Hans-Dieter Stölting · Eberhard Kallenbach · Wolfgang Amrhein

Handbook of Fractional-Horsepower Drives

Authors

Prof. Dipl.-Ing. Dr. Wolfgang Amrhein
　Johannes Kepler University Linz, Austria
Prof. Dr.-Ing. habil. Hans-Jürgen Furchert
　University of Applied Sciences Gießen-Friedberg, Germany
Prof. Dr.-Ing. Friedrich Wilhelm Garbrecht
　University of Applied Sciences Gießen-Friedberg, Germany
Dr.-Ing. Elmar Hoppach
　mecatronix GmbH, Darmstadt, Germany
Prof. Dr.-Ing. habil. Hartmut Janocha
　Saarland University Saarbrücken, Germany
Prof. Dr.-Ing. habil. Eberhard Kallenbach
　Steinbeis-Transfercenter Mechatronics Ilmenau, Germany
Prof. Dr.-Ing. habil Dr. h. c. Werner Krause
　Technical University Dresden, Germany
Dr.-Ing. habil. Andreas Möckel
　Technical University Ilmenau, Germany
Prof. Dr.-Ing. habil. Dieter Oesingmann
　Technical University Ilmenau, Germany
Dr.-Ing. habil. Christian Richter
　Saia Burgess Dresden GmbH, Germany
Dr.-Ing. Thomas Roschke
　Saia Burgess Dresden GmbH, Germany
Prof. Dr.-Ing. Wolfgang Schinköthe
　University Stuttgart, Germany
Dipl.-Ing. Dr. Siegfried Silber
　Johannes Kepler University Linz, Austria
Prof. Dr.-Ing. Hans-Dieter Stölting
　Leibniz University Hannover, Germany
Prof. Dr.-Ing. Heinz Weißmantel
　Technical University Darmstadt, Germany

Hans-Dieter Stölting
Eberhard Kallenbach
Wolfgang Amrhein

Handbook of Fractional-Horsepower Drives

With 462 Figures and 33 Tables

Springer

Prof. Dr.-Ing. habil Eberhard Kallenbach
Steinbeis-Transfercenter Mechatronics
Werner-von-Siemens-Str. 12
D-98693 Ilmenau
Germany
e-mail: eberhard.kallenbach@stz-mtr.de

Prof. Dr.-Ing. Hans-Dieter Stölting
Leibniz University Hannover
Institute of Drive Systems
and Power Electronics
Welfengarten 1
D-30167 Hannover
Germany
e-mail: stoelting@ial.uni-hannover.de

Prof. Dipl.-Ing. Dr. Wolfgang Amrhein
Austrian Center of Competence in Mechatronics (ACCM)
Johannes Kepler University Linz
Institute of Electric Drives
and Power Electronics
Altenberger Str. 69
A-4040 Linz
Austria
e-mail: wolfgang.amrhein@jku.at

Authorized translation of the German original edition, published at Carl Hanser Verlag, München
Stölting/Kallenbach (Hrsg.): Handbuch Elektrische Kleinantriebe, 3. Auflage
Copyright © 2006
Carl Hanser Verlag, München
All rights reserved.

ISBN 978-3-540-73128-3 Springer Berlin Heidelberg New York

Library of Congress Control Number: 2007929489

This work is subject to copyright. All rights are reserved, whether the whole or part of the material is concerned, specifically the rights of translation, reprinting, reuse of illustrations, recitation, broadcasting, reproduction on microfilm or in any other way, and storage in data banks. Duplication of this publication or parts thereof is permitted only under the provisions of the German Copyright Law of September 9, 1965, in its current version, and permission for use must always be obtained from Springer. Violations are liable for prosecution under the German Copyright Law.

Springer is a part of Springer Science+Business Media

springer.com

© Springer-Verlag Berlin Heidelberg 2008

The use of general descriptive names, registered names, trademarks, etc. in this publication does not imply, even in the absence of a specific statement, that such names are exempt from the relevant protective laws and regulations and therefore free for general use.

Typesetting and Production: le-tex publishing services oHG, Leipzig
Cover: WMXDesign GmbH, Heidelberg

SPIN 11532583   60/3180/YL - 5 4 3 2 1 0   Printed on acid-free paper

Preface

As a result of continually increasing mechanization and automation, electric fractional-horsepower drives have achieved immense application variety today. On the one hand, improvements made in material technology, microelectronics, power electronics, and automatic control science and technology have produced exceptional design diversity of drives; on the other hand, modern calculation, simulation and measurement methods lead to novel, improved drives. These developments allow us to avoid motors with expensive production engineering and to adjust instead the drive characteristics *via* easily variable control electronics. Even if the trend seems to converge toward simple motor designs, a generic motor will never be produced because the range of application demands is too heterogeneous and too contradictory.

Usually there are several solution possibilities for a drive problem. Therefore a user of small electrical drives must possess knowledge and judgment to find a solution to a problem with regard to function, form, integration in the driven apparatus, operation, maintenance, reliability, safety, economic efficiency, noise and balance quality, and especially costs. This is true above all for the assessment of competing offers and for discussions between users and manufacturers of small electrical drives.

This book strives to help in selecting a drive and in estimating which technical expenditures and which costs will be necessary to optimally fulfil the respective demands. In order to enable a complete solution, we provide an overview not only of motors but also of electronic open-loop and closed-loop control systems with low power and of the mechanical transmission elements. The latter is rising in importance as future motor manufacturers will increasingly supply not only the motor, the drive circuit and the gearing, if required, but also apparatus components. The development and production of electrically driven mechanical assembly groups are partly shifting from the apparatus manufacturer to the drive manufacturer.

In addition to rotational drives, this book covers linear-motion drives in a detailed description of the special qualities of the different drive possibilities for short motions; this is intended to alleviate the literature's omission in this area. Furthermore, these direct drives require very narrow adaptation to the driven mechanical system making it more important to know the different characteristics of the numerous design variants.

Above all, electromagnetic motors are presented here. However, we felt obliged to present the most important unconventional drives, the piezoelectric actuators, too. Finally, tips are given on drive configurations and on the solution of typical drive tasks, including some examples. Many bibliographical references and the most important regulations and standards help the reader to delve deeper into the technology, if necessary, or to use a drive that meets technical regulations.

This book is suitable for engineers who want to employ small electrical drives to solve configuration tasks, e.g. in automotive engineering, machine-tool engineering, household appliance engineering, office and data processing, medicine, laboratory engineering, robotics, and consumer electronics. It is also recommended for students in the fields of electrical engineering, precision mechanics, mechanical engineering and mechatronics.

The authors of this book come from industry and universities and have compiled their knowledge and experience in a concise form for a presentation as uniform as possible. Nevertheless, the foci have been set individually by the authors of the various subjects, which are also responsible for the content of their contributions. This book is based on the third edition of the *Handbuch Elektrische Kleinantriebe* published of Hanser-Verlag, Leipzig, in 2006.

The authors and editors thank Springer-Verlag for their support and the appealing layout of this book.

Hans-Dieter Stölting, Hannover, Germany
Eberhard Kallenbach, Ilmenau, Germany
Wolfgang Amrhein, Linz, Austria

Contents

1 Introduction
1.1 General View 1
1.2 The Electromagnetic Drive System 2
1.3 Drive Units 3
   1.3.1 Motors 3
   1.3.2 Electronic Circuits 8
References 11

2 Motors for Continuous Rotation
2.1 Motors with Commutator 13
   2.1.1 Electrically Excited DC Motors 13
   2.1.2 Permanent Magnet DC Motors 14
   2.1.3 Commutator Series Motor (Universal Motor) 33
   2.1.4 Contact System and Commutation 51
   2.1.5 Noise Behaviour 65
   2.1.6 Radio Interference Suppression 65
2.2 Brushless Permanent Magnet Motors 66
   2.2.1 Introduction 66
   2.2.2 Design Features 72
   2.2.3 Dynamic Model of the Brushless AC Motor 92
   2.2.4 Electronic Suppression of Torque Ripples 99
   2.2.5 Motor Characteristics 104
   2.2.6 Sensor Technology 105
   2.2.7 Special Designs of Brushless Permanent Magnet Motors ... 111
2.3 Switched Reluctance Motor 116
   2.3.1 Magnetic Circuitry 116
   2.3.2 Operation 117
   2.3.3 Commutation 118
   2.3.4 Control Strategies and Motor Performance 121
   2.3.5 Sensorless Modes of Operation 123
   2.3.6 Characteristics 124

VIII   Contents

2.4 Rotating-Field Motors 125
    2.4.1 Asynchronous Motors 125
    2.4.2 Synchronous Motors 144
References 155

3 Electromagnetic Stepping Drives
3.1 Overview 161
3.2 Classification of Stepping Motors 164
    3.2.1 Claw-Poled Stepping Motor 164
    3.2.2 Disk Rotor Stepping Motor 166
    3.2.3 Reluctance Stepping Motor 167
    3.2.4 Hybrid Stepping Motor 168
3.3 Control of Stepping Motors 170
3.4 Operating Behaviour of Stepping Motors 179
    3.4.1 Step Angle 179
    3.4.2 Torque 181
    3.4.3 Positioning Accuracy 182
    3.4.4 Step vs Time Characteristic 183
    3.4.5 Operating Characteristics 185
    3.4.6 Resonance Frequencies 188
    3.4.7 Micro Stepping 188
    3.4.8 Assigning Reference Position 190
3.5 Motion Sequences 191
    3.5.1 Classified Motion Sequences 191
    3.5.2 Closer Examination of Dynamic Operation 194
References 203

4 Drives with Limited Motion
4.1 Electromagnets 205
    4.1.1 Electromagnets as Drive Sections 205
    4.1.2 Direct Current Magnets 207
    4.1.3 Alternating Current Magnets 227
    4.1.4 Polarized Electromagnets 236
    4.1.5 Project Planning of Electromagnets and Magnetic Drives 240
    4.1.6 Design of Magnetic Drives 241
4.2 Electrodynamic Linear and Multi-Coordinate Drives 250
    4.2.1 Working Principle and Basic Structure 251
    4.2.2 Configurations of Electrodynamic Linear Motors 255
    4.2.3 Types of Construction of Integrated Electrodynamic
          Multi-Coordinate Drives With Single-Mass Armatures
          for xy-, $x\varphi$-, $xy\varphi$- and $xy\varphi z$-Travel 269
    4.2.4 Behaviour of Electrodynamic Linear and Multi-Coordinate
          Motors 285
    4.2.5 Control of Electrodynamic Linear
          and Multi-Coordinate Motors 288

4.2.6 Commercially Offered Systems 298
4.3 Linear and Planar Hybrid Stepping Motors 298
    4.3.1 Linear Hybrid Stepping Motors 298
    4.3.2 Multi-Coordinate Hybrid Stepping Motors 303
    4.3.3 Dynamic Properties of Linear Hybrid Stepping Motors 306
    4.3.4 Principle of Microstepping 309
    4.3.5 Linear Hybrid Stepping Motors
          as Non-Linear Magnetic Drives 310
References 311

5 Piezoelectric Drives
5.1 Physical Effect 317
5.2 Piezoelectric Components 318
    5.2.1 Piezoelectric Materials 318
    5.2.2 Piezoelectric Elements 320
5.3 Piezoelectric Actuators 322
    5.3.1 Stack Translator (Stacked Design) 322
    5.3.2 Laminar Translators 326
    5.3.3 Bending Elements 326
    5.3.4 Tube Elements 327
    5.3.5 Displacement Amplification 328
5.4 Piezoelectric Motors 329
    5.4.1 Inchworm and Walking Motors 330
    5.4.2 Ultrasonic Motors 332
5.5 Driving Electronics 338
    5.5.1 Power Amplifier 338
    5.5.2 Inverse Control 340
5.6 Implementation Examples 341
    5.6.1 Positioning System with Piezo Drives 342
    5.6.2 Clamping Elements for Inchworm Motors 343
References 345

6 Open-Loop and Closed-Loop Control
of Electric Fractional-Horsepower Motors
6.1 Introduction 347
6.2 Circuit Components and Pulse-Width Modulation 348
    6.2.1 Circuits for Indirect Coupling and Voltage Level Adapting 348
    6.2.2 Control Elements for DC and Rotating-Field Motors 352
    6.2.3 Control Elements for Stepper Motors 358
    6.2.4 Modulation Methods for Three-Phase Motors 359
    6.2.5 Control Elements 364
6.3 Open-Loop Control of Rotational Speed 367
    6.3.1 Control Elements and Power Sections for Universal Motors 367
    6.3.2 Open-Loop Rotational Speed Control of DC Motors 368

　　　　6.3.3 Open-Loop Rotational Speed Control
　　　　　　of Asynchronous Motors 371
　　　　6.3.4 Open-Loop Control of Motors with Synchronous Speed 373
　　　　6.3.5 Positioning and Rotational Speed Open-Loop Control
　　　　　　of Stepper Motors 376
　6.4 Positioning and Rotational Speed
　　　Closed-Loop Control 376
　　　　6.4.1 Closed-Loop Control of Universal Motors 376
　　　　6.4.2 Closed-Loop Control of DC Motors 377
　　　　6.4.3 Closed-Loop Control of Asynchronous Motors 380
　　　　6.4.4 Closed-Loop Control of Electronically Commutated Motors ... 390
　6.5 Information on the Practice-Oriented Adjustment
　　　of Controllers and Simulation of Rotating Speed
　　　Closed-Loop Control Circuits 395
　6.6 Structure of Drive Electronics
　　　for Three-Phase Alternating Current Motors
　　　Using Circuits with Large-Scale Integration 398
　6.7 Sensorless Determination and Closed-Loop Control
　　　of Rotational Speed, Rotor Position and Internal Torque 401
　References 414

7 Magnetic Bearing Technology
　7.1 Introduction 417
　7.2 Passive Magnetic Bearings 419
　　　　7.2.1 Permanent Magnet Bearings 420
　7.3 Active Magnetic Bearing Systems 426
　　　　7.3.1 Electromagnetic Bearings 427
　　　　7.3.2 Bearingless Motors 434
　References 445

8 Mechanical Transfer Units
　8.1 Gearings 450
　　　　8.1.1 Kinds of Gearings 450
　　　　8.1.2 Gear Trains 452
　　　　8.1.3 Flexible Drives 472
　　　　8.1.4 Screw Mechanisms 477
　　　　8.1.5 Coupling Drives 477
　　　　8.1.6 Cam Mechanisms 480
　　　　8.1.7 Stepping Gears 481
　8.2 Couplings 482
　　　　8.2.1 Firm Couplings 484
　　　　8.2.2 Self-aligning Couplings 485
　　　　8.2.3 Clutches 487
　8.3 Axles and Shafts 492
　　　　8.3.1 Calculation of Conceptual Design 492

8.3.2 Recalculation	493
8.4 Bearings	495
8.4.1 Sliding Bearings	496
8.4.2 Rolling Bearings	502
References	505
Standards and Directions	506

9 Project Design of Drive Systems

9.1 Requirements and Specifications	509
9.2 Approach to Drive Functions	511
9.3 Classification of Common Drive Functions	511
9.3.1 Classification According to Motion	511
9.3.2 Classification According to Operating Mode	514
9.4 Aspects of Motor Selection	517
9.5 Comparison of Position-Controlled DC Drives and Stepping Drives	519
9.6 Conversion of Mechanical Drive Parameters	524
9.7 Samples of Drive Functions	526
9.7.1 Direct Drive of a Disk Storage Device	526
9.7.2 Drive of a Dosing Piston Pump	532
9.7.3 Drive of a Drilling Machine for Printed Circuit Boards (PCBs)	539
9.7.4 Drive of a Light Pointer	546
9.7.5 Drive of a Flexible-Tube Pump or Peristaltic Pump	551
9.7.6 Bowden Wire Drive of a Recording Device	556
9.7.7 Drum Drive	566
Standards and Directions	576
Important Symbols	579
Index	585

1
Introduction

Hans-Dieter Stölting

1.1 General View

The economic importance of fractional-horsepower drives with a nominal power within the range of about 1 kW is considerable. A study of the market researcher Frost & Sullivan assesses a 5.4 billion dollar sale of small electric drives (<700 W) for 2006 in Europe, i. e. an increase of 23% compared to 1999. Whereas the market of AC motors is gradually saturated, brushless motors are increasingly in demand (Sect. 2.2.1.1).

Fractional-horsepower drives are marked by an unusual application variety. Drives used in consumer goods are generally produced in a large number of pieces, often more than 1 million pieces per year. In this case, the manufacturing costs have to be kept as small as possible. These low-cost drives not only have to fulfill the electromechanical conditions of the specific application case but they also have to be adapted constructionally as good as possible both to the driven mechanism and to the most economical production technology. Typical conditions are for instance:

- No excessive demands on the power/weight ratio and efficiency;
- Integration in the driven apparatus or taking over apparatus functions by motor components (e. g. the motor end shield is also part of a pump housing);
- Largely automatic mass production.
- Punch-bend-joint technique using components with commercial (standard) dimensions (no special design, e. g. for magnets, bearings, capacitors, etc.), stator and rotor core made of unalloyed sheet (so-called semi-finished sheet), if the speed is lower than about 4000 rpm. (tinplate, often not annealed used), ferrite magnets; rough dimension gradation of motor families (outer and inside diameter, core length), great width of air gap, shaft without steps (if possible), sleeve bearing, greater manufacturing tolerance range; low slot space factor, very easy winding, stoved-enamel wire; very easy electronic control.

Apart from low-cost drives there are high-grade drives, whose design depends on special (often extreme) demands:

- Optimal electromechanical and constructional adaptation to the apparatus;
- Small batch production: cutting, assembly by screwing, high-grade components: electrical or special low-loss sheet, rare-earth magnets (SmCo, today above all NFeB), rolling bearing;
- Four-quadrant operation, if required;
- Special properties related to power/weight ratio, efficiency (low energy demand, lower temperature rise), very high speed, true running, synchronous operation, dynamic (low mechanical and/or electrical time constant), positioning, overload capacity, lifetime, freedom of maintenance, noise and balance quality, electromagnetic compatibility, insensitivity towards environmental conditions (temperature, vibrations, accelerations, pressure, pollution (dust-, water-, gas-tight).

As a result of these different conditions, a nearly immense design variety has been created in the course of time, even further expanding due to new developments in the field of micro and power electronics as well as in the field of materials, above all with respect to magnetic materials.

1.2 The Electromagnetic Drive System

The electromagnetic drive system serves for the generation of motion. It consists of an information-processing subsystem, the control element, and energy-transmitting function elements, the controlled element, the actuator, the transfer elements and the effector (Fig. 1.1). According to the special task, this structure can become very complex (several controllers, several feedbacks, several observers, etc.).

Tasks of the information-processing subsystem (control element):

- Comparison of the given reference input variables R (setpoint values: torque, rotational speed, angle of rotation; force, speed, position) with the corresponding actual values A and forming the controlled variable C resulting from the control deviation;
- Registration of the disturbances D and monitoring as well as protection of the drive system;
- Output of message values M to higher-level monitoring or other systems.

Tasks of the electrical energy converter (controlled element, power section, output stage):

- Conversion of electrical energy (e. g. three-phase energy into direct current energy);
- Adaptation of the motor voltage to the supply voltage;

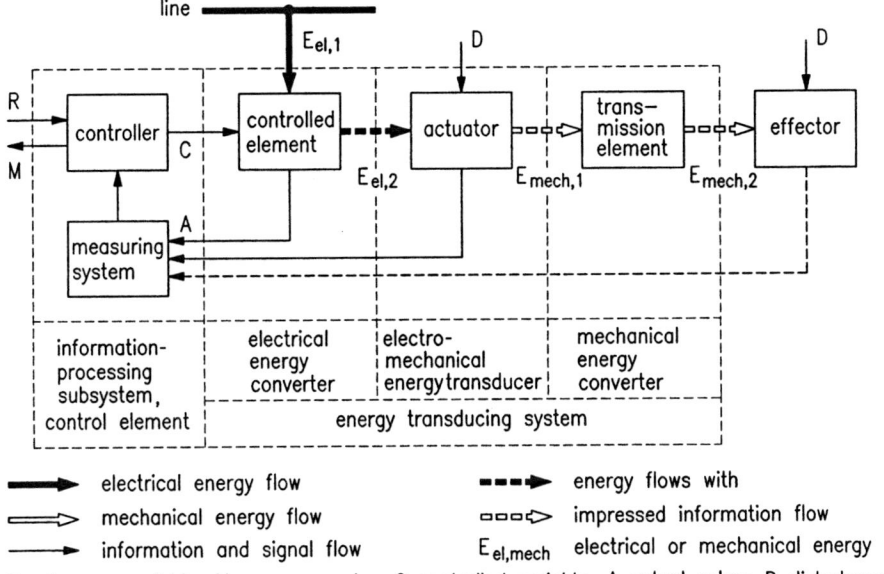

Fig. 1.1. The electromagnetic driving system

- Conversion of the input signals (manipulated variable) into currents useable by the motor, that is, in the controlled element the information flow meets the energy flow.

Tasks of the electromechanical energy transducer (actuator, motor):

- Torque or force generation;
- Generation of a steady or stepping motion, rotary or linear motion.

Tasks of the mechanical energy converter (transmission element, gear unit):

- Change of the torque and the rotational speed;
- Reduction of the load inertia torque;
- Conversion of a rotational motion into a linear motion.

1.3 Drive Units

1.3.1 Motors

1.3.1.1 Motor Systematic

Among other things, the fundamental characteristics of electrical motors depend on the method, how voltage is applied to their windings. The cyclic switching of the windings of self-commutated (self-clocked, position-switched) motors happens automatically, depending on the rotor position, the cyclic

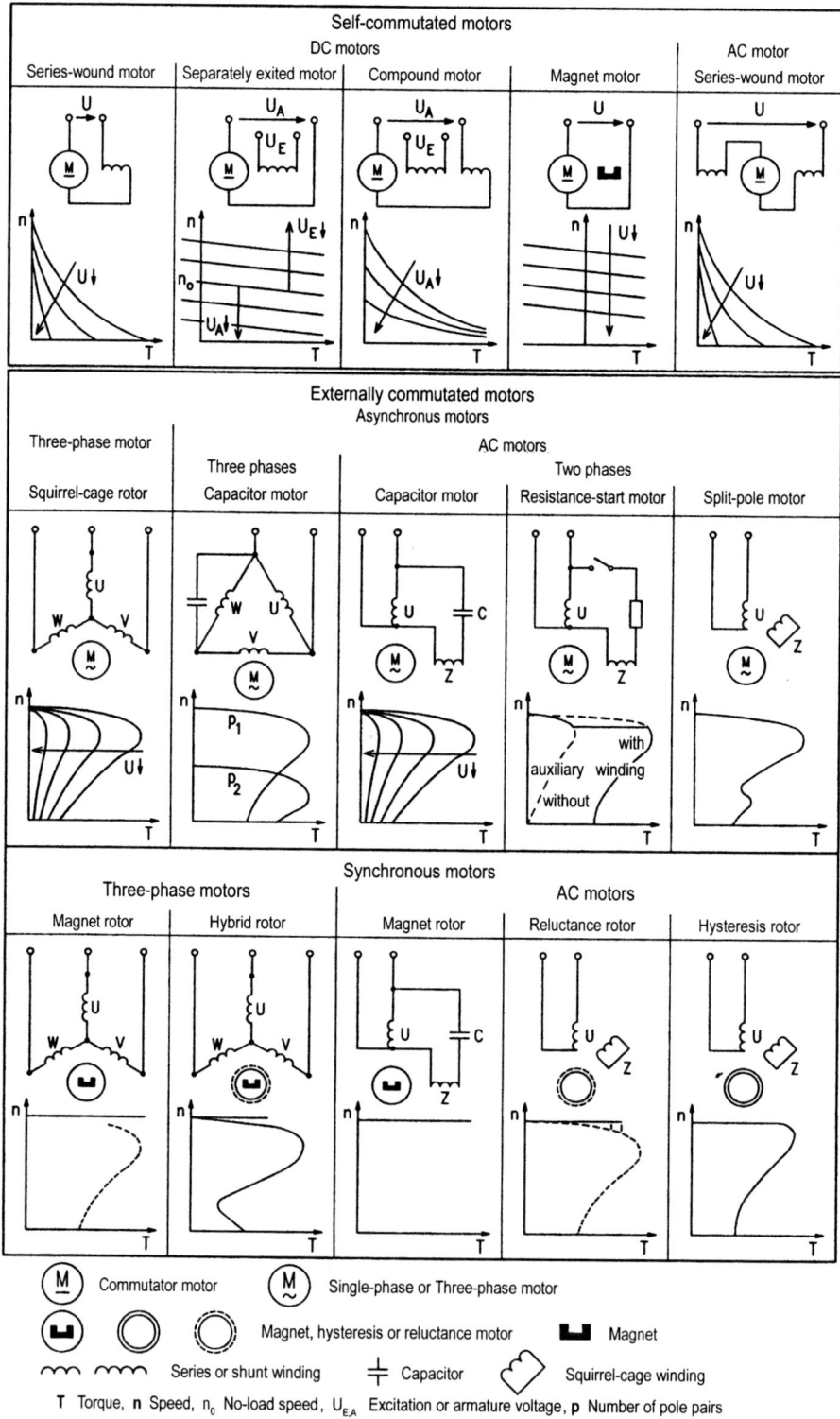

Fig. 1.2. Motors driven by direct line supply

Table 1.1. Motor systematics

self-commutated motors			externally commutated motors	
mechanical commutator		electronic commutator	load-dependent speed	frequency-locked speed
AC motors	DC motors	EC motors	asynchronous motors	synchronous motors
AC commutator-motor (universal motor)	series-wound motor	motor with magnet rotor and with square-wave or sinusoidal current (AC servo motor)	three-phase motor: cage rotor	three-phase motor: magnet rotor hybrid rotor
	shunt-wound motor		AC motors: capacitor motor, resistance-start motor, split-pole motor	AC motors magnet motor reluctance motor hysteresis motor
	permanent-magnet motor	switched reluctance motor		stepper motors: magnet, reluctance, hybrid motor
possible: $n > 3000$ min^{-1}; small, lightweight motors; easy open- and closed-loop control			$n \leq 3000$ min^{-1} (50 Hz line) expensive open- and closed-loop control	
motor: less robust, lower lifetime, noisy, expensive		motor: robust, quiet running; low cost with ferrite magnet or reluctance rotor	motor: robust, quiet running, low cost	
electronics: low cost		electronics expensive	electronics: very expensive	

switching of externally commutated (line-clocked) motors is carried out either by the supply system or the control electronics. Table 1.1 specifies the most important types of electromagnetic fractional-horsepower motors. The statements concerning their characteristics are valid only when comparing motor types of the same size or power.

The fundamental circuit configurations of small electric motors with direct line or battery supply are shown in Fig. 1.2 including the typical speed-torque characteristic and the open-loop speed control.

1.3.1.2 Basic Design Principles

In the following section, the basic design principles with their specific characteristics are described in shorthand. Since electric machines may be constructed in nearly all of the following design variants and combinations, this results in the enormous design variety as mentioned above.

a) Stator-rotor configurations (Fig. 1.3)
- Cylindrical rotor: the most widespread type of construction because of the low production costs, low torque of inertia (especially in case of slim rotors) and small diameter.
- Disk rotor: little overall length, often with ironless winding, danger of axial forces for rotors with magnetically hard and soft material (bearing defects), sometimes greater mechanical torque of inertia.
- Internal rotor: the most widespread type of construction because of good cooling of the stator winding, simpler bearing arrangement and easy mounting (no rotating housing).
- Intermediate rotor: bell rotor or disk rotor design; low electrical and mechanical time constants, better commutation and true running especially with moving coil; bell rotors for lower power only (gen. <100 W) or for higher power (<250 W) with a lower rotational speed only; brushless DC

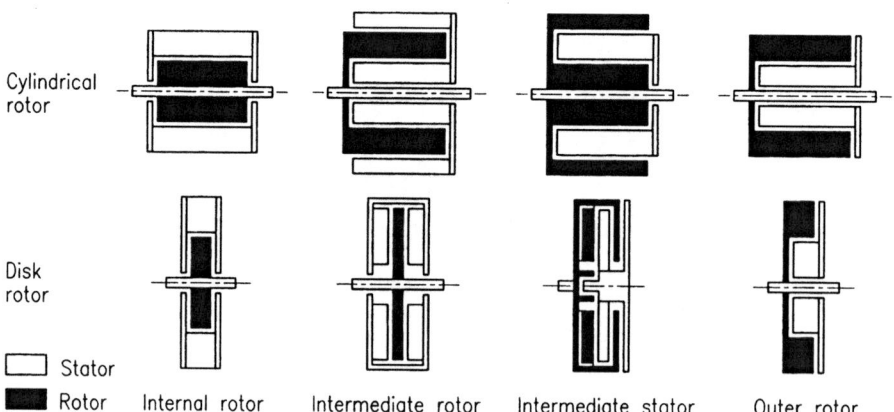

Fig. 1.3. Stator-rotor configurations

motors (electronically commutated motor, EC motor) and asynchronous motors with expensive stator production (core, winding).
- Intermediate stator: good true running especially with unslotted stator winding; no eddy-current loss with coupled magnetic return path.
- Outer (external) rotor: for special applications like fans and winders; good true running drives; often easier stator taping, but worse stator cooling.

b) Symmetry of cross section (Fig. 1.4)
- Biaxial symmetric cross section (stator and rotor are concentric): often easier mounting, better cooling of the stator winding.
- Uniaxial symmetric cross section (asymmetric or C-shape cross section): mostly low-cost production; sometimes noise problems because of alternating fluxes making the poles swing, which are mounted on one side.

c) Pole order (Fig. 1.5)
- Heteropolar motor: changing polarity along the circumference; good power/weight ratio because of the great magnetic flux ⇒ for this reason, this type is produced most.
- Homopolar motor: changing polarity along the axis; often unwound rotor with a great number of teeth ⇒ low but high-frequency flux oscillations because of the air-gap variation.

d) Winding type
- Slot winding: expensive but in general more favorable flux shape (lower loss).
- Concentrated winding on salient poles: easy production, electromagnetically more unfavorable for synchronous and asynchronous motors.

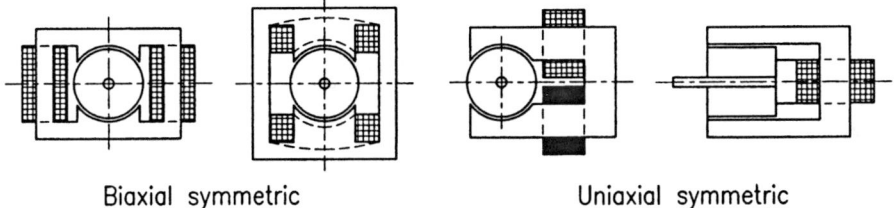

Biaxial symmetric   Uniaxial symmetric

Fig. 1.4. Die-set symmetry

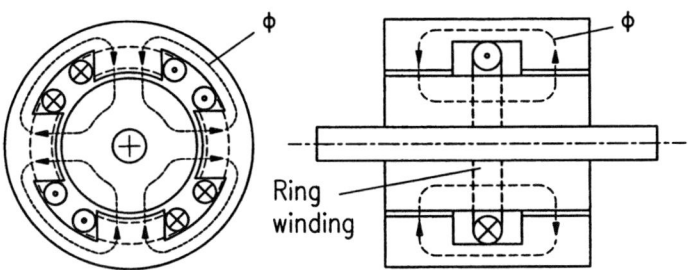

Fig. 1.5. Pole order

- Unslotted winding (air-gap, coreless, self-supporting winding): good true running, better commutation because of lower winding inductivity, lower motor utilization because of the greater air gap.
- Ring winding: for claw-pole systems (Sect. 2.4.2.2), the easiest construction for high numbers of poles; disadvantage of claw-pole systems: high flux leakage, high eddy-current losses in case of alternating fields.

e) Motion mode
- Rotating motors: This is the predominant type because of comparatively low costs.
- Linear motors: rarely used in the field of consumer goods because of higher costs (instead use of rotating motors with screw spindles (also as hollow shaft), toothed racks or toothed belts); limited motion, drives with individual adaptation to the apparatus (direct drive); advantages of direct drives: no gear problems, e. g. noise, loss, no play at positioning drives; disadvantage: increase of motor torque or decrease of the load torque of inertia is not possible because of the missing gear.
- Motors with very short motion: electromagnets, oscillating armatures and voice-coil motors.

f) Operating mode
steady, stepping or oscillating; continuous, short-time, intermittent operation.

1.3.2 Electronic Circuits

Figure 1.6 shows the possibilities of electronically driven fractional-horsepower motors, which are described in the following in brief to give a first overview.

1.3.2.1 Circuits for Self-Commutated Motors

DC Motor

AC operation: Rectifier is required; problem in case of permanent magnet motors: the armature winding has to be designed for line voltage, if there is no series resistor or voltage divider ⇒ expensive armature; simple drives without smoothing reactor (pulsating-current motor).

Operation with a linear transistor (transistor working in the linear range): Transistor acting as series resistor; lowest expenditure; high losses because of the difference between the line and the motor voltage at the collector/emitter line, especially for low rotational speed and high torques ⇒ low efficiency; therefore only for drives with very low power or short running time.

Chopper circuits: Switched-mode transistor; better efficiency.

Operation with one transistor: single-quadrant operation; if necessary, change of sense of rotation by a mechanical changeover switch; free-wheeling diode in parallel to the motor or to the armature winding to carry current during the blocking state.

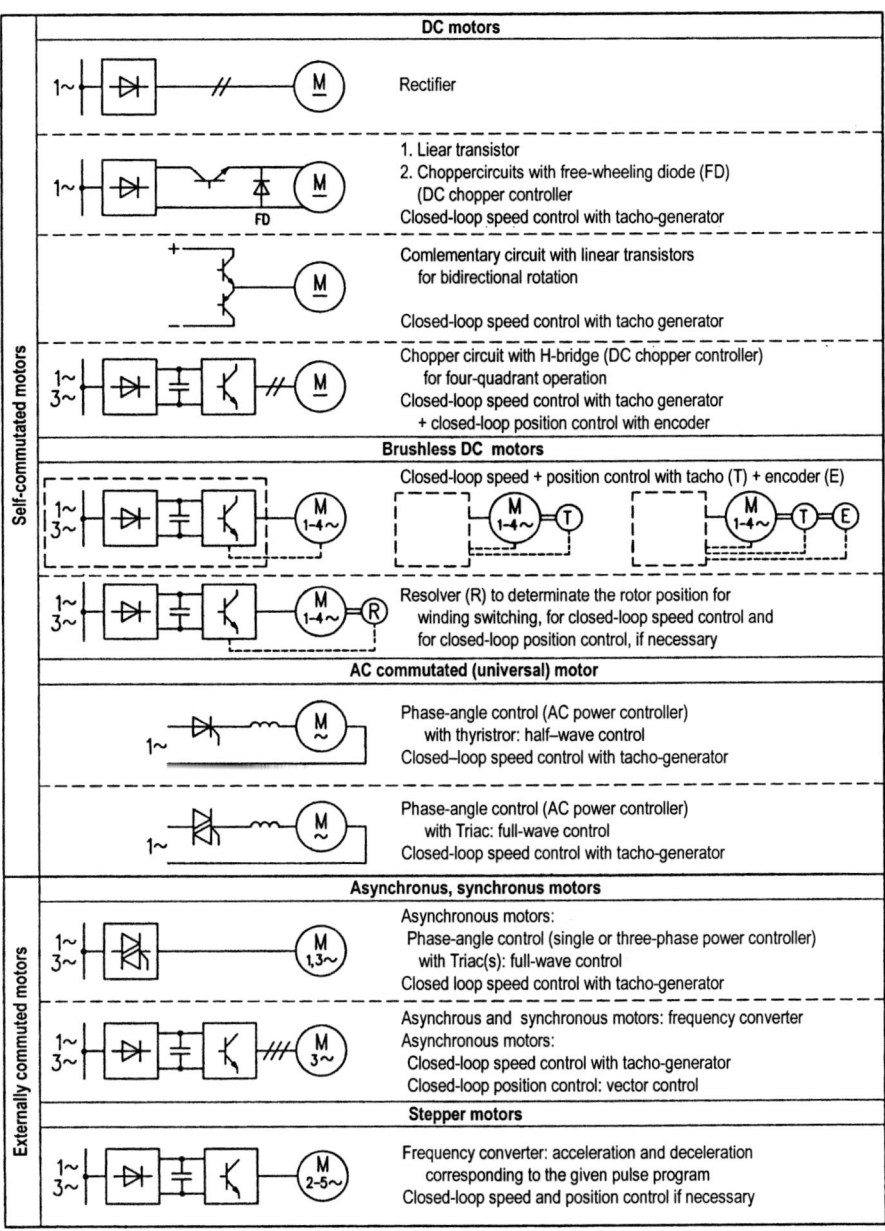

Fig. 1.6. Electronically driven fractional-horsepower motors

Full-bridge mode (H-bridge mode): four-quadrant operation; in case of AC supply rectifier and smoothing capacitor (DC link); adjustment of the voltage needed for the wanted speed *via* pulse-width modulation (variation of the operating interval/clock-time ratio);

Open-loop speed control: voltage input by a potentiometer; DC shunt-wound motor: expansion of the speed range by field weakening.

Closed-loop speed control: additional speed sensor (tacho-generator) and controller;

Closed-loop position control including the generation of a holding torque should the occasion arise: H-bridge and sensor for position detection (encoder); control mode is usually a cascade control with position regulator and inner speed and current control loop; advantage of DC motor: one current value has to be controlled only.

Brushless DC Motor (Electronically Commutated (EC) Motor)

AC operation: rectifier, DC link and H-bridge or three-phase bridge connection: the electronics cyclically switches the stator phases dependent on the rotor position ⇒ rotating field; detecting of the rotor position by a stator-integrated rotor position encoder or sensorless by analysis of the voltage induced in the present non-conducting phase.

Open-loop speed control by changing of DC link voltage.

Closed-loop speed control and position control like for DC motors, however, two or three currents are to be controlled; but a higher dynamic is possible because of the non- mechanical commutation.

Universal Motor (AC Commutator Series Motor)

Application above all in consumer goods, because AC operation is possible and electronic open-loop control (phase-angle control) is especially easy (cheap) to realize.

Half-wave control: lowest cost, but for low power only.

Full-wave control: both half waves are used, therefore for higher power application; triggering of the Triac increasingly realized by phase-angle IC; for higher demands, a rectifier is series connected with the phase angle control unit ⇒ ripple-free DC ⇒ improvement of the commutation;

Problem of universal motors: the speed decreases disproportionately to the increasing load ⇒ therefore often an open-loop control is necessary: expansion of the phase-angle control by a tachometer and a controller.

1.3.2.2 Circuits for Externally Commutated Motors

Asynchronous Motors

Phase-angle control: the simplest connection, but restricted because of the constant synchronous speed; in addition T (torque) $\sim U^2$ (voltage) and high power losses.

Frequency converter for three-phase motors only: speed control from zero to the rated speed; field weakening, i.e. further speed rise is possible (like DC motor); to get a constant motor utilization a constant induced voltage/frequency ratio is necessary; with increasing speed (frequency), the difference between the constant terminal voltage and the induced voltage decreases due to the decreasing voltage drop of the stator winding, and with it U_i/f; therefore, remedy, if necessary, e.g. by means of $I * R$-compensation or slip compensation: calculation of the slip frequency for the expected load based on the motor values; the sum of the expected speed frequency and slip frequency is the rotating-field frequency to be supplied by the frequency converter.

Closed-loop speed control: expansion of the frequency converter by a tachometer and a controller.

Closed-loop torque control: purpose is a fast and exact torque control like for DC motor; as the position of the rotating field with respect to the rotor position is variable, a conversion of the motor currents and voltages of the three-phase system into a coordinate system is necessary, geared to an electric variable (flux) ⇒ field-oriented control, vector control; calculation of the motor flux and the manipulated variable by means of a motor model.

Closed-loop position control: similar to DC motors: cascade control with position regulator and inner speed and current control loop with an additional closed-loop flux control.

Three-phase Synchronous Motor

Open-loop control with frequency converter: acceleration and deceleration along a frequency ramp; in contrast to DC and AC motors, speed extension by field weakening possible to a limited extension only.
Closed-loop position control: easier than the control of AC motors, as the field is fixed on the rotor; additional position sensor; controller type similar to DC and AC motors. cascade control with position regulator and inner speed and current control;

Open-loop and closed-loop control of synchronous motors used with high-power applications only.

Stepper Motors

Positioning with given step width *via* the number of pulses (Full- or half-step operation); control frequency given by electronics; open-loop control, therefore cost-effective; adapted pulse quantity for start and stop; precise positioning by microstep operation: subdivision of a full step in a sum of microsteps by pulse-width modulated triggering of adjacent windings.

Sometimes as servo-drive with encoder and controller type similar to the other drive variants.

References

[1.1] Pustola, J., Sliwinski, T.: Kleine Einphasenmotoren. Berlin: VEB Verlag Technik 1961
[1.2] Veinott, C.G.: Fractional- and Subfractional-Horsepower Electric Motors. New York: McGraw-Hill 1970

References

[1.3] SEW EURODRIVE: Handbuch der Antriebstechnik. München: Hanser Verlag 1980

[1.4] Phillipow, E.: Taschenbuch Elektrotechnik, Bd. 5. München: Hanser Verlag 1981

[1.5] Nasar, S.A. (Ed.): Handbook of Electric Machines. New York: McGraw-Hill 1987

[1.6] Stölting, H.-D., Beisse, A.: Elektrische Kleinmaschinen. Stuttgart: B.G. Teubner-Verlag 1987

[1.7] Kallenbach, E., Bögelsack, G.: Gerätetechnische Antriebe. München/Wien: Hanser-Verlag 1991

[1.8] Brosch, P.F.: Moderne Stromrichterantriebe. 3. Auflage. Würzburg: Vogel-Verlag 1998

[1.9] Vogel, J.: Elektrische Antriebstechnik. 6. Auflage. Heidelberg: Hüthig-Verlag 1998

[1.10] Dittmann, F., Stölting, H.-D.: Alles bewegt sich – Beiträge zur Geschichte elektrischer Antriebe. Geschichte der Elektrotechnik Bd. 16. Berlin/Offenbach: VDE-Verlag 1998

[1.11] Janocha, H. (Ed.),: Actuators. Berlin/Heidelberg/New York: Springer-Verlag 2004

[1.12] Moczala, H (Hrsg.): Elektrische Kleinmotoren. 2. Aufl. Ehningen: expert-Verlag 1993

[1.13] Richter, Chr.: Servoantriebe kleiner Leistung. Weinheim/New York/Basel: VCH 1993

[1.14] Wehrmann, C.: Elektronische Antriebstechnik. Dimensionierung von Antrieben mit Mathcad. Braunschweig/Wiesbaden: Vieweg 1995

[1.15] Brosch, P.F.: Moderne Stromrichterantriebe. 4. Auflage. Würzburg: Vogel-Verlag 2002

[1.16] Vogel, J.: Elektrische Antriebstechnik. 6. Auflage. Heidelberg: Hüthig-Verlag 1998

[1.17] Dittmann, F., Stölting, H.-D.: Alles bewegt sich – Beiträge zur Geschichte elektrischer Antriebe. Geschichte der Elektrotechnik Bd. 16. Berlin/Offenbach: VDE-Verlag 1998

[1.18] NN: Marktübersicht. Sonderausgabe der Zeitschrift antriebstechnik. Mainz: Vereinigte Fachverlage (erscheint jährlich)

[1.19] NN: Automation & Drives Kompendium. München: publish-industry Verlag (erscheint jährlich)

2
Motors for Continuous Rotation

2.1 Motors with Commutator

Heinz Weißmantel, Dieter Oesingmann, Andreas Möckel

2.1.1 Electrically Excited DC Motors

DC motors using electrical excitation do not face relevant market share. This decrease was due to the decline of the national grid based on DC and the advantage of DC permanent magnet motors. The majority of electrically excited DC Motors use neither a compensation winding nor an interpole winding, as these would require too much space and cost.

The field winding is connected in parallel with the armature winding on shunt wound motors. The rotational speed is a linear function of the load and decreases with torque. This type of motor is not often used. This type of motor is limited to specific applications, such at as very low temperatures, where permanent magnets may not be used. The speed can be set cheaply using different taps on the field winding.

The field winding is connected in series with the armature winding on the series wound motor (universal motor). The series wound motor is especially suited for applications requiring high pullout torque. The applicable torque increases disproportionately. The speed torque diagram of this type of motor shows the so-called series characteristic. This type of motor is predominantly used on the AC-line.

The compound motor combines the benefits of the latter two types, high pullout torque and linear torque over speed characteristic. The disadvantage is the complex and expensive manufacturing, limiting the use to a few applications.

2.1.2 Permanent Magnet DC Motors

2.1.2.1 Introduction

Characteristics

Permanent magnet DC motors, most often called DC motors, are the most popular fractional-horsepower drives, especially at low voltages (≤ 42 V). The efficiency is high, and this type of motor can be manufactured rather cheaply and the design offers a wide range of variations.

Permanent magnet DC motors are often an alternative to universal motors in 230 V AC line applications. From the technical point of view some aspects should be heeded.

The DC motor requires for operation on the AC line:

- Line rectification
- An armature winding with high resistance, leading to higher costs (as the motor has no field winding, the line voltage is applied to the brushes)
- An expensive EMC filter
- Complex electronics, as there is no way to change speed by different taps on the field winding

The series wound motor:

- Requires complex electronics for speed control, as speed is not a linear function of torque
- Commutates not very well, as voltages are induced on commutating the inductances

Compared to equal cost, both types of motors offer the same torque to volume ratio. For applications requiring constant speed, the DC motor is in most cases the better solution, whereas in applications requiring high starting torque and a wide speed range the universal motor is often more appropriate.

The expected lifetime of the commutator is in the range from a few hundred to several thousand hours of continuous duty, depending on the application. Due to wear of the commutator or the brushes, the lifetime is shorter than that of brushless DC motors. Table 1.1 shows the most relevant advantages and disadvantages of the motors with commutator and brushless DC motors.

Examples for Applications

DC motors are used in a wide variety of applications:

- Auxiliary drives in vehicles (wipers, blowers for air conditioning and motor coolers, gasoline pumps, water pumps for window cleaning, power windows, sun roof, seat adjustment, anti skid braking and door lock actuators)

- Domestic and hobby appliances (blowers, hair dryers, barbecues, lawn trimmers, tooth brushes, can openers, massage devices, model railroads, model cars)
- Home automation (heating, air conditioning and ventilation, door lock actuators, window shutters)
- Workshop (cordless power tools, handling devices, welding grips)
- Office automation (printers, copying machines, plotters, fax machines)
- Photo and video (cameras and camcorders, video recorders, tape recorders, projectors, cutters)
- Medical devices (X-ray radioscopic equipment, wheelchairs, dialysis devices, centrifuges, mixers, blood pressure measurement)
- Vending machines, cash machines, golf carts, packing machines, pneumatic delivery
- Printing machines, label printers

As opposed to larger DC motors, small permanent magnet DC motors are used in a wide variety of applications in control engineering. They provide low-cost drive with direct connection to DC power supply, low inertia (ironless rotor) low-cost driver and electronics (speed control by setting the motor current). Brushless motors require a resolver and more complex electronics, as two-phase or three-phase currents must be controlled.

2.1.2.2 Permanent Magnets

AlNiCo Magnets

Aluminium-Nickel-Cobalt magnets offer a high remanent flux density, and the coercive field strength is rather poor (Fig. 2.1, Table 2.1). This is critical due to the risk of partial demagnetisation. On AlNiCo magnets this risk is higher than on other magnet materials when taking the magnet out of the magnetic circuitry. Figure 2.2 shows the demagnetising curve and two load lines. The intersection of the load line (A) with the demagnetisation curve is the operating point of the magnetic circuitry. The operating point flux density and field strength can be denoted from the chart. The angle α from ordinate to load line depends on the ratio of the width of the air gap to the length of the magnet

Table 2.1.

	B_r	$-H_{cB}$	$(BH)_{max}$	T_k für B_r/H_{cB}
	T	kA/m	kJ/m^3	%/K
AlNiCo	0.8 ... 1.3	40 ... 150	10 ... 60	0.02/-0.02
Ferrite	0.2 ... 0.44	120 ... 260	6 ... 40	-0.2 /+0.4
SmCo	0.8 ... 1.2	400 ... 900	140 ... 320	-0.04/-0.2
NdFeB	0.6 ... 1.5	700 ... 1100	100 ... 420	-0.12/-0.6

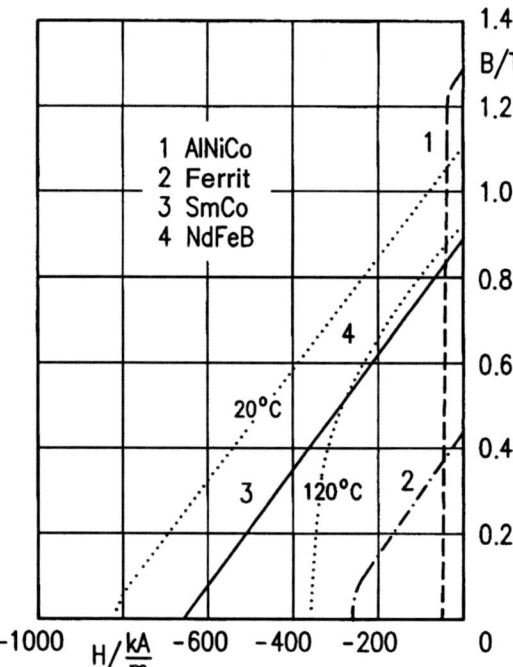

Fig. 2.1. Principle characteristic curves of different kinds of magnetic materials

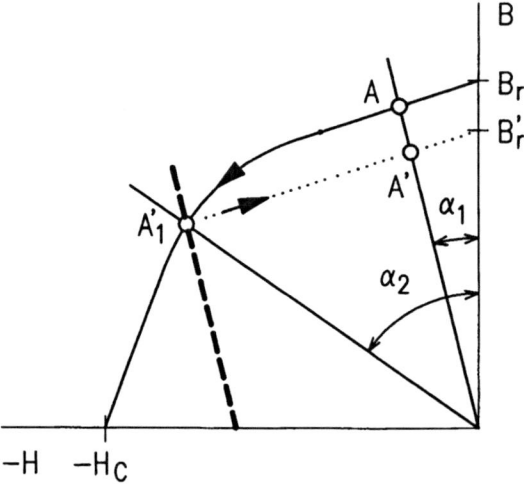

Fig. 2.2. Irreversible demagnetisation

in the direction of the magnetisation. The load line with the angle α_1 results from motor designs with long magnets and/or small air gaps. By opening the magnetic circuitry, for example taking the rotor from the stator, the air gap widens and so does the angle, for example up to α_2. Minimising the air gap does not bring the operating point back to the starting point. The new demagnetising curve is A1′A′Br′ resulting in a magnet that shows demagnetisation.

Load lines intersecting the point of origin are only valid at an ideal no-load condition; the permanent magnet DC motor does not draw any current from the supply. The load line moves in parallel dependent on the current in the armature winding to the left in section where the magnetic field is opposed to the excitation of the permanent magnet. At very high armature currents in consequence of high torques, e. g., during start up, operating points below the load line (dashed line in Fig. 2.1) may occur, leading to demagnetisation. Finally, it has to be heeded that saturation of the magnetic circuitry leads to bending of the work line. The operating point will be closer to the ordinate. The cases denoted are valid for all kind of magnetic materials, but are most visible on AlNiCo magnets. In a nutshell, the AlNiCo magnets shall be long, the iron parts shall not saturate and the air gap shall be as small as possible.

The magnetisation of AlNiCo is constant over a wide temperature range, as it offers the lowest temperature coefficient T_k. The maximum energy, the product of $(BH)_{max}$, is rather low when compared to rare earth magnets. AlNiCo magnets are available as cast or vacuum sintered raw material that will be milled to the desired dimensions.

Ferrite Magnets

Ferrite magnets are the most common on the market due to the low-cost. Their remanent flux density is rather poor, but the coercive field strength is so large, that demagnetisation is not relevant on proven designs. Ferrite magnets are brittle, hard and sensitive to temperature. Operation at low temperatures may lead to demagnetisation, as the demagnetisation curve is moved as shown in Fig. 2.3. The load line of the motor is moved to the left, as the current is increasing to the falling resistance of the armature winding. Ferrite magnets are sintered or die-cast to the desired dimensions using ferrite powder. If this is done in a magnetic field, magnets with a higher energy product will result. These magnets offer a higher flux density in the direction of magnetisation.

Rare-Earth Magnets

Rare-earth magnets are gaining a high market share as they drive high-quality AlNiCo magnets out of the market even though the price of rare earth magnets is higher and the material is harsher. DC motors using rare-earth magnets require less space as they offer a higher power density. Rare-earth magnets are often magnetised before mounting into the motor, requiring special tooling. Samarium-cobalt magnets (SmCo) offer a five times higher energy density when compared to AlNiCo magnets, and the temperature coefficient is slightly higher. SmCo magnets are quite robust to demagnetisation. Neodymium-Iron-Boron magnets offer even higher values for coercive field strength and remanent flux density than SmCo. These magnets are robust against demagnetisation at room temperature. A big disadvantage is the effect of the large temperature coefficient on the coercive field strength. Careful design is mandatory

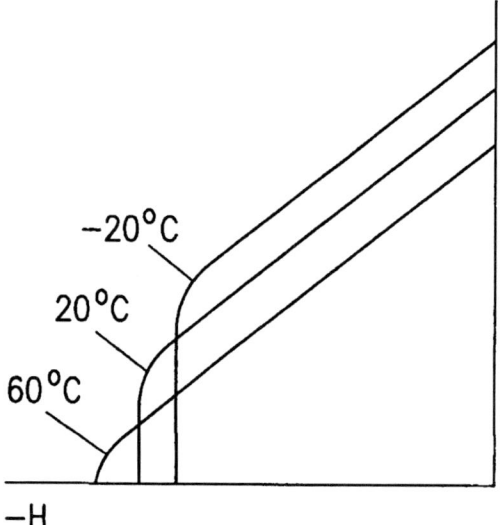

Fig. 2.3. Demagnetisation of hard ferrite magnets

to guarantee high temperature operation without demagnetisation. In general, SmCo magnets should be preferred at temperatures in the 150–200 °C range over their NdFeB counterparts. Resin-bound magnets are on the rise in various applications. These magnets offer a higher resistance which reduces eddy currents and therefore reduces temperature. A further benefit is the improvement against corrosion due to the encapsulation of NdFeB in the resin. These magnets are cheaper compared to SmCo due to widely available raw materials.

Assembly of Permanent Magnets

Permanent magnet are delivered ready for assembly. They may be fixed into the motor using springs, a very cost-sensitive practice, or by adhesive, bolts or screws. Adhesives offer noise damping. As permanent magnets are generally quite harsh, care must be taken to prevent damage in the assembly of the magnets and the completion of the motor. Particles from the magnet will cling to the magnet and may cause noise or even block the motor when entering the air gap. Resin-based magnets and flexible magnets go for a cost-sensitive assembly process, as they are softer and cling better to round shapes. Unfortunately, these magnets offer only lower energy densities.

2.1.2.3 Structure

Stator

Many different variations of the stator match different demands of different applications. Figure 2.4 shows variations of low-cost motors. The housing is often made of deep-drawn or rolled steel. The magnets are usually made of fer-

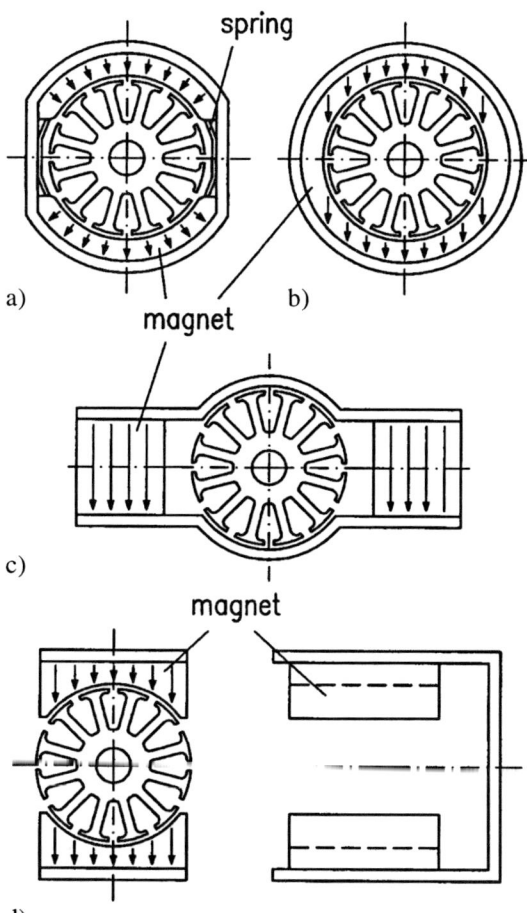

Fig. 2.4. Principle design of cost-efficient motors

rite and shaped to cylinders, blocks or pods. Radial magnetisation (Fig. 2.4a) results in a constant or trapezoidal magnetic field in the air gap. In two-pole motors a diametric magnetisation (Fig. 2.4b) results in a sinusoidal magnetic field in the air gap. A flat design is shown in Fig. 2.4c, d. The bearing bracket is usually made of stamped steel and holds the bearing. Alternatively, injection moulding forms bearing brackets that allow integration of the brushes. The parts are assembled by welding, adhesive or caulking.

High grade motors usually incorporate rare-earth or – with diminishing market share – AlNiCo magnets. Figure 2.5a shows an example of a motor using AlNiCo magnets, Fig. 2.5b an example of a motor using rare-earth magnets. The latter shows the usage of a flux concentrator for higher flux density. Low time constants for high dynamic drives and better commutation and higher lifetime require magnets in the air gap. Housing and bearing brackets are milled to measure and mounted by screws. High grade motors use ball bearings.

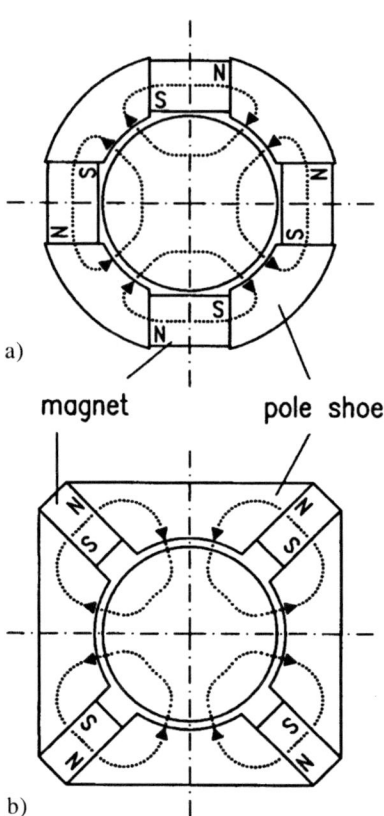

Fig. 2.5. Principle design of high grade motors

Rotor

The rotor is laminated (Fig. 2.6 left). To minimise cost, simple tin plated sheet metal is used, often up to 1 mm in thickness. At higher rotational speeds, roughly more than 3000 rpm, annealed or silicon iron offers lower iron losses. At least three commutator poles and windings allow for a reliable start up (Fig. 2.6). Five poles are state of the art for motors in the 10-W class (Fig. 2.7). Motors with higher power use more poles and benefit from lower cogging torque.

The armature winding is usually carried out as a lap or wave winding. Figure 2.8 shows the scheme of the winding with one coil for each layer. The number of coils is equal to the number of sections on the commutator. When using two coils for each layer, the number of sections on the commutator is twice the number of slots in the rotor. As the manufacturing of the windings is the most time consuming step in the production of a motor, the number of windings should be kept small. For lap windings, the number of brushes is equal to the number of poles, for wave windings two brushes are sufficient. The selection of the winding is based on the application and the demands of manufacturing and construction. A lap winding where the number of parallel

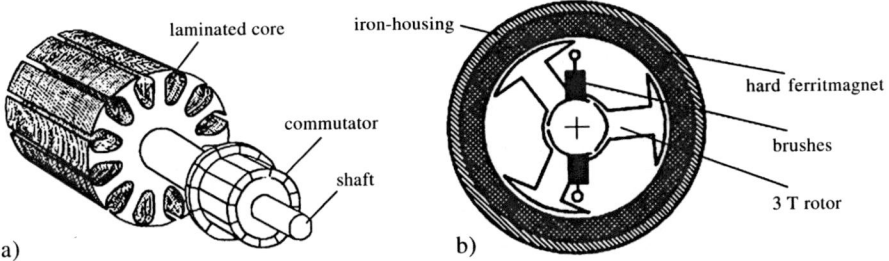

Fig. 2.6. Types of slotted rotors. a Drum-type rotor with 12 slots. b 3-slot rotor

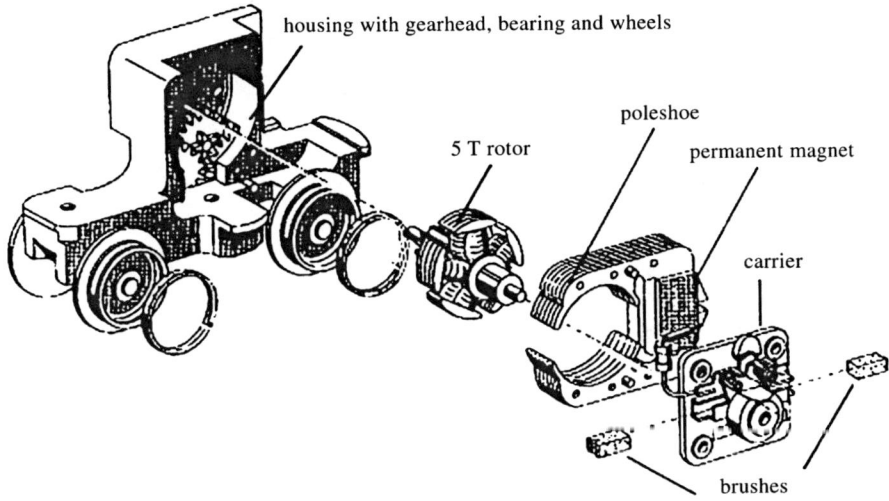

Fig. 2.7. Exploded view of a PM DC motor (model of a locomotive Märklin)

laps is equal to the number of poles is preferred with low voltages and higher currents. A wave winding where two laps are connected in parallel is used on higher voltages and lower currents. The number of windings can be even or odd. On even numbers of windings the so-called H-armature is applicable, leading to a well balanced rotor. An odd number of windings improves commutation, as only one winding is commutated at a time. Ripple torque is also minimised. Slot skewing at a ratio of one slot is often used to minimise noise, as the slot is not parallel to the pole. This is also a means to minimise torque ripple [2.1].

Commutator

Brush, spring and commutator allow for the application of electrical energy to the rotating armature winding. The brushes usually made of carbon, metal or mixtures of both, slide on a cylindrical or flat commutator. The brushes are guided and pressed on to the commutator by springs. The brushes conduct

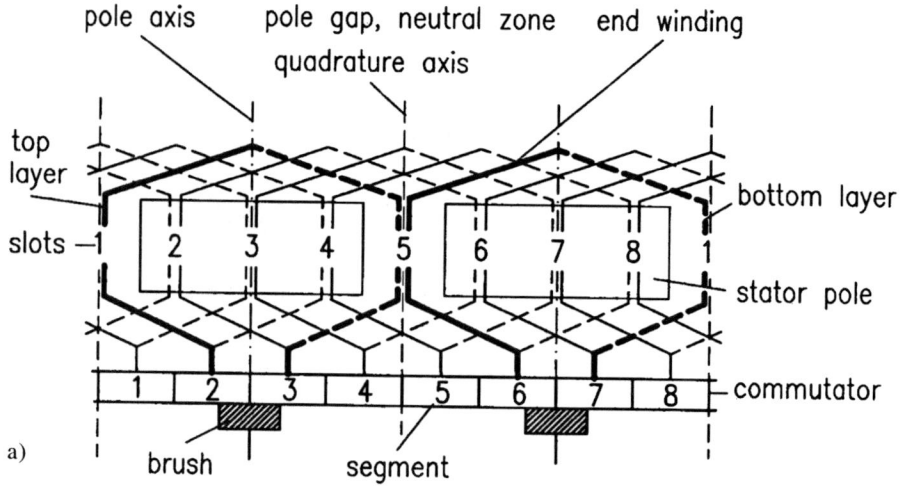

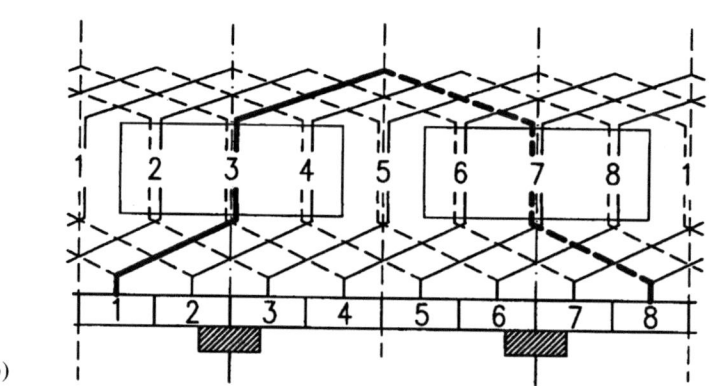

Fig. 2.8. Schema of windings. **a** Lap winding. **b** Wave winding

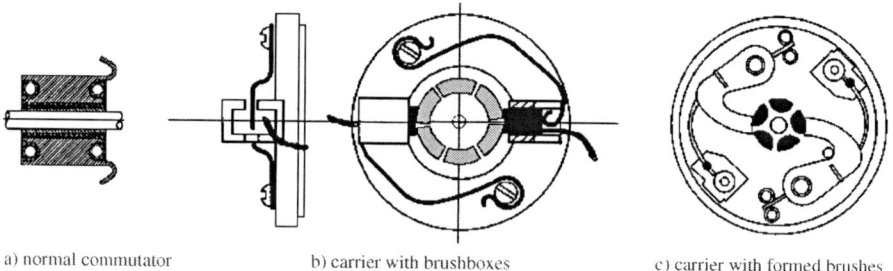

a) normal commutator   b) carrier with brushboxes   c) carrier with formed brushes

Fig. 2.9. Brushes, commutators, brush boxes and carrier

the current to the fins of the commutator. The latter are usually integrated on a carrier (Fig. 2.9b).

Commutation System

The commutation system enables energy transmission to the winding of the rotating armature from a resting source. For that reason, a commutator with copper sections is attached to the shaft. Each section is electrically connected to the armature winding. The copper sections form a cylindric (cylindric commutator) or a circular (flat commutator) surface on which the brushes glide along. The shape of the brushes is geometrically adapted to the construction. They are stabilized by a brush holder and pressed by springs on to the surface of the commutator. Brushes, brush holder, springs, elements for radio interference suppression, strip conductors and connectors are frequently combined into one assembly unit.

Bearings

Low-cost motors use bearings made of sintered alloys or plastic. Cylindrical and spherical porous bronze bearings usually fitted with an oil supply are state of the art. Their lifetime can reach up to 10 000 h or even more. Lubricants matched to operation temperatures of $-60\,°C$ to $+150\,°C$ are used. Hydrodynamic lubrication is available a few milliseconds after the start up of the motor. Porous bronze bearings are a low-cost solution but are unable to bear any axial forces. Ball bearings can sustain axial forces and the additional wear due to belt drives or friction gearboxes. On a well designed motor the lifetime is not determined by the bearing but by the commutation.

Motors with Coreless Rotors

Bell rotors with self-carrying winding are available with power ratings up to 250 W. The rotor is rotating between a two-pole AlNiCo or rare-earth

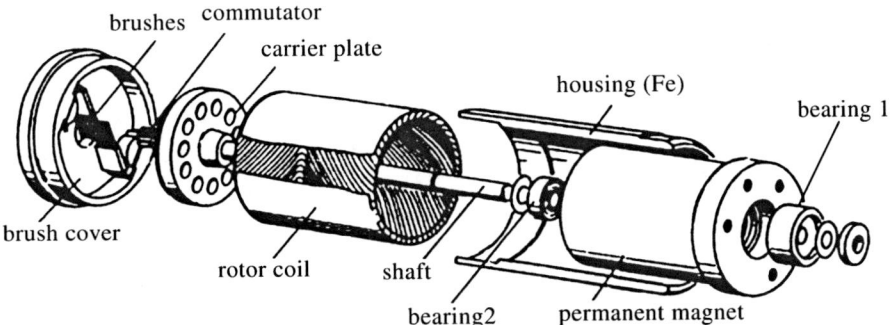

Fig. 2.10. Exploded view PM DC motor with coreless rotor

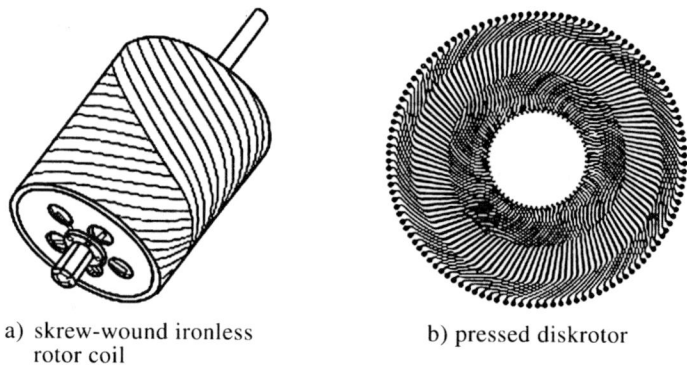

a) skrew-wound ironless rotor coil

b) pressed diskrotor

Fig. 2.11. Rotorforms of ironless PM-DC motors

magnet and the housing (Fig. 2.10). The rotor integrates seven to nine windings, carried out according to the maxon (Fig. 2.10) or faulhaber (Fig. 2.11a) patent. The wires overlap and yield a higher mechanical stiffness. Small motors use precious metal brushes for low resistance. The fins are often made of free wires, larger motors use conventional commutators and brushes. Bell rotors offer the lowest mechanical time constants and are very well suited for applications in control technology. The electrical time constant is usually lower when compared to slotted rotors leading to a better commutation. Disk or pancake motors with coreless rotor are available for various power ratings, overcoming the mechanical restrictions of bell rotors. The rotor is positioned between the permanent magnets and the steel housing. Either a magnet ring or separate magnets depending on the number of poles are used. Small motors with a diameter of 20 mm and beyond use four poles, three windings and a flat commutator. The armature winding is often stamped or etched (Fig. 2.11б). The winding is made of two parts, mounted on an isolator. The conductors are soldered on the inside and outside. The brushes glide on the armature winding; in larger motors

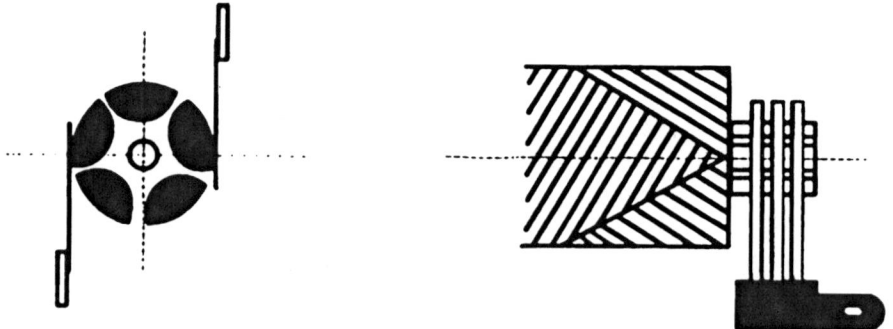

Fig. 2.12. Noble metal commutator and brushes of skew-wound ironless rotor coils

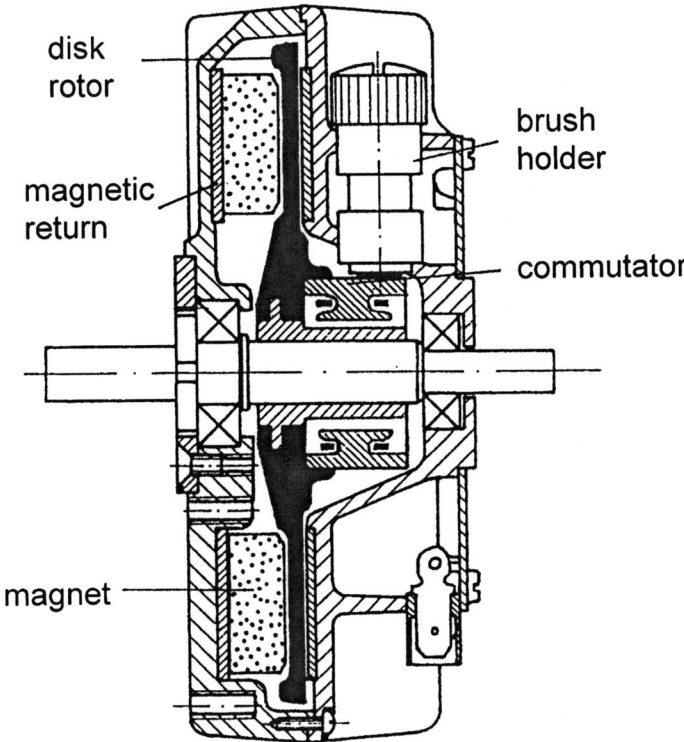

Fig. 2.13. Cross section of a motor with disk rotor

a conventional commutator is used (Fig. 2.13). The electrical time constant and the quality of commutation is better than that of slotted rotors. Large pancake motors offer a comparatively low mechanical time constant.

2.1.2.4 Principle of Operation

Torque Speed Function

Apart from the reluctance motors (see Sect. 2.3) the torque of electrical motors is proportional to the excitation flux ϕ and the armature current I. This leads to the equation for the magnetic torque

$$T_\mathrm{i} = c\phi I = k_\mathrm{T} I \:, \tag{2.1}$$

where c is the motor constant, that sums up all the geometrical data of the motor. The commutation system switches the terminal voltage in such a way that the current and direction are appropriate to the poles. The current in the windings alternates at a frequency proportional to the rotational speed. The higher the number of windings, the more constant the torque of the motor.

Due to friction, the torque on the shaft is minimised to the output torque

$$T = T_i - T_R \tag{2.2}$$

available from the motor.

The term $k_T = c\phi$ in Eq. (2.1) has the unit Vs and is called the flux linkage or torque constant. Flux linkage multiplied by angular velocity $\omega = 2\pi n$ leads to the induced voltage

$$U_i = k_T \omega \tag{2.3}$$
$$U = IR + U_i . \tag{2.4}$$

Taking the voltage loss on the winding into account, the equation for the armature voltage U gives the nominal voltage of the motor. The armature resistance R is the sum of the resistance of the armature winding and the conduction resistance between brushes and commutator. From equations (2.3) and (2.4) the speed can be derived.

$$n = \frac{U}{2\pi k_T} - \frac{IR}{2\pi k_T} , \tag{2.5}$$

The motor does not draw any current on ideal no-load operation. The first term of Eq. (2.5) is the ideal no-load speed n_0. In the case where the current is replaced by the torque leaving the friction aside, the torque speed curve is denominated as

$$n = n_0 - R\frac{T}{2\pi k_T^2} . \tag{2.6}$$

The rotational speed falls with increasing torque, determined by the resistance R. At standstill, $n = 0$, the equation shows the stall torque.

This might be written as

$$T_H = \frac{2\pi n_0 k_T^2}{R} = \frac{U^2}{2\pi n_0 R} \hat{=} T_S . \tag{2.7}$$

This equation denotes neither the warming up of the armature winding or the magnets nor

$$n = n_0 \left(1 - \frac{T}{T_H}\right) \tag{2.8}$$

the armature magnetisation. It is a guide for the selection of the motor in the design process.

Motor Current

At standstill ($U_i = 0$) using Eq. (2.4) the locked rotor currents are respectively

$$I_H = \frac{U}{R} \tag{2.9}$$

$$I_H = \frac{T_H}{k_T} . \tag{2.10}$$

The resistance R should be corrected to the value at operation temperature. On no-load the motor draws the current I_0, to compensate for the friction torque. The friction, which is a function of speed, is usually set to a constant value leading to the current

$$I = I_0 + I_H \frac{T}{T_H} . \tag{2.11}$$

Power

The output power P, that the motor develops at the shaft, can be calculated by Eq. (2.8) to be

$$P = \omega T = \omega_0 \left(T - \frac{T^2}{T_H} \right) . \tag{2.12}$$

Differentiated to T and set to zero this delivers the maximum power:

$$P_{max} = \frac{\omega_0 T_H}{4} = \frac{U^2}{4R} . \tag{2.13}$$

The power changes to the square of the voltage applied (Fig. 2.14). The maximum power is available at $T_H/2$ and $n_0/2$ (Fig. 2.15).

It should be noted, that at the point of maximum power the efficiency η is lower than 0.5. This causes an increase in temperature of the winding and the motor. For this reason DC motors are used in the first third of the torque speed diagram. This is also beneficial to the lifetime, as wear of the brushes is minimised and the efficiency is close to the maximum.

Efficiency

Efficiency is defined as

$$\eta = \frac{P}{P_1} = 1 - \frac{P_v}{P_1} \tag{2.14}$$

using input power $P_1 = UI$.

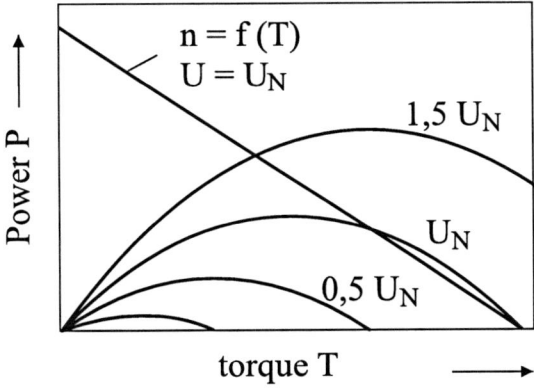

Fig. 2.14. Power $P = f(T, U)$

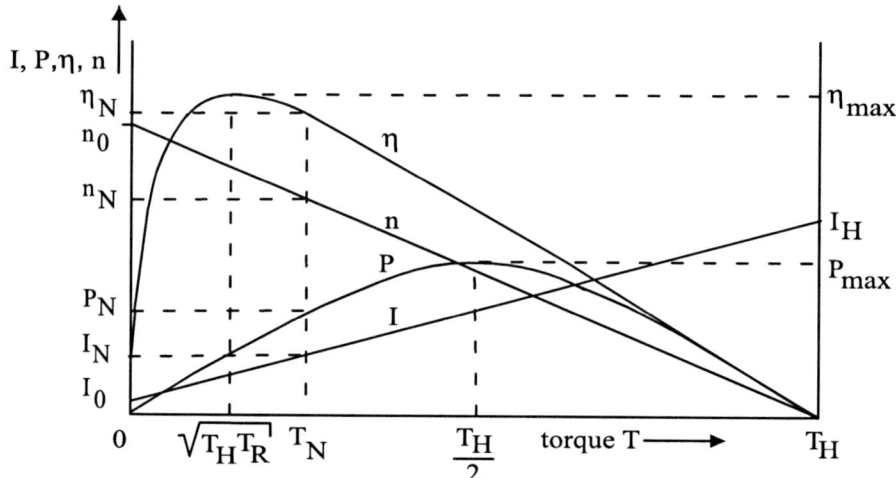

Fig. 2.15. Principle characteristic curves of PM DC motors

The losses in the motor are caused by:

- Iron losses, due to hysteresis and eddy currents
- Ohmic losses on the winding and brushes
- Friction on the bearings, air and commutator

The largest amount relates to the losses on the winding. Mohr [2.3] denotes a ratio of 1:4:5:8:10:15 to air, bearing, resistance of the brushes, friction of the brushes, iron losses and losses in the winding, depending largely on the design and application. On coreless motors the losses in the winding are dominant, with the rest the effect is less.

The efficiency is shown in Fig. 2.15. When operating on battery, the torque at maximum efficiency is of interest. The efficiency can be approximated to:

$$\eta = \left(1 - \frac{T_i}{T_H}\right)\left(1 - \frac{T_R}{T_i}\right) . \tag{2.15}$$

By deriving and setting to zero, the optimum torque can be found, leading to maximum efficiency:

$$T_{opt} \approx T_{i,opt} = \sqrt{T_R T_H} . \tag{2.16}$$

Using $T_{i,opt}$ in Eq. (2.15) the maximum efficiency is approximated to

$$\eta_{max} = \left(1 - \sqrt{\frac{T_R}{T_H}}\right)^2 . \tag{2.17}$$

Bell rotors using precious metal brushes offer a ratio of $T_R < 0.03\, T_H$. On other types of motor the ratio is lower.

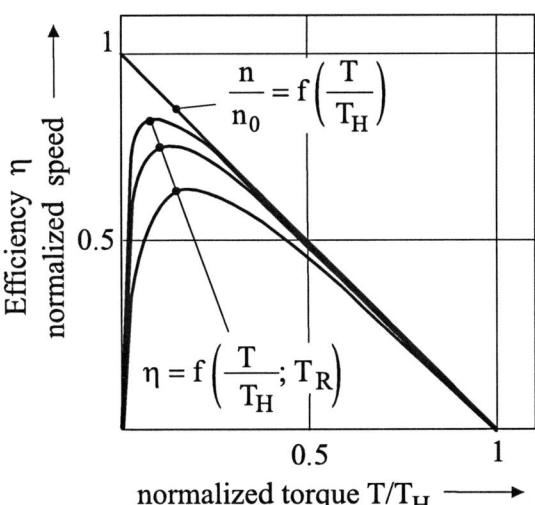

Fig. 2.16. Efficiency η and normalized speed $= f$ (normalized torque)

In Fig. 2.16, $T_{R1} < T_{R2} < T_{R3}$.

The most important technical data of DC PM motors are defined in DIN 42025, part 2. The motors are usually specified by their nominal data, but may also be used in the whole range of the diagram (Fig. 2.17).

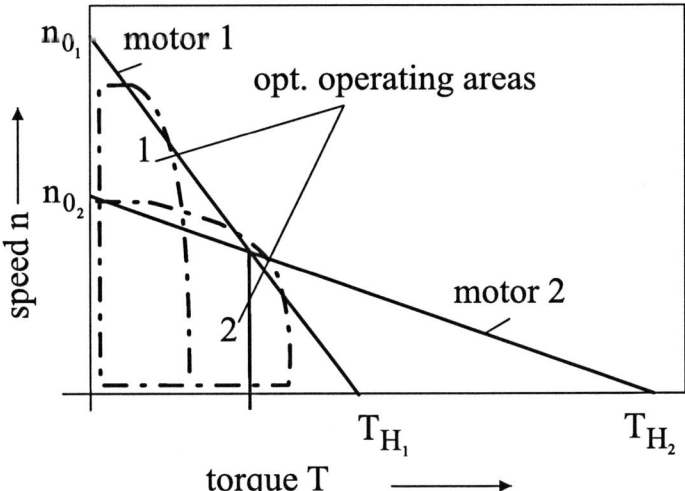

Fig. 2.17. n–T characteristic curves of two PM DC motors with recommendable operation areas

2.1.2.5 Speed Control

In principle, every point in the speed-torque diagram is a valid operation point. Power losses and lifetime have to be considered, so that the operation point can be chosen dependent on the application and the demands, if there is no special demand on price and lifetime.

The speed-torque diagram is moved in parallel to higher speeds on applying higher voltages to the motor, or *vice versa* (Fig. 2.18a).

Torque and power will rise at higher voltages, but so does overtemperature. The duty cycle decreases and so does lifetime. Still, this is acceptable in quite a few cases of short-term duty. Lifetime usually increases at lower voltages, but care must be taken using field tests on the application (Fig. 2.18). Resistors in series to the motor terminals will increase the gradient of the speed torque characteristic and will generate additional losses. Modern speed control is based on switch-mode controllers, thereby avoiding these losses.

Motors for wipers in automobiles often use a third brush (Fig. 2.19) to change the gradient of the speed torque characteristic. The motor is stable at an additional no-load speed [2.2, 2.3]. The commutation is getting worse, as the shortened windings remain in the field of the stator. More and more electronic speed control is applied to give a more suitable speed of wiper.

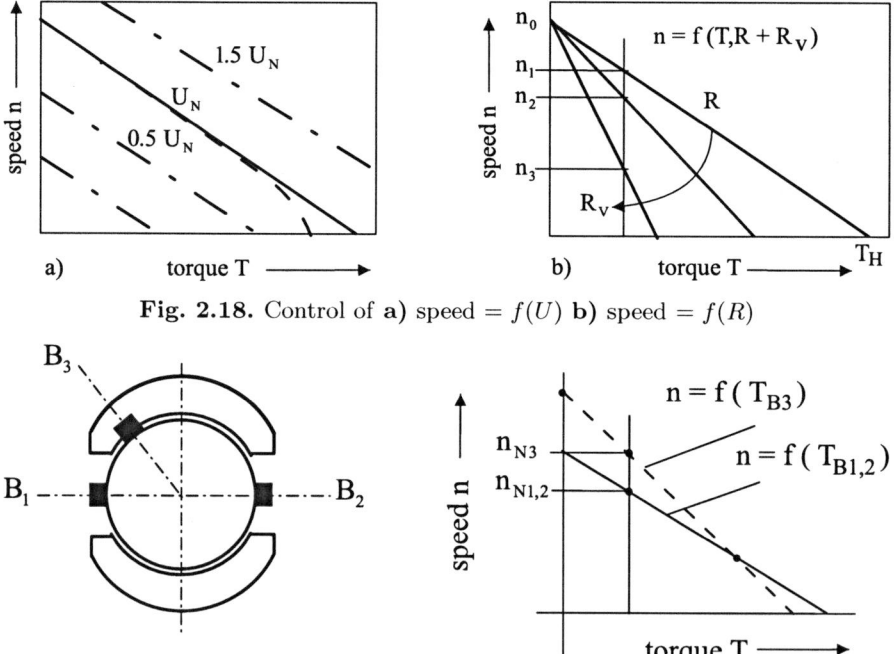

Fig. 2.18. Control of **a)** speed $= f(U)$ **b)** speed $= f(R)$

Fig. 2.19. Third brush and the influence to the n/T characteristic

2.1.2.6 Dynamic Operation

The power rate is sometimes used to characterise the dynamic performance of a motor. At start up with no external inertia, $T_L = 0$, the angular acceleration is $d\omega/dt = T/J_M$.

According to Eq. (2.2) the torque of the motor $T = T_i$ is assumed without friction and J_M denotes the inertia of the motor. The power rate indicates the increase in power over time denoted as

$$\dot{P} = \frac{dP}{dt} = T\frac{d\omega}{dt} = \frac{T^2}{J_M}. \qquad (2.18)$$

The mechanical time constant of the motor might also be considered (Eq. 2.23).

The gradient of the speed torque diagram, $\Delta n/\Delta T = n_0 R/nk_T$.

The calculation of the dynamic behaviour is usually not relevant, apart from some servo or stepper motors.

In the forthcoming section all time variant data is written in lower case letters. In Eq. (2.2) for the calculation of the torque the inertia J has to be added. When using a gearbox, the inertia of the load has to be transformed to the shaft of the motor. The same applies for the torque T_L. The friction torque, assumed to be constant is also included. Using Eq. (2.1) for the torque delivers

$$T_L = k_T i - J\frac{d\omega}{dt}. \qquad (2.19)$$

The current is not constant on dynamic operation, so to the equation for the voltage (Eq. 2.4) another term for the inductance has to be added. Using Eq. (2.3) the voltage is

$$u = Ri + L\frac{di}{dt} + k_T \omega. \qquad (2.20)$$

The Laplace Transform is a way to solve these equations and to calculate the speed over time (Eq. 2.16). To calculate current and speed on changes of voltage and load, the equations for continuous operation (Eqs. (2.2) and (2.4)) also have to be transformed. Subtraction of the two equations for voltages and torques and the solving in the Laplace domain delivers, after transformation, the equation for the change in current at constant torque with changing voltage:

$$\Delta i(t) = \frac{\Delta U}{R}\frac{1}{\sqrt{1 - 4\frac{\tau_e}{\tau_m}}}(e^{P_1^t} - e^{P_2^t}) \qquad (2.21a)$$

using the time constant of the rotor:

$$\tau_m = \frac{RJ}{k_T^2} \qquad (2.22)$$

and with

$$P_{1,2} = -\frac{1}{2\tau_e}\left(1 \pm \sqrt{1 - \frac{4\tau_e}{\tau_m}}\right)$$

and with the root of the solution of the characteristic equation in the Laplace domain.

The change in speed is described as

$$\Delta n(t) = \frac{\Delta U}{2\pi k_T} \left[1 + \frac{1}{\sqrt{1 - 4\frac{\tau_e}{\tau_m}}} \left(p_2 \, e^{p_1^t} - p_1 \, e^{p_2^t} \right) \right]. \qquad (2.23a)$$

In the case that $\tau_m > 4\tau_e$, current and speed will relate to the stationary value aperiodically. This is usually the case for small motors with relatively high armature resistance. In a lot of cases the armature winding is dimensioned in such a way that $\tau_m > 4\tau_e$ leads to a much simpler form for current and speed:

$$\Delta i(t) = \frac{\Delta U}{R} e^{-\frac{t}{\tau_m}} \qquad (2.21b)$$

$$\Delta n(t) = \frac{\Delta U}{2\pi k_T} \left(1 - e^{-\frac{t}{\tau_m}} \right). \qquad (2.23b)$$

In Fig. 2.20a the change in voltage is shown to increase speed at a given torque. The current increases for a short period of time and returns to the stationary value. Motors with a higher time constant show a lower peak in current, the risk of a high current is lower. On the other hand, a rise in speed will take more time, leading to poorer dynamic behaviour.

For $\tau_m < 4\tau_e$ damped oscillations will arise:

$$\Delta i(t) = \frac{\Delta U}{\omega_k L} e^{-\frac{t}{2\tau_e}} \sin \omega_k t \qquad (2.24)$$

$$\Delta \omega(t) = \frac{\Delta U}{k_T} \left[1 - \left(\cos \omega_k t + \frac{1}{2\tau_e \omega_k} \sin \omega_k t \right) e^{-\frac{t}{2\tau_e}} \right]. \qquad (2.25)$$

With $2\tau_e$ and ω_0 we get

$$\omega_k = \sqrt{\frac{1}{\tau_e \tau_m} - \frac{1}{4\tau_e^2}}. \qquad (2.26)$$

On a change of torque at constant value the current is

$$\Delta i(t) = \frac{\Delta T_L}{k_T} \left[1 + \frac{\tau_e}{\sqrt{1 - 4\frac{\tau_e}{\tau_m}}} \left(p_2 \, e^{p_1^t} - p_1 \, e^{p_2^t} \right) \right] \qquad (2.27a)$$

$$\Delta n(t) = -\frac{\Delta T_L R}{2\pi k_T^2} \left\{ 1 + \frac{\tau_e}{\sqrt{1 - 4\frac{\tau_e}{\tau_m}}} \left[\left(p_2 + \frac{1}{\tau_m} \right) e^{p_1^t} - \left(p_1 + \frac{1}{\tau_m} \right) e^{p_2^t} \right] \right\}, \qquad (2.28a)$$

simplified to

$$\Delta i(t) = \frac{\Delta T_L}{k_T} \left(1 - e^{-\frac{t}{\tau_m}} \right) \qquad (2.27b)$$

$$\Delta n(t) = -\frac{\Delta T_L R}{2\pi k_T^2} \left(1 - e^{-\frac{t}{\tau_m}} \right). \qquad (2.28b)$$

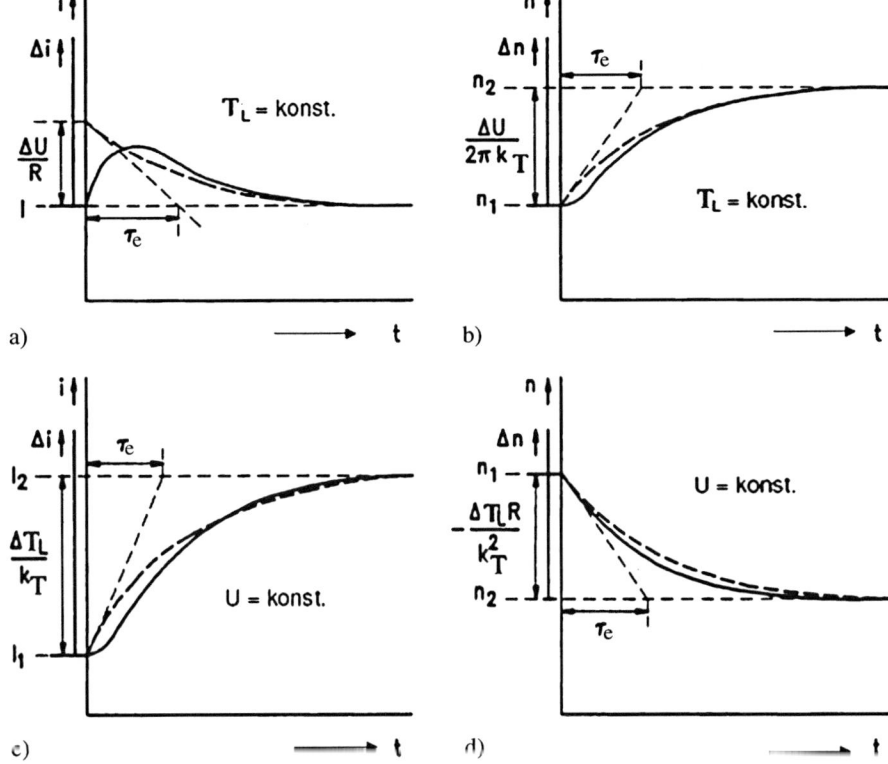

Fig. 2.20. Characteristic curves of current and angular velocity, voltage jump and torque jump

Figure 2.20c, d shows the effect of the change of load. The motor draws a higher current on a higher torque, the rotational speed falls. A motor with a higher time constant shows poorer dynamic behaviour.

The start up of servo drives is often an issue. Most applications start with a given inertia, friction being the only load. On neglecting friction, Eqs. (2.20) to (2.22) can be used. The terminal voltage U replaces the step in the voltage ΔU. The rotational speed reaches the ideal no-load speed $\omega_0 = U/R$ after about $4\tau_m$; the current will decrease to I_0 or ideally zero after an initial overshot.

2.1.3 Commutator Series Motor (Universal Motor)

2.1.3.1 Description

Commutator series motors in the power range up to 2.5 kW are termed universal motors, or AC–DC motors. Their development is closely related to household appliances and electric power tools. These devices are produced in high quantities for a broad number of applications and users. They had to be operated with both AC and DC during the first decades of their more than hundred year's long history. The power range is adapted to mobile use of the equipment, which only requires a conventional power outlet to be available. The possibility of using both current supplies is achieved by a serial connection of exciter and armature winding. In case of an alternating current source, the magnetic field of the stator winding and the magnetic flux of the rotor change sign simultaneously. Thus, the direction of the torque remains unchanged.

The majority of commutator series motors have a stator lamination cross section as shown in Fig. 2.21a, independent of specific application purposes, operation parameters and design aspects. It is frequently referred to as universal motor lamination cross section.

Due to the multitude of applications of these motors compared to other motor types, a much larger effect of reducing material and production costs is achieved by specific modifications of the lamination cross section.

An important aspect of cross section design is the width of the handle bar of a single-handed device. This leads to lamination contours with one main extension that is not much longer than the diameter of the rotor. The cross sections shown in Fig. 2.21b–e are examples of different design criteria.

Cross section b) is used for a simultaneous winding of both stator windings in case of large-scale production numbers. The stator yoke in d) was split for a specific assembly of exciter winding. The two separate lamination stacks are combined after the exciter coils have been inserted. The air-gaps between the stator yokes are problematic, since the complete magnetic flux is carried through them with high induction. Cross section b) shows the hollows which are characteristic for interlocking. Apart from symmetrical stators, unsymmetrical stators are also produced (c and e). They have the stator yoke located only on one side of the pole pair. With this design, particular assembling requirements of the construction are fulfilled, and simple technological solutions for the production of exciter windings can be realized.

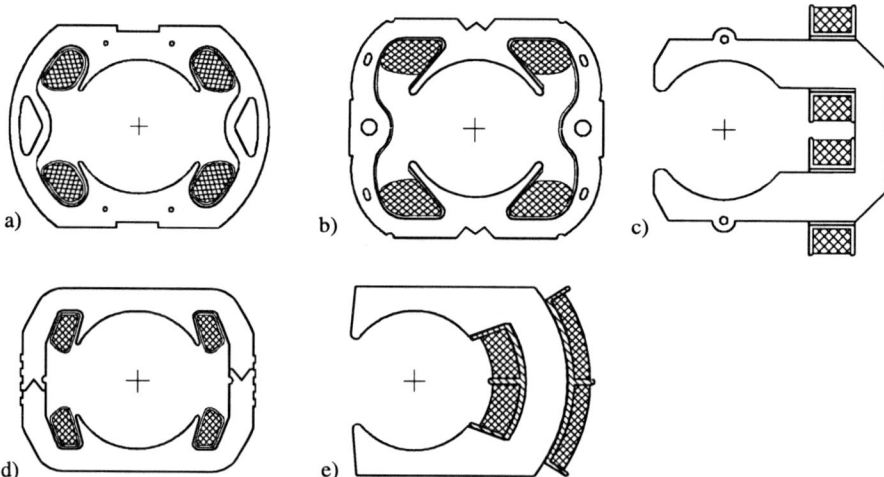

Fig. 2.21. Various cross sections of stator laminations

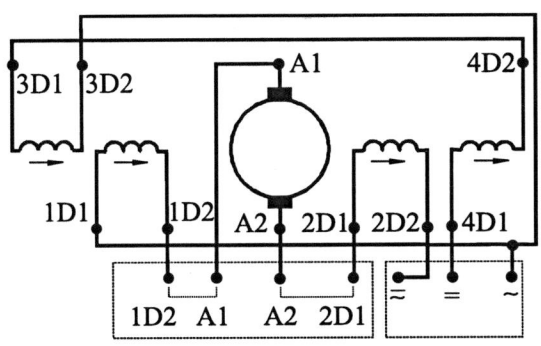

Fig. 2.22. Electrical connections of the commutator series motor

Generally, it has to be considered that a stator shape is evaluated not only by material usage, production costs and device design. The stator construction is also influenced by the lifetime of carbon brushes and radio shielding.

The worldwide installation of an AC supply requires the optimal specification of the motors as either an AC motor or a DC motor with rectifier. The naming of the commutator series motors as "universal motors" can thus be considered to have developed over time. Without any technical modifications, the same working point for both current sources is only achieved, if the power factor $\cos\phi$ is larger than 0.95.

The adaptation to operation with both current sources is realized by a change of the number of exciter windings of the rotor. It is increased when switching from AC to DC operation mode. The pole winding consists of two partial coils, which are combined depending on the kind of voltage (Fig. 2.22).

Three of the ten available connections must be accessible. A further four connections have to be added in order to facilitate a reversal of rotation. Therefore, the dashed connections 1D2-A1 and A2-2D1 can be opened and the contacts 1D2-A2 and A1-2D1 are closed. The direction of the armature field is hence rotated by 180°, while the direction of the excitation field is maintained.

2.1.3.2 Characteristic Features of Commutator Series Motors

Commutator series motors for use with low power are characterized by a few peculiarities compared to larger motor constructions that are used, for example in railway locomotives. Those specifics are determined by cost constraints, application purposes, and manufacturing processes. The following factors can be emphasized:

- The motors are exclusively manufactured bi-polarly.
- The armature winding is primarily connected between the two excitation windings in order to minimize radio interference. Nevertheless, an additional component for radio shielding is required.
- The stator has no commutating and compensating winding.

- Apart from a few exceptions, there is only one direction of rotation permitted. In order to improve the commutation, a tilt of the carbon brush bridges or a specific connection of the coils is applied, which causes the commutation zone not to be located in the middle of the pole gap.
- The resistance of the armature windings varies up to 10% due to the difference in winding length.
- The windings of the armature remain in the same layer, and do not change from upper to lower layer.
- Tang commutators are used almost exclusively.
- The rotor is not equipped with special balancing fixtures. Instead, balancing is achieved by milling holes into the rotor or by attaching epoxy putty to the windings.
- The working point is located far in the saturation region of the magnetization characteristic.
- The typical speed range is between $4000\,\text{min}^{-1}$ and $40\,000\,\text{min}^{-1}$.
- The primarily used principle of cooling is self-ventilation by a fan attached to the shaft.
- The motors are mainly used for inside appliances. Hence, the electrically accepted power is declared, not the mechanical power provided at the shaft.
- The inspection requirements and the achievable life cycle depend on the specific motor.
- Different to larger machines, the no-load speed is limited by the friction of brushes, bearings and air.
- The hysteresis and eddy-current loss in the lamination stack of the stator depends on the frequency of the source voltage. The loss in the lamination stack of the rotor is correlated to the rotation speed.

Advantages of the commutator series motors result from a relatively large locked-rotor torque in comparison to the rating torque. The short duration of the currents does not require changes in construction for thermal reasons that might increase the size of the components. The torque pulsation during operation is transmitted automatically without special controllers or regulators. The motor reacts accordingly with a change of rotation speed and current consumption. Any rotation speed can be realized between standstill and no-load operation. The actual speed is determined by the load only. Because of this property, several functions of operation can be realized by one speed-torque characteristic, as for example in a kitchen appliance with exchangeable tools (Fig. 2.23).

In general, the stationary characteristics of small series motors can be experimentally measured in the complete speed range between stand-still and no-load operation. Only a limited speed range is accessible for motors with high power that are conceptually attached to a gearing. Small rotation speeds cannot be held stable due to the temperature increase by large currents, and the no-load speed is basically limited by the gearing. Higher rotation speeds can only be realized without the gearing. However, the higher perpetual forces can exceed the permitted limits and will lead to damage of commutator and winding.

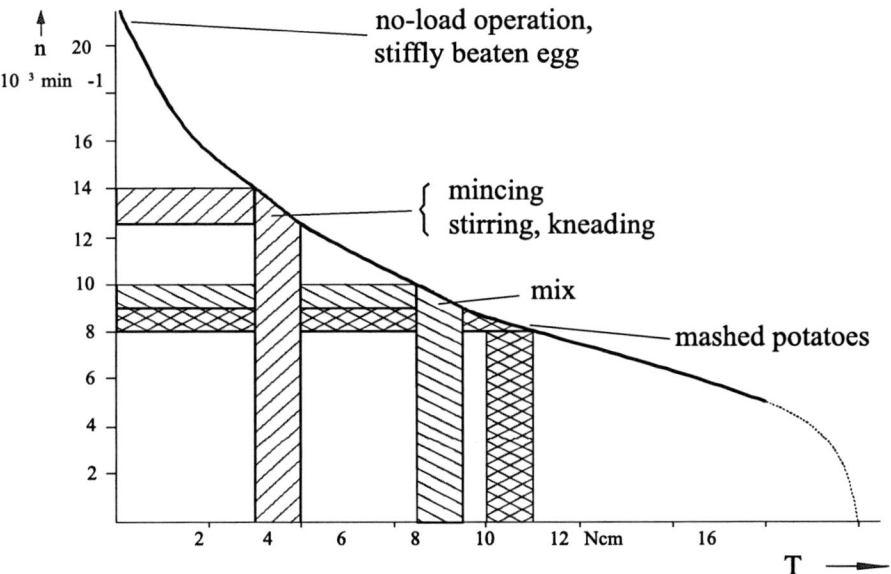

Fig. 2.23. Variable working points and load conditions of a food processor

Figure 2.24 shows the measurable interval of the speed-torque characteristic of an electric power tool with the power rating of $P_{el} = 2000\,\text{W}$. Additionally, the total input power P_{el}, the mechanically collected power P_{mech} at the shaft, the current I, the efficiency factor η and the power factor $\cos\varphi$ are depicted as a function of the torque at the shaft.

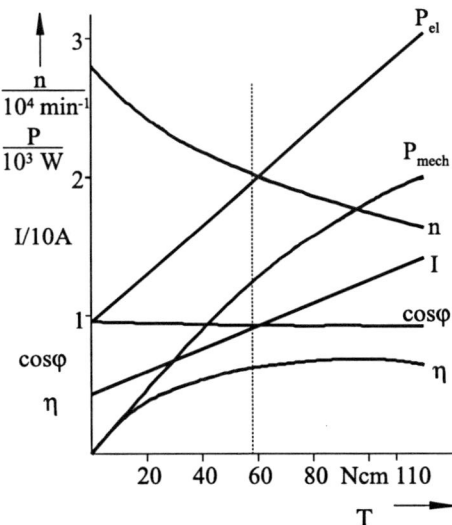

Fig. 2.24. Measured speed-torque characteristic of electric power tools with a power rating of $P_{el} = 2000\,\text{W}$

Table 2.2. Life ranges and brush endurances of selected appliances

Appliance	Life time in h
Chopper	30–60
Coffee grinders	50–100
Hammer drills	50–150
Compass saws	100–250
Circular saws	120–350
Angle grinders	200–600
Hand mixers	250–500
Vacuum cleaners, pumps	700–1500
Washing machines	2000–3000

Commutator series motors are not suitable for permanent operation due to the limited lifetime of the carbon brushes. The conventional life range requirements depend on the application. They are valid for an operation period of 10 years and last for between 15 and 3000 h. Occasionally, the operation cycles are referenced, as in washing machines (between 2000 and 4000 washing cycles). Life ranges or brush endurances of selected appliances are listed in Table 2.2.

2.1.3.3 Principal Operating Behavior of Commutator Series Motors

Description of the Simplified System

Simple circuitry is used to explain the differences of the stationary operation behavior that is valid for both AC and DC operations. It is characterized by the perpendicular orientation of armature and exciter fields. There is no transformatory coupling of the two windings. The magnetization losses are neglected and problems with commutation and radio interference are disregarded.

The partition of the exciter winding for radiation interference damping can be simplified by the depiction of one exciter coil only (Fig. 2.25). The connection between the total excitation flow Φ_E, that is linked up with excitation winding and the current is given by the magnetization characteristic.

Its particular trace can be applied to the dependency of the flow through the brush level Φ_B as function of the current I. In order to explain the oper-

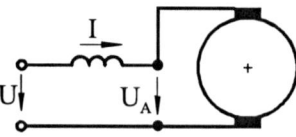

Fig. 2.25. Elementary diagram of the commutator series motor

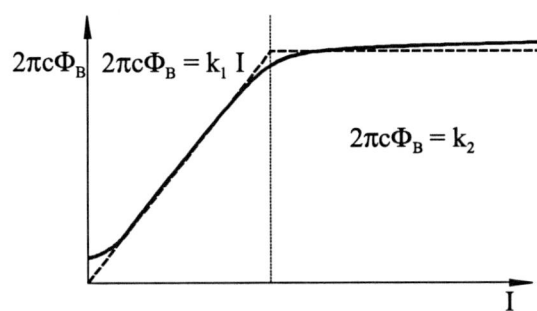

Fig. 2.26. Flow through the brush level as a function of the current

ation mode of the series-motor, the graph $2\pi c\Phi_B = f(I)$ is divided into the non-saturated part with $2\pi c\Phi_B = k_1 I$ and the saturated part with $2\pi c\Phi_B = k_2$ (Fig. 2.26).

The remanent flux at $I = 0$ can be neglected, since it is not significant for motor operation. The non-saturated range is most important for no-load operation. The saturated region determines the properties of the motor during start and operation rating.

DC Operation

The exciter and armature winding are understood as ohmic resistances R_E and R_{AA} in the voltage equation of the DC operation. This mode is characterized by the steady-state operation point with $I = $ constant and $\Phi_B = $ constant. The brush contact resistance R_{BC} has also to be considered. Hence, the total motor resistance R is equal to

$$R = R_E + R_{AA} + R_{BC} = R_E + R_A \ . \tag{2.29}$$

The voltage equation can be derived from the equivalent network of the DC series motor (Fig. 2.27). Using the positive counting of voltages and currents as depicted, the voltage is

$$U = IR + U_i = IR + 2\pi c\Phi_B n \ . \tag{2.30}$$

The operation behavior of the motors is characterized by rotation speed and current as a function of the torque. The number of revolutions can be obtained

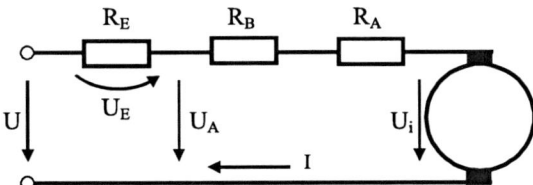

Fig. 2.27. Equivalent network of the DC series motor

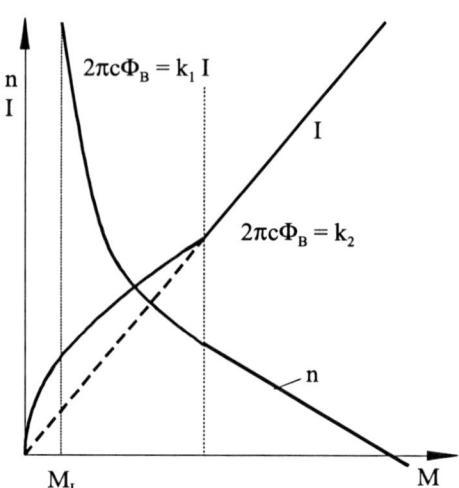

Fig. 2.28. Current and number of revolutions as function of torque in both regions of the magnetization characteristic

from the voltage equation by

$$n = \frac{U}{\sqrt{2\pi k_1}} \frac{1}{\sqrt{T}} - \frac{R}{k_1} . \qquad (2.31)$$

This expression is composed of a hyperbolic term $n_1 = \frac{U}{\sqrt{2\pi k_1}} \frac{1}{\sqrt{T}}$ and a constant term $n_2 = -\frac{R}{k_1}$. For a vanishing torque, i. e. $T = 0$, the number of revolutions is infinitely large. It is limited in practice by the torque of the no-load operation that is determined by the friction moments of brushes, bearings and ventilation. Hence, the number of revolution approaches a finite value while reducing the load.

The series motor progresses to a shunt motor in the second range of the magnetization characteristic with $2\pi c\Phi_B = $ constant. The dependency of the rotation speed from the torque is described by

$$n = \frac{U}{k_2} - \frac{2\pi}{k_2^2} RT . \qquad (2.32)$$

Theoretically, a change of the magnetization regions in the function $I = f(T)$ (Fig. 2.28) causes a significant alteration of the graph. The transition from a series to a shunt characteristic cannot be clearly distinguished in real motors, the more so as the ohmic resistance usually varies during a series of measurements.

AC Operation

In comparison to the stationary operation of the DC motor, the induction voltage drop of the exciter and armature windings of the AC motor have to be considered in the voltage equation (a pure sinusoidal voltage source is assumed here).

The rated operating point is located in the saturation region of the magnetization characteristic. This is described as one specific feature of the series motor in Section 2.1.3.2. The assumption of a constant inductance of the exciter winding is hence an approximation. The average inductances in no-load and rated operation can be different by more than a factor of 2. The inductance of the armature is considerably smaller compared to the stator and can be set constant.

Large errors occur if the mutual inductance between exciter and armature winding is measured with a tilted brush bridge and zero current through the exciter winding. However, the rated operation behavior with linear conditions and constant parameters permits a fairly good estimation of the motor properties.

With the use of this simplification, vector graphs of the voltages and the current phasor diagram can be developed for constant voltage of the source net. The voltage equation in abbreviated form is

$$\underline{U} = U_\mathrm{E} + U_\mathrm{A} = R_\mathrm{E}\underline{I} + jX_\mathrm{E}\underline{I} + R_\mathrm{A}\underline{I} + jX_\mathrm{A}\underline{I} + \underline{U}_{\mathrm{ir}} . \tag{2.33}$$

It leads to the equivalent network description of Fig. 2.29. The ohmic voltage drops and the armature voltage $\underline{U}_{\mathrm{ir}}$, induced by the rotation, are in phase with the current.

The voltage across the exciter winding is composed of the ohmic voltage drop and the inductive components across the leakage and air-gap reactances ($X_{\sigma E}$, $X_{\delta E}$). Due to the transformatory coupling with the armature field, an additional component appears if the commutation zone is different from the center of the pole gap:

$$U_E = \underline{I}R_\mathrm{E} + j(X_{\delta E} + X_{\sigma E})\underline{I} - jX_\mathrm{EA}\underline{I} \quad \text{with} \quad X_{\delta E} + X_{\sigma E} - X_\mathrm{EA} . \tag{2.34}$$

In order to improve the commutation, the brush bridge is tilted away from the pole gap center such that a field with opposite orientation to the air-gap field of the exciter winding is created. This can be experimentally observed very well: When the exciter winding conducts no current and a current flows through the armature winding, the rotor moves against the motor rotation during normal operation. Hence, the voltage drop across the counter inductivity $jX_\mathrm{EA}\underline{I}$ has a negative sign.

The following terms have to be considered in the voltage equation of the armature:

- The ohmic voltage drops at the contacts brush-commutator ($R_{\ddot{\mathrm{U}}}$) and at the armature winding resistance (R_AA)
- The induced electromagnetic force ($\underline{U}_{\mathrm{ir}}$)

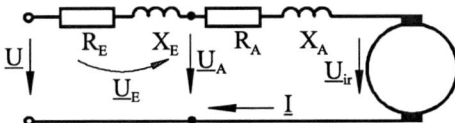

Fig. 2.29. Equivalent network of the AC commutator motor

- The voltage drops across the self-inductance of the armature (X_{AA}) and across the mutual inductance between armature and exciter winding (X_{AE})

$$U_A = (R_{AA} + R_{\ddot{U}})\underline{I} + \underline{U}_{ir} + j(X_{AA} - X_{AE})\underline{I}. \qquad (2.35)$$

Since the induced electromagnetic force is proportional to the magnetic flux through the brush level, which is directed by the current, the following expression holds:

$$\underline{U}_{ir} = 2\pi cn\underline{\Phi}_B = \frac{\Omega}{\omega} X_{AEr}\underline{I},$$

where Ω is the angular velocity of the rotor and ω is the angular velocity of the source net. The voltage equation of the motor has the simple form

$$\underline{U} = \left(R + \frac{\Omega}{\omega} X_{AEr}\right)\underline{I} + jX\underline{I}. \qquad (2.36)$$

The two terms of the terminal voltage have a 90° phase shift relative to each other and describe a circle as phasor diagram (Figure 2.30a) with the magnitude of the terminal voltage as diameter.

It is impossible to measure the circular phasor diagram of the voltages with the ratio of the mechanical and electrical angular velocities as parameter. Only the armature and exciter voltages can be measured. The end-points of their phasors are located within the circle, and the components U_E and U_A are not perpendicular to each other.

The current is determined with the use of the voltage equation (Eq. 2.36):

$$\underline{I} = \frac{\underline{U}}{R + \frac{\Omega}{\omega} X_{AEr} + jX}. \qquad (2.37)$$

The resulting current phasor diagram is a circle with the parameter Ω/ω. The maximum of the current is given at negative rotation speed Ω with a parameter value of

$$\frac{\Omega}{\omega} = -\frac{R}{X_{AEr}}.$$

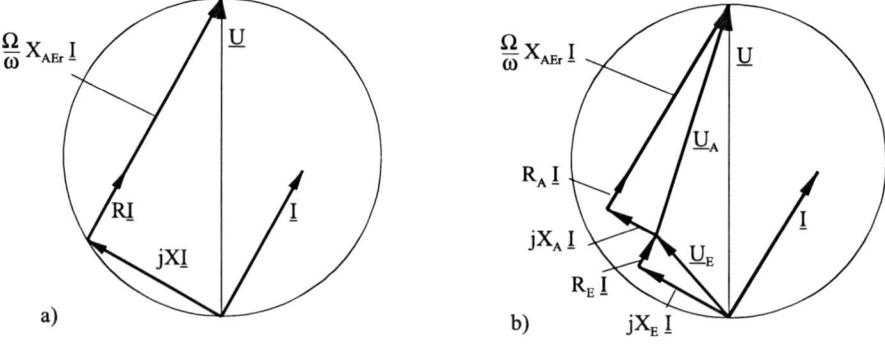

Fig. 2.30. Phasor diagrams of the voltages according to: **a** Eq. (2.38); **b** Eq. (2.33)

A variety of characteristic features of the AC series motor can be deduced from the phasor diagram depiction (Fig. 2.31). The phase shift (phase displacement) between current and voltage decreases with the increasing ratio of the angular velocities Ω/ω. The power factor $\cos\varphi$ becomes >0.9 at rotation speeds larger than $20\,000\,\text{min}^{-1}$, and it is almost 1 at speeds above $35\,000\,\text{min}^{-1}$.

The straight line of standstill is defined by the link of the coordinate origin with the point $\Omega/\omega = 0$ of the current phasor diagram. It divides the effective component of the current into the segments $\overline{AB}$ and $\overline{BC}$. Segment $\overline{AB}$ is proportional to the winding loss, and segment $\overline{BC}$ corresponds to the mechanical power. The tangent to the current phasor diagram parallel to the straight line of standstill defines the operation point at which the highest mechanical power is achieved.

The distance $\overline{DA}$ is proportional to the reactive component of the current. Verifiably, it is a measure of the torque. In the high speed range, small changes of the torque cause large variations of revolutions. Small differences in the frictional torques cannot be prevented in series production. Appliances are accepted as error-free with a deviation of up to $\pm 10\%$ from the revolution average after final assembly.

The current characteristic allows for the deduction of relations for the torque

$$T = \frac{1}{\omega} \frac{X_{\text{AEr}} U^2}{(R + \frac{\Omega}{\omega} X_{\text{AEr}})^2 + X^2} \tag{2.38}$$

and for the ratio of angular velocities

$$\frac{\Omega}{\omega} = \frac{X}{X_{\text{AEr}}} \sqrt{\frac{T_{\max}}{T} - 1} - \frac{R}{X_{\text{AEr}}} \tag{2.39}$$

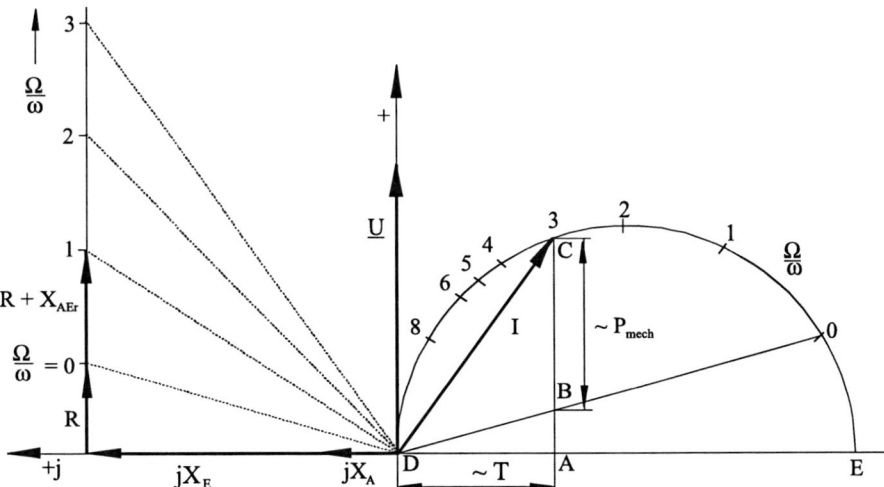

Fig. 2.31. Current phasor diagram of the AC series motor for constant inductances

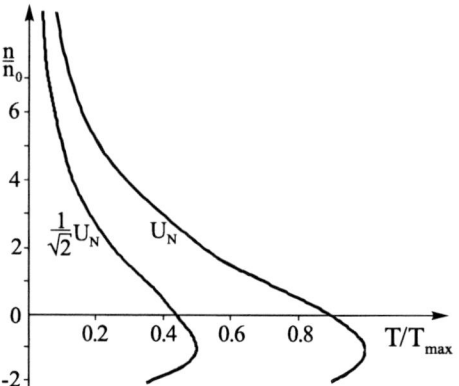

Fig. 2.32. Speed-torque characteristic of the AC series motor

with

$$T_{\max} = \frac{1}{\omega}\left(\frac{U}{X}\right)^2 X_{\text{AEr}} \,. \tag{2.40}$$

The resulting speed-torque diagram is shown in Figure 2.32 with the relations $\Omega = 2\pi n$ and $\omega = 2\pi n_0$. It is characterized by an infinite large revolution for an ideal no-load operation and by a maximum of the torque in the range of negative speed.

Considering the fact that the current is a time-dependent quantity, described by the fundamental wave equation $i = \hat{I}\cos(\omega t + \varphi_i)$, the temporal development of the torque is expressed by

$$T_g(t) = \frac{1}{\omega} X_{\text{AEr}} \hat{I}^2 \frac{1}{2}(1 + \cos(2\omega t + 2\phi_i)) \,. \tag{2.41}$$

The torque oscillates around the average value $T = \frac{1}{2\omega} X_{\text{AEr}} \hat{I}^2$ with the amplitude of the average and a frequency that equals twice the frequency of the source supply. It hence varies between zero and the maximum value $T_{\max} = 2T$. The number of revolutions is therefore not constant. However, the range of the oscillations is small due to the moments of inertia of the rotor and the attached load and the efficiency of the AC commutator series motor is not impaired.

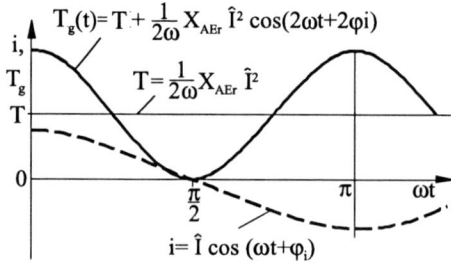

Fig. 2.33. Current and torque as function of time in stationary operation mode

2.1.3.4 Design of the Stator Winding of the Commutator Series Motors Used as Universal Motor

The functions of operation have to be maintained when connecting the universal motor to one or other voltage source. In order to compare both variations of operation, the voltage equations are simplified by neglecting the ohmic resistances. Two questions can be discussed with the derived equations:

1. How does the number of revolutions change if the voltage source is switched at constant torque and unchanged motor parameters?
2. What number of exciter windings has to be selected if the same operation point needs to be realized for both sources?

The answer to the first question is obtained with the use of the conditions $R = 0$, $U_\sim = U_=$ and $T_\sim = T_=$. The number of revolutions is reduced by $\cos\varphi$ for AC voltage operation:

$$n_\sim = n_= \cos\varphi . \tag{2.42}$$

The second question is answered with the additional assumption of $n_\sim = n_=$. For AC voltage operation, the number of exciter windings is reduced by the factor $\cos^2\varphi$:

$$w_{E\sim} = w_{E_=} \cos^2\varphi . \tag{2.43}$$

The power factor, which explains the differences of the exciter winding numbers quantitatively, deviates only slightly from 1 at revolutions above $20\,000\,\text{min}^{-1}$. Hence, the motors can be operated with DC or AC voltage. This conclusion needs to be restricted in case of differences in the commutation of the two operation modes, which affect the live cycle of the contact system.

2.1.3.5 Speed Control

Speed Control Requirements

The applications of the commutator series motor cause different speed requirements. Among those are:

- Compliance with a constant number of revolutions within a variation limit of ±1%;
- Soft starting (current and torque limitation during start);
- Braking from the speed of no-load and rated operation mode to standstill within the shortest possible time (for example in choppers, angle grinders or lawn mowers);
- Switch of revolution direction within a short duration of operation or for the same operating point in both directions;
- Speed control range up to the ratio of 1:80.

The speed control range can be realized by a variety of measures. Handling, costs and reliability determine the most suitable solution.

Voltage Control with Transformer

The change of the terminal voltage affects the number of revolutions as shown by the relation of speed and torque (Equation 2.30). The terminal voltage can easily be modified by the use of a variable transformer. The speed torque characteristics for a decreasing voltage are characterized by a smaller starting torque and the reduction of the achievable number of revolutions in no-load operation. The maximum of the torque declines, while the corresponding number of revolutions remains constant. Nowadays, transformers are rarely used in practice due to high costs and space requirements.

Ohmic Resistances for Speed Control

Ohmic resistors can be connected in series or in parallel to the armature. Both options can be combined in the Barkhausen circuit (Fig. 2.34). According to Eq. (2.31), the number of revolutions can be reduced by an ohmic resistor within the electric circuit. Compared to the voltage control with a transformer, the reduction of the achievable number of revolutions is smaller and the magnitude of rotation speed at maximum torque is higher. This option for speed control is rarely used in equipment due to its limited efficiency. Nevertheless, it is used in combination with centrifugal governors for speed stability control.

The connection of an ohmic resistor in parallel to the armature is possible, because the phase shift between armature voltage and current is small. The additional current parallel to the armature through the exciter winding causes a negligible phase shift between exciter field and current. The exciter current is made larger than the armature current. Therefore, the current through the exciter winding proceeds during load relieving and the no-load speed is limited. The situation becomes critical from a thermal point of view, if the parallel resistance is zero and the exciter winding alone determines the magnitude of the current.

This disadvantage can be avoided by the Barkhausen circuit, if series and parallel resistance are changed in the opposite direction. The series resistance is increased, when the parallel resistance is reduced, and vice versa. None of the circuits are common anymore.

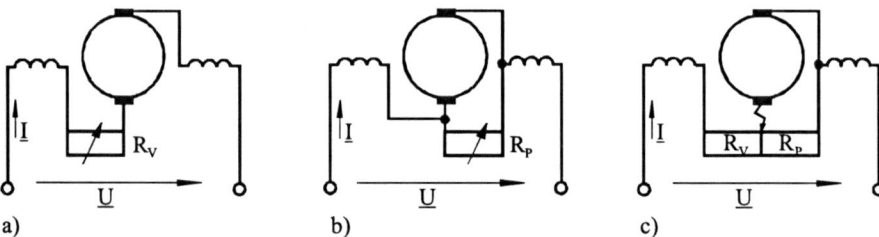

Fig. 2.34. Ohmic resistors for speed control. **a** Series resistor. **b** Parallel resistor. **c** Barkhausen circuit

Tapping of Winding

The construction of the exciter winding with tapping appears well suitable for speed control, as inferred from the comparison of DC and AC operation. A tap changer is required that provides the corresponding number of settings. It is common practice to preserve a symmetric circuit. For production reasons, however, a complete pole coil is occasionally switched off if large speed differences need to be realized. In this case it is important that the number of exciter windings roughly corresponds to the number of armature windings. Frequently, a diode is engaged in order to realize a low speed level (Fig. 2.35). Figure 2.36 shows the set of characteristics $n(T)$, $P_{el}(T)$, $I(T)$ and $\eta(T)$ of a kitchen appliance with a rated power of $P_{el} = 150$ W in comparison to an increase of the number of exciter windings by 23%.

Speed Control by Electronic Controlling Elements

The effective voltage at the terminals of the motor and hence the speed can be controlled by the implementation of electronic elements into the circuitry

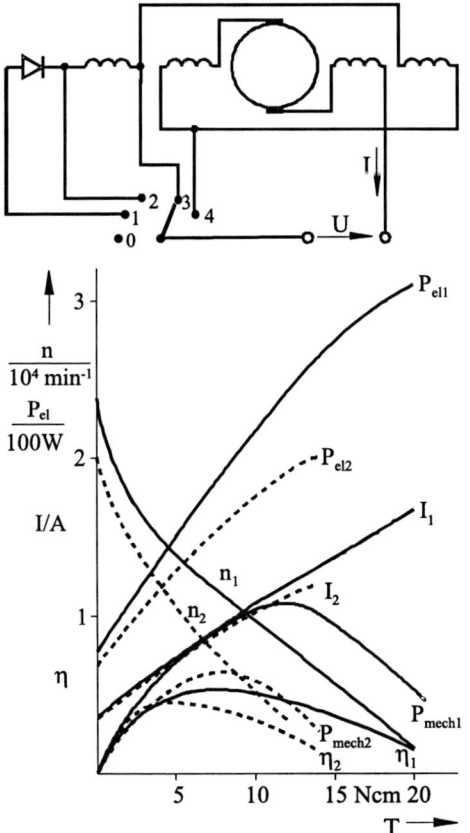

Fig. 2.35. Four speed levels with series diode and tapping of the exciter winding

Fig. 2.36. Set of characteristics of a kitchen appliance for two speed levels caused by a 23% increase of the number of exciter windings.
——— Smaller number of windings
- - - - Larger number of windings

of the commutator series motor (compare Chapter 5). This allows for both AC and DC operation.

The terminal voltage is mainly determined by the ground wave or the static amplitude of the source voltage. It additionally shows significant components with higher frequencies that deteriorate the quality of the network, and must not increase above a certain threshold. The electronic control elements also affect the lifetime of the contact system and the radio interference. The design of the circuitry and the motor rating are therefore not independent of each other.

The speed control in AC operation mode is accomplished by the power controller and the multi-cycle control. The AC controller is designed with a TRIAC (Fig. 2.37). It controls the average of the voltage by a change of the firing point during both half waves of the net voltage.

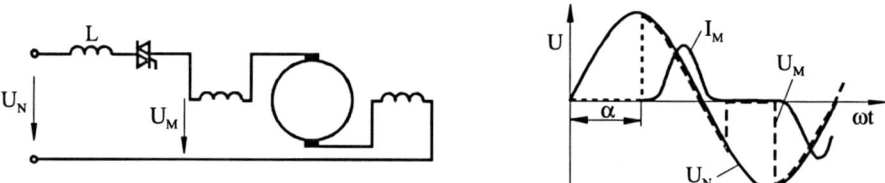

Fig. 2.37. Circuit diagram of the AC controller with power control, and the corresponding graphs of voltage and current (α = firing point)

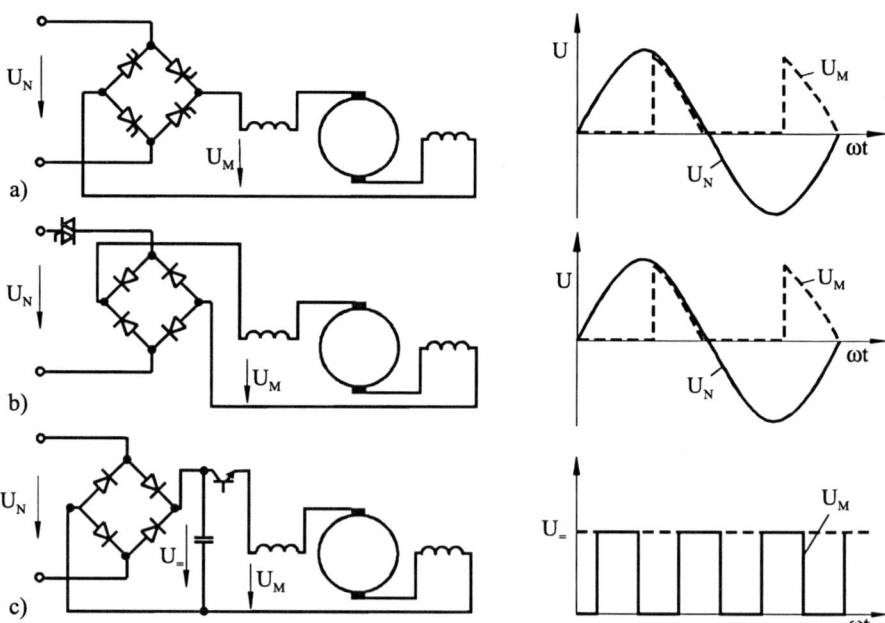

Fig. 2.38. Circuits for direct voltage control. **a** Completely controlled rectifier. **b** AC controller with uncontrolled rectifier. **c** Uncontrolled rectifier and DC controller

The source voltage of the net is switched on and off over several wave periods, if the wave package control is used. The advantage of low radio interference is opposed by a pulsation of the power and by a rough motor rotation. Three methods (Fig. 2.38) can be used for direct voltage control:

- Complete controlled rectifier;
- AC controller with non-controlled rectifier;
- DC controller with non-controlled rectifier.

The induction of the power current circuit is dimensioned such that no current gap occurs. With circuit a), similar to the power control, each half-wave will be completely or partly switched through to the load according to the adjusted trigger angle. The motor is driven by half-waves of identical polarity only.

Therefore the stator does not follow the hysteresis loop with the frequency of the net. This is an advantage compared to the alternating voltage control. The magnetic flux oscillates between a small value corresponding to the remaining field and a maximal value. This reduces the losses due to the magnetization changes. The circuit in b) shows an AC controller with an uncontrolled rectifier. This does not provide a different time dependency of the motor voltage compared to the version in a).

The selection of the implemented circuit is usually based on cost-minimization. Different voltage dependencies are obtained with an uncontrolled rectifier according to circuit c) together with a DC controller (Fig. 2.39).

The rectifier provides an intermediate circuit voltage that can be used by a transistor for motor control. Since the average voltage value is relevant for the speed, it can be adjusted by a control of the pulse width with constant pulse frequency or by the pulse-frequency control and constant pulse width. The advantage of this circuit is the improved dynamic of the drive compared to the other suggested ones. Also, the noise reaction can be reduced to a certain extent by adjusting the circuit parameters.

The direct voltage control becomes only widely accepted for superior appliances due to the higher costs. The power system disturbances are an additional drawback of all mentioned circuits. In particular the rectifier with filter capacitor is problematic. In certain circumstances, the compliance with DIN EN 61000-3-2 is facilitated only by the implementation of an additional inductivity in order to dampen the harmonics. Other methods for a reduction of the power system disturbances cause additional efforts, too. The application of an AC pulse controller is a very exclusive variant.

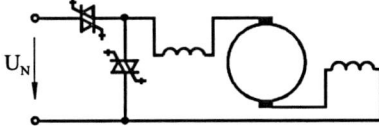

Fig. 2.39. AC pulse control

Stopping the Motor

The rotor of an appliance has to be stopped after switching off, before the user can touch it unintentionally. Mechanical brakes like one- and two-face brakes are difficult to incorporate by design, since the required space is not affordable. Additionally, they cause significant noise during usage. Therefore, the armature is short-circuited by one pole coil during switch off. The remanent flux is sufficient to induce a voltage that causes an armature current with opposite direction compared to the operation mode. The exciter coil has to be connected with the armature such that this current leads to an increase of the remanent pole flux. The required contacts for switching are quite expensive and have to be carefully checked by reliability tests. The safety requirements of choppers and angle grinders include the braking.

Speed Reversal

The reversal of the rotational direction is considered to be problematic, since it places extraordinary demands on the contact system brush-commutator. The brushes have to adapt quickly to the varying direction of rotation without a decrease in life cycle and an increase of the radio interference level. The brush displacement cannot be used to improve the commutation for fixed brush bridges if the demand is similar for both directions of rotation. Tilting of the brushes is avoided during the change of rotational direction by mounting the brush fixation with an angle of up to 30° to the surface of the commutator. This requires brushes with a corresponding inclination of the contact surface and the head. In some manual appliances, as for example in power screwdrivers or tappers, the improvement in commutation is achieved by pivoted brush bridges. These are mechanically connected to the rotational direction switch.

Realization of a Constant Speed

Centrifugal governors (Fig. 2.40) can be used to stabilize a particular speed.

They are used to switch on a series resistance if the speed exceeds a certain value and to bypass a resistance if the speed gets too low. This can be realized by a revolving switch which contacts two slip rings.

The connection between the slip rings is established through the brushes by an ohmic resistance. The switch opens due to the centrifugal force if the

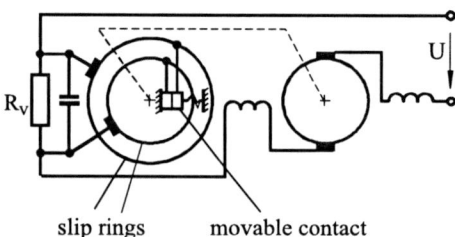

Fig. 2.40. Speed control with centrifugal governor

angular velocity is too high. This causes an ohmic voltage drop and a subsequent decrease of the speed, which leads to the closing of the switch again. A capacitor is used to extinguish the spark of the switch. The slip rings can be omitted if the centrifugal governor is integrated between commutator and armature winding. In this case, two centrifugal governors are required, one for each branch of the armature winding. The effort of the mechanical production, the required adjustment and the sensitivity of the contacts are reasons for the development of a more robust method of speed control. This is provided by the measurement of the speed with an AC tachometer generator and the electronic regulation of a constant value.

2.1.4 Contact System and Commutation

2.1.4.1 Constituents of the Contact System

The contact system brush-commutator is requested to fulfil two inseparable tasks. It provides the galvanic contact of the static energy source to the rotating armature winding, and it allows the alteration of the current direction in different armature coils by the commutation. The design of this system is linked to a variety of physical, structural and economical challenges that have an impact on the lifetime of the motor. The almost countless number of variations of the contact system can be explained by the vast number of requirements that depend on speed, current, voltage, mechanical vibrations, temperature, and installation conditions. They are furthermore affected by the minimization of material usage and production costs.

The contact system (Fig. 2.41) consists of commutator, brushes, brush holder, brush springs, and contact elements, which provide the current transition to the brushes and to the voltage source. Its construction is determined by the concept of the motor. It is also controlled by the production costs, the required lifetime and the permitted radio interference.

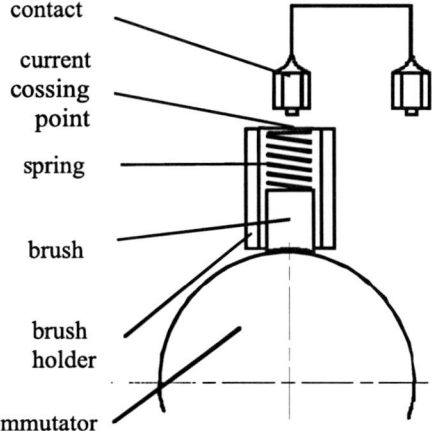

Fig. 2.41. Contact system

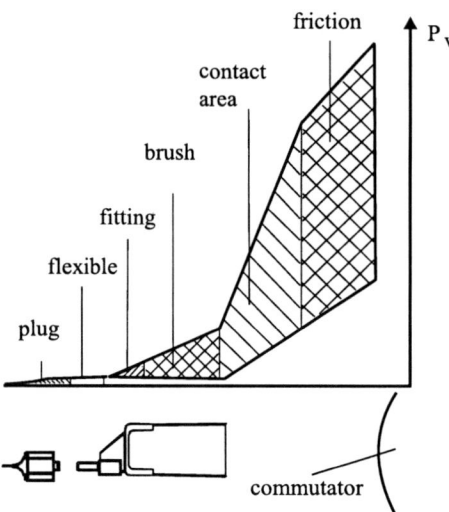

Fig. 2.42. Loss distribution at the contact system

The challenges to be solved are partially reflected by the losses within the contact system. Those are depicted principally for an armed brush in Fig. 2.42. The largest losses and the largest tolerance range occur at the transition between brush and commutator. The losses are modified by the brush material, the spring force and the state of the commutator surface. Together with the electromagnetic characteristics, these factors significantly determine the commutation processes. The restriction of the temperature at the commutator surface is one design criterion. In case of lifetime failure or radio interference, various brush materials are tested before the motor construction is changed. Moreover, the mechanical and electrical properties of the commutator have to be adjusted to the whole system and concept.

2.1.4.2 Configuration of Commutators

Commutator Shapes

Apart from special cases, commutators are individual components and supplier parts. They are manufactured as cylinder or flat commutators. Most motors are equipped with cylinder commutators that have a longer lifetime compared to flat commutators due to better mechanical characteristics. Design and dimension are determined by constructive measures, environmental influences and power requirements of the particular application. The commutator material is mainly copper, even though metal and carbon commutators are also used. The disk rotor with flat coils mounted on the disk constitutes a special case, since the brushes slide directly on the coil itself. Motors with high power are implemented using dovetail and shrink collar commutators. Up to now, the expensive carbon commutators are only used in petrol or gas pumps when immersed into the liquid fuel. Since certain components of the fuel react with copper, the lifetime of the electric motor would be reduced significantly. The implementation of a pricy component is hence justified.

Configuration and Properties of Copper Commutators

The configuration and properties of the commutator are the result of considerations in respect of function and construction. The main components of commutators are the copper sections. In order to separate electrically the individual sections, uniform thin isolating ligaments are inserted that consist of mica chippings, epoxy resin or shellac. Because of cost pressure and in case of mass production, the isolating sections of commutators for motors with small electrical power are made from duro-plastic molding compounds that hold the sections together. The mechanical stability is increased by steel or brass bushes and ring enforcements, respectively.

The fabrication of the galvanic contacts between armature winding and commutator sections requires specific attention. Each section receives the end of an armature coil and the beginning of the neighboring coil. Hence, two wire ends have to be soldered, welded, or calked to each commutator section. The isolating varnish has to be removed from the wires prior to processing, either mechanically or thermally. Commutator sections with tangs or slots can be applied.

The slot-commutator is characterized by short and sufficiently deep slots of the sections on the side of the lamination stack. The wires are cut at the end after winding the corresponding coil. The tang commutator requires

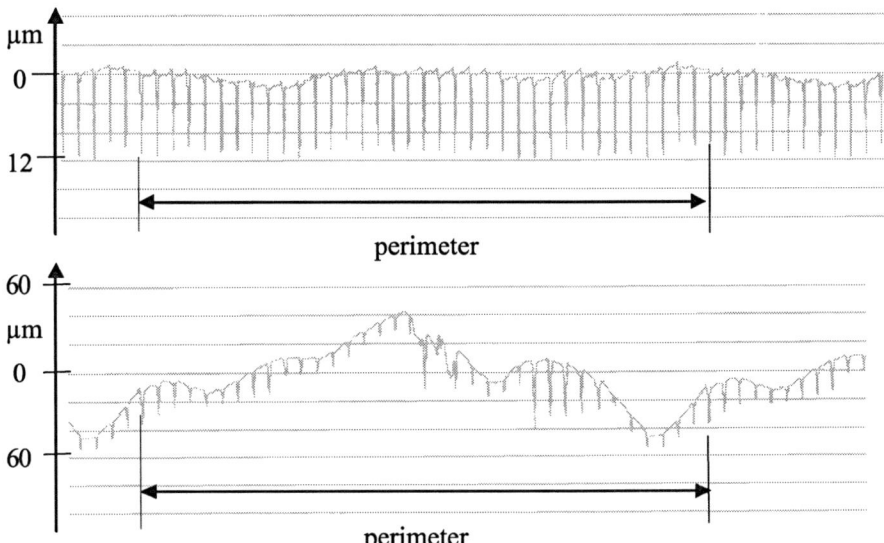

Fig. 2.43. Concentricity diagram prior and after the endurance test

less working steps, since the wire is guided around the tangs at the end of the sections. The winding process proceeds continuously without cutting the wire. The galvanic contact of wire and section is established by bending the tang to the section surface while hot.

The mechanical strength of the tangs and the space between the tangs of neighboring sections determine the applicable wire strength.

Shape variations of the commutator are visible in the diagram of concentricity.

Figure 2.43 shows the comparison of the diagrams of a manufactured commutator prior and after the endurance test that led to a failure. The changes are caused by the mechanical, thermal and electrical processes during normal operation.

2.1.4.3 Constitution and Properties of Brushes

The majority of brushes consists of carbon. Metal brushes made from brass, bronze and precious metals, respectively, are implemented only in low voltage motors with low power, since the less expensive carbon brushes do not provide a similar high performance during operation.

Commonly, the producers of motors do not specify the brush characteristics sufficiently accurately and the manufacturer of brushes do not know the exact implementation details of their product. Hence, the data sheets of the vendors provide information about the approved applications of brushes based on decades of operating experience. An ample lifetime of the contact system brush-commutator is achieved only by a careful manufacturing process and an adaptation to the specific application. This requires an intensive cooperation between the producers of motors, the designers of appliances, and the manufacturers of brushes.

The operating properties of the brushes are affected by brush material, current load, radial speed of the commutator, surface contour of brush and commutator, patina on the commutator surface, coefficient of friction, temperature of the commutator, contact pressure of the brush, mechanical vibrations of the appliance, and commutation characteristics.

It is commonly aimed to use a brush pressure that causes the least possible mechanical and electrical wear. The brush gets detached from the commutator by mechanical vibrations if the pressure is too low, and the current flow is maintained by electric arcs. The copper section of the commutator gets eroded and little parts of the brush burn away. An increasing pressure reduces the electro-eroding wear and enhances the friction loss. This leads to an increase in temperature of the commutator and enforced mechanical brush wear (Fig. 2.44).

The optimal brush pressure can only be determined experimentally during realistic operating conditions, since it is impossible to calculate all factors that influence the brush wear. This requires an extensive analysis considering that each machine type has its own characteristic of vibration. The investigation

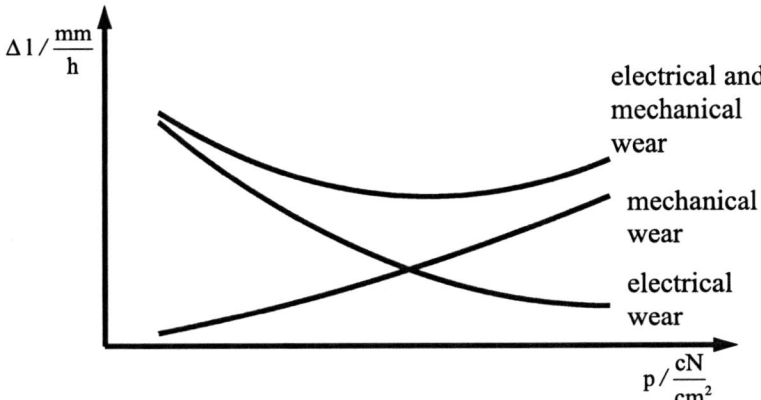

Fig. 2.44. Electrical and mechanical wear of the brushes

is advisably initiated by the recommended pressure values. The uniformity of the brush pressure needs to be checked closely if multiple brushes are switched in parallel, because an uneven pressure distribution causes an under- or an overload of single brushes.

In some motors, the brushes are tilted at an angle of up to 20° from their radial orientation to a trailing (following the rotation) or reacting (against the rotation) position in order to improve the operating characteristic.

Due to friction, the brush is pressed against the brush holder in the rotating direction. This causes a tangential deviation of the brush axis within the holder. The development of mechanical vibrations leads to an increase in noise and brush wear that can be reduced by a beveled edge at the top of the brush (Fig. 2.45). The spring force on the slanted brush face consists of a radial and a tangential component. The brush runs significantly smoother if

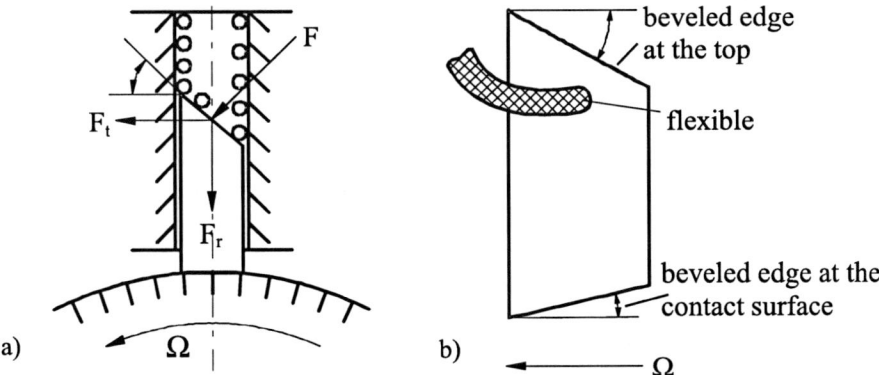

Fig. 2.45. a Brush with beveled edge at the top inside the brush box. b Brush with flexible wire, beveled edge at the top and at the contact surface

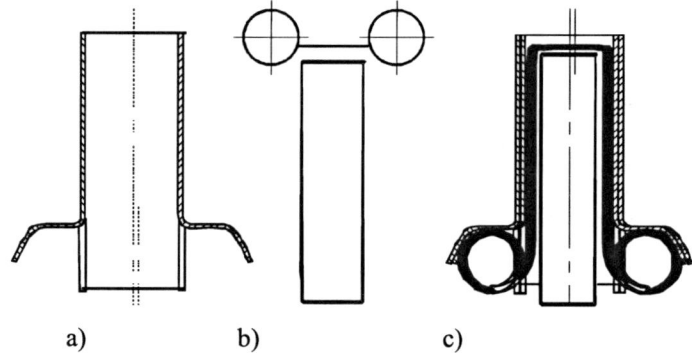

Fig. 2.46. Brush-holder for constant force spring. **a** Brush box. **b** Relaxed constant force spring and brush. **c** Brush, inserted into the box

the tangential component and the friction force are parallel. Simultaneously, the electric contact between brush and brush holder is improved. That is particularly important for using brushes without flexible wire.

The brush pressure can be created by helical (Fig. 2.45a) and band (Fig. 2.46) springs, as well as by spiral or leaf springs that are also partially used for current transmission. A secure fit at the brush head and a constant spring pressure must be realized over the total wear length. The contact faces of virgin brushes frequently receive a surface structure in order to reduce the initial running-in period. In addition, they obtain an initial curvature that corresponds to a 10% larger radius of the commutator.

A good electric contact between carbon brush and voltage source is established by a flexible cord that is made of fine copper wires (flexible wire) (Fig. 2.45). The contact at the brush must be sufficiently robust against mechanical vibrations. It needs to have a low ohmic resistance to permit a current transmission with low losses. Apart from stamped contacts that have proved of value, riveted and soldered joints are also possible. The cross section of the flexible wire depends on the current that needs to be transmitted with a maximal current density of $20\,A/mm^2$.

If the lifetime of the appliance with commutator motor exceeds the lifetime of the brushes, detecting devices are used to signal the wear limit (service signal device). Self-switching coal brushes (carbon brushes with cut-off devices) cut the energy flow before a pressure spring drags on the commutator. The lifetime of a machine is extended by timely change of the brushes.

2.1.4.4 Commutation Procedure in General

The control of the commutation is closely related to the lifetime of the commutator motor. A two pole commutator armature, the cross section of the magnetic circle of the motor, and the unrolling of the commutator armature are depicted in Fig. 2.47. The unrolling is used to describe the commutation.

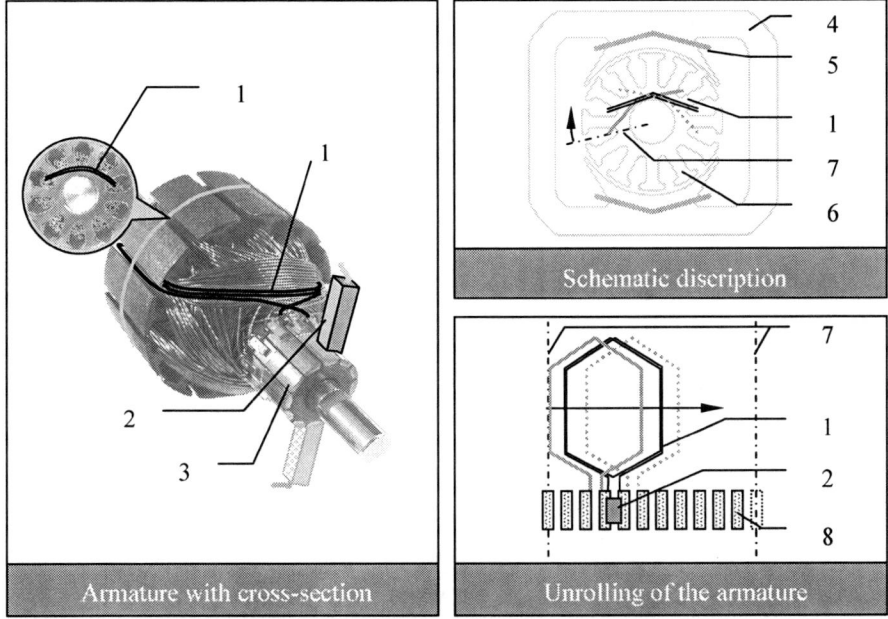

Fig. 2.47. Schematic description of the commutator armature. *1* commutating coil, *2* brush, *3* commutator, *4* stator, *5* excitation winding, *6* armature (rotating), *7* cross section of the unwinding, *8* commutator section

The schematic magnetic circle shows the orientation of the commutating coils to the excitation field.

The number of commutating coils is determined by the number of poles and the width of the brush. The number of brushes is equal to the number of poles in the case of lap winding. Independent of the number of poles, it can be reduced to two brushes for wave winding. If the armature is rotating, the coils change from one armature branch to the next within one short-circuit phase. This is shown by three positions of a two-pole armature in Fig. 2.48. The processes under a brush of a two-pole motor are considered as an example in order to explain the commutation.

The current through the brush is equal to the armature current in two-pole machines. It splits into two armature branch currents with equal magnitude at the transition from brush to commutator section (Fig. 2.48 left). At a particular time, one armature coil that carried the current so far is short-circuited by the rotation (Fig. 2.48 center). When the brush is leaving the corresponding commutation section, the short-circuit is removed and the armature branch current runs in the opposite direction in the armature coil (Fig. 2.48 right).

The available time for commutation T_K (see also Fig. 2.49) is given by

$$T_K = \frac{b_b - b_s}{\pi \cdot D_K \cdot n} \quad \text{for} \quad b_b \geq b_s \, .$$

Fig. 2.48. Movement of the two-pole armature by one commutator section

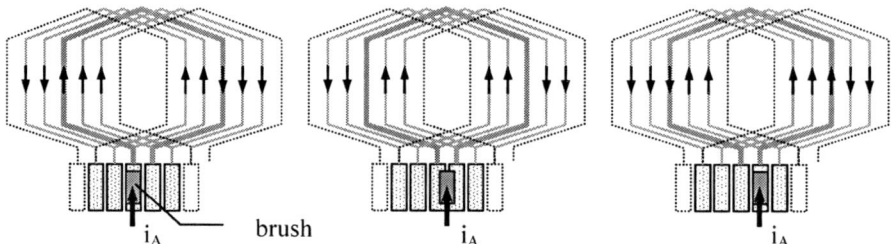

Fig. 2.49. Cycles of commutation process (commutating coil is *highlighted*, *arrows* depict the direction of the current)

The brush width is denoted by b_b, and the width of the section gaps by b_s. Alternating one and two coils are short-circuited by each brush, if the brush width is larger than one and smaller than two section intervals.

The shorted-circuited coils are magnetically interlinked. The electromagnetic interaction between neighboring commutating circuits have an extended influence on the commutation at the beginning and the end of the commutation process.

2.1.4.5 Electrical Parameters

A commutating armature coil can be described by a series circuit of ohmic resistance, leakage inductance and voltage source. The leakage inductance is determined by the number of windings, the geometry of the slots and the magnetic conductivity of the ferromagnetic sections at the boundaries of the slots. The voltage source is a sum of several components that can be separated according to their causation. These are in particular a rotational voltage from the excitation field, a counter emf from the armature field, a counter emf from commutating poles, a transformational component of the excitation field and one component induced by current alterations of additional magnetically linked current circuits. The ohmic resistance of the commutation circuit consists of the resistance of the winding, the resistance of the contacts at the copper sections, the cross-resistance of the brush and the resistance of the transition between brush and commutator. Those values depend on many variable parameters, and the fluctuation of the final resistance is large. This represents one of the reasons why the calculation of the commutation process is based on assumptions. An unerring selection of brushes and a reliable estimation of the lifetime remain impossible, though principal characteristics can be shown, and aspects for the design of the contact system can be derived.

2.1.4.6 Qualitative Description of the Commutation Process

Constant Current Density Distribution Underneath the Brush

A strongly simplified arrangement is employed to explain the commutation process, where all induced voltages in the commutating coils and the ohmic resistance of the winding are neglected. The current density distribution on the contact surface of the brush is set constant. Figure 2.50 shows the process in three steps with a brush width that is equal to the segmentation distance. The brush covers initially one section (Fig. 2.50a). If the brush contacts one additional section in the direction of motion, the armature current through the brush is split between the two commutator sections.

The ratio of the two currents is proportional to the ratio of the correspondingly covered areas A_1 and A_2 (Fig. 2.48). The current through area A_1 is decreasing steadily from the total current in the beginning, while the current through area A_2 is increasing accordingly.

The current i_k of the commutation circuit changes linearly because of its distribution according to both contact areas. It is zero if both areas are covered equally by the brush. The current increases in the opposite direction until the armature branch current i_{AZW} is reached, and the edge of the running down brush leaves the corresponding commutator section simultaneously. In this case, the opening of the commutation circuit happens at the exact point of time when no current runs through the running down edge of the brush and the respective commutator section. This is the criterion of a complete commutation. Neither electric discharge at the brush nor an additional load of the brush contact that is caused by the commutation process is observed. The current i_k of the commutation circuit is calculated by the following equation:

$$i_k(\varGamma) = -\frac{i_A}{2}\left(\frac{2\varGamma}{\gamma} - 1\right) \quad \text{with} \quad \varGamma = 0\ldots\gamma.$$

The covered distance within the commutation cycle is represented by $\varGamma$ and γ is the total short-circuit path bridged by the brush. The area A_1 in Fig. 2.50c equals zero and the armature branch current of the corresponding coil flows into the opposite direction compared to the beginning. The armature has turned by one section distance and a new commutation cycle starts.

Position of the brush

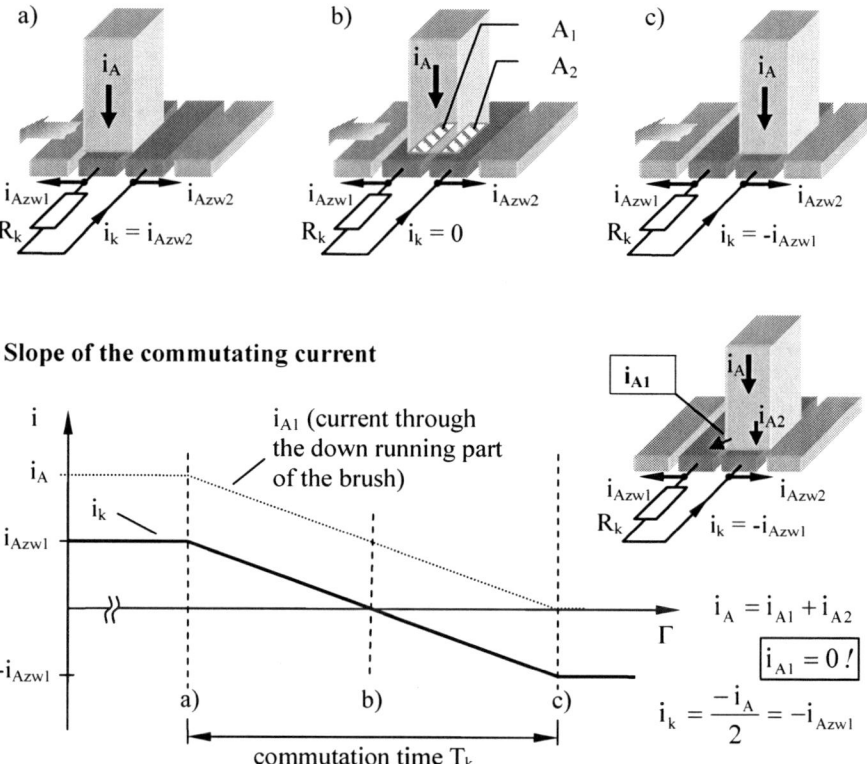

Fig. 2.50. Slope of the commutating current of an ohmic commutation circuit with the labeling **a** beginning **b** center and **c** end of commutation

Influence of the Leakage Inductance

The current change in the commutating coil is linked to a change of the leakage field that is related to the leakage inductance. This is causing the reactance voltage of commutation u_σ:

$$u_\sigma = L_\sigma \frac{di_k}{dt} .$$

The true time dependency of the current is not known. Hence, the following equation is used to estimate the reactance voltage of commutation U_σ with the assumption of a linear commutation:

$$U_\sigma = L_\sigma \Delta I \frac{2\pi n}{b_L} \frac{D_K}{2} .$$

However, this is an approximation of the real reactance voltage.

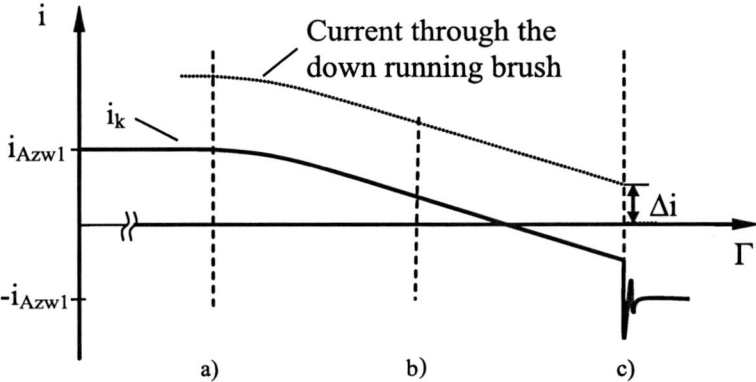

Fig. 2.51. Slope of the commutating current of an ohmic-inductive commutation circuit with the labeling **a** beginning **b** center and **c** end of commutation

The leakage inductance of the commutating coil causes the commutating current to decline with retardation. At the end of the commutation, the current suddenly assumes the value of the branch current (Fig. 2.51) if the given commutation time is too short. This depends on speed and brush coverage. A rapid change of the leakage flux interlinking leads to an induction of voltages that force currents through the opening brush contact. If the limiting voltage is exceeded, the appearing electric arc (spark) erodes the surfaces of brushes and commutator segments. The lifetime of the contact system gets reduced.

Effect of the Interpoles

The armature field induces a voltage in the commutating coils that contributes to the retention of the current through the commutating coil, similar to the reactance voltage of commutation. Both voltages can be compensated by the implementation of interpoles that create a field with opposite direction to the armature field within the pole gap. The commutating field accelerates the commutation process. The commutating winding carries the armature current. Its dimension can be adjusted to create a complete commutation, an under-commutation or an over-commutation (see Fig. 2.52).

The commutation characteristics of over-commutation, complete commutation and under-commutation differ in the current of the short circuit at the end of the commutation time (Fig. 2.52). The complete commutation is characterized by no current flowing cross the down running edge of the brush at the end of the commutation period. Instead, over- and under-commutation is causing the disruption of the current with a formation of electric arcs or sparks that wear away material from brush and commutator. Both commutation characteristics differ by the sign of the current that must be switched off, which is causing a switching of the direction of the electric arc or spark.

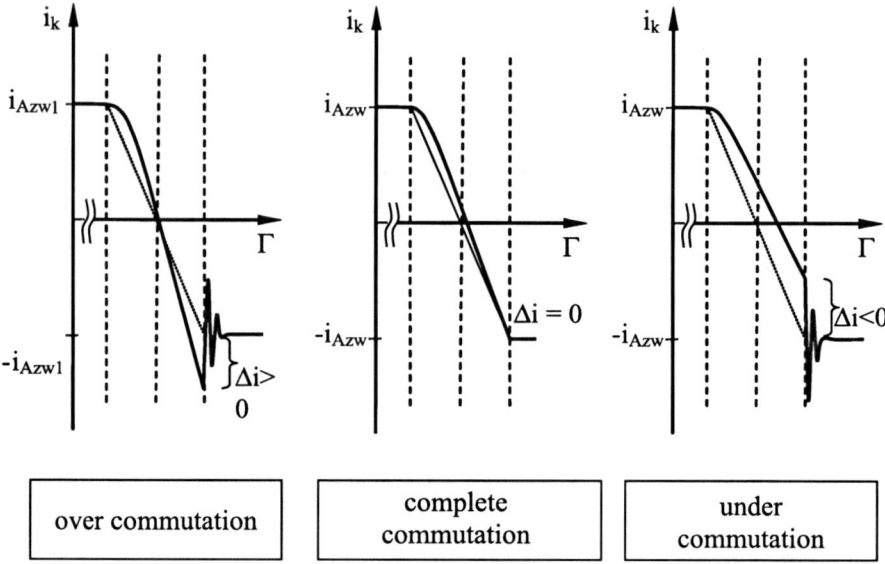

Fig. 2.52. Overview of commutation characteristics and their identification

Tilted Brush Bridge

Commutator motors up to 3 kW rated input are built without interpoles.

The desired lifetime can only be achieved by implementing a tilted brush bridge that causes the induction of electromotive force in the commutating coils by the exciter field. The induced voltages accelerate the commutation process. This requires a rotation of the brush bridge in motors, which operate in both directions of rotation.

No electromotive force is induced in the commutating coils by the exciter field if the commutating armature coils reside in the pole gap (Fig. 2.53).

The voltage induction is increased with opposite sign if the brush bridge is tilted in or against the direction of rotation of the armature (Fig. 2.54). In order to compensate the reactance voltage of commutation, the brush bridge

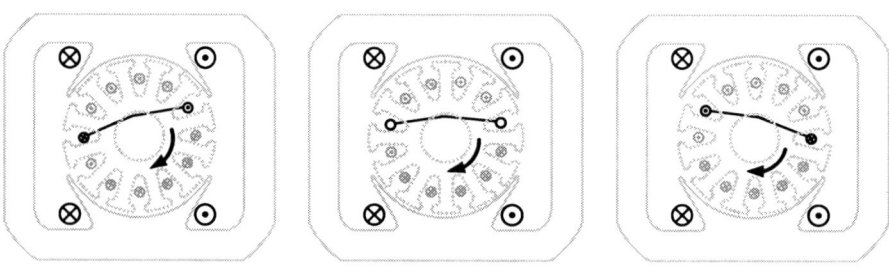

Fig. 2.53. Commutating coil with a brush bridge in neutral position

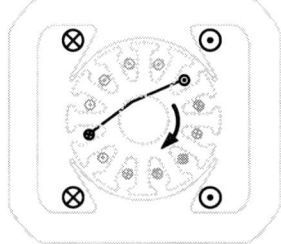 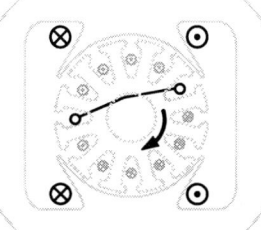 

Fig. 2.54. Commutating coil with a brush bridge tilted against the direction of the motor rotation

must be tilted against the direction of rotation in motor operation. For generator operation mode it needs to be tilted in the direction of rotation. The figures depict the positions of one particular coil prior, during and following commutation. Tilting the brush bridge has the disadvantage of a weaker field inside the air-gap.

Influence of the Transformational Voltages that are Induced by the Exciter Field

In AC commutator series motors, an alternating voltage is induced in the commutating coils by the magnetic flux through the brush level. This voltage has a phase shift relative to the armature current. It cannot be compensated by a specific construction and must hence be limited by an adequate choice of the number of armature windings.

Voltages Induced by Current Changes of Additional Magnetically Linked Current Circuits

Apart from the already mentioned flux linkage of the commutating coils, further magnetically linked current circuits exist dependent on the motor design. Among these are the simultaneously commutating armature coils and eddy current circuits of the bulk armature shaft, as well as eddy current circuits in the conductive ties of the lamination stack formed by riveted, interlocked and welded joints. This coupling partially helps to unload the commutation process in the switch off moment, since it is able to take over the common field from the commutating circuit by a current change. An overview of the voltage components of the commutating circuit is given in Table 2.3.

2.1.4.7 Difference in DC and AC Operation Mode

As discussed when considering the transformational voltages that are induced by the exciter field, the alternating voltage causes a phase-shifted voltage component in the commutating circuit. This voltage approaches its maximum at the zero crossing of the armature current. The electromotive force and the re-

Table 2.3. Overview of voltage components that are relevant for commutation (symbols according to Fig. 2.46)

	Schematic representation		Effect
Reactance voltage (self-inductance)		leakage field	Leakage field of the commutating armature coil is decreased and increased in opposite direction, delays commutation
Electromotive force by interpole field		interpole field	EMF accelerates commutation
Electromotive force by exciter field		exciter field	EMF increases due to tilt of brush-bridge, diminished exciter flux, accelerates commutation
Electromotive force by armature field		armature quadrative-axis field	EMF delays commutation
Transformational voltage		exciter field	Phase-shifted voltage with respect to other voltage components, accelerates or delays commutation depending on the appearance time
Voltage, induced by current changes of additional magnetically linked current circuits		bulk armature shaft	Field transfer to linked short circuit or neighboring coils; unloads the switch-off process at the end of commutation

actance voltage of commutation are zero at that particular point in time. The current of the short circuit is hence caused by the transformational voltage.

Starting from the zero crossing of the armature current and towards increasing armature currents, the induction of the transformational voltage components by the exciter field leads to over-commutation. If the armature current reaches a particular value, the complete commutation is realized for a short period in time after it gives way to under-commutation. Those are converted to an over-commutation at the next zero crossing of the armature current due to the reversed sign of the electromotive force and the reactance voltage. Over- and under-commutation alternate periodically. The maintenance of a complete commutation is impossible over the complete length of the alternating voltage period. The optimization aims for the best compromise between over- and under-commutation that allows for the requested lifetime. In direct voltage operation mode, a complete commutation can be realized for a specific operating point. The best brush position needs to be determined again, if the load cycles change.

In DC motors with a constant direction of rotation, the polarity of the brushes remains fixed. Therefore, the direction of the electric arc at the end of an incomplete commutation period is defined. This causes a stronger wear of brushes with one polarity in comparison to the brushes with the opposite polarity. However, it is impossible to determine the brush with a higher wear by the polarity alone. Also the prevailing commutation characteristic contributes to the higher wear of one brush by an electric arc. The brush wear is almost uniform in AC motors, since the polarity of the commutation characteristic permanently changes in alternating voltage operation. Here, differences in wear are mainly caused by irregularities of the mechanical construction.

2.1.5 Noise Behaviour

Every drive system generates noise. Users assess it differently.

Results of noise-measurement research (Eqs. (2.4) and (2.19)) recommend carrying out an assessment including the effect on the psyche. At least an assessment has to include the noise effect on man. Besides loudness, shrillness and tone of the noise, dissonance and modulation have an influence on the perceived sound. Very seldom is the small drive the disturbing component alone. The drive system has to be assessed together with the other components at the place of origin. Drive manufacturers and the automotive industry have drawn up an assessment method of noise for fractional-horsepower drives in connection with the end product place of manufacture (Eq. 2.4).

2.1.6 Radio Interference Suppression

The average values of direct and alternating values are sufficiently accurate for a discussion of energy conversion and speed control. However, the actual function of time of the electric parameters is mainly influenced by harmonics due to the geometric design of the magnetic circuit and the commutation of the currents in the armature winding. This can be verified by an oscillographic measurement of voltages and currents.

The commutator series motor can be understood as a bandwidth generator. The frequency bandwidth ranges from the third harmonic of the net frequency to the radio frequency range of several MHz. The radiation and conduction of higher frequency components cause radio interferences that have to be confined to the permissible limits by appropriate measures. The manifold mechanic and electro-magnetic causes have to be investigated during the development of the motor.

They have to be avoided or their negative effects have to be restricted. In this context, the contact system brush-commutator plays a central role. The elements for radio interference suppression are clearly visible and frequently dominate the visual appearance of the commutator series motor. Among these are spark suppression impedances (pole- or ring-core impedances), spark suppression capacitors and tube cores. A typical suppression circuit is shown in Fig. 2.55. The suppression effort has to be in due proportion, since motor and appliance represent a unit.

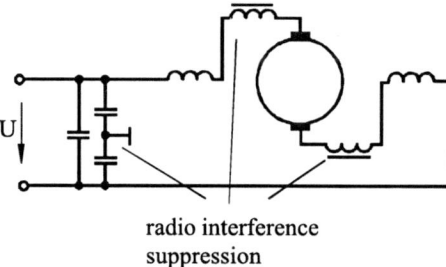

radio interference suppression

Fig. 2.55. Typical circuitry for radio interference suppression

2.2 Brushless Permanent Magnet Motors

Wolfgang Amrhein

2.2.1 Introduction

2.2.1.1 Operation Modes and Definitions

Operation Modes

Depending on the respective design, brushless permanent magnet motors are typically operated as single-phase or three-phase drives directly at AC mains using a start-up auxiliary device, or predominantly, with electronic power converters providing square-wave or sinusoidal currents or voltages [2.42–2.45].

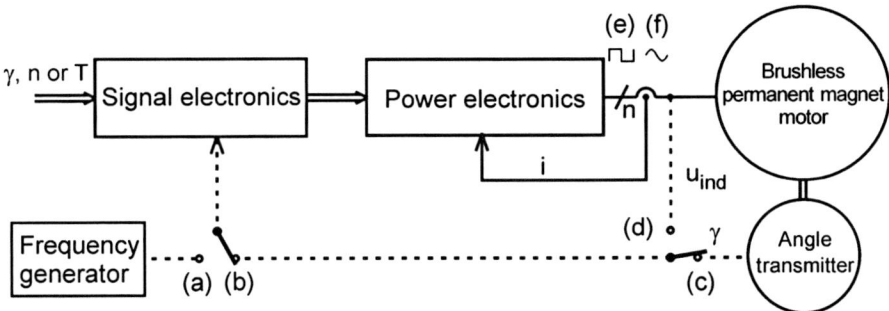

Fig. 2.56. Operating modes of converters of brushless permanent magnet motors. **a** Frequency-controlled. **b** Position-controlled or field-controlled. **c** Sensor-controlled. **d** Sensorless. **e** Square-wave. **f** Sinusoidal voltage or current feed

Operation with converters (see Fig. 2.56) can be done applying either frequency control or field control. The former method is used, e.g. for the group of stepper motors.

In the case of field-controlled permanent magnet drives, with the outputs of the converter either the motor voltages or the motor currents are adjusted in accordance with the flux linkage of the armature windings. With motors with surface mounted permanent magnets of low permeability and adequately large magnet heights, the angle of the armature linkage flux is largely fixed to the rotor position and thus is approximately independent from the mechanical load. Here the converter outputs are adjusted as per the following rules:

$$a_i(\gamma_{el}) = \hat{a}\,\text{sign}\,[\sin(\gamma_{el} + \varphi_i)] \qquad (2.44)$$

as a square-wave or

$$a_i(\gamma_{el}) = \hat{a}\sin(\gamma_{el} + \varphi_i) \qquad (2.45)$$

as a sinusoidal waveform.

In this connection γ_{el} denotes the electrical angle of the rotor position, and in the case of multiphase motors, φ_i denotes the phase shift of the associated phases i. The mechanical rotor angle is represented by γ.

Figure 2.56 illustrates that brushless permanent magnet motors can also be controlled without a sensor, e.g. with the help of the measurement of the electromotive forces (square-wave operation), by the application of test pulses or high-frequency signals or by using an observer to calculate the rotor or linkage flux position from the current and voltage outputs of the power converter (particularly with sinusoidal operation). Sometimes, the starting process is carried out with a frequency generator that causes the rotor speed to assume a value in the range of assessable induced voltage signals. But, it has to be taken into account that in the case of a simple control mechanism, on the one hand the motor can take the first step in the wrong direction on startup, depending on the respective rotor position, and on the other hand the rotor can fall out of step at high acceleration rates or loads. Meanwhile there are also some alternative methods to start from the very first under field-oriented control without these disadvantages [2.66–2.73].

Definitions

The term brushless permanent magnet motors is commonly used to denote the subgroup of field-controlled permanent magnet motors and is rather rarely used to refer to AC mains-driven synchronous motors or stepper motors. The following sections of Sect. 2.2 are restricted to the description of brushless field-controlled DC and AC permanent magnet motors.

Brushless DC motors are used for operation with block-shaped commutation in accordance with Eq. (2.44). Due to the principle of operation derived from commutator motors, such motors are also called electronically commutated motors or electronic motors.

The second group of motors fed with sinusoidal quantities is in part very similar to AC motors in stator design as well as in their magnetomotive force distribution and is thus often referred to as the group of brushless AC motors.

Both motor groups draw power from a DC link. Brushless DC motors use switching power converters generating square-wave or trapezoidal voltage or current waveforms whereas brushless AC motors are controlled by sinusoidal waveforms. In both cases the motor phases are fed with alternating quantities, i.e. even in the case of brushless DC motors. The term can thus be a little bit confusing.

As described in detail in Sect. 2.2.2, brushless DC and AC motors differ from one another primarily in their design with respect to the formation of the permanent magnetic field, the design of the windings and the requirements of the resolution of the angle transducers.

2.2.1.2 Relationship to Other Types of Motors

We also distinguish between brushless DC and AC motors in the course of the representation of the relationship of field-controlled permanent magnet motors to other types of drives.

Brushless DC Motors

Brushless DC motors and DC commutator motors are very similar in their design and principle of operation. Both have permanent magnet excitation, an armature winding and a mechanical or electronic commutation arrangement for feeding the windings with square-wave voltages or currents. The alternation of current in both cases occurs such that the armature current distribution keeps aligned relative to the permanent magnet poles; the local current distribution is merely subjected to small oscillatory movements due to the finite number of phases and coils, respectively.

Figures 2.57 and 2.58 illustrate the design principle of the DC commutator motor and the electronically commutated DC motor. Apart from the magnetic

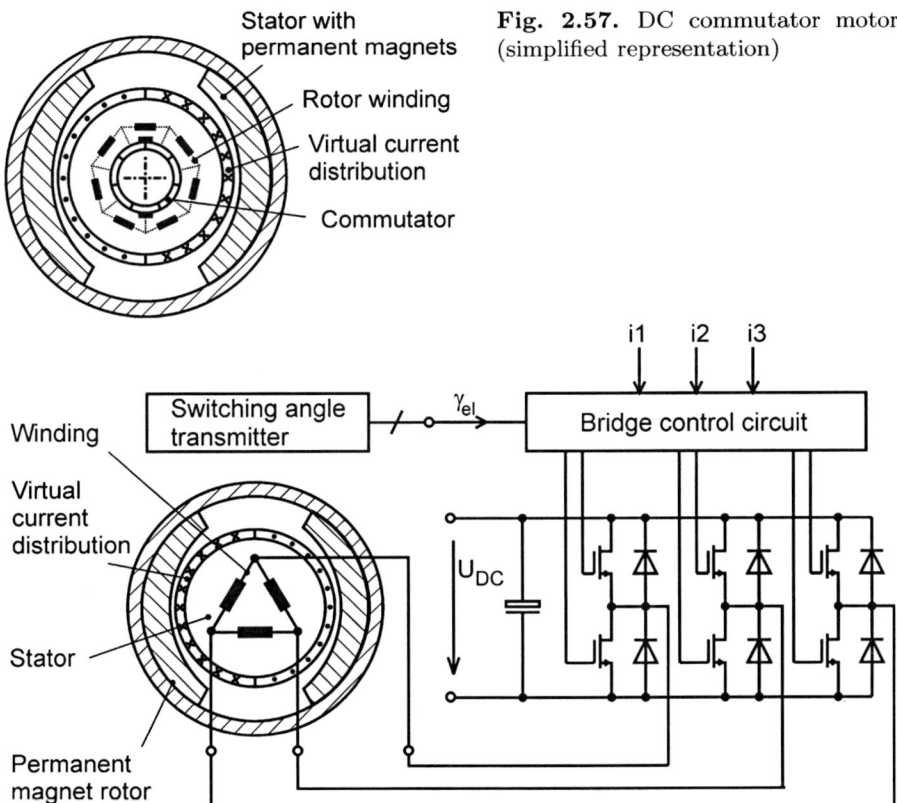

Fig. 2.57. DC commutator motor (simplified representation)

Fig. 2.58. Electronically commutated DC drive (simplified representation of an external rotor motor)

motor component, the brushless design calls for an additional rotor angle transducer, an electronic converter with EMC filter, over-voltage and over-current protection, inverse-polarity protection, as well as multi-pole cable and connectors in the case of non-integrated electronics, which can also represent a considerable cost factor.

In the case of both mechanically and electronically commutated motors, the stator often comprises full-pitch windings or slightly chorded windings so that, as far as possible, constant torque can be generated independently of the rotor angle in combination with the square-wave or trapezoidal distribution of the air-gap flux density and block-shaped currents. In contrast to brushless motors having a limited number of phases due to the increasing electronic complexity (usually $m \leq 3$), a substantially higher number of phase coils can be cost-efficiently implemented in commutator motors, so that even in the case of unpropitious sinusoidal flux linkages the resulting torque remains largely constant, e.g. in bell-type armature commutator motors with a great number of ironless rhombic winding coils.

The affinity of both types of motors in design and principle of operation also results in comparable operational characteristics. For example, with constant motor temperature, negligible armature reaction and magnetic saturation, in both cases there is generally a linear correlation between speed and torque and between torque and motor current.

Brushless AC Motors

For permanent magnet excited AC motors, there are clear differences with respect to brushless DC motors, particularly in mechanical design and magnetic circuitry. For example, the excitation flux density in the air-gap due to the corresponding magnetization or shape of the magnets as well as the armature linkage fluxes resulting from pitching, distribution or skewing of the phase windings, and also the current waveforms provided from the power converters should be sinusoidal in shape. Frequently, the design of the stator is very similar to that of asynchronous motors applicable for field-oriented control operation. Even the mathematical calculation methods used for the motor design are similar as in both cases complex space vector theory is applied in connection with the design of control algorithms [2.46, 2.47, 2.48, 2.49, 2.50].

While the angle ϑ between the space vectors of the stator current $\underline{i}_s$ and the stator linkage flux $\underline{\psi}_s$ adjusts itself according to the mechanical load in the case of frequency-controlled synchronous motors, the angle ϑ is electronically adjusted with brushless AC motors and field-controlled asynchronous motors. In contrast to asynchronous machines that require an additional direct-axis component of current to create the excitation field, a quadrature axis component is adequate for permanent magnet excited AC motors to generate the torque as illustrated in Fig. 2.59. Thus, when weakening of field is not required, the stator current and linkage flux space vectors are orthogonally oriented towards each other for all rotor positions.

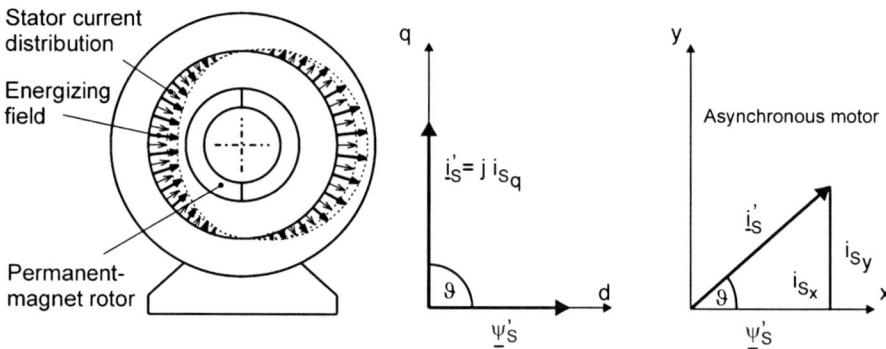

Fig. 2.59. *Left*: Sinusoidal stator current and excitation field distributions of the brushless AC motor (schematic representation). *Diagrams*: Stator current and flux space vectors of the brushless AC motor and the asynchronous motor (flux-oriented coordinates)

Table 2.4. Comparison of typical characteristics and applications of permanent magnet motors with and without brushes

DC commutator motors
Features: Direct connection to DC mains possible; partly high efficiency achievable (in particular when using SmCo or NdFeB magnets and also with bell-type armature motors having ironless windings and metallic brushes); limited dynamic operation (durability of brushes) electrical and thermal overload for short durations possible; substantial influence of the speed and torque on the life of the motor; good running performance with large numbers of armature coils and commutator bars, respectively; largely linear characteristics; relatively low-costs for electronic control (full bridge circuit), inevitable brush noises and wear.
Costs: Often the most cost-effective solution for simple drive tasks, particularly when low voltage connection is available.
Typical applications: Positioning drives in automobile technology (power windows, seat adjustments, flap controllers, wipers); drives for medical applications (dosage apparatus, hose pumps); measurement technology (tacho-generators, torque measurement devices); toys industry (drives for automobiles, trains, aircrafts and ships).
In combination with power controllers: position-controlled and speed-controlled drives (industrial servo drives without very high demands on operational life).

Brushless DC and AC motors
Features of motors equipped with SmCo or NdFeB magnets: high power densities achievable (frequently, good carrying-off of stator heat possible); high efficiency values reachable; large power factor under full-load operation; high dynamic operation possible (precondition: permanent magnets must be protected mechanically and magnetically); in contrast to DC motors no speed-dependent current limit necessary; suitable for high nominal speeds (in connection with tape binding of the permanent magnets); application in explosive environments basically possible; capable of low noise levels and negligible torque ripples (particularly for motors designed for sinusoidal commutation; brushless DC motors, as a rule, require special measures such as soft commutation, skewing of the slots or of the permanent magnet poles, etc.); largely linear characteristic; long life and high reliability (largely determined by the bearings and the electrolytic capacitors and power switches of the electronics).
Costs: Simple and very cost-effective solutions are possible, particularly in the area of low voltage applications, with single-phase drives having integrated electronics (note: low starting torque!). Three-phase designs are often used for sophisticated position-controlled or speed-controlled drive systems. Brushless DC motors are generally more cost-effective than those designed for sinusoidal commutation. In general brushless AC motors have more expensive windings, need sinusoidal airgap flux densities and thus adequate magnetic circuit design, have more stringent demands on the angle resolution of the sensors, but feature improved operational characteristics with respect to torque performance, positional accuracy and noise.
Typical applications: Single-phase drives: ventilating and air-conditioning systems (fans, blowers and ventilators) and heating systems (blowers for oil and gas burners).

Table 2.4. (continued)

Three-phase brushless DC drives: Computer systems (hard disk drives, CD-ROM and DVD drives), medical applications (pumps, centrifuges, agitators), audio and video systems (CD and DVD), machine tools and robotics.
Three-phase brushless AC drives: sophisticated, industrial position-controlled and speed-controlled drives combined with resolvers, high-resolution optical or magnetic transducers or sensorless control methods (angle observer).

Table 2.4 presents a comparison of brushless permanent magnet motors and DC commutator motors. Brushless motors with internal rotors, external rotors or even disk-shaped rotors are often used in applications that demand long life and reliability (fans), high power density (hard disk drives), high efficiency (battery operated drives), good dynamics in combination with low torque ripples and cogging torque (servo drives) or low noise levels (audio applications).

2.2.2 Design Features

It is necessary to match the permanent magnet excitation field, the arrangement of the armature winding, and the motor voltage or current supply carefully in order to achieve proper motor performance [2.43, 2.44]. As already preintimated in the previous section, in principle there are two basic designs: the brushless permanent magnet motor appropriate for (a) square-wave and (b) sinusoidal commutation.

Basic Brushless DC and AC Motor Concepts

Figure 2.60 presents a simple illustration of the design of a brushless DC motor with a radially magnetized, two-pole permanent magnet ring in combination with full-pitch windings. Under ideal conditions (ignoring flux leakage, influences of the slots, saturation and iron losses), triangular rotor angle-dependent linkage fluxes result from the square-wave distribution of the air-gap flux density.

Basically the resultant torque T of the m phases

$$T = \sum_{i=1}^{m} \frac{d\psi_i}{d\gamma} i_i \qquad (2.46)$$

with $\gamma = \gamma_{el}/p$ is independent of the rotor angle for a square-wave current feed with a duty cycle of $2\pi/3$ as shown in Fig. 2.62. The prerequisite for this behavior is a constant slope of the linkage fluxes during the on period of the associated winding currents.

Figure 2.61 illustrates the corresponding motor design for sinusoidal commutation. The cylindrical two-pole rotor magnet is magnetized diametrically

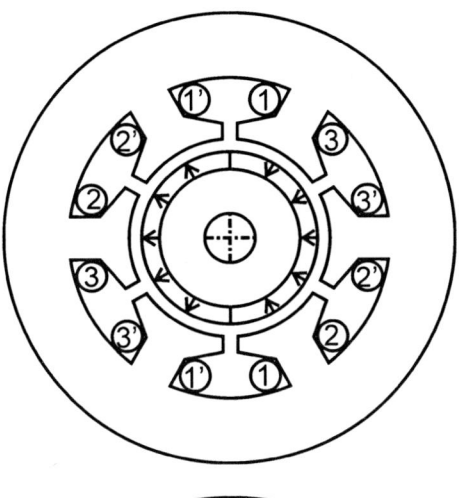

Fig. 2.60. Brushless DC motor with radially magnetized, two-pole permanent magnet ring and full-pitch windings

Fig. 2.61. Motor design for sinusoidal commutation with diametrical two-pole permanent magnet cylinder

and creates a sinusoidal air-gap field. The fluxes linked with the full-pitch windings are thus also sinusoidal in nature and hence, in conjunction with the sinusoidal shapes of the winding currents as shown in Fig. 2.63, also lead to a total torque that is independent of the angle.

In order to obtain uniform torque the winding flux pattern of the brushless DC motor may even deviate from the sketched shape beyond the switching period of the windings assigned (cf. Fig. 2.62). As a result, as explained in the subsequent section, an additional degree of freedom becomes applicable for the design of the magnets and the windings.

Similarly, many options are available for motor designs with sinusoidal commutation to optimize the operational characteristics. For example, the negative influence of a non-sinusoidal air-gap field can be corrected by means of appropriate winding design. The following section contains examples for this purpose.

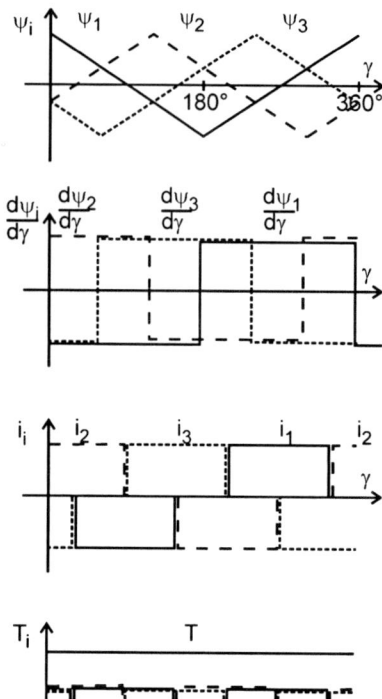

Fig. 2.62. Idealized representation of the linkage fluxes and their derivatives, the winding currents and phase torques for the brushless DC motor in Fig. 2.60

2.2.2.1 Design and Selection of the Stator Winding

Beyond the selection of suitable magnets (see Sect. 2.2.2.2), the design of the armature winding is also significant from the point of view of optimizing the motor operation characteristics. With the help of illustrative examples, this section presents an insight into numerous design variants and their influence on motor performance.

Brushless Motors for Square-Wave Commutation

In order to obtain as uniform a torque as possible, the armature linkage fluxes, the current waveforms and the winding design must be carefully matched as mentioned before. An example of a brushless DC motor design that basically fulfils these requirements has already been presented in Figs. 2.60 and 2.62.

Given Eq. (2.46), Fig. 2.62 clearly shows that the torque curves remain unchanged even when the values of the derivatives of the linkage fluxes change between the angular regions of $\pm\pi/6$ related to the discontinuity of the curves. Such disturbing effects can be, for example, the result of non-ideal magnetizations of the permanent magnets, gaps between the magnetic poles or the

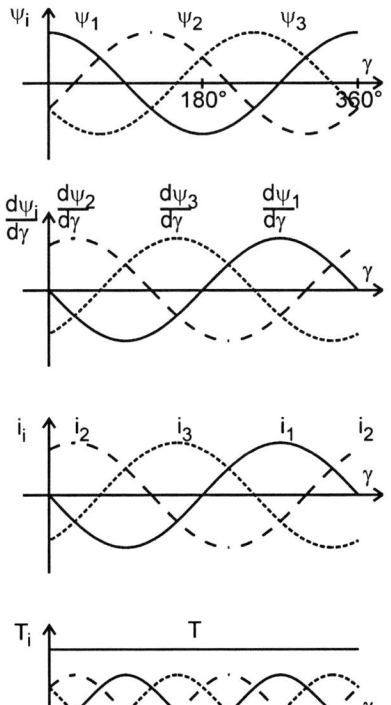

Fig. 2.63. Idealized representation of the linkage fluxes and their derivatives, the winding currents and phase torques for the brushless AC motor in Fig. 2.61

influence of magnetic flux leakages. Thus flux curves or their derivatives deformed in the border regions are not really critical as long as these discrepancies don't fall together with the duty cycle of the associated square-wave phase currents.

However, due to induced voltages, particularly in medium to high speed ranges, the practical realization of extremely steep current slopes requires unreasonably high-voltage reserves on the part of the power electronics. As a result, this leads to higher costs for power converters, and in general also to higher commutation switching noises that are not permissible for many applications. Thus in many practical applications the current curves are allowed to deviate from the ideal square-wave pattern, leading to an increase in the torque ripple in the interest of reducing costs and switching noise levels. Particularly in the case of square-wave voltage control, highly deviating current curves can arise to some extent, resulting from the influence of the induced voltages.

In general, in order to minimize the cost of the power electronics in the case of multiple phase motors, the windings are interconnected using a delta connection or a star connection without mid-point conductor. For these cases the sum of phase currents becomes zero:

$$i_1 + i_2 + i_3 = 0 \, . \tag{2.47}$$

The motor design using delta-connected windings requires special care. An example is given in Fig. 2.60 where the electromotive forces would cause disturbances in the case of delta-connected windings [2.51]. The condition for the voltage sum

$$u_1 + u_2 + u_3 = 0 \tag{2.48}$$

is strongly violated by the electromotive forces in this case so that a circular current is superimposed on the line-to-line currents which can result in substantial torque disturbances, losses and consequently unacceptable heating and noise (see Fig. 2.64). Here, the odd multiples of the third harmonic of the induced voltages are responsible for the interferences. The effect described can be eliminated or substantially reduced by means of suitable fractional pitch windings, the distribution of the winding coils (e.g. as shown in Fig. 2.65) or by shortening the pole width to a pole angle of $2\pi/3$.

The interconnection of the windings influences the motor characteristics, especially in the case of non-sinusoidal motor quantities [2.51]. As with the number of turns, the selection of the type of interconnection can be used for the adaptation of the motor to the power supply, in particular in conjunction with the prevention of manufacturing problems like wire break or winding problems with stiff wires as a result of extremely small or large wire cross sections. Frequently, in spite of the danger of short-circuited currents, delta-connected windings are preferred to star-connected windings, since they permit the uninterrupted and therefore cost-saving continuation of the winding manufacturing process.

Reasons for the use of chorded windings in brushless DC motors could be the specific influence on harmonics, the reduction of copper losses or the attenuation of cogging torque effects. The latter can be achieved by selecting an appropriate slot to pole number ratio as described in a later section.

Figures 2.65 and 2.66 illustrate examples of fractional pitch windings with a coil width of $2\pi/3$ for a four-pole motor with square-wave air-gap field distribution. The first figure shows an arrangement of concentrated windings

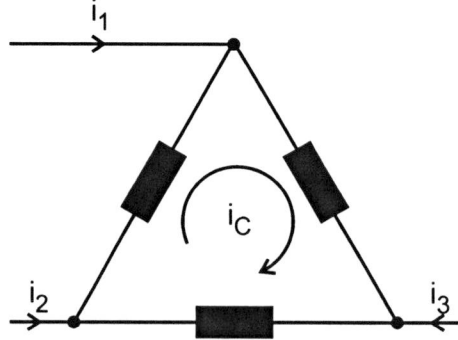

Fig. 2.64. Build-up of a circular current i_C in delta winding connections when the sum of the electromotive forces is not equal to zero in all operation modes

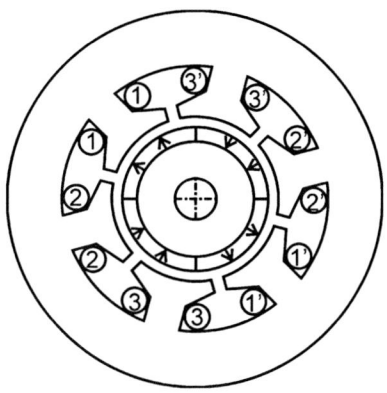

Fig. 2.65. Four-pole brushless DC motor with concentrated windings (coil width = $2\pi/3$)

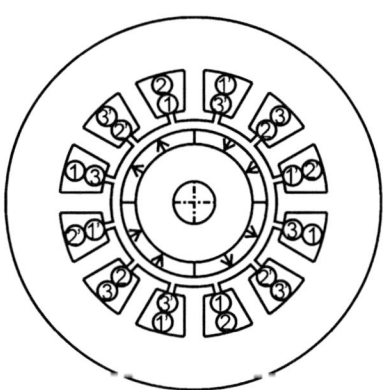

Fig. 2.66. Four-pole brushless DC motor with a comparable two-layer winding concept (coil width = $2\pi/3$)

(tooth coil windings); the second figure illustrates a corresponding two-layer winding. The flux linkage of one phase, the associated square-wave phase current, and the resulting constant torque curve are illustrated in Fig. 2.67. Under ideal considerations, e. g. neglecting of additional slot effects, the curves for both winding designs are identical. However, actually the short pitching of both winding concepts leads to torque fluctuations as a result of the reduction of the constant $d\psi_i/d\gamma$ section due to magnetic leakages at the pole and tooth edges. Thus motor windings with larger coil width are recommended for applications with stringent requirements on the torque performance.

Brushless Motors for Sinusoidal Commutation

Figure 2.69 illustrates a chorded and also twofold distributed winding with coil widths of $2\pi/3$. The series connection of the adjacent phase coils as illustrated in Fig. 2.69 leads to an approximately sinusoidal curve for $d\psi_i/d\gamma$. Thus disturbing torque ripples would occur in this motor configuration in connection with a square-wave current feed.

The previous considerations indicate that the waveform of the linkage flux, the electromotive force and the magnetomotive force can be brought close to

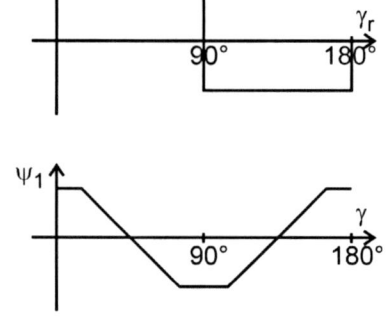
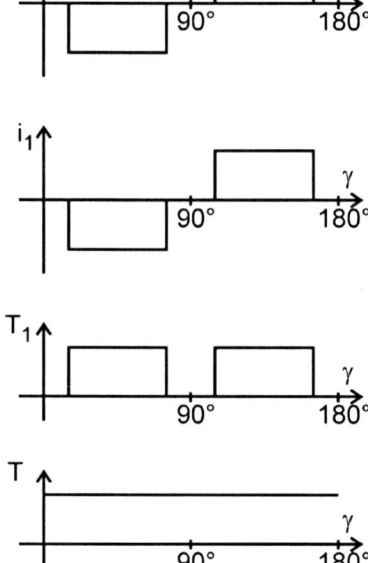

Fig. 2.67. Idealized representation of the radial air-gap flux density, the flux linkage, its derivative, the appropriate phase current, the phase torque, and the total torque for the brushless DC motor design shown in Figs. 2.65 and 2.66

a sinusoidal waveform even with a square-wave air-gap flux density by means of short pitching and distribution of the phase coils. Further it is possible to improve the results by additional means of skewing the slots or the permanent magnet poles. Windings of the type as illustrated in Fig. 2.69 featuring a specific damping or suppressing of harmonics are thus preferred for permanent magnet excited AC motor designs with special applicability of sinusoidal commutation.

In order to achieve a better basis for the generation of sinusoidal waveforms, these motors are often provided with diametrically magnetized or arc-shaped formed permanent magnets. The amplitude reduction of the harmonics can be calculated with the help of the winding factor. This takes into

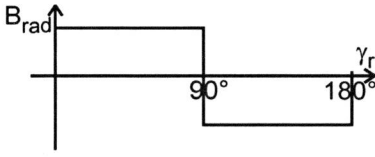
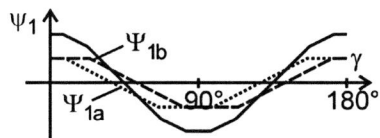
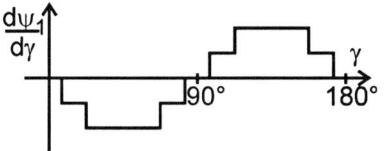
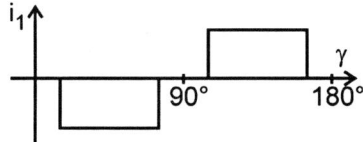
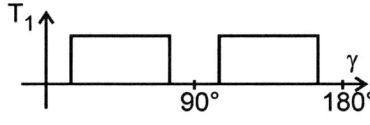
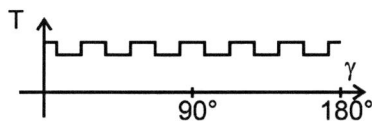

Fig. 2.68. Idealized representation of the radial air-gap flux density, the flux linkage, its derivative, an (inappropriate) square-wave phase current, the resulting phase torque, and the total torque for the brushless AC motor design shown in Fig. 2.69

account the influence of short pitching as well as of the distribution. Compare [1.8, 2.52, 2.53, 2.54] for further information.

Reduction of the Cogging Torque

The cogging torque can be damped by skewing either the stator lamination stack or the poles of the permanent magnets. The waveform of the linkage flux is affected by skewing the slots in a similar way as by chorded or distributed windings, which in turn also affects the electromotive force. The references at [2.44, 2.52] provide detailed calculations for this purpose.

The manufacture of the lamination stack with skewed slots and fitting the winding over it, in general, requires more effort and involves higher costs com-

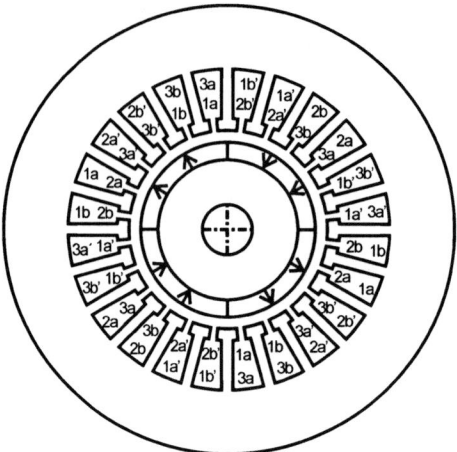

Fig. 2.69. Four-pole brushless permanent magnet motor with a distributed two-layer winding (coil width = $2\pi/3$)

pared to non-skewed windings. Whether skewing the magnet poles or an axial arrangement of angularly offset magnet segments offers more cost-effective solutions is a case-by-case decision. When magnet rings are applied a skewed magnetization of the poles can be made practically without additional production costs. For this purpose only the development of a custom-built magnetization tool becomes necessary.

Another method to suppress substantially the cogging torque is the selection of a fractional value for the slot to pole number ratio. Figures 2.70 and 2.71 illustrate two stator winding designs for a four-pole permanent magnet motor with a slot to pole number ratio of 3 and 2.25, respectively. While in the first case, for every turn of the rotor by one slot pitch, four pole edges simultaneously face four slot openings and thus generate correspondingly large changes in magnetic energy and associated high cogging torque values, in the

Fig. 2.70. Four-pole brushless permanent magnet motor with an integral slot to pole number ratio $z = 3$

Fig. 2.71. Four-pole brushless permanent magnet motor with a fractional slot to pole number ratio $z = 2.25$

second case only one such instance occurs. However, the significant reduction in the cogging torque is accompanied by a weakening of the fundamental waveform of the armature linkage fluxes as a result of the short pitching and distribution of the winding. The approximately sinusoidal waveform of the linkage fluxes and the electromotive forces in this arrangement results in good torque characteristics for operation with sinusoidal current feed.

In addition, selecting the slot to pole number ratio requires consideration of the fact that in the case of asymmetric winding arrangements like in Fig. 2.71 there are radial forces acting on the rotor, depending on the strength of the asymmetric armature field.

The motor design in Fig. 2.65 illustrates a symmetric winding arrangement with a fractional slot to pole number ratio of 1.5.

2.2.2.2 Design and Selection of the Permanent Magnet Shapes

Various magnetic materials such as ferrite, AlNiCo, SmCo and NdFeB (Sect. 2.1.2.2) in combination with numerous permanent magnet designs can be used for developing the excitation field in brushless permanent magnet motors. Figure 2.72 illustrates a few basic design examples. The respective specifications and the desired mode of motor operation (square-wave or sinusoidal control) help to determine the magnet geometry and the type of magnetization to match the stator winding design (cf. Sect. 2.2.2.1).

In support of the motor design, Table 2.5 contains some notes on the suitability and use of variants for internal and external rotors.

While radial magnetization of the magnets is preferred in the case of brushless DC motors, in the case of brushless AC motors diametrical magnetization with parallel magnetization alignment is often the choice, due to the desired sinusoidal flux linkage.

For both applications, simple relations for approximate determination of the torque-related radial component of the air-gap flux density B_δ can be specified using the basic equations for the permanent magnet circuit. The following specifications ignore both voltage drops in iron and magnetic leakage.

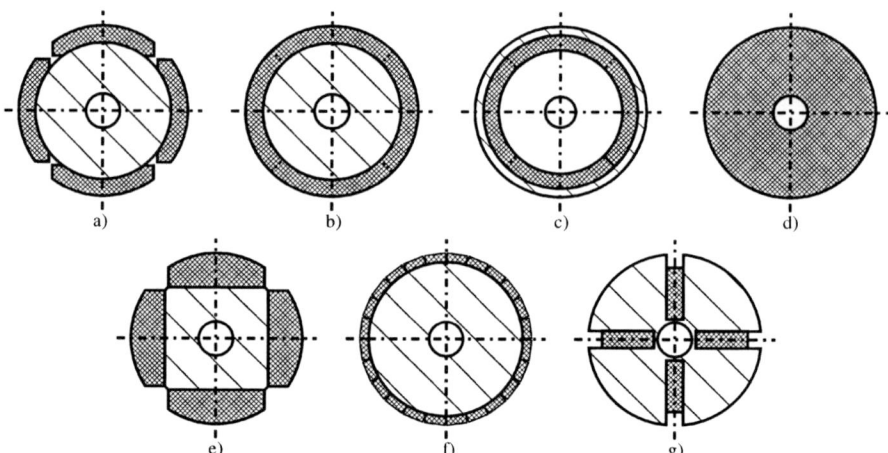

Fig. 2.72. Different design forms of rotors. Permanent magnets as: **a** Ring segments. **b** Rings. **c** Flexible bands. **d** Cylinders. **e** Arc-shaped block segments. **f** Partial blocks of magnetic poles. **g** Embedded disks

With the help of Ampere's law (assumption: $\boldsymbol{J} = 0$)

$$\oint_C \boldsymbol{H}\,\mathrm{d}\boldsymbol{s} = \int_A \boldsymbol{J}\,\mathrm{d}\boldsymbol{A}\,, \qquad (2.49)$$

of the relation

$$\boldsymbol{B}_\delta = \mu_0 \boldsymbol{H}_\delta\,, \qquad (2.50)$$

of the flux density equations (assuming a small air-gap in comparison with the medium air-gap radius r_δ and magnet radius r_m)

$$r_\delta B_\delta = r_\mathrm{m} B_\mathrm{m}\,, \qquad (2.50\mathrm{a})$$

and

$$r_\delta B_\delta = r_\mathrm{m} B_\mathrm{m} \cos\varphi_\mathrm{r} \qquad (2.50\mathrm{b})$$

for (a) the radial magnetization as shown in Fig. 2.73 or (b) the diametrical magnetization as shown in Fig. 2.74, and furthermore using the (linear) magnetization characteristic

$$B_\mathrm{m} = \mu_0 \mu_\mathrm{r} H_\mathrm{m} + B_\mathrm{r}\,, \qquad (2.51)$$

the radial component of the air-gap flux density B_δ yields the following approximation for radially magnetized ring-shaped magnets:

$$B_\delta = \frac{r_\mathrm{m}}{r_\delta} \frac{B_\mathrm{r}}{1 + \mu_\mathrm{r} \frac{\delta}{h_\mathrm{m}}}\,, \qquad (2.52)$$

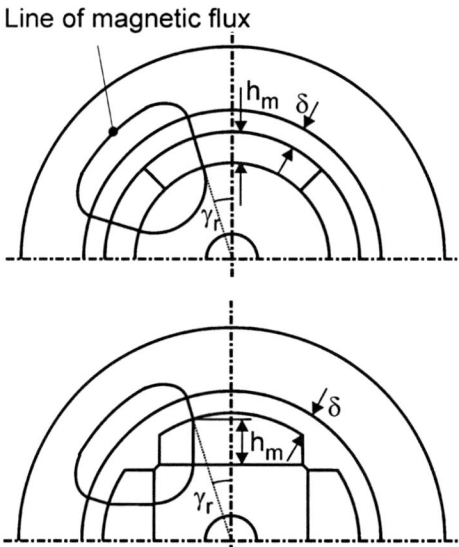

Fig. 2.73. Model for calculating the air-gap flux density for radial magnetization of ring magnets

Fig. 2.74. Model for calculating the air-gap flux density for diametrical magnetization of block magnets

or for diametrically magnetized, arc-shaped formed block magnets

$$B_\delta(\gamma_r) = \frac{r_m}{r_\delta} \frac{B_r \cos\gamma_r}{1 + \mu_r \frac{\delta}{h_m(\gamma_r)}} \ . \tag{2.53}$$

With a pole pair number $p > 1$, due to the reduced pole angle it becomes increasingly difficult (even with diametrical magnetization) to achieve a satisfactory approximation of a sinusoidal waveform of the air-gap field and the linkage fluxes. It can thus become necessary either to increase the air-gap between the magnet and the stator towards the direction of the pole edges, to skew the magnets or the stator lamination stack or to use methods like short pitching and distribution of the armature phase windings.

Basically the magnetic volume is put to better use by employing radial magnetization with isotropic materials that basically permit both radial and diametrical magnetization. This is particularly appropriate to constructions with a very small number of pole pairs (cf. Eq. (2.53)). Thus, in the case of unsaturated iron, the fundamental wave amplitude is $4/\pi$ larger for a square-wave air-gap field than for a sinusoidal air-gap field of the same magnetic material.

A large number of brushless permanent magnet motors is fitted with surface-mounted magnets (Fig. 2.72). Due to smaller armature inductivity values, smaller electrical time constants and reduced armature reaction, there are benefits in the motor characteristics with respect to dynamics and linearity. The disadvantage, however, is that, in contrast to embedded magnets in case of strong opposing fields, there is no special protection against demagnetization and in view of the mechanical robustness there is no protection against centrifugal forces. Magnets mounted on the outer surface in the case

Table 2.5. Use of various permanent magnet design shapes for brushless motors with square-wave and sinusoidal commutation

PM Design	Typical features	Typical applications (square-wave commutation)	Typical applications (sinusoidal commutation)
Segment	Magnetic orientation: isotropic, anisotropic mostly in diametrical direction Reduction of the cogging torque: axial arrangement of angular offset segments, angle-dependent segment height Mechanical characteristics: two-line contact (risk of fracture, gap filling by the use of adhesives), with two-pole design reduced pole width ($\ll 180°$ geom.) Applications: internal rotors, external rotors	Radially or diametrically magnetization	Diametrically magnetized
Ring	Magnetic orientation: isotropic, partly anisotropic in the radial direction for thin rings Magnetic characteristics: all pole configurations can be implemented; special magnetizing equipment necessary in case two-pole diametrical magnetization is not desired. Reduction of the cogging torque: magnetization of skewed poles Mechanical characteristics: gaps as a result of manufacturing tolerances, gap compensation using adhesives Applications: internal rotors, external rotors	Radial magnetization	Diametrical or radial, pole-oriented in the case of high ring strength and large number of poles (no back iron required)
Flexible strip	Magnetic orientation: mostly anisotropic perpendicular to the strip plane Magnetic characteristics: all pole configurations can be implemented; special magnetizing equipment necessary Reduction of the cogging torque: magnetization of skewed poles Mechanical characteristics: length tolerance compensation by means of crushing Applications: external rotors	In built-in (rolled-in) condition: radial magnetization	
Cylinder	Magnetic orientation: isotropic, anisotropic in axial or diametrical direction Magnetic characteristics: isotropic and anisotropic: pole-oriented magnetization (no back iron required); diametrical two pole magnetization	Axial, diametrical (two-pole), pole-oriented magnetization	Diametrical (two-pole), pole-oriented magnetization

Table 2.5. (continued)

	Reduction of the cogging torque: pole-oriented magnetization; magnetization of skewed poles possible; diametrical magnetization: axial arrangement of angular offset partial cylinders Applications: small internal rotors, motors with air-gap windings between magnet and back iron	
Block segment	Magnetic orientation: isotropic, anisotropic in diametrical direction Magnetic characteristics: sinusoidal field distribution in the air-gap can be achieved by means of angle-dependent adaptation of the magnet height (variable air-gap) Reduction of cogging torque: angle-dependent adaptation of magnet height Applications: internal rotors	Diametrical magnetization / Diametrical magnetization
Partial blocks (subdivided poles)	Magnetic orientation: isotropic, anisotropic parallel to the block height Magnetic characteristics: approximation of radial magnetization with blocks magnetized in parallel; multiple adjacent blocks make up a pole Reduction of the cogging torque: axial angular offset blocks Mechanical characteristics: automatic assembly is recommended for highly subdivided poles, thin magnetic layers can also be achieved in connection with large pole widths Applications: internal rotor, not usually used for external rotor	Parallel magnetization / Parallel magnetization
Embedded disks	Magnetic orientation: isotropic, anisotropic parallel to the disk height Magnetic characteristics: magnets protected from demagnetization by the armature field, sinusoidal field distribution in the iron pole shoe geometry achieved by angle-independent adaptation of the iron pole shoe geometry; larger armature reaction and electrical time constants (thus reduced dynamics), ferromagnetic linkages between the rotor poles: local magnetic short circuits in the rotor (under saturation) Reduction of the cogging torque: skewing of the lamination stack in several steps, pole shoe geometry Mechanical characteristics: protected permanent magnets Applications: internal rotor	Parallel magnetization, adapted pole shoe geometry (angle-dependent air-gap)

of internal rotors with large radii or high speeds are thus often encased using carbon, optical fiber or non-ferromagnetic sleeves.

Not only due to the better demagnetizing protection, but also by reason of better field weakening possibilities and favorable manufacturing costs with the block-shaped forms, rotors based on embedded magnet design are more and more used in small motor applications.

2.2.2.3 Design and Selection of the Type of Motor

As illustrated by the application examples in Sect. 2.2.1, the requirements of brushless permanent magnet drives are multi-faceted. The various technical problems arising from the applications are solved with the help of very individual types of motor designs, especially in the case of large serial productions with extreme low-cost demands.

This is why the choice of the motor design is very important at the concept stage. Basic criteria should be used, such as the space available (length, diameter), the required type of operation (constant speed, speed adjustment range, motor dynamics, running performance) and requirements for mechanical stability, vibration and noise or the capability of integrating the motor into an existing application design.

In order to aid the decision-making process, the following sections present significant characteristics of internal rotor, external rotor and disk rotor designs of brushless motors taking into consideration slot-based as well as slotless winding arrangements.

Brushless Motors with Internal Rotor Design

The brushless motor with an internal rotor represents an all-purpose design to some extent which is adequate for a wide range of applications.

The length to diameter ratio of rotor design varies substantially in practice from very slim to disk-shaped rotor constructions. In general, however, a slim design permits better utilization of the magnetic circuit since the magnetic flux leakage on the axial front sides and the losses at the end windings are of less consequence in this case. Particularly in conjunction with SmCo or NdFeB magnets, high torque to moment of inertia ratios and thus good dynamic characteristics can be obtained with this constructional concept.

Figure 2.75 illustrates the principal design of a motor fitted with an internal permanent magnet rotor. In order to protect the permanent magnets from destruction by high external stresses resulting from large centrifugal forces, it can be necessary to encase the segments or rings in epoxy resin reinforced with glass or carbon fibers. For this purpose, thinly drawn non-ferromagnetic steel or aluminum sleeves are also used to some extent [2.55].

In the case of small stator slot openings and large magnet heights, it is generally not necessary to use a laminated stack as magnet carrier for the conducting of the magnetic flux in the rotor. Nonetheless, this is the most cost-effective solution in many cases, especially since the rotor laminations

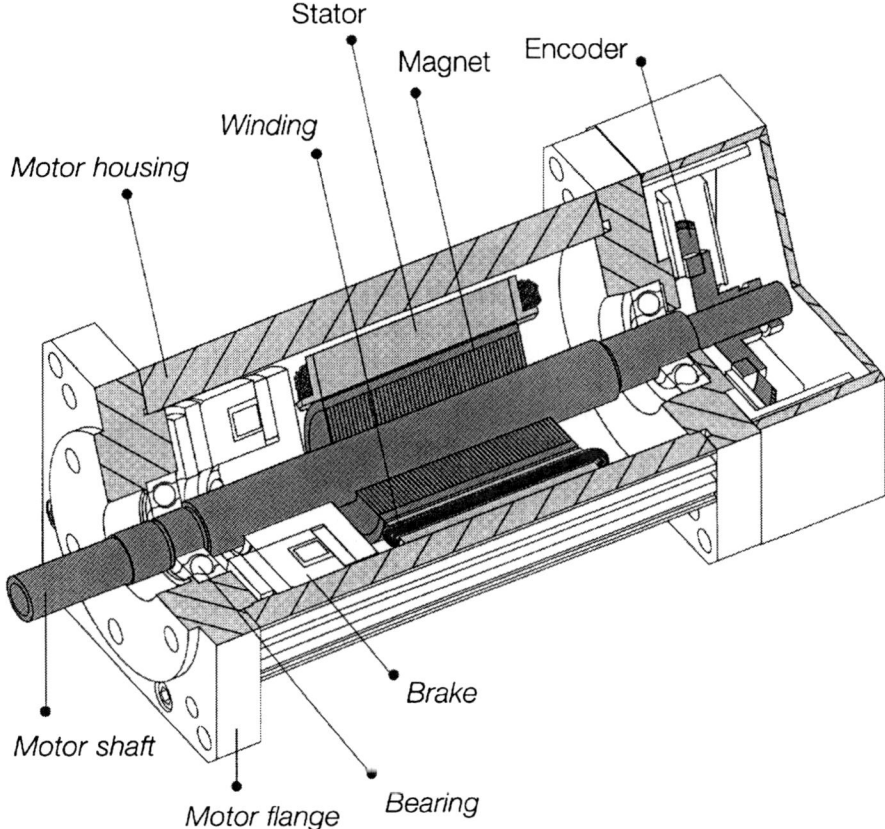

Fig. 2.75. Cross section of a brushless permanent magnet motor with internal rotor design (Source: © ebm-papst)

can be punched from the core of the stator lamination sheet in the same punching process. When there are space constraints on the rotor side, the shaft is also used to carry the magnetic flux. However, in contrast to the soft magnetic laminations, in general, there are no stringent magnetic tolerance values available for the shaft material. Due to their poor magnetic characteristics, shafts are basically suitable only to a limited extent for conducting the magnetic flux.

The arrangement of the elongated rotor lamination stack and the shaft leads to a rigid rotor construction. As a result the self-resonance frequencies and thus the vibration-related critical speeds are substantially higher in the case of internal rotor designs compared to corresponding external rotor alternatives.

The motor losses (winding losses, hysteresis losses, eddy current losses) occur primarily in the stator. The resulting heat can be dissipated very effectively by means of the external stator cooling devices with the help of suitable housing and flange design. Thus with proper heat dissipation the power density may attain very high values. At times, the eddy current losses of the magnets are not negligibly small. They are influenced by the electric conductivity and the magnetic permeability (increased armature reaction) of the permanent magnetic material, the slot openings of the stator, the transient behavior of the armature field, and its relative oscillating movement with respect to the rotor.

Brushless Motors with External Rotor Design

The greatest number of brushless permanent magnet motors is manufactured with external rotor design. The most significant application fields of this group are fans, blowers and ventilators and also to a great extent, applications of the growing computer industry, e. g. hard disk drives, CD-ROM and DVD-drives.

Apart from the typical characteristics of brushless motors such as long life and high reliability, the basic demands on these applications include a bundle of special requirements like low flutter, low noise and low manufacturing costs, and, particularly for computer applications, high torque densities and low thermal losses. The external rotor design offers a few special benefits in this respect as compared to the internal rotor design. The high moment of inertia of the rotor has a very positive effect on the running smoothness of the drive. Speed variations caused by torque ripples or cogging torque are damped with increasing rotor speed. Other technical benefits include the larger surface area available for the permanent magnets resulting in concentrated air-gap fields due to the radially inward oriented flux directions, and also the shorter end windings on both sides of the stator leading to less copper losses.

The external rotor is very suitable from the design point of view for large series production (Fig. 2.76). The motor housing is simultaneously built from both the stator (connecting flange) and the rotor (rotor bell). In contrast to internal rotor motors, the stator typically has only one flange with an integrated bearing pipe. This dispenses with the need for precise manufacture and the flush mounting of two separate flange pieces. The two ball bearing hubs can be lathed precisely with a single clamping of the work piece. Flyer winding machines are available with very low manufacturing times for cost-effective production. Moreover, it is possible to wind various phase windings simultaneously in case of concentrated windings. The latter also applies to motors with internal rotor design.

In fans, blowers and ventilators, the motor electronics is usually located on a PCB integrated on the inner side of the motor flange and protected by the overlapping rotor. Typically the rotor housing consists of a drawn soft magnetic bell that is riveted, welded or molded to the shaft. Small sheet thickness and thus a light weight of the rotor housing can be achieved by including the bell base in the flux distribution. The magnets, which are frequently plastic-bound elastic anisotropic ferrite magnet strips in the case of ventilation devices or NdFeB magnet rings in the case of hard disk drives, are stuck to the rotor housing. Encasing the magnets is not necessary. In the case

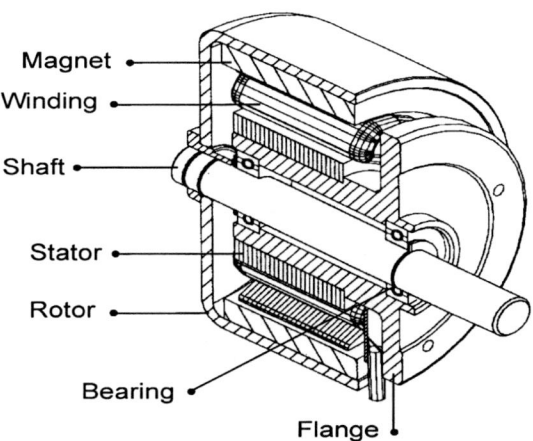

Fig. 2.76. Cross section of a brushless permanent magnet motor with external rotor design (Source: © ebm-papst)

of extreme centrifugal forces, merely the axial faces of the magnets must be additionally supported.

Brushless Motors with Disk Rotor Design
(Slotless Discoidal Winding Design)

Figure 2.77 illustrates a preferred direct drive variant for audio and video devices comprising a PCB to hold the winding and electronics, one ferromagnetic back iron fixed behind the PCB and an axially magnetized ferrite magnet rotor disk also equipped with a back-mounted iron sheet as back iron.

A simple solid steel sheet can be adequate as the ferromagnetic stator part for applications with low speed and flux densities. For more stringent requirements, the ferromagnetic stator part has instead to be built to limit the eddy current losses with a plastic-bound soft magnetic iron powder composite (SMC) or, rather rarely, as a spirally laminated disk.

Very good values can be obtained for the flutter due to the slotless armature design and the high moment of inertia of the rotor (cf. Fig. 2.77). The stator can also be made completely of nonferrous material when the winding, for instance embedded in a plastic disk, is placed between two rotor disks connected over the shaft. In this case the rotor can be built symmetrically with permanent magnet rotor disks on both sides or asymmetrically with a ferromagnetic rotor plate combined with another rotor plate assembled with permanent magnets. In the latter case the first rotor plate serves only as the ferromagnetic back iron. In contrast to the single disk design, there are no technical problems encountered here with respect to the three-dimensional flux orientation in the stator, the associated hysteresis and eddy current losses and also the high bearing load due to axial magnetic forces between the stator and the rotor.

In slotless designs as shown in Fig. 2.77 the eddy current losses within the windings resulting from the permanent magnet excitation field also have to

Fig. 2.77. Brushless permanent magnet motor with a single disk rotor design (Source: © Sony)

be taken into account. This applies in particular in case of high speeds and numerous rotor poles, and for PCB, foil or tape windings with large conductor cross-section dimensions.

As illustrated in Fig. 2.77, the electric motor with disk-shaped rotor is very suitable for integration of the motor electronics and facilitates a very compact flat design.

Brushless Motors with Slotless Cylindrical Winding Design

Brushless permanent magnet motors with slotless cylindrical windings are based on the design of bell-type armature commutator motors (Sect. 2.1.2.3). An example is shown in Fig. 2.78. The generally three-phase rhombic-wound or skew-wound winding is placed in the motor air-gap. It is enclosed on the inside with a cylindrical permanent magnet placed on the shaft and enclosed on the outside by a solid bell-type back iron also fixed to the shaft. As a result of this construction there are practically no iron losses in the motor. This has a particularly good effect on the efficiency at high speeds.

Figure 2.79 illustrates another design variant with a laminated ferromagnetic stack in the stator. In contrast to the first design the permanent magnet circuit is no longer loss-free as a result of the rotating field in the stator. How-

Fig. 2.78. Design principle of a slotless EC motor with a solid rotating back iron and a self-supporting cylindrical skew-wound winding as armature (Source: © Faulhaber)

ever, the moment of inertia is substantially smaller in this type of motor and provides high dynamic performance.

The life of brushless motors with self-supporting windings as compared to bell-type armature commutator motors is determined merely by the bearing and the commutation electronics. The electronics as well as the angle transducer are integrated into the rear portion of the motor housing in certain designs (refer to Figs. 2.78 and 2.79).

Both the winding design and the diametrical magnetization of the two-pole cylindrical magnets lead to a sinusoidal flux linkage and thus to sinusoidal electromotive forces. Therefore sinusoid-shaped control is recommended from the viewpoint of torque uniformity.

In contrast to bell-type commutator motors that achieve very low torque ripple values due to the high number of coils, the three-phase brushless design can be troublesome for highly demanding servo applications with square-wave commutation. However, in some applications square-wave commutation is preferred over sinusoidal commutation due to the cost and constraints on the space available for mounting the adequate electronics and sensors. The

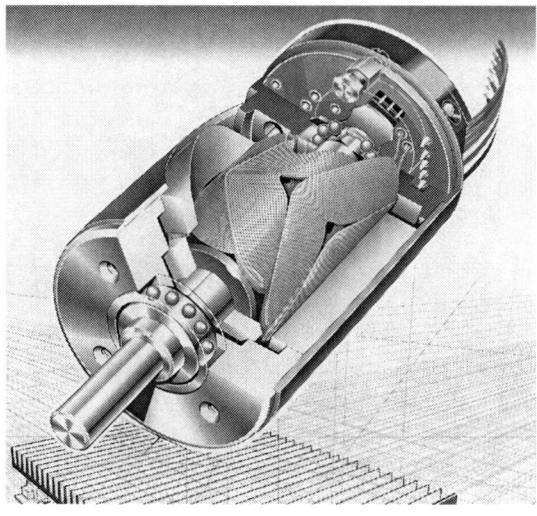

Fig. 2.79. Design principle of a slotless EC motor with a stator lamination stack and a self-supporting cylindrical rhombic-wound winding as armature (Source: © maxon)

differences between square-wave and sinusoidal commutation reduce with increasing speeds due to the smoothing influence of the rotor moment of inertia.

2.2.3 Dynamic Model of the Brushless AC Motor

Brushless, field-oriented controlled permanent magnet motors enjoy an increasingly significant trend of ever-growing automation of technical processes with increasing requirements on dynamics, power density, efficiency, reliability and life. This also applies to the field of servo applications, where with the availability of high-energy demagnetization-proof and temperature-resistant permanent magnet materials the brushless AC motor has become an interesting alternative, both technically and commercially, to asynchronous motors. The brushless permanent magnet excited servo motor is an excellent option in general due to its robustness, good dynamics, reduced power to weight ratio and good controllability.

Simulation of the dynamic processes has meanwhile become almost indispensable for the selection and optimal adaptation of the components and also for computer-aided design of advanced speed or position control algorithms. When the mathematical models of the motors and the associated drive components have been defined in a general form and the specific parameters and characteristics of the components considered have been obtained, then significant statements can be made regarding the dynamic operational behavior of the system with a bare minimum of cost and time. That way cumbersome test runs with iterative mechanically or electrically adapted functional samples can usually be omitted.

2.2.3.1 Motor Model

Both the mechanical and electromagnetic dynamic behavior of the motor must be taken into consideration for developing a mathematical model.

We assume that the inductivity values are independent of the rotor position, that the iron losses are negligibly small, and that there is no damping influence as a result of the eddy currents in the magnetic material for the determination of the motor model. The voltage equations for the motor with $i = 1 \ldots m$ phases can be specified with the relation

$$u_i = R_i i_i + \sum_{j=1}^{m} L_{ij} \frac{\mathrm{d}}{\mathrm{d}t} i_j + \Omega \frac{\mathrm{d}}{\mathrm{d}\gamma} \psi_{pm_i} , \qquad (2.54)$$

whereby, the last term in the equation represents the electromotive forces and Ω refers to the mechanical angular velocity of the rotor.

For a symmetric three-phase motor, the values for the individual self-inductances $L_{ki,\, k=i} = L_\mathrm{s}^*$ as well as the mutual inductances $L_{ki,\, k \neq i} = L_\mathrm{g}^*$ are the same in each case.

With the equation for the sum of the currents

$$i_1 + i_2 + i_3 = i_0 \tag{2.55}$$

for a mid-point connection and equal phase resistances $R_i = R$, Eq. (2.54) can be rewritten in a general from as

$$u_i = Ri_i + L_s \frac{\mathrm{d}}{\mathrm{d}t} i_i + L_g^* \frac{\mathrm{d}}{\mathrm{d}t} i_0 + \Omega \frac{\mathrm{d}}{\mathrm{d}\gamma} \psi_{pm_i} \tag{2.56}$$

with

$$L_s = L_s^* - L_g^* . \tag{2.57}$$

When the mid-point is not connected ($i_0 = 0$), the voltage equation at Eq. (2.56) simplifies to

$$u_i = Ri_i + L_s \frac{\mathrm{d}}{\mathrm{d}t} i_i + \Omega \frac{\mathrm{d}}{\mathrm{d}\gamma} \psi_{pm_i} . \tag{2.58}$$

Figure 2.80 illustrates the corresponding equivalent circuit of the brushless permanent magnet motor. The last term of Eq. (2.58) corresponds to the electromotive forces E_{A_i}.

Apart from the voltage equation, the mechanical movement equation is required to describe the mathematical motor model. For the internal motor torque this is

$$T(\gamma) = J\ddot{\gamma} + D\dot{\gamma} + T_l . \tag{2.59}$$

The calculation of the internal motor torque can be analytically performed based on the angle-dependent change of the magnetic co-energy of the motor. Thus the following applies to the torque generated electromechanically:

$$T(\gamma) = \frac{\partial W_{mag}^{co}}{\partial \gamma} . \tag{2.60}$$

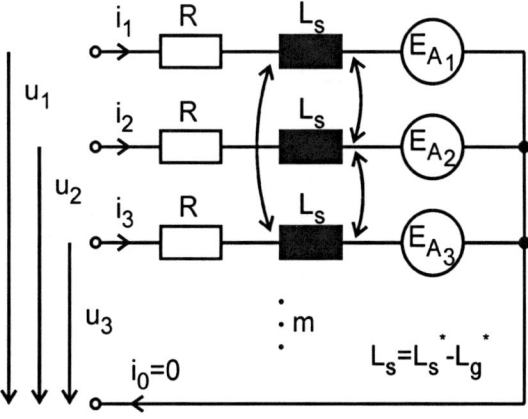

Fig. 2.80. Electrical equivalent circuit of the brushless AC motor

For linear magnetic conditions with marginal magnetic potential differences in the iron and negligible cogging torque the equation for the magnetic co-energy can be derived as follows:

$$W_{\text{mag}}^{\text{co}} = \sum_{k=1}^{m} \psi_{pm_k}(\gamma)_{i_k} + \frac{1}{2} \sum_{k}^{m} \sum_{i=1}^{m} L_{ki}(\gamma) i_i i_k . \quad (2.61)$$

From Eqs. (2.60) and (2.61) we obtain the internal torque of the machine for the general case of angle-dependent inductances:

$$T(\gamma) = \sum_{k=1}^{m} \frac{\mathrm{d}\psi_{pm_k}(\gamma)}{\mathrm{d}\gamma} i_k + \frac{1}{2} \sum_{k}^{m} \sum_{i=1}^{m} \frac{\mathrm{d}L_{ki}(\gamma)}{\mathrm{d}\gamma} i_i i_k . \quad (2.62)$$

The internal torque of the brushless AC motor comprises a component that arises from the interaction of the phase currents with the permanent magnet linkage fluxes (first term), and the reluctant component (second term).

Angle-dependent inductance values occur, for example, as a result of non-circular ferromagnetic rotor cross sections or embedded magnets or with the use of permanent magnets with high reversible permeability values if the magnet alignment does not result in a closed ring-shaped distribution. The latter can arise from gaps between the pole edges or in case of angle-dependent magnet heights.

The derivative of the linkage fluxes can be obtained either from measurements of the angular speed-dependent electromotive forces

$$E_{A_k} = \frac{\mathrm{d}\psi_{pm_k}(\gamma)}{\mathrm{d}\gamma} \frac{\mathrm{d}\gamma}{\mathrm{d}t} , \quad (2.63)$$

or by analytical or finite element calculation methods.

Equation (2.62) can be extended by the addition of a permanent magnet generated cogging torque $T_{\text{cog}}(\gamma)$ that occurs even without armature currents (see Sect. 2.2.4).

In many cases, particularly with rotors having rotational symmetry and surface-mounted permanent magnets, the reluctant component can be largely ignored on account of the angle-independent inductance values. Also the component $T_{\text{cog}}(\gamma)$ can be ignored in the case of the slotless motors and also for those motors with special arrangements for cogging torque suppression by means of slot skewing, pole skewing or a favorable slot to pole number ratio (compare Sect. 2.2.2.1).

The equations presented describe the dynamic operational behavior of brushless permanent magnet AC motors and can be combined to summarize a simple motor model under the assumption of angle-independent inductance values. As an example, Fig. 2.81 illustrates a mathematical model of a single-phase motor with linear behavior. The functions $T_{\text{cog}}(\gamma)$ and $\mathrm{d}\psi_{\text{pm}}(\gamma)/\mathrm{d}\gamma$ can be obtained analytically, numerically with a finite element program, or by means of measurements on a test bench (cf. Eq. (2.63)), and can be implemented into the simulation program analytically as equations or in the

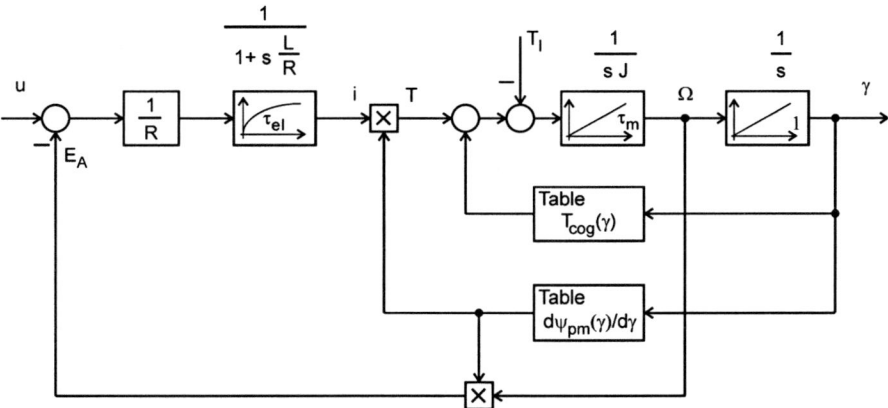

Fig. 2.81. Motor model of a brushless single-phase motor

form of tables. The model of the single-phase motor in Fig. 2.81 can also be extended with the equations presented in this section for multiple phase motors.

2.2.3.2 Two-Phase Equivalent Motor Model for Field-oriented Control

It is beneficial to use a two-phase equivalent motor model with flux-oriented coordinates for field-oriented control of three-phase permanent magnet AC motors with sinusoidal voltage or current supply [2.46–2.50,2.56]. In this manner the stator current can be specified with two orthogonal components that determine the torque and the magnetization independently of one another.

For highly dynamic operation it is desirable to be able to change very quickly the current component controlling the torque, as in the case of the adjustment of the armature current of DC commutator motors. The magnetization component used to build the excitation field can be omitted for brushless motors in many applications due to the available permanent magnet excitation (exception: field weakening operation). In order to ensure decoupling of both the current components, it is necessary to control the stator currents in value and phase depending on the flux orientation. For this purpose, particularly when the control is to be done digitally with the help of a microprocessor or signal processor circuit, implementing the required control and transformation algorithms in complex space vector representation is recommended. This leads to a clear and simple program structure and permits the realization of short computation times and thus high sampling rates.

In order to represent the space vector components of the electrical and magnetic parameters (such as the phase currents, phase voltages or linkage fluxes) in an appropriate manner, as a rule, the three-phase system of the motor is converted to a two-phase equivalent system as shown in Fig. 2.82.

The transformation equation for transforming the above variables into the stator α, β coordinate system can be specified with

$$\begin{pmatrix} x_\alpha \\ x_\beta \\ x_0 \end{pmatrix} = \frac{2}{3} \begin{bmatrix} 1 & -\frac{1}{2} & -\frac{1}{2} \\ 0 & \frac{\sqrt{3}}{2} & -\frac{\sqrt{3}}{2} \\ \frac{1}{2} & \frac{1}{2} & \frac{1}{2} \end{bmatrix} \begin{pmatrix} x_1 \\ x_2 \\ x_3 \end{pmatrix}. \qquad (2.64)$$

The inverse transformation is described by

$$\begin{pmatrix} x_1 \\ x_2 \\ x_3 \end{pmatrix} = \begin{bmatrix} 1 & 0 & 1 \\ -\frac{1}{2} & \frac{\sqrt{3}}{2} & 1 \\ -\frac{1}{2} & -\frac{\sqrt{3}}{2} & 1 \end{bmatrix} \begin{pmatrix} x_\alpha \\ x_\beta \\ x_0 \end{pmatrix}. \qquad (2.65)$$

The brushless permanent magnet motor can now be modeled on the basis of the two-phase equivalent winding system. For the following analysis we assume that the motor has the following characteristics:

- Magnets mounted on the surface with sinusoidal flux density distribution in the air-gap
- Sinusoidal phase current feed
- Sinusoidal magnetomotive force distribution of the armature winding
- Magnetic materials having linear characteristics
- Negligible influence of saturation

The magnetic behavior of the permanent magnets can be modeled approximately with an equivalent electrical coil enclosing the magnet contour. Accordingly, the following equation applies to the space vector of the rotor current

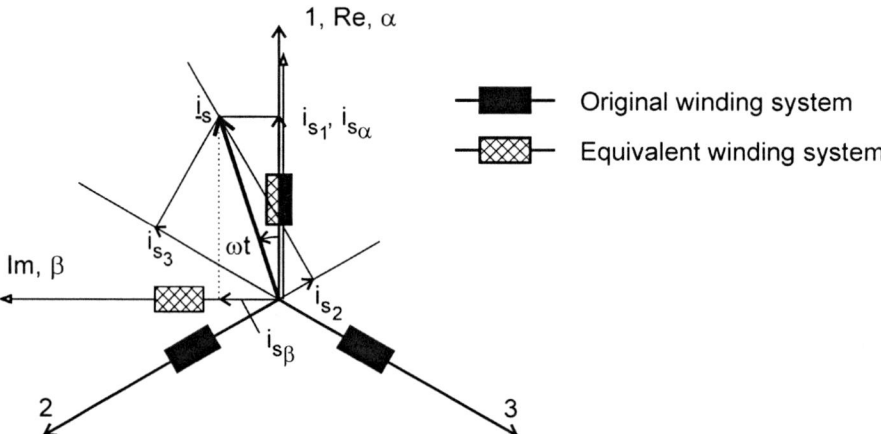

Fig. 2.82. Conversion of the three-phase system into a two-phase equivalent system (Clarke's transformation between stator-fixed coordinate systems)

in the flux-oriented (rotor-fixed) coordinate system (d, q):

$$\underline{i_r} = (i_{r_d} + j i_{r_q}) = \begin{pmatrix} i_{r_d} \\ i_{r_q} \end{pmatrix} \qquad (2.66)$$

with the direct current component $i_{rd} = I_r$ and the quadrature current component $i_{rq} = 0$.

The magnetic flux linked with the stator winding is required to determine the motor torque. The following simple relation applies to the stator linkage flux space vector:

$$\underline{\psi_s} = L_s \underline{i_s} + L_m \underline{i'_r} \qquad (2.67)$$

with $L_s = L_s^* - L_g^*$ and $L_m = \frac{3}{2} L_{sr}^*$, whereby L_s^* is the self-inductance and L_g^* the mutual inductance of the stator phases and L_{sr}^* is the coupling inductance between the stator phases and the permanent magnet equivalent winding. The fictitious rotor current space vector $\underline{i'_r}$ is specified in the stator coordinate system (marked with an apostrophe) and corresponds to the rotor to stator transformation $\underline{i'_r} = \underline{i_r} e^{j\gamma_{el}}$. The permanent magnet linkage flux of the stator winding is defined by $\psi_{s_{pm}} = L_{m_d} i_{r_d}$.

In the flux-oriented rotor coordinate system the flux equation is

$$\underline{\psi'_s} = (L_s \underline{i_s} + L_m \underline{i'_r}) e^{-j\gamma_{el}} = (L_s \underline{i'_s} + L_m \underline{i_r}) \qquad (2.68)$$

or, expressed in the complex d and q components, this is

$$\underline{\psi'_s} = \psi_{sd} + j\psi_{sq} = L_{s_d} i_{s_d} + jL_{s_q} i_{s_q} + L_{m_d} i_{r_d} \ . \qquad (2.69)$$

The first two summands reflect the flux linkage within the three-phase stator winding. This is significant for small slot openings, for low magnet heights or for permanent magnet materials with high permeability values, e. g. AlNiCo magnets. The third summand describes the flux linkage of the permanent magnet field with the stator winding.

With the relation

$$T = \frac{3}{2} p \underline{\psi_s} \times \underline{i_s} \qquad (2.70)$$

and its transformation to flux-oriented rotor coordinates

$$T = \frac{3}{2} p \underline{\psi'_s} \times \underline{i'_s} \qquad (2.71)$$

the following motor torque is obtained from the vector product:

$$T = \frac{3}{2} p \left(\psi_{sd} i_{s_q} - \psi_{sq} i_{s_d} \right) = \frac{3}{2} p \left[L_{m_d} i_{r_d} i_{s_q} + \left(L_{s_d} - L_{s_q} \right) i_{s_d} i_{s_q} \right] \ . \qquad (2.72)$$

If the permanent magnet segments are arranged on an iron cross-section having the shape of a regular polygon as shown in Fig. 2.83, then the inductances

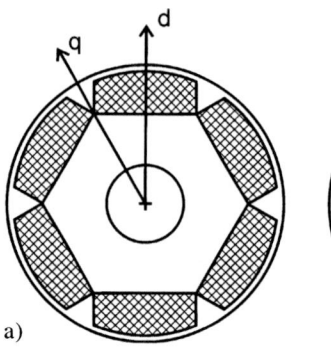 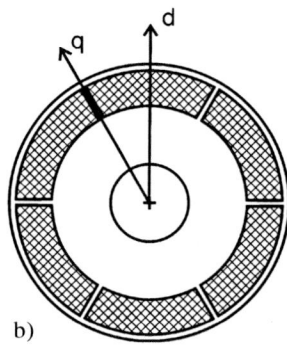

Fig. 2.83. Geometrical designs of surface-mounted permanent magnet rotors with: **a** $L_{s_d} < L_{s_q}$; **b** $L_{s_d} = L_{s_q}$

in direct and quadrature directions have the relation $L_{s_d} < L_{s_q}$. In other cases, due to the rotational symmetry and negligible pole gaps or in the case of magnets with larger distances and very low reversible permeability values close to 1, there is no difference between the direct and quadrature inductance values.

If the space vector of the stator current (in flux-oriented rotor coordinates) is selected as

$$\underline{i}'_s = j i_{s_q} \tag{2.73}$$

without any *field weakening* component, then the equation for the torque (Eq. 2.72) can be simplified as follows:

$$T = \frac{3}{2} p L_{m_d} i_{r_d} i_{s_q} = \frac{3}{2} p \psi_{s_{pm}} i_{s_q} . \tag{2.74}$$

Fulfilling this torque equation, which is very similar to that for DC commutator machines, requires field-oriented control of the stator current. Since the stator linkage flux of the permanent magnet field is coupled in a fixed manner to the d-axis of the rotor, the current control can easily be derived from the rotor position signal.

For applying the currents to the power electronics the stator current space vector must be available in stator coordinates. With the complex description of the space vector

$$\underline{i}_s = i_{s_\alpha} + j i_{s_\beta} \tag{2.75}$$

the transformation from flux-oriented rotor to stator-fixed coordinates results in

$$\underline{i}_s = \underline{i}'_s e^{j\gamma_{el}} . \tag{2.76}$$

From the previous two equations, the separation of the real and imaginary parts leads to following rules for the control of the stator current space vector:

$$\begin{aligned} i_{s_\alpha} &= i_{s_d} \cos \gamma_{el} - i_{sq} \sin \gamma_{el} \\ i_{s_\beta} &= i_{s_d} \sin \gamma_{el} + i_{s_q} \cos \gamma_{el} . \end{aligned} \tag{2.77}$$

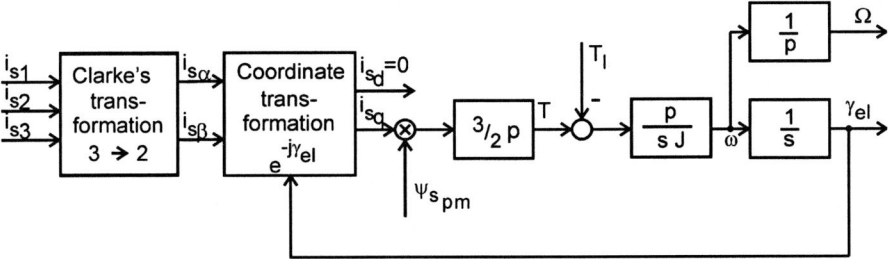

Fig. 2.84. Simple model of the brushless permanent magnet motor for sinusoidal field-oriented stator current feed

The trigonometric functions compose the rotation matrix for the transformation. The rotation is also defined by the factor $e^{j\gamma_{el}}$ in Eq. (2.76).

The model of the permanent magnet motor can be developed for dynamic simulation of the drive system by taking the previous equations into account. Equations (2.76) and (2.77) have to be replaced by an inverse formulation of the transformation. Equation (2.59) for the mechanical movement is also included in this model.

The d-component of the stator current need not compulsorily be selected as zero as represented in Eq. (2.73) and Fig. 2.84. Field weakening operation can be set with it to a limited extent. Thus, for example, speeds beyond the nominal operating range can be achieved without an increase of the nominal voltage and the nominal power of the power converter. However, as a rule, this option is limited, especially in the case of arrangements with surface-mounted magnets.

For integrating brushless motors into speed- or position-controlled applications, see Sect. 6.4.4.2. Significant components of the associated signal flow diagram are the position, speed and torque controllers which are frequently arranged in a cascading structure, the torque determination, the transformations of the stator currents between the stator-specific and rotor-specific (field-oriented) coordinates, and the transformations between the three-phase motor and the equivalent two-phase system. Time delays like the processor computing time or the A/D and D/A conversion time periods can be compensated by an appropriate angle correction of the stator current space vector.

2.2.4 Electronic Suppression of Torque Ripples

Various options have already been considered in Sects. 2.2.2.1 and 2.2.2.2 with which the torque performance of the motors can be influenced by constructional design measures. This section deals with the electronic options for smoothing the torque [2.57–2.61].

The following torque influences are primarily responsible for an inadequate torque behavior:

- Ripples in the electromagnetic torque component (first term in Eq. (2.62)), caused by inadequate matching of the interaction between the armature

phase currents, the winding arrangement and the permanent magnet fluxes linked with the phases (torque ripple)
- Ripples in the reluctant torque component (second term in Eq. (2.62)), caused by angle-dependent changes in the winding inductances as they can occur with non-uniform air-gaps or non-cylindrical ferromagnetic rotor back iron (reluctance torque)
- Permanent magnet specific torque fluctuations also generated in the zero current state as a result of the interaction between the permanent magnet fields and the winding slots in the ferromagnetic armature (cogging torque).

However, torque ripples of a drive can also have purely mechanical causes such as the axis offset of inflexible couplings, the bending of the motor shaft as a result of eccentrically manufactured or mounted bearing carriers or external load-specific disturbance torques periodical recurring within one revolution.

In all these cases the disturbances occur cyclically over a period of one rotation of the rotor during stationary operation, so that basically it is possible to suppress the torque fluctuations by means of suitable current shape control. In such cases no superimposed torque controller will be necessary for smoothing the torque.

Since the reluctant torque ripple of brushless permanent magnet motors plays only an insignificant role in many cases, close attention is directed to the electronic reduction of the torque ripple of the first part of Eq. (2.72) and in a second step to the cogging torque caused by the permanent magnet influence.

Equation (2.62) yields the following torque function taking Eq. (2.63) into consideration, and ignoring the reluctant component of the internal motor torque:

$$T = \frac{1}{\Omega} \sum_{k=1}^{m} E_{A_k}(\gamma, \Omega) i_k(\gamma) \,. \tag{2.78}$$

The associated optimal current waveforms shall be obtained for the requirement of constant internal torque T for a motor with a star connection of the winding. This requires a distinction between motors with and without mid-point conductor (sum of currents $i_0 \neq 0$ or $i_0 = 0$).

2.2.4.1 Motors with Mid-point Conductor

Equation (2.78) in its present form is under-determined. To obtain the optimal phase current waveform, we need to impose an additional requirement.

As an example, the condition of minimum resistive winding losses can be imposed under the assumption of a symmetric motor design with identical phase windings and, to begin with, without the limitation of a zero sum of the currents

$$P_v = \sum_{k=1}^{m} R i_k^2(\gamma) = \min \,. \tag{2.79}$$

To solve Eq. (2.78) with m variables and the additional condition as given in Eq. (2.79), first the necessary condition for the independent control of the phase currents can be determined with the help of the Lagrange multiplier λ:

$$\frac{\partial P_v}{\partial i_k} + \lambda \frac{\partial T}{\partial i_k} = 0 \, . \tag{2.80}$$

The solution to the set of equations consisting of Eqs. (2.78–2.80) leads *via*

$$\lambda = -\frac{2\Omega^2 T}{\sum_{k=1}^{m} \frac{E_{A_k}^2(\gamma,\Omega)}{R}} \tag{2.81}$$

to the definition of the current curve for constant internal torque T

$$i_k(\gamma) = \frac{T \Omega E_{A_k}(\gamma,\Omega)}{\sum_{k=1}^{m} E_{A_k}^2(\gamma,\Omega)} \, . \tag{2.82}$$

Similarly, correction currents can be derived for linear relations to suppress the permanent magnet cogging torque. Here the current components are determined in such a manner that an angle-dependent electromagnetic countertorque is created to the cogging torque T_{slot}. The current components $i_{c_k}(\gamma)$ are obtained with the help of Eq. (2.82), in which the constant torque T must be replaced with the negative cogging torque $-T_{\text{slot}}(\gamma)$.

For the unsaturated linear magnetic circuit, the torque components of the two previous equations are independent and can be controlled by means of superimposition of the current components $i_k(\gamma)$ and $i_{c_k}(\gamma)$ of the phases $k = 1\ldots m$.

2.2.4.2 Motors Without Mid-point Conductor

Under the restricting condition of a zero sum of phase currents

$$\sum_{k=1}^{m} i_k = 0 \tag{2.83}$$

the following current equation system can be obtained for constant internal torque for the three-phase motor in a similar way:

$$i_{[1,2,3]}(\gamma) = \frac{T\Omega(2E_{A_{[1,2,3]}} - E_{A_{[2,1,1]}} - E_{A_{[3,3,2]}})}{(E_{A_1} - E_{A_2})^2 + (E_{A_1} - E_{A_3})^2 + (E_{A_3} - E_{A_2})^2} \, . \tag{2.84}$$

whereby the electromotive forces E_{A1}, E_{A2}, E_{A3} are functions of the angle γ and the time derivatives. Equation (2.84) describes the waveform of the three phase currents. The phase variables are differentiated with the indices $[1, 2, 3]$.

2.2.4.3 Electronics Design

The methods presented for an optimized current feed, with or without a midpoint conductor, can easily be implemented by means of software on electronic circuits as they are commonly used in servo drive technology.

The calculation of the current waveforms in the microprocessor is performed based on the angular position of the rotor and the desired motor torque in accordance with Eqs. (2.82) and (2.84), respectively. For this purpose, the use of standardized curves $i_k^*(\gamma)$ is recommended that are either stored in the form of tables or analytical functions in the memory.

The current amplitudes are not dependent on the angular speed, as appears to be the case when we consider the two equations. This becomes evident when we substitute the electromotive forces with Eq. (2.63). Thus the following relation emerges from Eq. (2.82), which is basically dependent only on the angle and the linkage fluxes:

$$i_k(\gamma) = \hat{i} i_k^*(\gamma) = T \frac{\frac{\mathrm{d}\psi_{pm_k}(\gamma)}{\mathrm{d}\gamma}}{\sum_{\kappa=1}^{m}\left(\frac{\mathrm{d}\psi_{pm_\kappa}(\gamma)}{\mathrm{d}\gamma}\right)^2} . \qquad (2.85)$$

The current amplitudes in Eq. (2.84) can be represented in a similar manner independently of the angular speed.

In the block diagram in Fig. 2.85, a current curve $i_{c_k}(\gamma)$ is superimposed on the operational current curve $i_k(\gamma)$ for suppressing the cogging torque under the assumption of linear magnetic circuits. The amplitude of the operational current curves is set either in a direct manner by the reference value for the

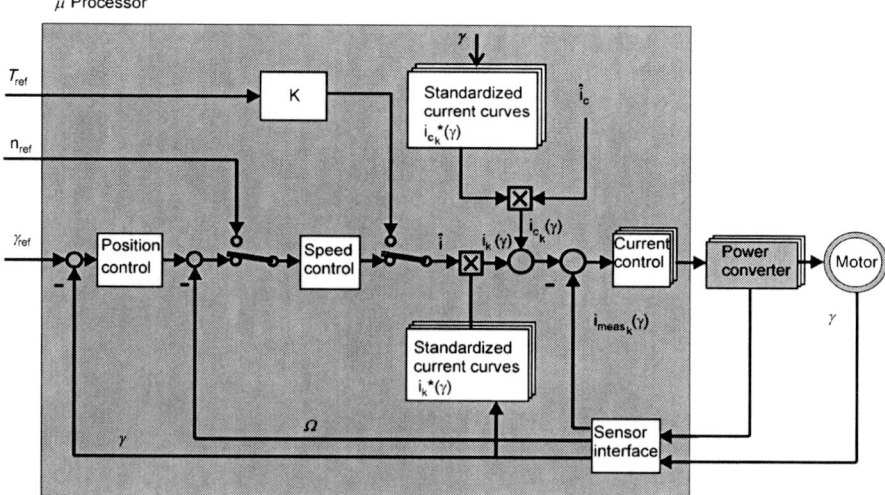

Fig. 2.85. Electronics design for suppression of torque ripples and cogging torque

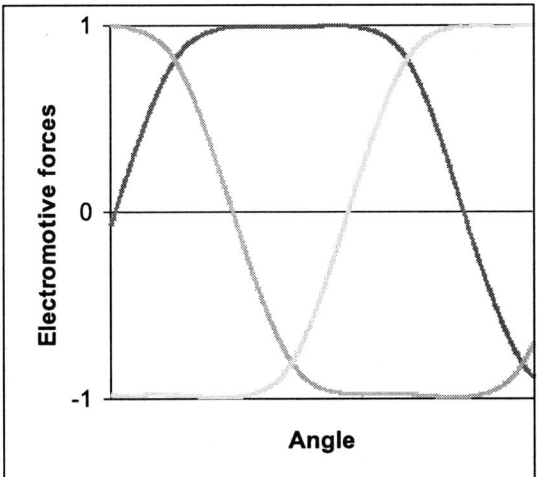

Fig. 2.86. Electromotive forces of a brushless permanent magnet motor with star-connected winding

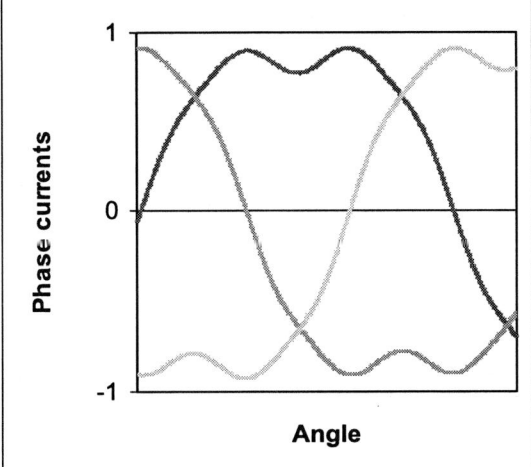

Fig. 2.87. Optimal phase current waveforms for a winding with mid-point connection (with reference to Fig. 2.86)

torque or in the control loop by the output of the speed controller, whereas the amplitude $\hat{i}_c$ of the cogging torque current curves always has a constant value.

The torque uniformity that can be achieved depends to a large extent on the accuracy of the simulation or measurement curves obtained for the electromotive force and the cogging torque, the dynamic characteristics of the power electronics, and the resolution of the angle transducer. Thus torque disturbances with harmonics of a higher order can be suppressed adequately only when the required current rise and drop times can be realized within the required extent. In particular, control problems can occur with the suppression of high-frequency cogging torques. However, since smooth running is increasingly supported by the rotor moment of inertia at higher speeds, elec-

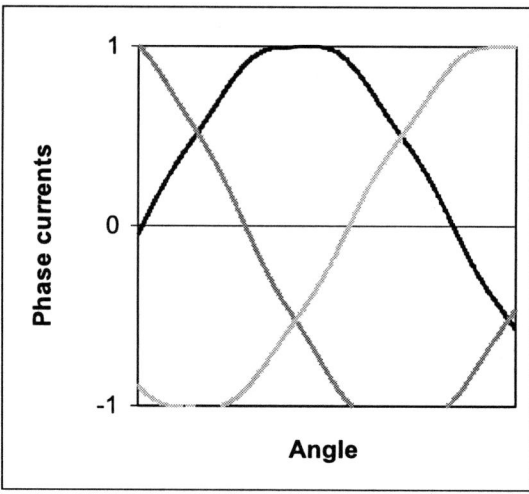

Fig. 2.88. Optimal phase current waveforms for a winding without mid-point connection (with reference to Fig. 2.86)

tronic suppression of the torque ripple and cogging torque is usually significant mainly at lower ranges of speed.

Figures 2.86, 2.87 and 2.88 illustrate examples for the optimized currents $i_k(\gamma)$ for a three-phase motor with approximately trapeze-shaped electromotive forces for two winding arrangements with and without mid-point connection. As illustrated in Figs. 2.87 and 2.88 the phase current curves for constant torque, adapted to the motor characteristics, deviate substantially from the conventional sinusoidal or square-wave feed.

2.2.5 Motor Characteristics

The load characteristics of brushless permanent magnet motors are comparable to a large extent with those of the permanent magnet excited commutator motors. This characteristic can also be seen in the equations in Sects. 2.2.3 and 2.2.4. Figure 2.89 illustrates typical curves for the speed, current, power and efficiency depending on the motor torque. The broken lines in the graph display curves which increase under the influence of temperature which affects the amplitude of the permanent magnet field and the armature resistances.

Magnetic saturation in the iron caused by high mechanical loads can result in a further drop of the speed *vs* torque characteristic. Variations of the supply voltage lead to parallel speed *vs* torque characteristics provided that the motor temperature remains constant and the magnetic circuits do not saturate.

Figure 2.90 illustrates the working range that occurs typically for permanent magnet excited motors. For highly dynamic accelerations, the nominal motor torque can be exceeded many times for short durations. This is particularly applicable to motors fitted with SmCo or NdFeB permanent magnets whose coercive field strengths have very high values. Protection against demagnetization can be improved with the help of design measures such as implementation of greater magnet heights, higher number of poles or ferromagnetic

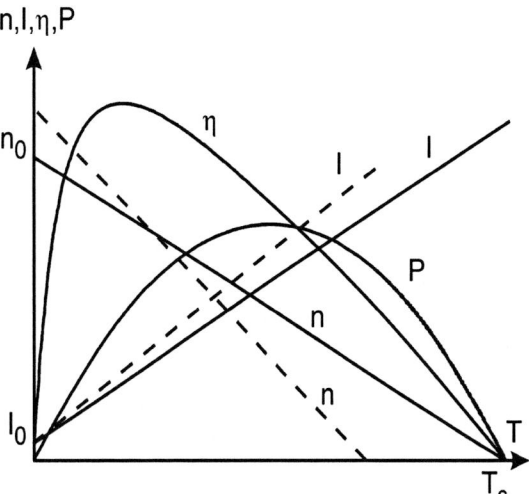

Fig. 2.89. Typical load characteristics of permanent magnet excited brushless motors (*broken lines*: curves at higher temperatures)

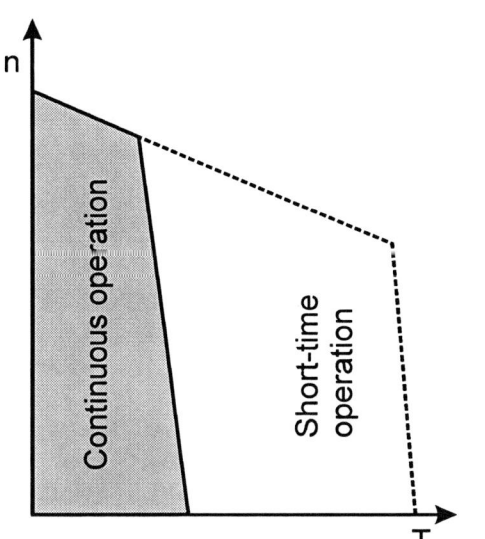

Fig. 2.90. Continuous and short-time operation range

pole shoes which is the case with rotors having embedded magnets. Other limits are specified by the robustness of the mechanical design of the motor, the current and voltage limits of the power controller, and the thermal losses.

2.2.6 Sensor Technology

For the operation of field-controlled brushless motors, it is necessary to know the direction of the flux orientation at any time. In many cases outlined in Sects. 2.2.1 and 2.2.3 the armature linkage flux is largely fixed to the rotor. The rotor position can be determined, for example, optically using incremental or absolute angle encoders, magnetically using magnetic resolvers, Hall sensors or magneto-resistive sensors, or without the use of sensors by measuring and evaluating electrical motor parameters. The selection of the angle measurement method is influenced by the type of design of the motor and also by the respective application.

Thus for the square-wave commutation corresponding to the brushless DC motor designs in Sect. 2.2.2, the angle information has only to be updated in the switching regions. Thus, depending on the number of phases, up to three switching Hall sensors (three-phase motor) are sufficient to provide the signals for commutation. If the brushless DC motor is used in servo drive applications, then the angular resolution of the digital outputs of the switching Hall sensors is not adequate for high-precision position demands as a rule. Frequently, in these cases high-resolution angle transducers such as resolvers, optical or magnetic incremental transducers or absolute transducers, analog Hall sensors or magneto-resistive sensors are used.

Similarly, a high angle resolution is also required for the operation of permanent magnet excited brushless AC motor designs for sinusoidal voltage or current supply. In applications with primarily stationary operation with largely constant speeds, interpolation of the angle between the switching transitions of the Hall signals by means of timer control in the microprocessor could be a good choice in order to save on the costs of a high-resolution angle transducer.

2.2.6.1 Resolvers

Resolvers are often used for high-class servo applications in combination with brushless AC motors due to their mechanical and thermal robustness and their high angle resolutions. The typical temperature range of such systems range from approximately -55 to $+155\,^\circ\mathrm{C}$.

The resolver is built in accordance with the principle of a three-phase variable transformer. A high-frequency voltage signal u_r is transmitted to the rotor with the help of two ring coils mounted opposite each other in the stator (stator winding 0) and the rotor:

$$u_\mathrm{r} = \hat{u}_\mathrm{r} \sin \omega t \,. \tag{2.86}$$

As illustrated in Fig. 2.91 the induced rotor voltage u_r feeds a second (usually) two-pole distributed winding which generates a sinusoidal air-gap field. Depending on the rotor angle, voltages u_{s_1} and u_{s_2} are induced in the stator windings 1 and 2, offset to each other by 90°; their envelope takes on the shape of a sine and cosine function, respectively:

$$\begin{aligned} u_{s_1} &= k\hat{u}_\mathrm{r} \sin \omega t \cos \gamma \\ u_{s_2} &= k\hat{u}_\mathrm{r} \sin \omega t \sin \gamma \,. \end{aligned} \tag{2.87}$$

The high-frequency carrier can be eliminated with the help of a conventional demodulation circuit (see Fig. 2.92). The envelopes of the sine and cosine

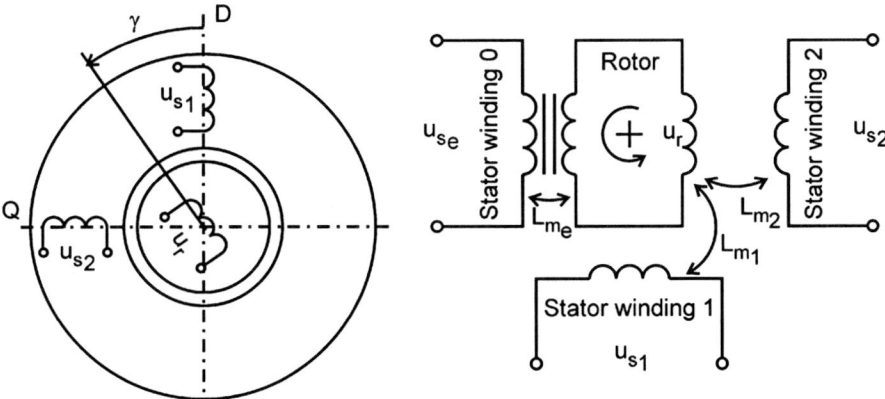

Fig. 2.91. Electrical equivalent circuit diagram of the resolver

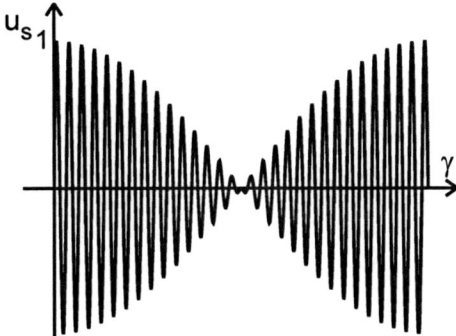

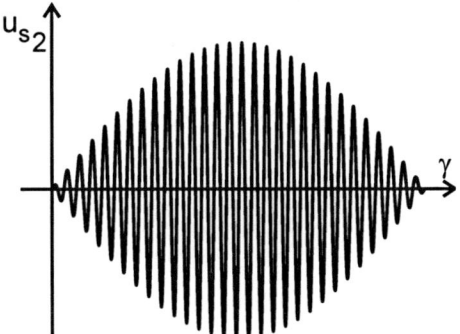

Fig. 2.92. Stator voltages u_{s1} and u_{s2} of the resolver

functions represent the angle information which is usually evaluated digitally by a microprocessor. Resolutions of 12 bits or more can be achieved with the help of commercially available resolvers and integrated signal processing circuits.

2.2.6.2 Incremental Optical Encoders

Incremental optical encoders can be a cost-effective and compact alternative to resolver systems, particularly in the lower and medium angle resolution range. Typical resolutions for optical encoders are 64 to 1024 impulses per revolution. But higher resolutions are also available. Due to the integrated electronics, the permissible temperature range is substantially lower than in resolvers (typically $-40/+100\,°C$). Other disadvantages are the sensitivity to mechanical vibrations and dust and the absence of absolute angle information at the time of switching on. This information is not available until the index impulse occurs; this is the case at the latest after one revolution.

Frequently, the angle signal is generated by means of two tracks of an encoder disk that are scanned by three optical channels A, B (track 1) and C (track 2), usually using an aperture located close beneath it. The channel C delivers an index signal per revolution. The two other channels generate two incremental signals offset by 90°. The output signals of the associated photo diodes, which can be of sinusoidal shape due to the scattering of the light, are converted by means of integrated comparator circuits to square-wave output signals.

Figure 2.93 illustrates the design principle of an encoder which exemplifies the signal generation of channels A and B. The angle signals of all three channels are shown in the sketch in Fig. 2.94. *Via* evaluation of the slopes of the signals from channels A and B, the impulse rate per revolution can be easily quadrupled. The direction of rotation is obtained from the phase sequence of these two channels.

Many digital signal processors support fast angle and speed evaluations by the integration of special logical evaluation units usable for standard two- and three-channel incremental encoders. The particular pulse rate is selected *via* the software program.

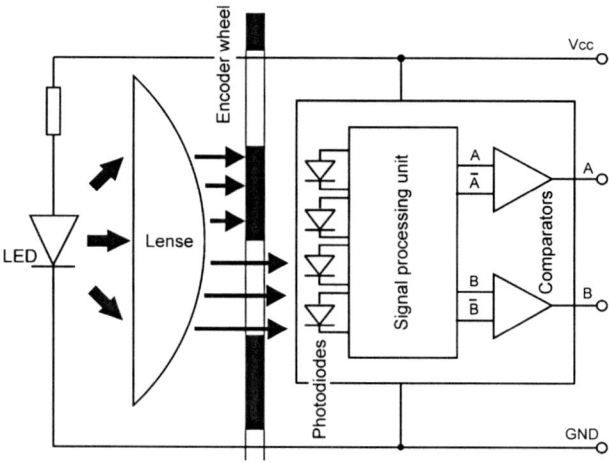

Fig. 2.93. Design principle of an incremental encoder (two-channel encoder)

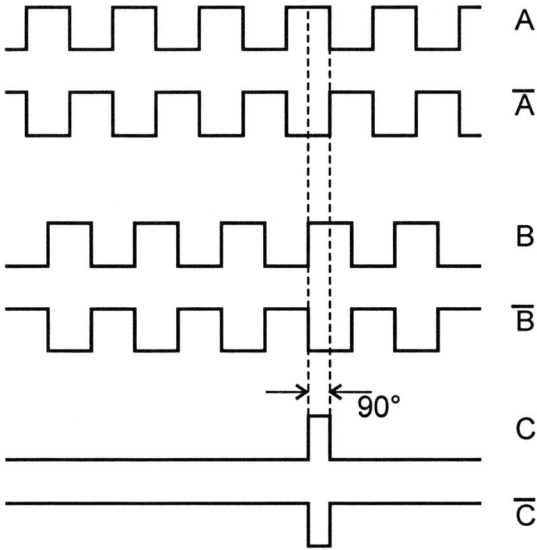

Fig. 2.94. Output signals of index channel C and channels A and B offset by 90°

2.2.6.3 Magnetic Integrated Circuit Sensors

Magnetic integrated circuit sensors offer very cost-effective and compact solutions for manifold measurement applications and are preferentially applied in large serial productions. They can be used as switching devices as well as for analog measuring tasks.

The component that has been applied the most over many years is the Hall sensor, which is used in brushless DC motors normally in an integrated version with digital comparator amplifier output in order to facilitate direct control of the commutation electronics. In many cases the evaluation of the lateral magnetic leakage field of the permanent magnet rotor is adequate to obtain its electrical angular position.

Magneto-resistive sensors, also known as MR sensors, are being used increasingly in brushless AC applications and for continuous angle measurement in servo applications. Generally they not only provide high measurement sensitivity but are also suitable for use in higher temperature ranges.

The anisotropic magneto-resistive effect was discovered as long ago as 1857 by William Thomson. However, it was only much later, at the end of the 1960s, that MR sensors came to be used in technical applications. Many of the principles known today have been the subject of international research activities in the last few years and a few of them have already been introduced in series production. Some of the most significant principles are the AMR effect (anisotropic magneto-resistive effect), the GMR effect (giant magneto-resistive effect), the CMR effect (colossal magneto-resistive effect) and the GMI effect (giant magnetic inductance effect).

AMR Sensors

AMR sensors use ferromagnetic materials (nickel-iron) that have the characteristic of changing their electrical resistance depending on an external magnetic field. Figure. 2.95 illustrates the basic principle of operation.

Without an external field, the magnetization orientation of the anisotropic material is in the direction of the current. On creation of an external field H not aligned to the current, the magnetization vector M turns by an angle φ. The magnetization vector orients itself completely in the direction of the external field with a correspondingly strong field (typically $H > 100\,\text{kA/m}$) and follows this in case of rotational movement. The relation for the change of the electrical resistance can be expressed as

$$R = R_0 + \Delta R_0 (\cos \varphi)^2 \ . \tag{2.88}$$

Here ΔR_0 is typically about 2% to 3% of R_0.

In order to achieve both temperature compensation and a larger evaluation range, two magneto-resistive bridge circuits offset by 45° are arranged in the integrated angle sensors. The Wheatstone bridges deliver two temperature-dependent output signals A_1, A_2:

$$\begin{aligned} A_1(\varphi, T) &= A_0(T) \sin 2\varphi \\ A_2(\varphi, T) &= A_0(T) \cos 2\varphi \end{aligned} \tag{2.89}$$

from which, assuming identical bridge behavior, the rotor angle ϕ can be determined with the function

$$\varphi = 0.5 \arctan(A_1/A_2) \ . \tag{2.90}$$

The angle evaluation range is 180° and can be considered as largely temperature compensated by reason of the quotient A_1/A_2 in Eq. (2.90).

GMR Sensors

Quantum mechanics effects form the basis of GMR sensor technology. The effect was first discovered independently by Peter Gruenberg and Albert Fert.

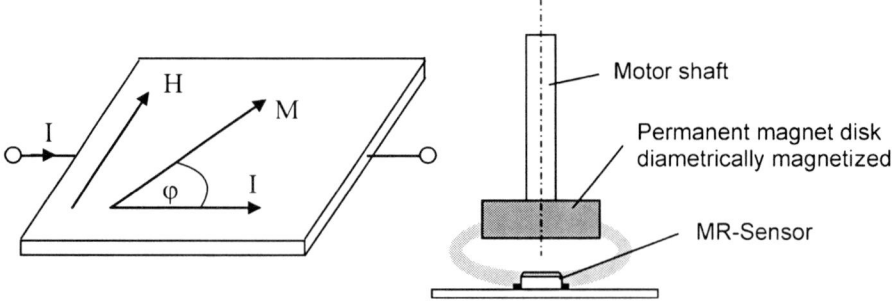

Fig. 2.95. The anisotropic magneto-resistive effect

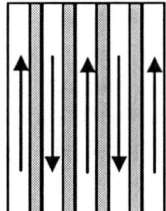 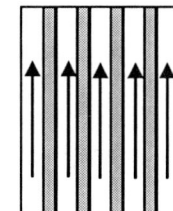 

Fig. 2.96. Direction of magnetization in the ferromagnetic layer of the GMR structure

Fig. 2.97. Magneto-resistive 180° absolute angle sensor in use as an automotive steering angle sensor (Source: © Sensitec)

The structures used consist of successive alternating ferromagnetic (nickel-iron, cobalt-iron) and non-ferromagnetic (copper, ruthenium) layers on top of one another. Here the layer thicknesses are only of the order of nm. The magnetization of neighboring ferromagnetic layers is anti-parallel to one another without the effect of an external magnetic field (opposing electron spin). If an external field is generated, the magnetization is oriented in a parallel manner (matching electron spin). Here the electrical resistance of the substrate structure reduces by up to 17% (nickel-iron) or 50% (cobalt-iron). Due to the large change as compared to AMR materials, the effect is termed giant. As in the case of the AMR effect, the GMR effect is independent of the sign of the magnetic field.

2.2.7 Special Designs of Brushless Permanent Magnet Motors

The requirements of brushless permanent magnet motors can vary greatly depending on the particular application. In practice this leads to a high number of constructional variants. As a rule the design process of fractional horse power motors is affected by technical and economic optimization steps, in many cases resulting in relatively simple mechanic motor constructions refined by integrated control electronics responsible for the high grade of functional performance and excellent operational and control behavior.

The extremely large number of variants of brushless motors cannot be handled completely within the framework of this handbook. Thus the following examples serve merely to trigger thoughts and to provide an insight into design variants that deviate from standard solutions. In this context the following sections present a low-noise, low-cost compact blower and three design principles of brushless direct drives for low speeds. These examples are supplemented by related bearingless motor constructions and applications presented in Chapt. 7.

2.2.7.1 Single-Phase Motor with Elliptical Coil

Figure 2.98 illustrates a single-phase radial blower designed for measuring the temperature in passenger compartments of automobiles. The blower is fitted behind small slots of the dashboard in order to guide a specific amount of air flow to the surface of a temperature sensor. The important requirements here are low manufacturing costs and low air-borne as well as structure-borne sound emissions.

The design of the motor can be seen in the cross section of Fig. 2.98. A plastic carrier contains the sintered slide bearing, the auxiliary magnet to ensure blower startup, the bifilar wound coil, the ferromagnetic back iron, and the contact tabs and snap-fit connectors for automatic PCB mounting. The blower can be plugged in and soldered in a wave soldering bath like an electrical component. The blower wheel is also an injection-molded plastic part that includes the four-pole main magnet, a ferromagnetic back iron disk and the shaft.

The radial component i_{rad} of the current i following the elliptical winding curve is responsible for the motor torque. All regions of the coil contribute to the torque generation with the exception of the zones in the main elliptical axes. Figure 2.99 illustrates a schematic that represents the principle of operation.

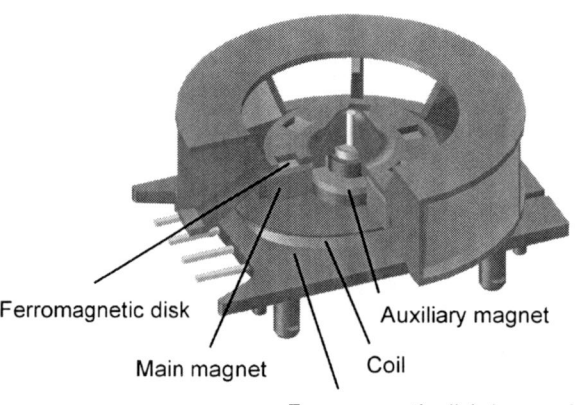

Fig. 2.98. Construction of a compact blower for PCB mounting

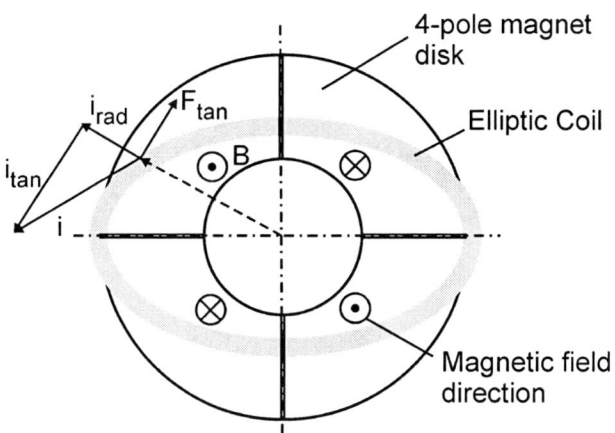

Fig. 2.99. Torque build-up in the disk rotor by means of the radial component of the winding current

The motor shown comprises only one phase coil. Thus it is necessary to provide an auxiliary torque for startup that overcomes the four dead centers. The auxiliary torque is generated by means of a second four-pole permanent magnet ring fitted within magnetic range of the rotor magnets on the bearing pipe of the stator. The commutation signal is obtained from the second coil of the bifilar winding. Due to the slotless, low inductance design of the motor, very low levels of commutation noise are generated during blower operation.

2.2.7.2 Direct Drives for Low Speeds

As a rule, mechanical gear solutions are used for adapting the motor characteristic to the low speed requirements of the load. In certain applications, however, direct drives are preferred in order to avoid disturbances caused by the gears, e.g. backlash, noise, wear or high-frequency torque ripples.

Usually, electrical motors have the disadvantage that the efficiency is very poor in the range of very low speeds. An improvement can be achieved with adapted drives comprising a large number of poles and permanent magnet materials with high energy density, such as the transverse flux motor in Fig. 2.100 or the hybrid motor illustrated in Fig. 2.101. Both can be used either as stepper motors or as field-controlled motors in connection with a suitable position sensor. While the stator of the transverse flux motor consists of only one ring winding per phase and many permanent magnetic poles and ferromagnetic flux conducting parts, the large number of poles is achieved by only two fine toothed lamination stacks enclosing a thin permanent magnet intermediate layer in the case of the presented hybrid motor. In return a larger number of coils is needed for the arrangement of the phase windings. For the generation of a rotary field the winding legs of the individual motor phases are angularly offset by a fraction of the tooth pitch depending on the number of phases.

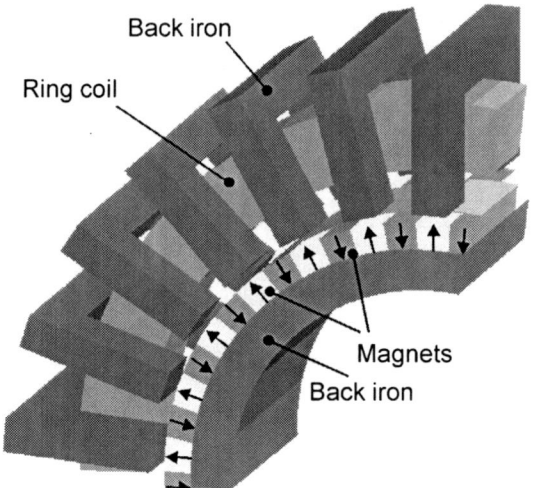

Fig. 2.100. Design principle of the transverse flux motor

Figure 2.102 illustrates another variant of a direct drive for low speeds. This is a design with different number of poles for the stator and the rotor. The switching cycle of the three-phase stator winding comprises six pulses. Figure 2.103 represents the various switching states of the winding. While the magnetomotive force revolves around the motor once, the rotor moves by only two-pole pitches. Various combinations can be worked out for the stator legs and the rotor poles. The design of the motor can thus be adapted to the requirements of the respective application.

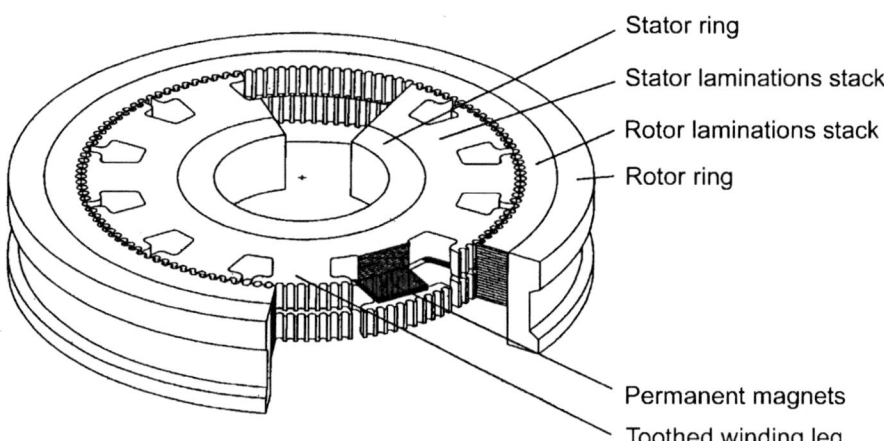

Fig. 2.101. Hybrid motor with permanent magnet bias of the air-gap field (Source: © Institute for Drive Systems and Power Electronics, Leibniz Universität Hannover)

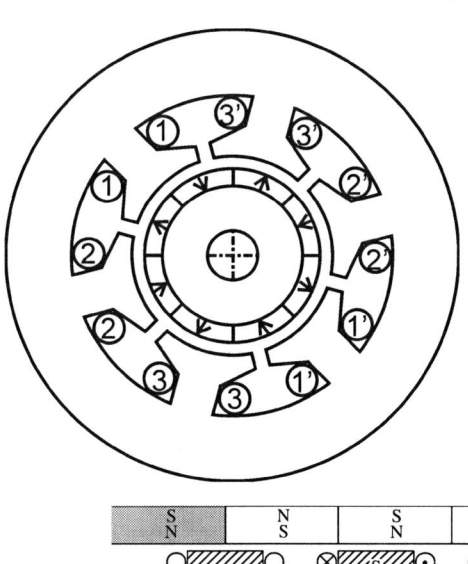

Fig. 2.102. Permanent magnet motor with different number of poles in the stator and rotor

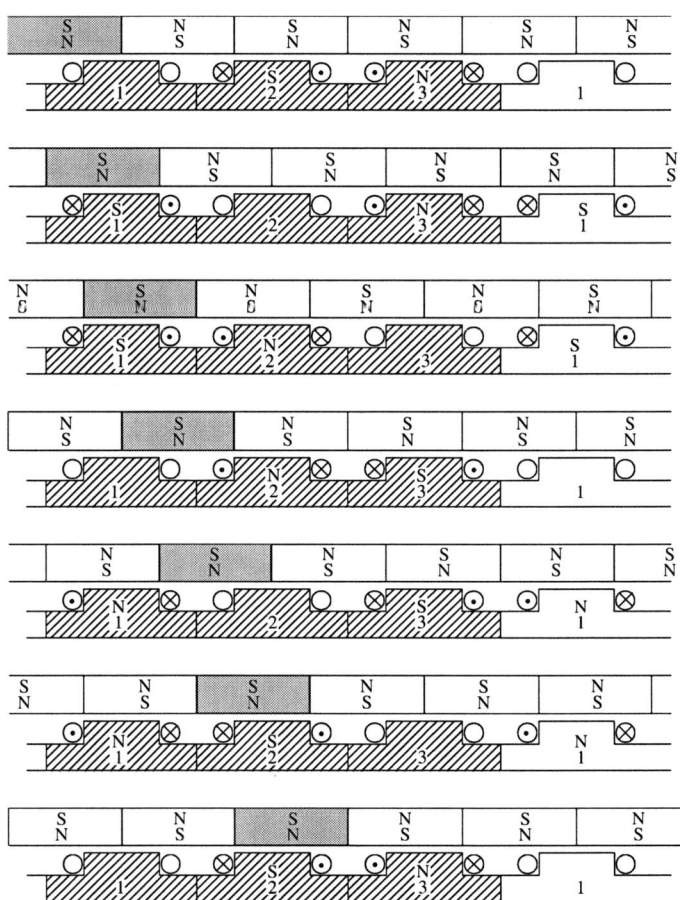

Fig. 2.103. Commutation sequence for the motor in Fig. 2.102. (Source: © Institute for Drive Systems and Power Electronics, Leibniz Universität Hannover)

2.3 Switched Reluctance Motor

Elmar Hoppach

The switched reluctance motor is based on a quite well understood operating principle. Due to the advances on sensors, power electronics and signal processing in recent years this type of motor has gained a lot of commercial and academic interest.

The operating principle of the magnetic structure is comparable to the reluctance stepper discussed in Chap. 3. It differs from the reluctance stepper in the control of the phase current. As the stepper is operated open loop, the switched reluctance motor requires control of the currents oriented on the position of the rotor. From the point of commutation and control of the phase currents the motor has to be classified similar to the brushless DC motor of Sect. 2.2.

The design for continuous operation has been carried out in recent years with the advent of fast electronic power switches and electronics. The motor is referred to as switched reluctance or switched reluctance synchronous motor.

2.3.1 Magnetic Circuitry

The magnetic circuitry of the switched reluctance motor neither requires permanent magnets nor moving windings in the rotor. Hence, the switched reluctance motor is simple and robust in construction. It offers the possibility of low-cost manufacturing.

The stator carries designated poles with concentrated coils. The rotor does not require a winding and also carries designated poles. Depending on the design the number of poles on the rotor is either higher or lower than that of the stator. A larger number of rotor poles is often found on outer rotor type designs. Two examples of commercial magnetic circuitries are shown in Fig. 2.104. On the left there is a three-phase motor using 12 poles on the stator and 10 on the rotor, a 12/10 switched reluctance motor. On the right is the nowadays most common switched reluctance design using four phases, eight poles on the stator and six on the rotor, the popular 8/6 switched reluctance motor.

Switched reluctance motors of the three-phase design are often 6/4 or 6/8 designs. The phase current is commutated depending on the position of the

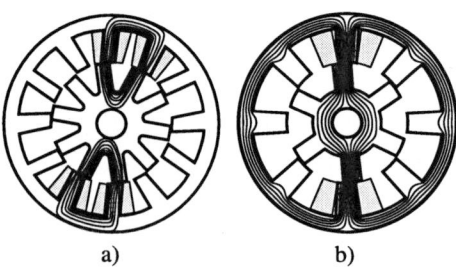

Fig. 2.104. Field distribution in short flux 12/10 switched reluctance motor (**a**); 8/6 switched reluctance motor (**b**)

rotor. To detect the rotor position feedback is used, broadly based on Hall sensors or optical encoders.

2.3.2 Operation

The generation of force in the magnetic circuitry of a switched reluctance motor is based on the operating principle of electric magnets (Sect. 4.1.2.2). The voltage across winding 1 is derived using the winding resistance R, the flux linkage of this winding ψ and the position of the rotor φ:

$$U_1 = i_1 R + \frac{\mathrm{d}\psi_1(\varphi, i_1 \ldots i_n)}{\mathrm{d}t} . \qquad (2.91)$$

The coupling between the phase windings of the switched reluctance motor is kept small by design. The switched reluctance motor can be thought of as a system of independent electric magnets that are switched on and off in synchronization with the rotor position. The independence is rather obvious in designs using a short flux path, i.e. Fig. 2.104a; it is not that obvious in the most common 8/6 design, as shown in Fig. 2.104b. Even in this design, the assumption of independent electric magnets is acceptable in small electrical drives where the magnetic structure is usually not saturated. The error in torque due to coupling between the phase windings in fractional horsepower drives is much lower than those due to tolerances in manufacturing. Deriving and splitting the term into inductance and change of position leads to the voltage of a winding comparable to the electric magnet (Sect. 4.1.1):

$$U_1 = i_1 R + \frac{\mathrm{d}\psi_1(\varphi, i_1)}{\mathrm{d}i_1}\frac{\mathrm{d}i_1}{\mathrm{d}t} + \frac{\mathrm{d}\psi_1(\varphi, i_1)}{\mathrm{d}\varphi}\frac{\mathrm{d}\varphi}{\mathrm{d}t} . \qquad (2.92)$$

Torque over position is periodic to the poles of the rotor. A linear plot of one half of an 8/6 switched reluctance motor (Fig. 2.105) shows the three periods.

Two positions in each period do not deliver any torque. The aligned position, rotor and stator are face to face, and the unaligned position, where the active stator pole is situated in the middle between two rotor poles.

On normal motor operation the rotor moves towards the aligned position. The motor of Fig. 2.104a will be commutated to the next stator pole on the left. The rotor will move one quarter of a pole subsection clockwise. The

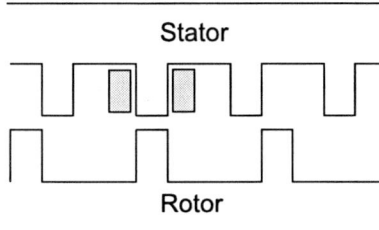

Fig. 2.105. Rotor and active phase in the stator in the aligned position

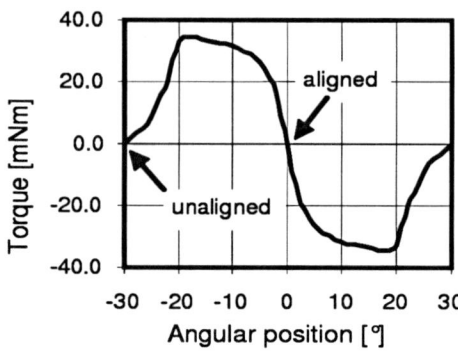

Fig. 2.106. Torque over angular position

rotor of the 8/6 switched reluctance motor is therefore moving in the opposite direction to the magnetic field of the stator. If the number of phases is m and $2p$ the number of pole pairs on the rotor, the rotor will proceed by a number of degrees given by

$$a_\mathrm{s} = \frac{360°}{2pm} . \qquad (2.93)$$

The 8/6 switched reluctance motor requires 24 commutations for 1 revolution of the rotor. The 6/4 switched reluctance motor will require 12 commutations for one revolution. At 6000 rpm this requires a frequency for commutating an 8/6 switched reluctance motor of 2400 Hz. Iron losses have to be considered when designing a switched reluctance motor.

The aligned position will be denoted by φ_0. The period of the 8/6 switched reluctance motor is determined by the six rotor poles to $60°$. The torque over the angular position is shown in Fig. 2.106.

In small switched reluctance motors the ψ–i-transfer function is almost linear. Therefore the torque can be determined by the same assumption as for the electrical magnet and expressed in the formula at Eq. (4.11). Comparably the torque will be described in Sect. 2.3.4 as a function of current and angular position:

$$M(\varphi, i) = \frac{1}{2} i^2 \frac{\mathrm{d}L}{\mathrm{d}\varphi} . \qquad (2.94)$$

To maximize torque, the change in the inductance must be as high as possible, whereas the bias is not evident. In normal operation as a motor the phase is switched on in the unaligned position and switched off in the aligned position. This corresponds to switching on at a low inductance and therefore a shorter time constant of the magnetic circuit and switching off at a higher time constant. This will be analysed in the following section.

2.3.3 Commutation

The switched reluctance motor must operate closed loop with position feedback for continuous rotation. The controller of most appliances is implemented on a microcontroller or digital signal processor. The power stages are commonly based on a DC link.

According to Eq. (2.94) the switched reluctance motor does not require inversion of the phase current for two or four quadrant operation.

Recent publications have been showing a varied number of topologies for power electronics [2.89]. The focus shall be drawn on some topologies that are well suited for fractional-horsepower drives, and the demands for current control.

An obvious disadvantage of switched reluctance motors is the fast switch off at the end of the commutating cycle. As mentioned before the winding has to be switched off during the state of highest inductance and electrical time constant. This disadvantage leads to a poor power factor; switched reluctance motors are typically well beyond a $\cos\varphi$ of 0.5. To reach this, the magnetic circuitry must be partially saturated. This calls for a good understanding of magnetic circuitry by means of analytic [2.81] and especially valuable numerical field computation and optimisation [2.82]. The latter are inevitable for acceptable overall performance of switched reluctance motors.

These demands call for an inverter that can handle the switch on and off of the phase current without lag. In order to minimise losses in the demagnetising circuitry, appropriate inverter design allows for feeding the energy of the demagnetised phase into the DC link. These demands are quite well satisfied using an inverter in bridge configuration with additional freewheeling diodes (Fig. 2.107).

The technical realisation requires one bridge for each of the m phases of the switched reluctance motor, requiring $2m$ power switches and diodes for the entire inverter. This type of inverter is therefore called a $2m$ inverter (Fig. 2.108). The high component count in discrete realisation will in most cases only be accepted at an overall power of the motor beyond 100 W.

In order to reduce the component count, less complex circuitries appeared on the market. The less demanding circuit uses one high side switch carrying the pulse-width signal (PWM); the low side switches carry the pattern for commutation. This topology paves the way to low-cost current sensing, i.e. using current sense MOSFETs or resistors. Figure 2.109 shows the circuit diagram of an $m+1$ inverter.

The predominant disadvantage of the $m+1$ topology is the inevitable short circuit of the phase which must demagnetise when the consecutive phase is

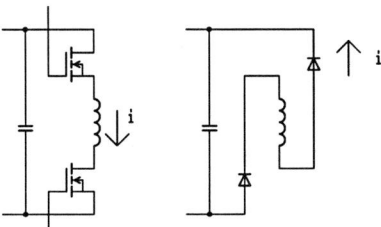

Fig. 2.107. Current during on state and in freewheeling condition

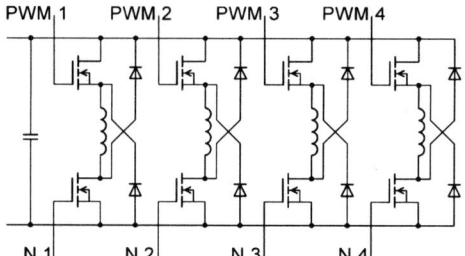

Fig. 2.108. $2m$-Inverter topology

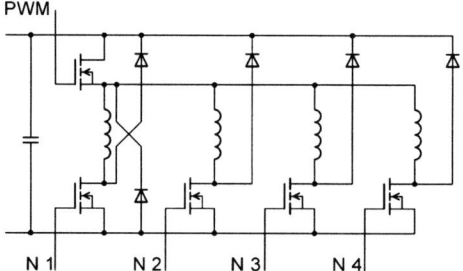

Fig. 2.109. $m + 1$ Inverter topology

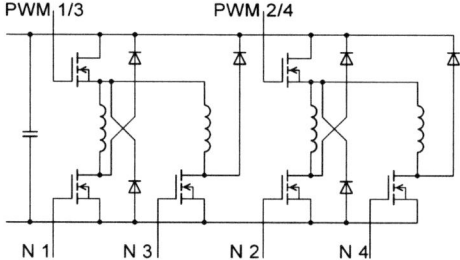

Fig. 2.110. $m + 2$ Inverter topology

magnetised. The consequence is a long current decay and self-excitation if the phase is not demagnetised when passing the aligned position. This disadvantage can be minimised by lowering the PWM modulation leading to a lower inverter utilisation. Even further, at high speed and high commutating frequencies, the demagnetisation is not sufficient and the speed range of the switched reluctance motor cannot be fully utilised.

The $m + 2$ topology offers a very good compromise on commutation and part count. In normal operation of an 8/6 switched reluctance motor, this topology has no drawbacks, as the commutation of two consecutive phases can take advantage of full PWM modulation during switch on and off (Fig. 2.110).

An 8/6 switched reluctance motor can thus be operated using two highside PWM switches and four lowside switches.

2.3.3.1 Control of the Inverter

The switched reluctance motor requires a controller for continuous operation; it is not a self-commutation system, like the widely used DC motor. The

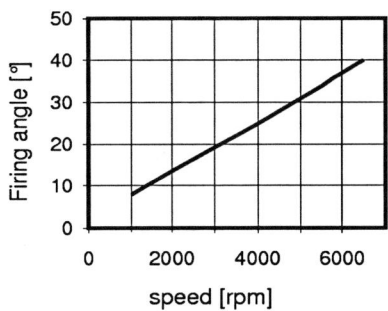

Fig. 2.111. Optimised pre-firing angle in degrees for an 8/6 switched reluctance motor

controller ensures a beneficial generation of torque by adequately feeding the motor phases with appropriate current patterns. The magnetic circuit of the switched reluctance motor itself should be optimised for continuous torque over the firing angle. This is 15° for an 8/6 switched reluctance motor. Figure 2.106 shows the usable firing angle of an 8/6 switched reluctance motor. The usable firing angle of an 8/6 switched reluctance motor is typically −20° to −5° for motor operation and 5° to 20° while braking. The gap between the aligned position and the start of the commutation is the pre-firing angle that is −5° at low rotor speed.

Higher rotation speed requires an earlier switch point to reach the nominal current in the firing angle. The current decay in the aligned position is too slow to switch the phase current to nominal value. The pre-firing angle must be increased when the motor is running at higher speeds. It is advisable to obtain the data on the real application. An example for an 8/6 switched reluctance motor is shown in Fig. 2.111.

2.3.4 Control Strategies and Motor Performance

The overall performance of a switched reluctance drive system is highly dependent on the control strategy chosen. The switched reluctance motor is a non-linear system for converting current to torque, requiring a controller to adopt the performance to the requirements. This can be met using three common control strategies for commutating the switched reluctance motor:

- Block voltage commutation
- Block current commutation
- Linearization using a pre-calculated current pattern

To compare these three control strategies, a setup of an 8/6 switched reluctance motor designed for 100 mNm nominal torque and 50 W nominal power will be used. All plots show the torque over time at start up.

Noise is a critical issue on switched reluctance motors. The control strategy has a large impact on the noise emission of the switched reluctance drive system. Even linearization (Sect. 2.3.4.3) of torque will not reach a comparable noise to a DC motor.

2.3.4.1 Block Voltage Commutation

Block voltage commutation offers the advantage of low demands on part count and computing power for the controller. The motor phases are switched directly to the supply voltage on the DC link. It is advisable to use a speed dependent pre-firing angle. The torque over time plot shows the large torque ripple and, on averaging the system, behaviour of a universal motor (Fig. 2.112).

The level of torque obtained using block voltage also shows the limit in torque over speed on a given DC link voltage. The more complex control strategies will not allow for a higher torque level.

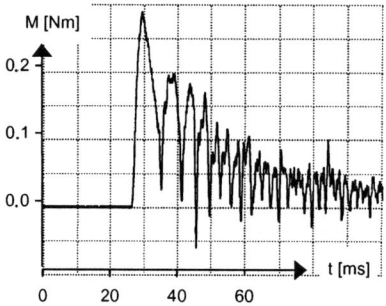

Fig. 2.112. Torque on start up using block voltage commutation

2.3.4.2 Block Current Commutation

Controlling the current in the winding is beneficial as it offers the protection of the power semiconductors against a failure due to over current. The torque remains constant up to the point where the switched reluctance drive system suffers from field weakening.

The torque ripple using block current commutation is directly dependent on the transfer function of the magnetic circuitry shown in Fig. 2.113. The demands on torque quality are usually met using a motor geometry with an optimised transfer function.

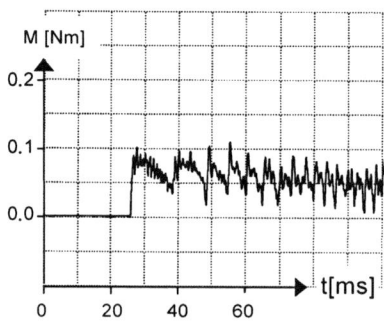

Fig. 2.113. Torque on start up using block current commutation

2.3.4.3 Linearization Using Current Pattern

Targeting at low torque ripple, linearization by using a pre-calculated current pattern is a way to go for open-loop control. The controller structure is the same as the well understood solution for brushless DC motors and uses a comparable circuit topology (Fig. 2.85). As long as the magnetic circuitry is not saturated, the torque of the switched reluctance motor will remain proportional to the square of the phase current. This is true for most fractional-horsepower drives. The phase current for linearization i_{lin} can be calculated at the average torque M_0 regarding the current for nominal torque i_0 and the torque $M_{i_{\text{const}}}$ at constant current using Eq. (2.95).

$$i_{\text{lin}}(\varphi) = i_0 \cdot \frac{M_0}{\sqrt{M_{i_{\text{const}}}(\varphi)}}. \tag{2.95}$$

In case of higher current densities when saturation occurs, a two-dimensional function for linearization must be used. Most fractional-horsepower drives do not require this computing intense solution as manufacturing tolerances usually count for much higher deviations.

The linearization of Fig. 2.114 is based on the average torque of all phases over one half of the pole angle of 30°. This is efficient in terms of memory usage. For an even higher level of linearization, compensation for manufacturing tolerances is necessary. This requires current patterns for each phase and pole; a high-resolution encoder is also mandatory. The minimisation of the manufacturing tolerances and the linearization of the magnetic circuitry is hence providing a more efficient solution for high-quality switched reluctance drive systems. It should be denoted that the utilisation of magnetic circuitry is kept at a reasonable level [2.87].

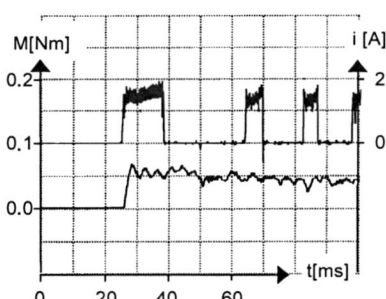

Fig. 2.114. Current in one phase and linearized torque during start up

2.3.5 Sensorless Modes of Operation

The switched reluctance motor offers high potential and low manufacturing costs. Opposed to this, the controller needs a high-quality position sensor. Further cost savings require the elimination of the position sensor, and a lot of control strategies on sensorless control have been published. The most common way to implement sensorless control is the measurement of the inductance of a phase [2.84, 2.85]. The gradient of the phase current in PWM operation is a way to calculate the inductance of a phase using Eq. (2.96):

$$L = \frac{U - Ri - U_i}{\frac{di}{dt}} \, . \tag{2.96}$$

At the start of the cycle the gradient of the current in unaligned position is larger than the inductance found in aligned position. The phase current is low; the second term, the generated voltage, is also low. By the end of the commutation cycle current increase and decay are almost the same, and the generated voltage is approaching the phase voltage.

To increase the precision of the sensorless controller, an observer based position estimator is widely used. An example of this is based on a Kalman observer [2.80] and requires a continuous state model of the plant.

2.3.6 Characteristics

A subsumption of the characteristic behaviour of switched reluctance drives systems is listed below.

Positive characteristics

- The switched reluctance motor offers the chance of low-cost manufacturing, as neither permanent magnets nor moving windings are required.
- At low speed losses are predominately ohmic losses in the winding of the stator. Cooling is efficient on inner rotor type motors.
- The rotor is outstandingly robust and offers a very low inertia.
- The switched reluctance motor is suited for high speed and high dynamic positioning drives.
- Torque is proportional to the square of the phase current leading to high pullout torque at moderate phase currents and moderate losses in the inverter.

Negative characteristics

- The width of the air-gap is rather small, and the switched reluctance motor is critical on manufacturing tolerances.
- The commutation causes audible noise due to high flux densities generating cyclic deformation of the stator. Audible noise is often not acceptable. The audible performance of switched reluctance motors with a higher amount of poles is often better.
- The motor suffers from a large torque ripple requiring complex and costly linearization.
- The four-phase motor often performs better than three-phase designs.
- The duty factor of the motor is rather poor due to magnetisation causing poor inverter usage.

2.4 Rotating-Field Motors

2.4.1 Asynchronous Motors

There also exist three-phase asynchronous motors in the low power range. Generally these are standard motors which are built according to IEC 60072 and EN 50347 from frame size 56, that is from a rated power of 60 W; these are in fact produced mostly for industrial applications. But AC asynchronous motors (single-phase induction motors) are mainly used for consumer goods.

2.4.1.1 Characteristics and Fields of Application

Characteristics

Advantages of asynchronous motors compared with commutator motors (Sect. 2.1):

- Robust (only the bearings are subject to wear and tear), low-maintenance, long lifetime
- Quiet running, low vibration
- Simple construction, cost-effective

Disadvantages:

- Limited maximum speed ($n_{\max} < 3000\,\mathrm{min}^{-1}$ with a line or rated frequency $f_\mathrm{N} = 50\,\mathrm{Hz}$), in fact a relatively high weight
- Expensive closed- and open-loop speed control
- Middling efficiency

Fields of application:

- Robust drives:
  - AC motors: heating-circulating pumps, garage-door drives, lawn mowers, fans and ventilators, copiers
  - Three-phase motors: high-frequency tools, cement mixers, grinding machines, circular saws, labelling machines
- Low-noise drives:
  - Heating-circulating pumps, ironing machines, fans and ventilators, copiers, high-frequency tools.
- Converter operation (three-phase motors with closed- and open-loop speed control):
  - Washing machines, main-spindle drives, servo drives

2.4.1.2 Types of Construction

As small asynchronous motors are generally built in the driven devices, they need no housings (cost benefit), but do need end shields or bearing brackets made of die-cast aluminium or punched and deep-drawn sheet metal

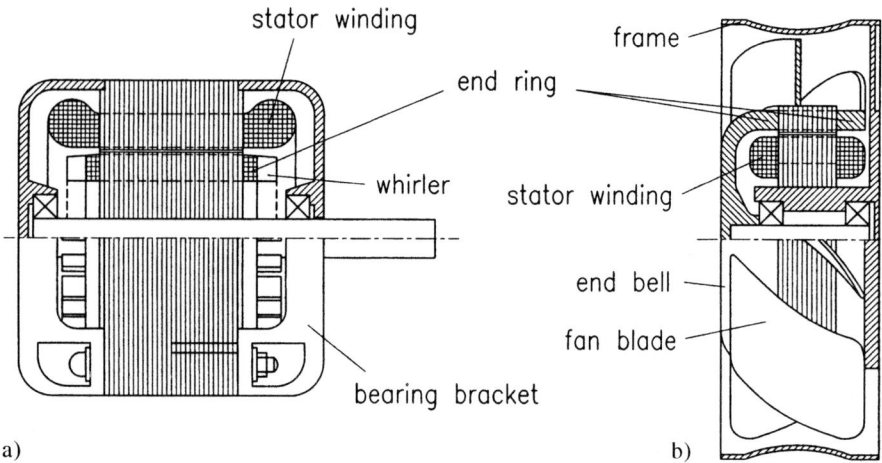

Fig. 2.115. Types of construction. a Internal rotor. b External rotor

(Fig. 2.115). Apart from motors with internal rotors, fans (e. g. device ventilators, extractor hoods) are built with external-rotor motors, where the blower blades are arranged on the rotor housing. Through this, a very compact construction and good fan efficiency are obtained. There also exist disk motors, but very few, with radial slots in the stator and the rotor. Despite their short axial length, their production is expensive, and they show a strong axial bearing pressure. Spherical rotors were suggested too, but their production is far too expensive. Split-cage motors are widespread as drives for pumps (e. g. heating-circulating pumps), where the rotors surrounded by a thin-walled tube of non-magnetic steel rotate in the delivery medium. Because of that, expensive seals are avoided.

Laminated Core

Often identical laminated cores are taken for several pole numbers to be cost-effective. In Fig. 2.116a, the cross section of internal-rotor motors with 24 stator slots and 32 rotor slots is shown. It can be used for two to six poles. The cross section of an external-rotor motor shown in Fig. 2.116b has partly slot openings which are slightly shifted out of the middle of the slot to simplify winding. The dashed lines of the end windings demonstrate this principle.

Apart from shaded-pole motors (Sect. 2.4.1.7) the stator winding is located in slots. Normally the two coil sides lie in slots, being connected to each other at the end faces (Fig. 2.117a). Even if rarely, there also exist coils surrounding the yoke which are located in slots on one side only (Fig. 2.117b). To be able to wind the stator, it has to be divided into two parts. After the winding, both parts are put together. The winding is embedded in synthetic resin at the outer housing. The advantage of a shorter overall length is opposed by the disadvantages of higher leakage and higher costs.

2 Motors for Continuous Rotation

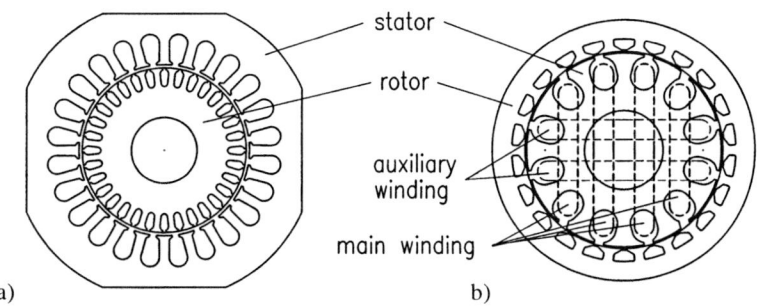

Fig. 2.116. Cross sections. **a** Internal rotor. **b** External rotor

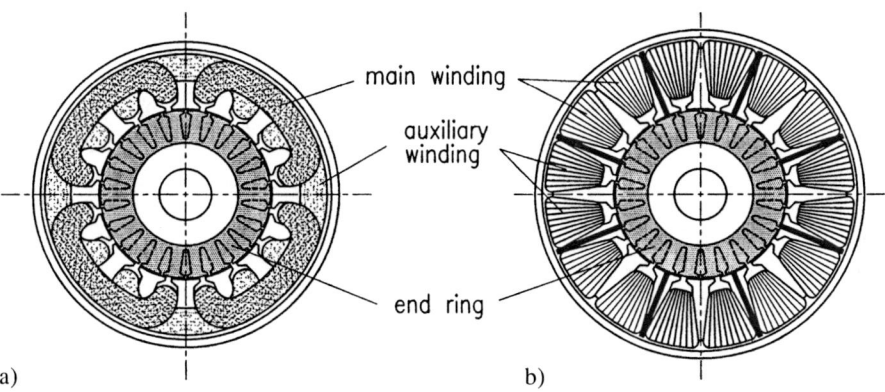

Fig. 2.117. End windings. **a** Conventional winding. **b** Winding across the yoke

The rotor always has a squirrel-cage winding. Every slot has one bar; all bars are short-circuited by rings at the end faces (Fig. 2.118). Bars and end rings are made of die-cast aluminium or aluminium alloys which possess a higher resistance (high-resistance cage rotor). Little blades for more intensive cooling of the end windings and small pins which can be shortened or equipped with rings for balancing are often fastened at the end rings. Against the risk of corrosion (e. g. split-cage motors in heating-circulating pumps) copper bars are

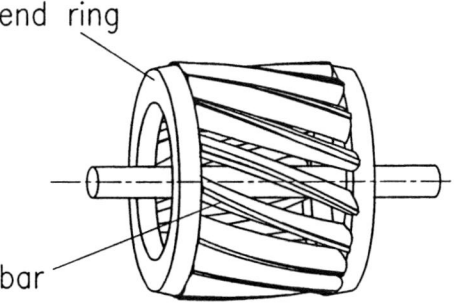

Fig. 2.118. Squirrel-cage winding (die-cast aluminium) without laminated core

driven into the slots and hard-soldered with copper rings at the end faces. The bars are axially twisted in order to minimize torque pulsations (slot skewing, Fig. 2.118).

2.4.1.3 Circuit Design and Types of Construction of the Stator Winding

Motors with Three Phases

These motors always have windings like three-phase asynchronous motors. The three phases are arranged symmetrically and shifted against each other along the circumference by the electrical degrees:

$$\alpha = \frac{2\pi}{3} \quad \text{resp.} \quad \alpha = \frac{360°}{3} . \tag{2.97}$$

Annotation: electrical degrees are often used, because they are valid for motors with any number of pole pairs. The corresponding geometrical degree is given by $\alpha_g = \alpha/p$. p is the number of pole pairs. Therefore, the phase displacement for two-pole motors is $\alpha = \alpha_g = 120°$ or $2\pi/3$, and for six-pole motors likewise $\alpha = 120°$, but $\alpha_g = 40°$.

Two of the three terminals are directly connected to the power supply. To achieve a phase displacement of the phase currents which is necessary to generate a rotating field, the third terminal connection is linked *via* a capacitor, sometimes also *via* a series connection of a capacitor and a resistor, to the power supply (Fig. 2.119). AC motors with three-phase windings require a greater and therefore more expensive capacitor than two-phase motors (Table 2.6). Therefore, the latter are generally employed only if no new motor shall be developed with a smaller number of pieces, but an existing three-phase motor can be used. Since three-phase motors are driven in star connection in a 400-V-three-phase system, the winding is switched to a delta connection for the operation at 230-V alternating voltage. This is possible

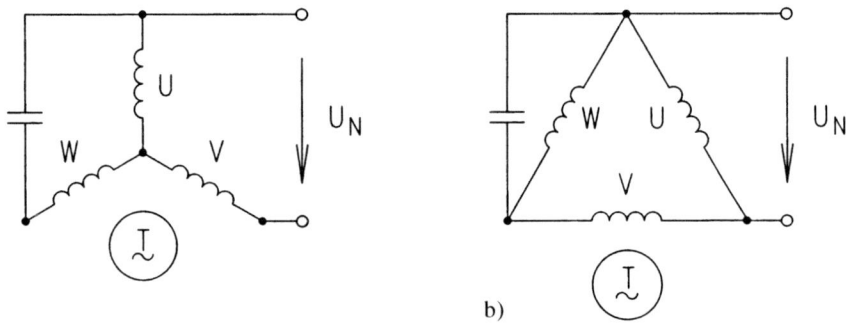

Fig. 2.119. Three-phase AC asynchronous motors – Steinmetz-connections. **a** Star connection. **b** Delta connection

without great effort by changing three links of the terminal board, where all conductor ends of the three phases are connected to six terminals. The star connection is only used in exceptional cases, if for instance the ring current which can arise in a delta connection generating undesirable magnetic fields of triple fundamental pole numbers might be avoided.

Motors with Two Phases

In these motors, the electrical degrees of the winding displacement amount to

$$\alpha = \frac{\pi}{2} \quad \text{or} \quad \alpha = 90°. \tag{2.98}$$

In two-pole motors, the winding axis is perpendicular, and in four-pole motors with a geometric angle of 45°. One of the windings, the so-called main winding, is connected directly with the power supply, the second phase winding, the so-called auxiliary winding, is connected to the power grid mostly *via* a capacitor, sometimes *via* a resistor connected in series.

If the motor shall generate a torque as high as possible during the operation, e.g. at rated load, the capacitor remains in action all the time (permanent-split capacitor motor with a running capacitor C_R, Fig. 2.120a). But if the motor shall generate a starting torque as high as possible, the capacitor must be considerably larger. After running up, the auxiliary winding is disconnected together with the capacitor by the starting switch (capacitor-start motor with a starting capacitor C_S, Fig. 2.120b). Otherwise inadmissibly high losses would arise during the operation. The starting switch may be centrifugal, magnetic, or static type. If a high continuous rating as well as a high starting torque is demanded, two capacitors are used in parallel – one for starting and one for running (two-value capacitor motor, Fig. 2.120c). For especially high-powered drives (e.g. in the case of compressor drives in refrigerators), due to frequent switching on and off, the resistance of the auxiliary phase is often enlarged instead of using a capacitor. This can be realized by using of a bifilar winding, an additional resistor connected in series or a wire with a higher specific resistance (resistance-start motor, split-phase motor).

There exist three types of two-phase windings each for special applications.

Symmetrical winding. This winding is used for reversal operation, because the motor thus has the same characteristics in both senses of rotation. Only a single-pole cost-effective switch is necessary (Sect. 2.4.1.6).

Quasi-symmetrical winding. The phases are placed in the same number of slots, but possess a different number of turns. In this way a balance operation of the two windings can be easier to achieve than with a symmetrical winding (Sect. 2.4.1.4). Choosing the number of turns of the auxiliary winding less than that one of the main winding – a typical ratio is 0.8 – a smaller and thus more cost-effective capacitor of the same ratio is sufficient in order to get the optimal phase displacement. Therefore, a quasi-symmetrical winding is the most used type of motor with one speed sense only (non-reversing motor).

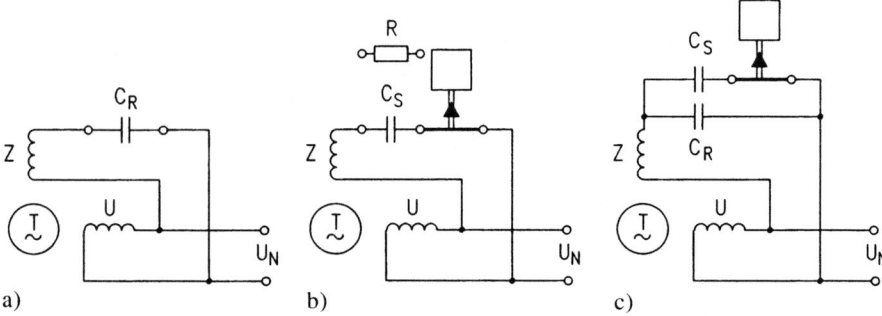

Fig. 2.120. Connection types of two-phase AC asynchronous motors. **a** Motor with running capacitor C_R. **b** Motor with starting capacitor C_S or resistor R. **c** Motor with starting and running capacitor

Asymmetrical winding. The most important version has a main winding lying in two thirds of the slots and an auxiliary winding in the remaining third (Fig. 2.116b). Through this, the main winding generates a magnetic field with less harmonic components or less losses. In contrast the auxiliary winding generates especially strong field harmonics, i.e. especially high losses. Therefore it is switched off after running up. The above-mentioned motors for refrigerator drives are typical applications, because these motors must have efficiencies as high as possible.

2.4.1.4 Mode of Functioning

The mode of functioning of AC asynchronous motors is similar to that of three-phase induction motors. In three-phase induction motors, identical currents which are phase-displaced by $2\pi/3$ (= one third period) towards each other flow in identical phase windings symmetrically arranged along the circumference. They generate a rotating field with an approximately constant size (circular field). With a line frequency $f_N = f_1$ (f_1 = frequency of the stator currents), it rotates with the synchronous speed

$$n_s = \frac{f_1}{p} \qquad (2.99)$$

and induces currents in the slower rotating rotor winding. The slip of the rotor, i.e. the speed difference of the rotating field and the rotor related to the synchronous speed is

$$s = \frac{n_s - n}{n_s} . \qquad (2.100)$$

So the rotor-current frequency is

$$f_2 = s \cdot f_1 . \qquad (2.101)$$

The rotor currents generate the torque together with the rotating field. The stator winding gets the active-input power from the power supply. Because of the power factor it is less than the complex power, the geometrical sum of the active power and the reactive power. In the stator, the currents provoke ohmic losses in the winding and the magnetic field causes losses in the ferromagnetic elements (core losses) as a result of hysteresis and eddy currents. Referring to the stator losses as P_{V1}, the so-called air-gap power is

$$P_\delta = P_1 - P_{V1} \qquad (2.102)$$

which is transferred to the rotor. P_δ is divided into the rotor losses

$$P_{V2} = s \cdot P_\delta \qquad (2.103)$$

and the output power at the shaft

$$P = (1-s)P_\delta \ . \qquad (2.104)$$

The torque results with $n = n_s(1-s)$ in

$$T = \frac{P}{2\pi \cdot n} = \frac{P_\delta}{2\pi \cdot n_s} \qquad (2.105)$$

and the efficiency in

$$\eta = \frac{P}{P_1} \ . \qquad (2.106)$$

All of the previous equations are also valid for AC asynchronous motors. As in these motors, phase windings and phase currents can be unequal and can deviate in position and phase angle from $\pi/2$; an elliptical rotating field generally arises in contrast to three-phase motors. Its quantity and speed waver with double line frequency during the rotation. For this reason, a pulsating torque superimposes onto the torque which is constant at a definite speed. This produces additional losses and a noise of twice the line frequency. But this noise is insignificant compared with the brush noises of commutator motors. Depending on the deviation of phase windings and currents from the optimal conditions, the elliptical field is more or less pronounced. In the extreme case, the ellipse becomes a line: an alternating field (spatially standing still) develops out of the rotating field. As mentioned in the preceding section, this is the case when the auxiliary winding is disconnected after the motor run up.

The alternating field of a single winding can be described as a superimposition of a field rotating in rotor direction (positive phase-sequence field) and an opposite rotating field (negative phase-sequence field). According to this one can imagine an AC motor with one winding switched on, replaced by two three-phase induction motors with opposite speed sense. As can be seen in Fig. 2.121a, both fictitious motors generate a similar torque of the same size in the opposite direction, i.e. the AC motor does not start. When starting

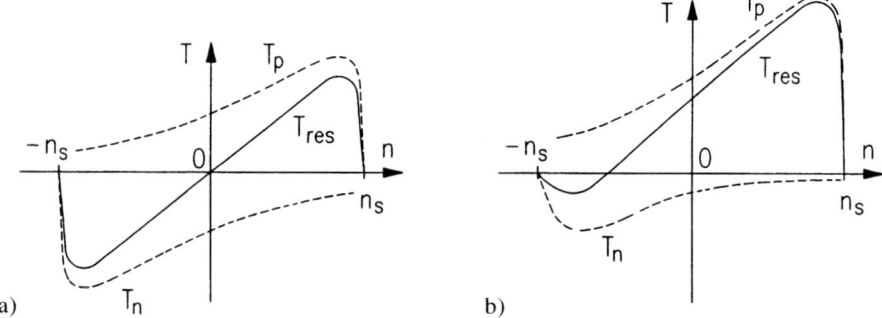

Fig. 2.121. Speed-torque characteristic. **a** Single-phase motor. **b** Two-phase capacitor motor

up in one specific direction, it runs up in this direction because the torque of the fictitious positive phase-sequence motor T_p is always higher than that of the negative phase-sequence motor T_n. Therefore, the resulting torque is $T_\mathrm{res} = T_\mathrm{p} - T_n \neq 0$, if $n \neq 0$. It must be added that the no-load speed n_0 is much lower than the synchronous speed n_s.

The positive phase-sequence field becomes greater than the negative phase-sequence field by using two windings with an electrical 90° space displacement and a current time phase shift which is greater than zero degrees. In this way, the fictitious motor rotating in the positive direction is stronger than the motor rotating in the opposite direction. In this way, the AC asynchronous motor generates a starting torque T_s (Fig. 2.121b). As the fictitious negative phase-sequence motor always works in a breaking manner, the utilization of an AC motor is worse than that of a three-phase motor of the same frame size. Winding and capacitor can be designed in such a way that an optimal phase displacement between the both currents, namely $\phi = 90°$, can be achieved. Through this, the negative phase-sequence field can be suppressed and its negative effects can be avoided: balanced operation, balanced motor. Both power factor and efficiency are high. But this is possible for one working point and by approximation only, because the values of the winding and the capacitor which are necessary, are practicable step by step only. As the efficiency possesses a very wide optimum in the field around the continuous rating (Fig. 2.123), a deviation from the favorable point of operation results in no essential change for the worse.

Harmonic fields. As the real air-gap field is not sinusoidal, there exist, apart from the basic field rotating with the synchronous angular velocity n_s, harmonic fields with rotational speeds departing from the base velocity n_s. These harmonic fields are undesirable, because they provoke losses, vibrations and noise. The number of pole pairs of all rotating fields, i.e. the number of their periods along the circumference, are to be calculated with

$$\nu = p(1 \pm 2m \cdot g) \qquad g = 0, 1, 2, 3, \ldots . \qquad (2.107)$$

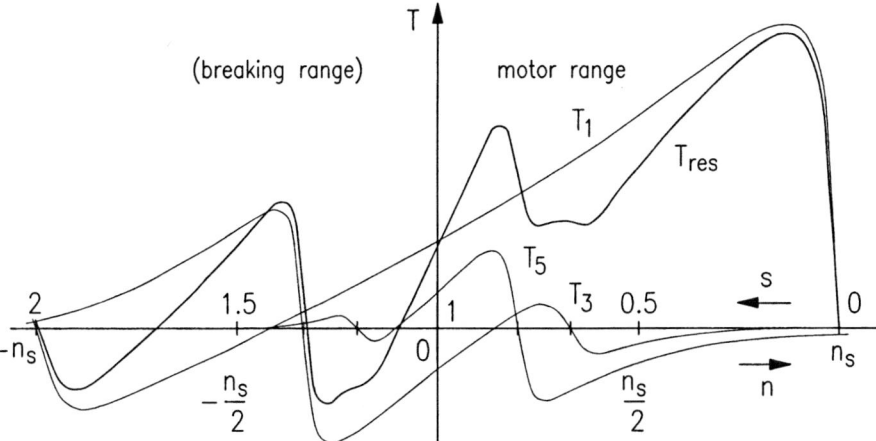

Fig. 2.122. On principle torque diagram of the basic field (T_1), the third field (T_3) or the fifth field (T_5) of an AC asynchronous motor

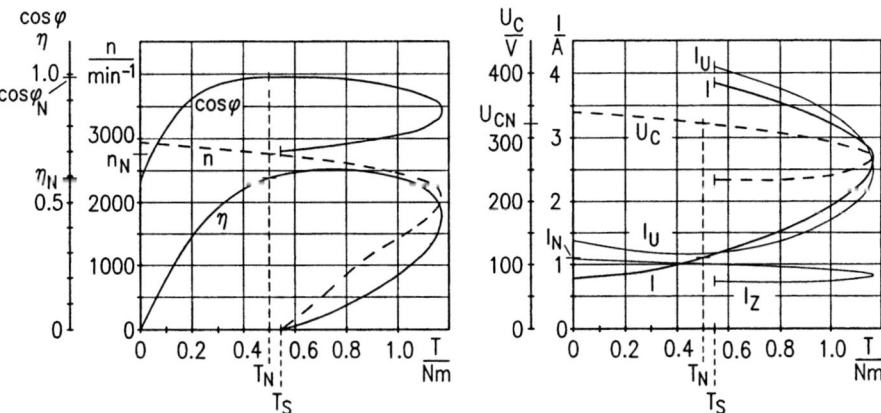

Fig. 2.123. Performance characteristics of a two-pole 150-W AC asynchronous motor. T_S = starting torque, I = total current, I_U = main current, I_Z = auxiliary current, U_C = capacitor voltage, η = efficiency, $\cos\phi$ = power factor, values of the continuous rating: T_N, n_N, I_N, U_{CN}, $\cos\phi_N$

In this equation, m is the number of phases of the stator winding. The sign declares the rotation sense. A field with a positive number of pole pairs rotates in the same direction like the basic field, a field with a negative sign in the opposite direction. The smaller the number of pole pairs, the stronger the corresponding harmonic field. While no numbers of field pole pairs divisible by three arise in three-phase motors and the lowest harmonic field is the fifth ($\nu = -5$ for $p = 1$), there exist third harmonic fields in AC motors causing higher harmonic torques than the fifth.

Corresponding to Eq. (2.99), the synchronous speed of the harmonic fields is with the speed of the basic field n_s

$$n_{\text{s}\nu} = \frac{f_1}{\nu} = n_\text{s}\frac{p}{\nu} \qquad (2.108)$$

and thus lower than that of the basic field. Figure 2.122 shows the effect of the 3. and the 5. harmonic field on the speed-torque characteristic of an AC asynchronous motor. It is an on-principle diagram with harmonic torques which are drawn exaggeratedly high compared to the basic torque. The figure shall explain the development of torque saddles in the lower speed range which are able to make the running up difficult. In the worst case, the drive gets stuck in the saddle and does not reach the desired operation speed.

This is valid especially for motors with a small number of slots per pole and per phase. The smaller it is, the more visible are the harmonic fields. Due to the costs and the space required, fractional-horsepower motors are designed with a minimum number. Often simple motors and motors with high speed have one slot per pole and per phase, only, and therefore especially deep torque saddles. In order to decrease the harmonic fields sometimes the conductors are distributed irregularly in the slots.

2.4.1.5 Operational Performance

Figure 2.123 shows the typical characteristic curves of a two-pole two-phase AC asynchronous motor. At a speed of $n \approx 1000\,\text{min}^{-1}$, i.e. at about a third of the synchronous speed ($n_\text{s} = 3000\,\text{min}^{-1}$) a little saddle arises in the speed-torque curve, as was mentioned in the preceding section. Such a hardly noticeable saddle can be obtained by corresponding dimensioning. The no-load losses are higher than those in the continuous rating, as can be seen in the curve shape. Therefore, a capacitor motor does not cool down by any means during no-load, because the capacitor does not fit this mode of operation. The capacitor voltage of AC motors is much higher than the line voltage. The capacitor motor takes hardly or no reactive power over a wide torque range from the power supply.

Comparison. Figure 2.124 shows speed-torque curves of capacitor motors compared with a three-phase motor (a) of the same frame size. A motor with a starting capacitor C_S possesses the curve b. In the range above the breakdown torque C_S is switched off. Then, the motor operates further on either two-phase *via* a running capacitor C_r (c: pointed line) or single-phase (d: continuous line). In Fig. 2.125, speed-torque curves are given for different values of capacity, verifying that the start and the breakdown torque can be increased with increasing capacities, whereas the continuous rating decreases, because the rotating field gets more and more elliptical in shape and the losses as well as the pulsating torques increase.

Typical values for running capacitors of two- and three-phase AC motors are given in Table 2.6. It shows that higher-pole motors need capacitors with

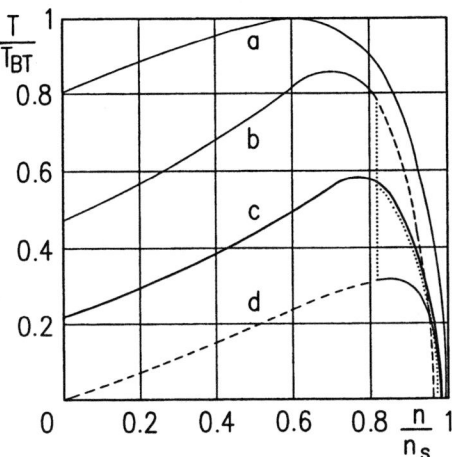

Fig. 2.124. Comparison of the speed-torque characteristics for different connection methods of capacitor motors. T_{BT} = breakdown torque of the three-phase motor

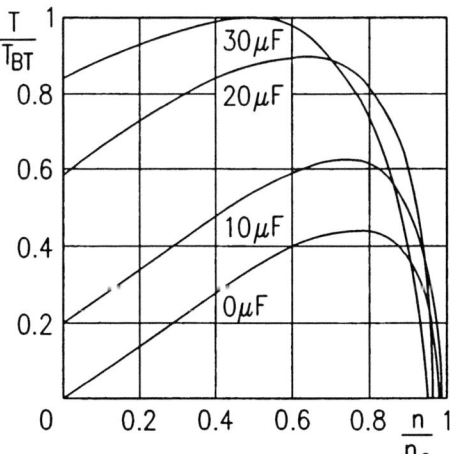

Fig. 2.125. Speed-torque characteristics of different capacitors

a higher capacity than two-pole motors. But it certifies especially that, for three-phase motors, much greater and thus more expensive capacitors are necessary than for two-phase motors.

The capacity of running capacitors for two-phase motors can be calculated approximately by

$$C_R = \frac{\ddot{u}}{2} \frac{P_1}{\omega \cdot U_N^2} \tag{2.109}$$

and for three-phase motors by

$$C_R = \frac{2}{\sqrt{3}} \frac{P_1}{\omega \cdot U_N^2} . \tag{2.110}$$

These equations are valid only if, at the wanted operation point, the conditions of a circular field come true which is not always the case. Approximate values

Table 2.6. Approximate values of the efficiency and the capacity C_R of running capacitors

P/W	η	$C_R/\mu F$			
		$m_1 = 2$		$m_1 = 3$	
		$p = 1$	$p > 1$	$p = 1$	$p > 1$
50	0.50	2	2	10	14
100	0.55	3	4	12	16
200	0.60	5	6	16	22
500	0.68	12	14	35	45
1000	0.72	18	25	80	100
1500	0.74	25	35	140	160

of the efficiency can be taken from Table 2.6, e. g. to calculate the input power P_1 from the output power P.

2.4.1.6 Control Methods

Reversing Duty

There exist two ways to change the speed direction:

- Change of main and auxiliary phase. Switching over the capacitor from one to the other phase (Fig. 2.126a, b). This method is more cost-effective, because only a single-pole switch is necessary. The stator winding however, has to be symmetrical if the same torque shall be supplied in both senses of rotation.
- Current reversal in one phase (Fig. 2.126c). This method is seldom used because, for this purpose, an expensive two-pole switch is required. But it is suitable for quasi-symmetrical windings.

In both cases the rotor should be stopped as a precaution before switching on the new sense of rotation. Otherwise, the rotor may remain in the previous sense of rotation.

Speed Control

The possibilities of speed control result from Eqs. (2.99) and (2.100):

$$\overset{(3)\downarrow \quad (2)\downarrow}{n = n_s(1-s) = \frac{f_1}{p}(1-s)}\underset{(1)\uparrow}{}. \tag{2.111}$$

(1) **Pole changing**. The stator has two or three separated windings with a different number of poles. The power-space ratio of the motor is worse,

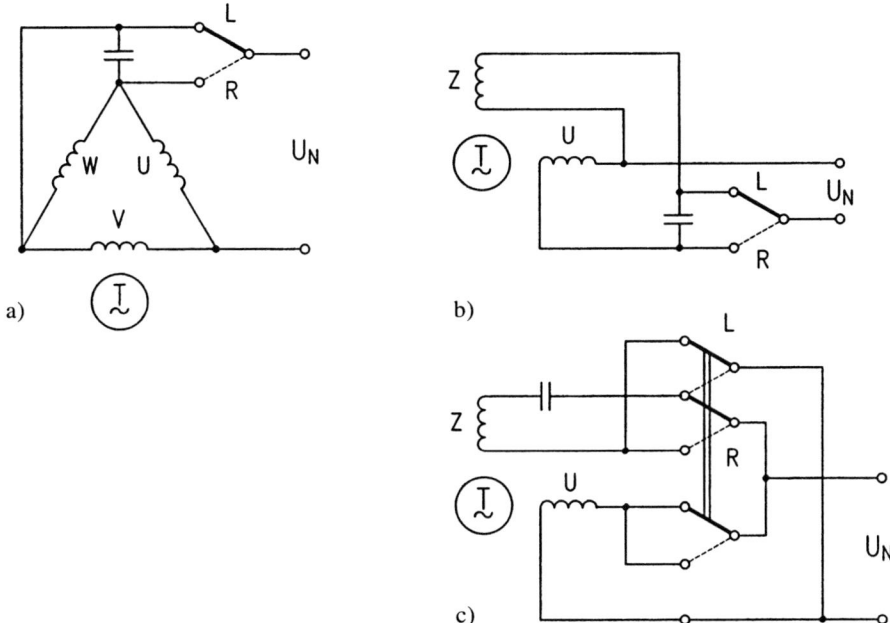

Fig. 2.126. Reversing connections: **a,b** change of main and auxiliary phase; **c** current reversal

but in contrast to Dahlander windings which are found in greater motors, any speed steps are possible and no expensive switches are necessary. The winding with the lower pole number can be single-phase, too. In this case the motor is started with the help of the higher-pole winding and then switched over to the lower-pole winding.

(2) **Slip changing**.
Tapped winding. The auxiliary winding consists of two parts which can be connected with one another, with the main winding and with the capacitor, by different methods. As the synchronous speed does not change, only a restricted speed variation is possible (approximately 1:2). With the artificial slip enlargement, considerable ohmic losses P_{V2} develop in the rotor winding according to Eq. (2.103). Besides this, a compromise has to be found with respect to the size of the capacitor, so that the rotating field is more or less elliptical depending on the different connection variants. Therefore, the use of this method is restricted to intensively cooled motors, as for example to split-cage motors of heating-circulating pumps (Sect. 2.4.1.2).

Motor-voltage changing. For this purpose, voltage regulating transformers or phase-angle controls, which are especially applicable for universal motors (Sect. 6.3.3.1), are used. With the help of the latter method, blocks with variable width are cut out of the sinusoidal voltage, and in this way the motor voltage is set to the mean value. In both cases, ohmic rotor losses increase

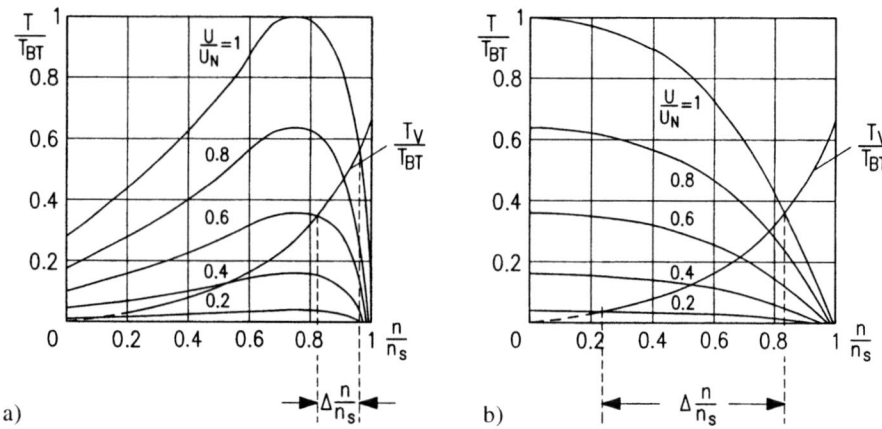

Fig. 2.127. Speed control by changing of the motor voltage. **a** Low-resistance rotor. **b** High-resistance rotor

considerably due to slip enlargement. With respect to phase-angle control, an enlargement of the phase angle increases the harmonics of the motor voltage and thus generates further losses. With this method, neither the synchronous speed nor the breakdown speed are modified in this way. The speed range is limited here, too. In addition, it must be considered that the torque decreases overproportionally with the voltage ($T \sim U^2$). For this reason, this method is suitable for intensively cooled motors, i.e. for ventilators and fans with a hyperbolic load characteristic allowing a greater speed difference. To extend the speed range to approximately 1:3, often a high-resistance squirrel-cage rotor is required. In this way, the speed-torque characteristic becomes more smooth and the locked-rotor torque increases (Fig. 2.127). It should be considered that the rated speed decreases; in AC motors, the breakdown speed as well as the breakdown torque decrease.

(3) **Frequency changing.** For this expensive method (Sect. 6.3.3.2), only three-pole motors supplied with three-phase current are used. Here, the synchronous speed is changeable by a frequency converter with a DC link (Fig. 2.128a), as it is principally used for larger motors. By this means, the speed can be regulated over a wide range, especially if the possibilities of field weakening are in demand (Fig. 2.128b). For permanent-magnet motors and EC motors, a field weakening is only possible due to special procedures, e.g. by a temporary reverse magnetization, and that only to a certain extent.

The torque remains nearly constant when varying the terminal voltage proportionally to the frequency. Nevertheless, it must be taken into consideration that the voltage drop at the stator resistance has an increasing influence with decreasing frequency or decreasing motor terminal voltage. This is especially valid the smaller the motor. If it is necessary the voltage drop has to be compensated. Variable-frequency motors are used sporadically in the lower-power range today, but they will compete with EC motors in the future.

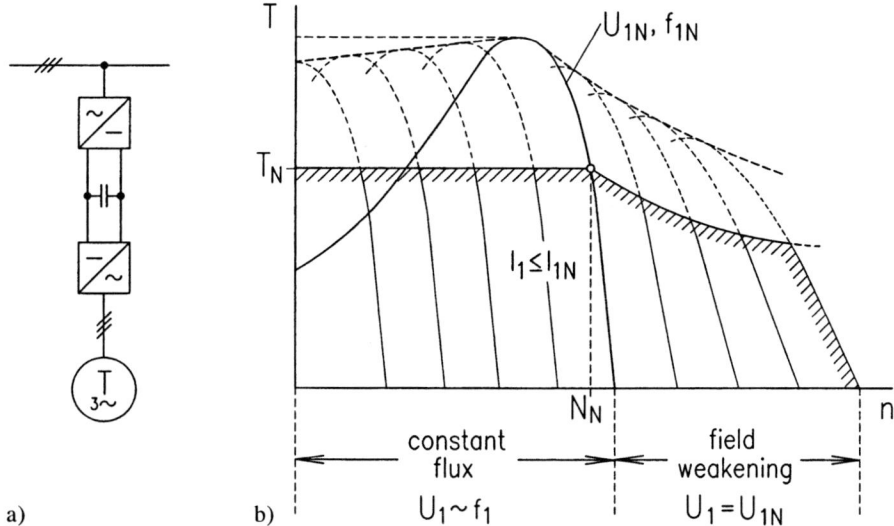

Fig. 2.128. Frequency-converter operation. a Block diagram. b Operating range with field weakening

Speed control

A vector control as required for greater drives hardly comes into question in the lower-power range today. Therefore, only conventional methods are applied, measuring the speed by tacho-generators and thus adjusting the frequency or the voltage (Sect. 6.4.3).

Dual-Voltage Motor

The phase windings of motors whose terminal voltage is switchable to half value, i.e. from 230 V to 115 V, consist of two parts. They can be connected once in series and once in parallel (series-parallel connection, Fig. 2.129a). As the capacity is inversely proportional to the square voltage according to Eqs. (2.109) and (2.110), it has to be the quadruple at the lower voltage. For this reason, the so-called T-connection is preferred (Fig. 2.129b), where the reversing takes place in the main winding only. The auxiliary winding is designed for the lower voltage only. The adjustment to the high voltage is more unfavorable. The motor characteristics are worse in the case of the higher voltage, but the changeover switch is simpler. However, both connection methods are seldom produced nowadays, because production engineering is much more flexible as a result of automation. Therefore, windings and capacitors can be easily designed for every value.

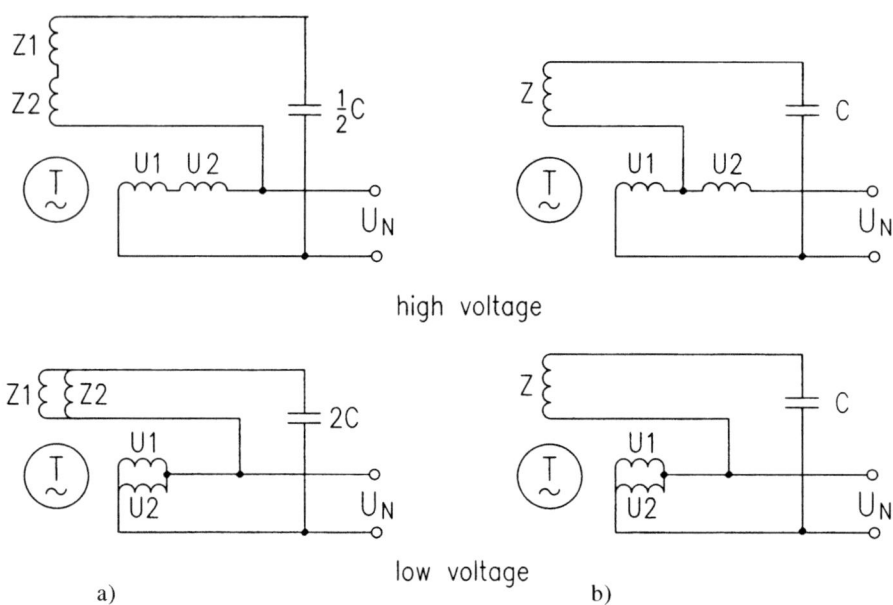

Fig. 2.129. Dual-voltage motor. **a** Series-parallel connection. **b** T-connection

2.4.1.7 Shaded-Pole Motor

Applications

Shaded-pole motors (split-pole motors) are the most cost-effective AC asynchronous motors due to their simple construction and simple production engineering. They are robust and quiet like all asynchronous motors, but the efficiency and the power factor are rather low. These motors are used for simple drive applications, e. g. for ventilators, fans, pumps, flap adjusting (heating).

Mode of Functioning

The main winding consists of concentrated coils on salient poles and is connected to alternating voltage. The auxiliary winding has one or several single turns, which are short-circuited (shading coils) and enclose a part of each pole. Opening the shading coil, like the auxiliary phase used in resistance-start motors after running up in order to reduce losses, is only possible with a complicated breaker unit. Doing so, the above-mentioned advantageous qualities of a shaded-pole motor will get lost. In addition the torque is reduced. The thermal advantages of this method are negligible.

Main winding and auxiliary winding work like a secondarily short-circuited transformer. On the one hand, both phase windings should have a common axis, on the other hand they should be shifted against each other about the electrical angle of $\pi/2$, as it is necessary in two-phase motors in order to generate a circular field (Sect. 2.4.1.3). The main winding induces a current in the auxiliary winding. The flux linked with the auxiliary current is out of phase with the main flux. Since the auxiliary flux is less than the main flux and the phase displacement is less than $\pi/2$, an elliptical field arises. Some improvement is reached by using magnetic bridges between adjacent poles. The flux linked with the shading coil increases by the leakage flux over these bridges. Through this, the effective angle between main and auxiliary winding increases. On the other hand, the torque developing fluxes linked with the rotor winding decrease. Therefore, the quantity of this flux linked with main and auxiliary winding has to be adjusted very carefully. This can be realized, e.g. by means of artificially saturated zones. At the same time, the shape of the air-gap field is improved by these flux bridges. The field can extend into the pole gaps, so that the trapezoidal field generated by a salient pole becomes a more sinusoidal shape.

The air-gap field gets a one-sided saddle by the slots of the auxiliary winding. In order to diminish the harmonics generated by the asymmetric field distribution, an air-gap grading is provided on the other side of the shaded pole (Fig. 2.130). The inductivity of this range decreases simultaneously, i.e. this part of the main flux leads the part across the small air-gap. Since on the other hand, the auxiliary flux across the shaded pole lags behind the main flux, the air-gap grading supports the sense of rotor rotation from the main pole to the auxiliary pole. In spite of these measures, the air-gap field contains a strong third harmonic causing a pronounced saddle in the speed-torque curve (Fig. 2.131). If the load characteristic runs through this saddle, the rotor only achieves the speed which results from the intersection of the load characteristic and the branch ascending from the saddle of the speed-torque curve: "The motor sticks fast in the saddle".

Types of Construction

The simplest motors have a two-pole asymmetrical laminated core (U- or C-core, Fig. 2.130, top left). The main winding is placed in a spool pushed over the yoke and is inserted into the stator together with it. The short-circuited winding consists of from one to three thick copper wires, each pole lying in drill holes of the poles. They are bent together, their ends being connected with one another by hard-soldering. These motors are produced most cost-effectively, but they have the worst efficiency (up to 15%) and are suited for low power only.

Motors with a symmetrical laminated core and a two-piece stator core (Fig. 2.130, top right) have efficiencies within the region of about 20% and a power up to about 30 W. First, the coils are pushed upon the poles from the outside, and then they are pressed together with the pole star into the yoke ring.

The greatest shaded-pole motors with a power up to approximately 150 W are built with a one-piece stator core (Fig. 2.130, below left). They can achieve efficiencies within the range of about 30%. Instead of a flux bridge, they have a leakage slot. The poles can be wound directly, which, however, is very costly. The short-circuited stator windings consist of copper or aluminium hollow sections which are pushed through the leakage slots over the pole horns, or they

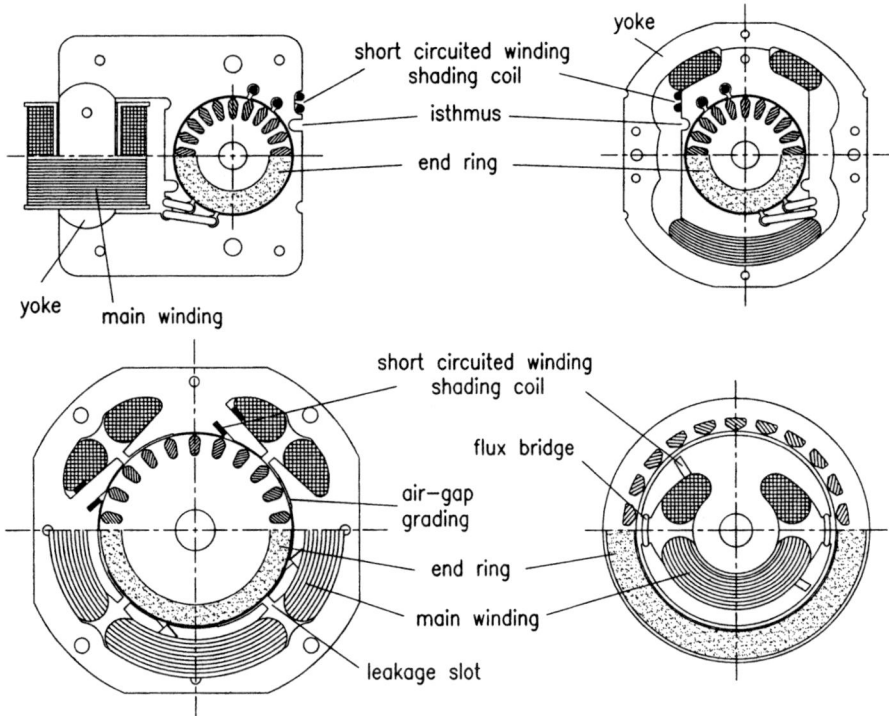

Fig. 2.130. Constructions of shaded-pole motors

consist of flat sections which are put into the slots, bent together and hard-soldered. The broad slot of the short-circuited winding is disadvantageous, because the air-gap field is, thus, very much saddled at this point, and therefore, the harmonic components are very high. If the leakage slot between the poles is too broad, it is closed by a squeezed-in sheet metal. If it is narrow, it normally remains open.

Slotted stator. A shaded-pole motor can principally be built with windings distributed in slots. The short-circuited winding is located in a part of the slots either at the slot ground or at the slot opening. Both winding axes enclose an electrical angle of about $\pi/3$. This design implicates no essential advantages, unless a laminated core of an existing three-phase motor is used, thus avoiding the development of a specially laminated core and the special production engineering of conventional shaded-pole motors. Since the production is much more expensive, these motors are only used very seldom.

Motors with external rotors are used especially for fans and ventilators (Fig. 2.130, bottom right). The fan blades are directly mounted on the outer rotor housing. By this means a shorter overall length is obtained, as is necessary, e. g. for cabinet fans. The worse heat discharge out of the stator, however, is disadvantageous.

Like all small asynchronous motors the rotor has a cage winding which is made of aluminium or silumin by die-casting. Slots are skewed here, too, for reducing position-dependent torques (Sect. 2.4.1.2). Nearly without exception oil-impregnated porous-bronze bearings are used, and often self-adjusting collar bearings. It must be considered that shaded-pole motors have to be driven in a horizontal position as those bearings are not able to cope with axial forces. On the one hand, the rotor is stabilized by electromagnetic forces, on the other hand axial oscillations might occur if they are excited by the driven apparatus. In vertical arrangement, a bearing must be provided for supporting, for example, a ball in the middle of the axis end. The rotor is mounted in bearing brackets consisting of aluminium or embossed sheet metal. In order to reduce the costs, one of the bearings can be made as a flange or the bearing can be integrated in the driven apparatus.

Performance Characteristics

Figure 2.131 shows the performance characteristics of a 50-W motor. The speed-torque curve shows the saddle which is typical for shaded-pole motors. At the ordinates, the rated values of the speed, the current and the efficiency are specified.

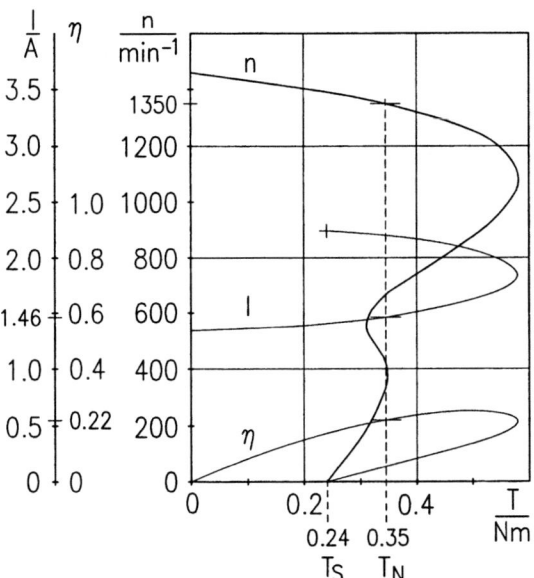

Fig. 2.131. Performance characteristics of a shaded-pole motor. T_S = start torque; T_N = rated torque; n_N = rated speed; I_N = rated current; η_N = rated efficiency

Control Methods

Reversing duty. As mentioned above, shaded-pole motors always rotate from the main to the auxiliary winding. The obvious possibility furnishing the

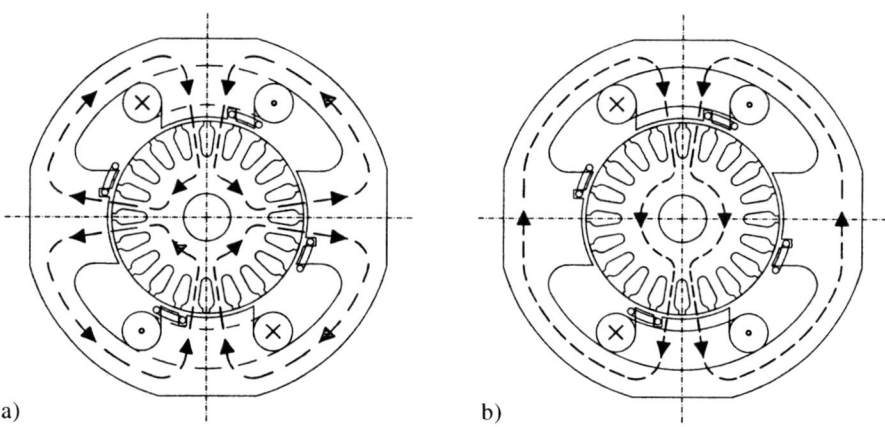

Fig. 2.132. Cross-pole connection. a Four-pole excitation. b Two-pole excitation

poles on both pole sides with short-circuited windings, which are switched on corresponding to the desired sense of rotation, proves to be problematic. The switches are worn out very fast in the case of the high currents, if they are not opened or closed exactly at the same time. In practice, this is not possible. Therefore, for reversing drive duties two motors are put together which are inversely mounted and are switched on mutually corresponding to the sense of rotation.

Speed control. Shaded-pole motors are built nearly exclusively for one speed. For fans, a speed control is occasionally used by means of a tapped winding. A two-step speed control is possible with a so-called cross-pole connection. Here, only every second pole is wound. Depending on the desired speed, adjacent coils are excited either equidirectionally or inversely. E. g. in a four-pole motor, only the two opposite poles have windings (Fig. 2.132). With an inverse excitation, a four-pole field is obtained. An equidirectional excitation results in a two-pole field if the third harmonic is as diminished as possible by leakage bridges or one-sided air-gap grading. A six-pole field occurs if the third harmonic and the torque saddle are considerably pronounced. Nowadays, this construction is also seldom in demand.

2.4.2 Synchronous Motors

2.4.2.1 Characteristics and Fields of Application

Characteristics

Speed. Just as in asynchronous motors, a rotating field is generated by the stator winding of synchronous motors. It is forced to rotate synchronously with the rotating field because the second torque-developing component is impressed in the rotor. Corresponding to Eq. (2.99) the speed is strictly proportional to the frequency of the power supply or of the control electronics and independent of the applied voltage and voltage fluctuations, too (except if the breakdown torque is exceeded). But the speed is constant only on average. In AC synchronous motors, pulsating torques arise due to the rotating field, which is generally elliptical as in AC asynchronous motors. These torques can spoil the true running. A symmetrization of the windings (balanced motor) as described in Sect.2.4.1.4 or an artificially increased torque of inertia can be a corrective measure.

Load angle. The pole axis of the rotor (direct axis) forms an angle with the pole axis of the rotating field dependent on the load, the so-called load angle β. At ideal no-load, i. e. if the motor is unloaded and the friction losses are negligibly low, the load angle is equal to zero: the rotor-pole axis lies in the direction of the field axis. At most, the pole axis can be perpendicular to the field axis, i. e. the electrical load angle can theoretically reach 90°. However, then the rotor is in an unstable position and can pull out of synchronism, i. e. the rotor is not able to follow the rotating field and comes to a standstill.

Besides rotor oscillations as a result of pulsating torques, oscillations arise in the case of state changes, i. e. in the case of load or voltage changes. The rotor does not suddenly reach the new position with regard to the rotating-field axis, but not until a transient reaction takes place. By the way, small synchronous motors are as robust, low-noise and cost-effective as asynchronous motors, and the same high expenditure is required for speed control.

Fields of Application

The above-mentioned peculiarities and the start problem described in the following have to be taken into consideration for the application of AC synchronous motors. They are suitable for frequency-proportional drives which need not be speed-controlled: clocks, control units, counters. Small synchronous motors can be constructed very simply and with good efficiency or with a small frame size. Therefore, there are application cases too for which synchronous speed is unimportant. On the one hand, these are two-pole motors for pumps in washing machines and dish washers, pond and aquarium pumps or citrus squeezers; on the other, high-pole motors for simple servo drives.

2.4.2.2 Types of Construction

In the lower range of performance no electrically excited synchronous motors are produced because the expenditure for production and maintenance of the slip rings and brushes would be too great. The rotors are constructed either as permanent-magnet, hysteresis or reluctance rotor, in the latter case occasionally in combination with magnets. There exist motors with internal and external rotors.

For small numbers of poles, the stator can be slotted and can have a two-phase or three-phase winding. Three-phase motors can be directly connected to the power supply. Operating at alternating voltage, a capacitor is used for phase displacement like AC asynchronous motors. There are also one-phase motors with and without split-pole (shaded-pole) winding (Sect. 2.4.1.7).

Motors with a high pole number are constructed according to the claw-pole principle (Fig. 2.133a). Metal flaps are bent around a ring winding alternately from right and from left, setting up alternately north and south poles. In this way, a high-pole alternating field is created, when connected to an alternating voltage. This construction (the so-called canned motor) is very simple, because it is produced in a punching-bending-fitting process without cutting operation and without screwed connections. But it causes high leakage fluxes between the flaps and eddy currents in the $1-2$ mm thick metal components.

In order to generate a rotating field there are the following possibilities:

- For three-phase operation three claw-pole systems are fitted together, which are twisted against each other by a third of a pole pitch.
- For AC operation, two systems are needed shifted by half a pole pitch (Fig. 2.133b). The winding of one stator is directly connected to the power supply, the winding of the other stator *via* a capacitor.
- One-phase motors get additional auxiliary poles with copper sheets across them. Currents are induced in these sheets by the ring windings as shown in Fig. 2.134 with broken lines. They react upon the fluxes of the auxiliary poles. In this way, a low phase displacement is achieved between the main and the auxiliary flux like for split-pole motors. Such motors have a distinctive elliptical rotating field (Sect. 2.4.1.3), but they are quite simple in their construction and therefore very cost-effective. Figure 2.134 only

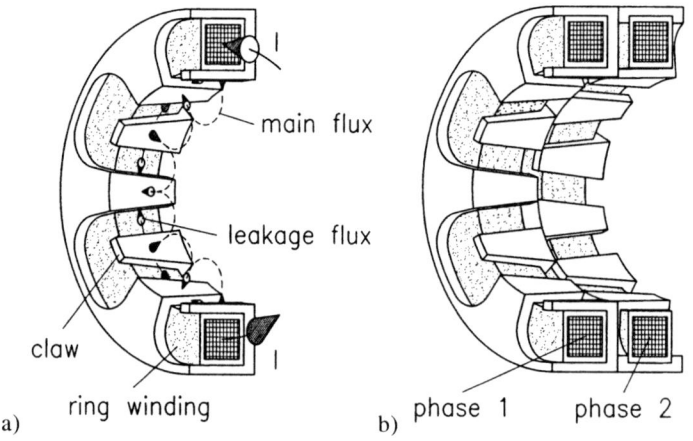

Fig. 2.133. Claw-pole motor. **a** Construction principle. **b** AC motor

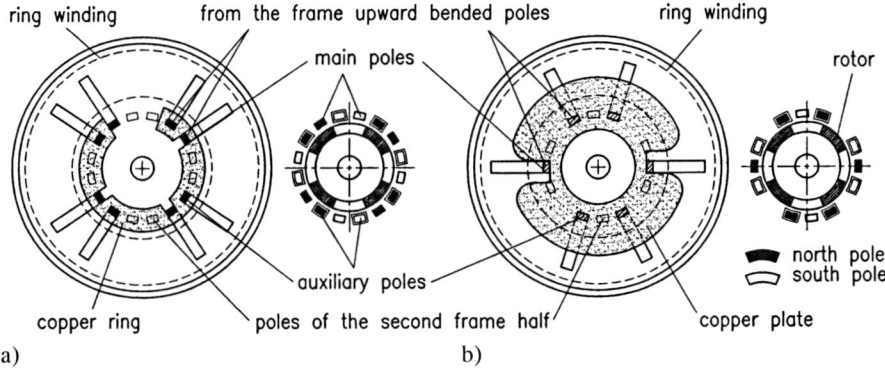

Fig. 2.134. Claw-pole motors with auxiliary pole. **a** Symmetrical pole arrangement. **b** Asymmetrical pole arrangement

shows one housing half for two eight-pole motors. The poles of the other half are indicated by dashed lines. Additionally to both housing halves, all stator poles are shown together with the rotor.

The rotation sense of the left motor with a symmetrical construction is not defined. If the rotor shall rotate in one direction only, a mechanical back stopping is provided. The rotor of the right motor rotates to the right due to the slightly asymmetrical pole arrangement. Sometimes, the pole fingers are asymmetrically constructed to prefer a rotation sense. Such synchronous motors are still in demand today in simple controls, as for instance in washing machines or in dish washers.

2.4.2.3 Synchronous Motors with Magnet Rotors

Synchronous motors are mostly built with permanent-magnet rotors, because they reach the highest efficiency or power-to weight ratio. The magnets are either produced as hollow cylinders, which are fastened on a plastic bearer, or as solid cylinders, in the bore of which the shaft is sticked in. Barium- or strontium-ferrite magnets are in demand as permanent-magnet material for these simple motors. It is disadvantageous that motors with magnet rotors cannot start without any difficulty, especially in the case of low pole numbers. Small motors reach the synchronous speed by a transient reaction. In order to support this, the load is not rigidly coupled with the rotor.

High-pole motors are built as one-, two and three-phase claw-pole motors, as described in the previous section.

Two-pole, single-phase motors have an uneven air-gap, as shown in Fig. 2.135. If the winding is not excited, the pole axis approximately adapts itself to the direction of the small air-gap parts. If the winding is switched on, the rotor swings to the right or to the left, depending on which direction of magnetization arises in the stator. One-phase motors generate an alternating

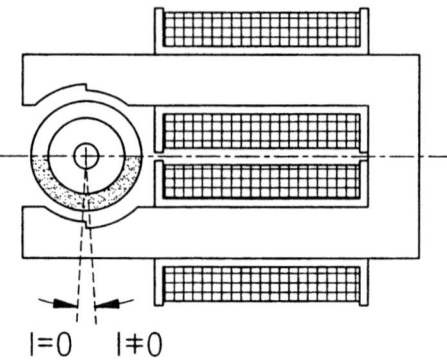

Fig. 2.135. One-phase synchronous motor

field, which can be imagined as two fields rotating in the opposite direction, as described in Sect. 2.4.1.4. The rotor pulls into step with one of these fields within one or a few periods. Therefore, the sense of rotation is not definite. If only one rotation sense is allowed, a mechanical backstop must be built in. Due to the wear, this is only sensible if the motor is rarely switched on. Such motors have mostly replaced shaded-pole motors as drives for pond and aquarium pumps or citrus squeezers, especially for draining pumps in washing machines and dish washers, because their construction and production is even more simple. In order to avoid a back stopping and to allow both senses of rotation, the streaming out of the pump takes place radially and not tangentially, leading to a more favourable pump efficiency. In order to avoid a special seal between motor and pump, the rotor rotates in the water (split-cage motor). As the air-gap of magnet motors can be essentially greater than that of an asynchronous squirrel-cage motor (Sect. 2.4.1.2), the split case in which the rotor is revolving and which protects the stator winding can be made of plastics. The efficiency is essentially greater than that of shaded-pole motors. Therefore, the synchronous motor with magnet rotor has a lower volume.

These one-phase motors can be built with a power up to about 50 W, only, because the torque of inertia of greater motors hinders the start. Therefore, there are efforts to start motors with higher power reliably and with a definite rotating sense by the simplest electronically starting controls. Motors of this kind could replace for instance the comparatively expensive capacitor motors, which are used in dish washers for driving the circulating pump.

Small synchronous motors are seldom driven with frequency converters (Sect. 5.3.4). An exception is the two-pole micromotor in Fig. 2.136. The voltage is applied in cyclic order to the three phase windings (slotless skew winding) by an electronic control. The motor is started with low frequency and can then be run up to the rated speed by raising the frequency. The rated speed is chosen as high as possible (e. g. $100\,000\,\text{min}^{-1}$), in order to a get power-for-size ratio as high as possible. A speed reducer is necessary, since such a speed is too high for practicable use.

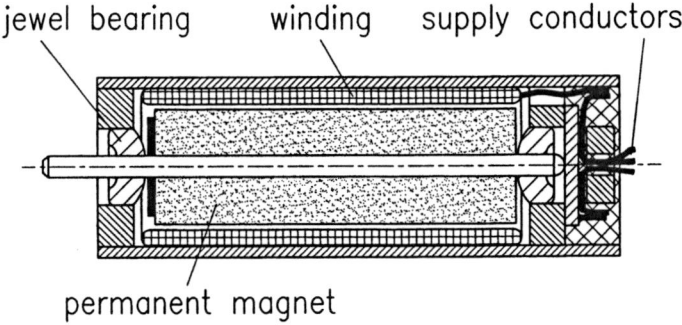

jewel bearing   winding   supply conductors

permanent magnet

Fig. 2.136. Micromotor (diameter: 1.9 mm; length: 5.5 mm)

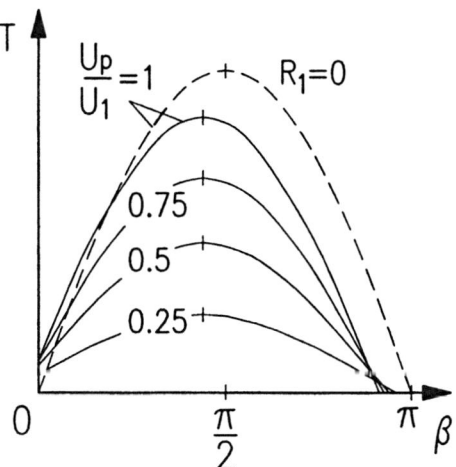

Fig. 2.137. Torque characteristics of synchronous motors with magnet rotors

Torque. Synchronous motors with permanent-magnet rotors where the ohmic resistance of the stator winding is negligible low compared with the reactance, possess a torque which is proportional to the sinus of the load angle β (Fig. 2.137, $R_1 = 0$). The maximum torque (breakdown torque) occurs at $\beta = \pi/2$. The maximum load torque should be clearly smaller, because this is an unstable operating point. In small motors, the resistance is not negligible with regard to the reactance in general. In the example shown in Fig. 2.137, the continuous curves are valid for a resistance R_1, which is one eighth of the synchronous reactance (the sum of the leakage and magnetizing reactance) in the pole or direct axis. It is shown that the breakdown torque and the maximum load angle decrease as a result of the resistance. The resistance can even be greater in small motors than assumed here. In addition, the torque depends on the so-called synchronous generated voltage (field e.m.f.) U_s, as shown in Fig. 2.137. This is the voltage generated in the stator winding by the rotor magnet.

2.4.2.4 Hysteresis Motor

Construction. The stator of hysteresis motors is produced in the same types of construction as motors with magnet rotor, i.e. with slots, with claw poles or with split poles. The rotor has a ring of hysteresis material. This is similar to permanent-magnetic materials; the hysteresis loop, however, is less broad and not prepolarized.

Mode of functioning. The ampere turns of the stator generate a field rotating with synchronous speed (Eq. 2.99). They permanently reverse the magnetization of the hysteresis rotor, so long as it stops or rotates slower as the rotating field. The magnetic state of the rotor elements runs through the whole hysteresis loop. The hysteresis losses P_H generated by that are proportional to the remagnetizing frequency

$$f_2 = s \cdot f_1 \qquad (2.112)$$

with the slip s of the rotor with regard to the rotating field according to Eq. (2.100). In the hysteresis material hysteresis losses arise analogously to the rotor losses in asynchronous motors (Fig. 2.138, left):

$$P_H = P_{V2} = s \cdot P_\delta \qquad (2.113)$$

and at the shaft, the output power is given by

$$P = (1-s)P_\delta \ . \qquad (2.114)$$

Neglecting first the eddy currents generated by the rotating field, no currents flow in the rotor, which are reacting upon the air-gap field. The stator current does not depend on the load, because there is no armature reaction. Therefore, the power losses of the stator P_{V1} and the input power coming from the power supply P_1 are constant. As a result, the air-gap power is constant, too:

$$P_\delta = P_1 - P_{V1} \ . \qquad (2.115)$$

Therefore, the torque generated by the hysteresis is constant according to Eq. (2.105):

$$T = \frac{P}{2\pi n} = \frac{P_\delta}{2\pi n_s} \ . \qquad (2.116)$$

The above statements are valid for ideal motors. In reality, eddy-current losses arise in every hysteresis rotor proportional to the square of the rotor frequency sf_1. The eddy-current torque T_E is inversely proportional to the slip. Therefore, it decreases linearly with the slip or with increasing speed (Fig. 2.138, middle). The torque characteristic of real hysteresis motors is

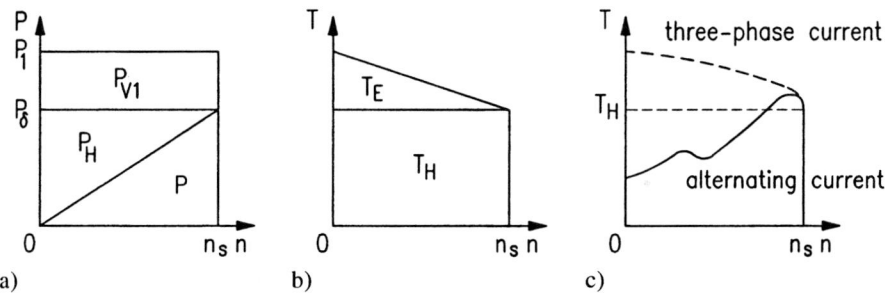

Fig. 2.138. Powers and torques of hysteresis motors. **a, b** Ideal motor. **c** Real motor

shown in Fig. 2.138 on the right side for three-phase or AC operation. Caused by leakage fluxes and a generally high harmonic content of the air-gap field, the real torque characteristic differs from that of an ideal motor. It is typical for a hysteresis motor that it starts independently without swinging in and pulls softly into synchronism. In the synchronism poles are formed. Now the hysteresis motor acts like a magnet-rotor motor, but it creates a torque, which is 20 to 30 times smaller. Therefore, gearings with a great reduction ratio are used, in order to get sufficiently great torques for the respective application case. Due to the low utilization hysteresis drives have considerably lost in importance today.

2.4.2.5 Reluctance Motor

The importance of reluctance motors has also greatly decreased in the last few years, so that they are hardly found in company catalogues any more. Therefore, their construction, mode of functioning and operational performance shall only be mentioned briefly.

Construction. Reluctance motors have rotors, whose permeance along the circumference varies depending on the number of poles. The rotor turns synchronously with the rotating field, because it always adjusts in such a way that the magnetic energy in the air-gap is a minimum. Like with the salient-pole motor a torque is generated on the basis of different reluctivity in the direct and in the quadrature rotor axis (reluctance torque).

Greater reluctance motors have a two- or three-phase winding distributed in the stator slots and in the rotor a squirrel-cage winding like an asynchronous motor (motor with starting cage). There are two rotor constructions:

- Rotors with salient poles (Fig. 2.139, left) have on average a greater air-gap. For this reason, they need a comparatively high magnetizing current in order to drive a sufficiently great flux across the air-gap. As a result, the power factor $\cos\varphi$, the ratio of the real input power to the complex input power, is low. As the high current generates high losses, the efficiency is

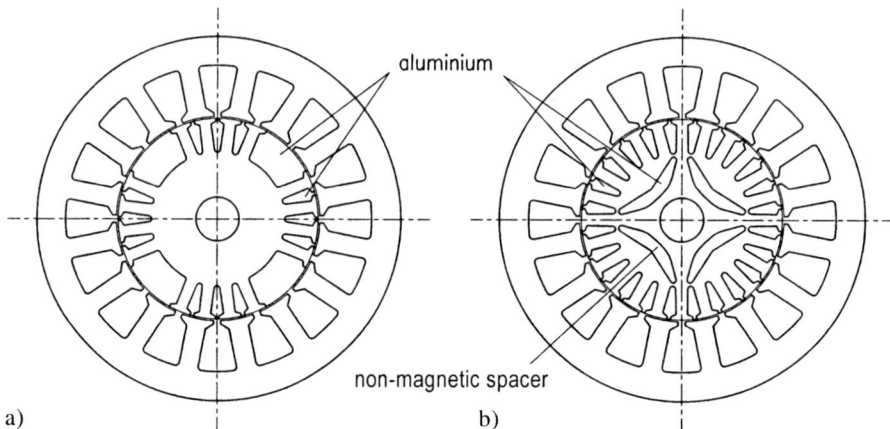

Fig. 2.139. Four-pole reluctance motors with squirrel-cage rotor. a Rotor with salient poles. b Rotor with sectional view of non-magnetic spacer

low, too. Therefore, the output power of these motors is about half of that of AC asynchronous motors with the same frame size.
- Rotors with non-magnetic spacers (Fig. 2.139, right) are more expensive in their production, but due to the constant air-gap width the output power of these motors is the same as that of AC asynchronous motors with the same frame size. The magnetic saturation should be not too high.

In the case of motors with external rotors, sometimes sectional views of asynchronous motors are used, in which little holes are punched, thus creating magnetic bottlenecks.

Operational performance. Motors with squirrel-cage winding run up automatically. After running through the asynchronous break-down torque, they suddenly pull in synchronism (Fig. 2.140). If the load torque becomes greater than the synchronous break-down torque, they pull out just as suddenly. But they do not stop if the load torque is lower than the asynchronous break-down torque. Motors without squirrel-cage winding but with salient poles, which can be toothed or not toothed, pull in by swinging like motors with magnet rotors. They have practically no importance, because the torque is too low.

For the torque behaviour at synchronous operation, the same is valid as for motors with magnetic rotors, but here it is to be considered that both the torque as well as the efficiency of reluctance motors are much lower. Figure 2.141 shows the torque depending on the load angle β for the same values of the winding resistance R_1 related to the direct reactance X_d. Motors with negligible resistance R_1 show the dashed line, which is hardly the case for small motors. Here it becomes especially clear how strongly the breakdown torque and the maximum possible load angle decrease with increasing resistance.

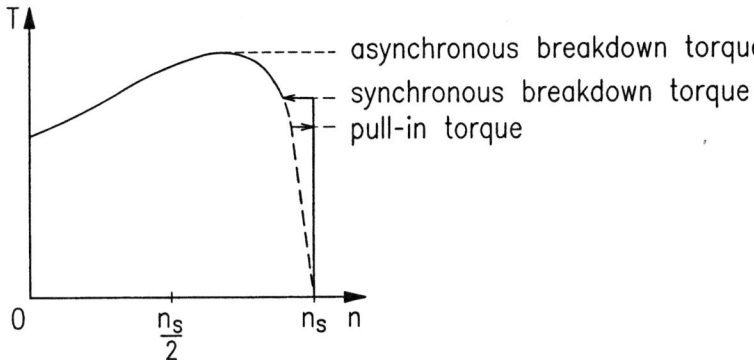

Fig. 2.140. Speed-torque curve

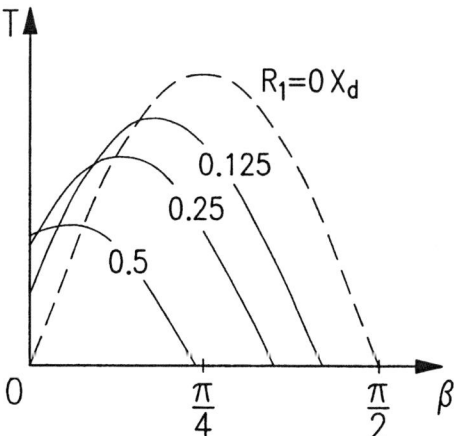

Fig. 2.141. Synchronous operation: torque depending on the load angle β

2.4.2.6 Permanent-magnet Motors with Anisotropic Rotor

Design. Permanent-magnet synchronous motors with anisotropic rotors, i.e. motors with magnets within a reluctance rotor are called hybrid-synchronous motors or synchronized asynchronous motors (Merrill motors). In general, these motors have a squirrel-cage winding for starting like asynchronous motors for starting (Fig. 2.142).

Without this winding, the motors have to be run up by frequency control (Sect. 6.3.4), because in the case of these expensive motors, a swinging in the synchronous speed is out of question, as is in case for the motors mentioned in Sect. 2.4.2.3. As magnets, both ferrite magnets and rare-earth magnets are used.

Mode of functioning. The motor with a squirrel-cage rotor generates a torque during running up like the asynchronous motor (Fig. 2.143). The permanent-magnet field which rotates together with the rotor excites currents in the stator winding with a frequency proportional to the speed. The line is a shunt fault for these currents.

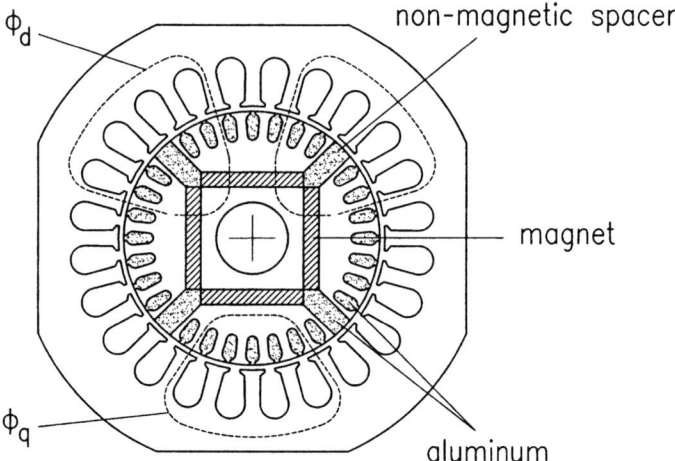

Fig. 2.142. Cross section of a permanent-magnet motor with anisotropic rotor. Φ_d direct-axis flux; Φ_q quadrature-axis flux

Therefore, they create a torque with the permanent-magnet field (magnet torque), which superimposes the asynchronous torque. In this way, the resulting torque has a deep saddle which can considerably hinder the running up. In synchronism, the slip-frequent pulsating torque generated by the rotor magnet changes into the synchronous torque. In Fig. 2.144 the principal synchronous torque/load-angle curve is shown, depending on the ohmic resistance and on the synchronous generated voltage U_s. The magnet plates and the non-magnetic spacers hinder the direct-axis flux, so that in contrast to electrically excited salient-pole motors, the quadrature-axis reactance is greater than the direct-axis reactance. As a result, the magnet torque is weakened by the reluctance torque in the range of low load angles and strengthened in the range of high load angles. Therefore, the maximum load angle increases.

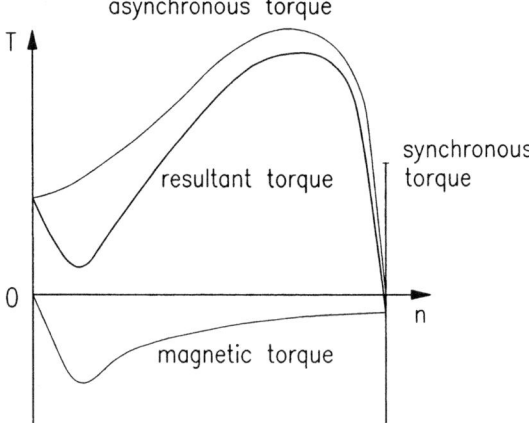

Fig. 2.143. Speed-torque characteristic

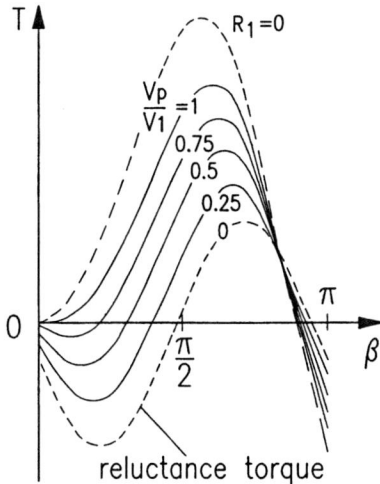

Fig. 2.144. Synchronous torque

Advantages of these motors in contrast to pure reluctance motors are their higher torque, their higher efficiency, and the fact that they need no reactive power for field generation. Therefore, corresponding converters can be designed with a lower power. Such drives are, however, expensive and therefore seldom used. They are applied in the synthetic-made fibre industry and for synchronous drives (roller table drives).

References

[2.1] Amrhein, W.: Motor-Elektronik-Rundlaufgüte. Zürich: Verlag Fachvereine an den schweizerischen Hochschulen 1989
[2.2] Beise, A., Lebsanft, J.: Betriebsverhalten permanenterregter Gleichstrommotoren bei Verschiebung einer der Kohlebürsten. etz-Archiv 7 (1985) H. 12, S. 389–394
[2.3] Braun, H., Ruschmeyer, K.: Polfühligkeit permanentmagnetisch erregter Gleichstrommotoren. F & M 104 (1996), H. 7-8. S. 562
[2.4] Dräger. 1.: Reduzierung von Reluktanzmomenten. F & M 103 (1995), H. 7–8, S. 433 DIN 45631. Berechnung des Lautstärkepegels aus der Lautheit und aus dem Geräuschspektrum. Beuth Verlag 1991 DIN FB 72, Geräuschqualität von Elektromotoren für KFZ-Zusatzantriebe. Beuth-Verlag 1998
[2.5] Gevatter, H.J., Ge, J.: Die Dynamik feinwerktechnischer Gleichstrommotoren. F & M 98 (1990). H. 5. S. 199
[2.6] Jucker, E.: Über das Physikalische Verhalten kleiner Gleichstrommotoren mit eisenlosem Läufer. La Chaux-de Fond/CH: Portescap 1975
[2.7] Jung, R.: Ein Beitrag zum Vergleich von fremd- und selbstgesteuerten Kleinmotoren mit Permanentmagneterregung. Diss. TH Darmstadt 1985
[2.8] Kenjo, T., Najomori, S.: Permanentmagnet and Brushless DC-Motors. Oxford: Clarendon Press 1985
[2.9] Koch, J., Plaumann, H.J., Ruschmeyer, K: Permanentmagnetisch erregte Kleinmotoren. Heidelberg: Alfred Hüthig Verlag 1986

[2.10] Lee, K.H.: Grenzen der technischen Miniaturisierung von permantmagneterregten Gleichstromkleinstmotoren mit Hilfe der Ähnlichkeitstheorie. Diss. U/GH Duisburg 1985

[2.11] Marinescu, M.: Einfluß von Polbedeckungswinkel und Luftspaltgestaltung auf die Rastmomente in permantmagneterregten Motoren. etz-Archiv 10 (1988), S. 83–88

[2.12] Merz, D.: Ein Modell für permanentmagnetisch erregte Kleinstmotoren mit wenigen Nuten und Metallbürsten. Diss. TH Darmstadt 1993

[2.13] Mohr, A.: Kleinmotoren mit Permanentmagneterregung. Bd. I u. Bd. 2. Bühlertal: Robert Bosch GmbH 1987

[2.14] Seinsch, HO.: Ausgleichsvorgänge bei elektrischen Antrieben. Stuttgart: B.G. Teubner 1991

[2.15] Siekmann. H: Gleichstromkleinstmotoren mit eisenlosem Läufer. F & M 85 (1977), H. 3, S. 2

[2.16] Stemme, 0., Wolf, P.: Wirkungsweise und Eigenschaften hochdynamischer Gleichstromkleinstmotoren. Sachseln/CH; Interelektrik AG 1994

[2.17] Volkmann, W.: Kohlebürsten. Gießen: Schunk u. Ebe 1980

[2.18] Weißmantel, H.: Einige Grundlagen zur Berechnung bei der Anwendung schnell hochlaufender trägheitsarmer Gleichstromkleinstmotoren mit Glockenanker. F & M 84 (1976) H. 4, S. 165–174

[2.19] Zwicker, E.; Fastl, H.: Psychoacoustic Facts and Models. Springer 1992

[2.20] Berghänel, D.: Das Verhalten elektrischer Kommutatormaschinen für Haushalt und Gewerbe bei Speisespannungen mit Frequenzen >50 Hz. Diss. der TU Dresden 2000

[2.21] Doppelbauer, M.: Oberfeldtheorie zur Berechnung der Kommutierung und des Betriebsverhaltens von Universalmotoren. Electrical Engineering 78 (1995), S. 407–416

[2.22] Figel, M., Labahn, D.: Fortschritte in der Konstruktion von Universalmotoren. Siemens-Z. 46 (1972), H. 9, S. 761–766

[2.23] Fujii, T.: Study of universal motors with lag angle brushes. IEEE PAS 1982, p. 1288–1296

[2.24] Kuhnle, H.: Die Ständerjochentlastung bei zweipoligen Universalmotoren. Diss. U Stuttgart 1969

[2.25] Metzler, K: Entwurf von unkompensierten Reihenschlußmotoren kleiner Leistung zum Anschluß an Gleich- und Wechselstrom gleicher Spannung. Leipzig: Verlag von Oskar Leiner 1925

[2.26] Möckel, A.: Analyse der Erregerspannung hochtouriger Reihenschlussmotoren kleiner Leistung im Hinblick auf Kommutierung und Fehlererkennung. ISLE Verlag 2001

[2.27] Moser, H.: Zur Konstruktionssystematik der Gehäuse kleiner elektrischer Maschinen. Konstruktion 20 (1968), H. 12, S. 465–477

[2.28] Oesingmann, D.: Neue Ständerkontur elektrisch erregter Kommutatormotoren. Elektrie 41 (1987), H. 5, S. 174–175

[2.29] Oesingmann, D., Siebenhaar, V.: Einfluss des Ankerfeldes auf die Auslegung von Kommutatorreihenschlußmaschinen. Neue Entwicklungen bei Elektrischen Kleinmaschinen. Kolloquium U/GH Paderborn, Meschede 1991, S. 7–16

[2.30] Oesingmann, D., Schuder, R.: Stromanalyse zur Fehlererkennung bei Kommutatormaschinen kleiner Leistung. Neue Entwicklungen bei Elektrischen Kleinmaschinen. Kolloquium U/GH Paderborn, Meschede 1991, S. 17–25

[2.31] Oesingmann, D.,. Schuder, R., Siebenhaar, V.: Einflußgrößen auf die Berechnung der Hauptabmessungen von Kommutatorreihenschlußmaschinen kleiner Leistung. Int. Wiss. Kolloquium TU llmenau 1992, Band I, S. 390–395

[2.32] Oesingmann, D., Siebenhaar, V.: Wechselstromkommutatormaschinen kleiner Leistung. Int. Wiss. Kolloquium TU Ilmenau 1995, Band 4, S. 521–526

[2.33] Pfeifer, R.: Beitrag zum Betriebsverhalten und zur lastgerechten Berechnung des magnetischen Kreises von Universalmotoren mit Phasenanschnittsteuerung. Diss. U Stuttgart, 1983

[2.34] Roye, D., Poloujadoff, M.: Contribution to the study for commutation in small uncompensated universal motors. IEEE PAS (1978), p. 242–250

[2.35] Scheffold, E.: Universalmotoren mit Hauptschlußcharakteristik. EMA 41 (1962), H. 10, S. 261–268; H. 11, S. 293–298; H. 12, S. 325–335

[2.36] Schroeter, W.: Die Berechnung des magnetischen Kreises von Universalmotoren. Diss. U Stuttgart 1956

[2.37] Stölting, H.-D.: Meßtechnische Wicklungsauslegung von Universalmotoren. F & M 92 (1984), H. 4, S. 182–184

[2.38] Stölting, H.-D.: Berechnung von Gleich- und Wechselstromkommutatormotoren. Neue Entwicklungen bei Elektrischen Kleinmaschinen. Kolloquium U/GH Paderborn, Meschede 1991, S. 71–80

[2.39] Volkmann, W.: Kohlebürsten. Gießen: Schunk & Ebe GmbH 1980

[2.40] Weinert, H.: Kommutierung und Bürstenverschleiß als Optimierungsproblem bei Universalmotoren. Techn. Mitt. der Ringsdorff Werke (1978), H. 9, S. 3–23

[2.41] Wiegel, M.: Prüfung von Universalmotoren-Ankern. F & M 96 (1988), H. 3, S. 91–93

[2.42] T. Kenjo, S. Nagamori: Permanent-Magnet and Brushless DC Motors, Clarendon Press, Oxford, 1985

[2.43] J. R. Hendershot Jr., TJE Miller: Design of Brushless Permanent-Magnet Motors, Magna Physics Publishing and Clarendon Press, Oxford, 1994

[2.44] D. C. Hanselman: Brushless Permanent-Magnet Motor Design, Mc. Graw, Inc., New York, 1994

[2.45] Y. Dote, S. Kinoshita: Brushless Servomotors, Clarendon Press, Oxford, 1990

[2.46] P. Vas, Vector Control of AC Machines: Clarendon Press, Oxford, 1990

[2.47] R. Schönfeld: Digitale Regelung elektrischer Antriebe, Verlag Technik GmbH Berlin, 1990

[2.48] P. Vas: Sensorless Vector and Direct Torque Control, Oxford University Press Inc., New York, 1998

[2.49] V.R. Stevanovic, R. M. Nelms: Microprocessor Control of Motor Drives and Power Converters, IEEE Industry Applications Society, 1993

[2.50] I. Boldea, S. A. Nasar: Vector Control of AC Drives, CRC Press Inc., Boca Raton, 1992

[2.51] H. Kleinrath: Stromrichtergespeiste Drehfeldmaschinen, Springer Verlag, Vienna, 1980

[2.52] G. Müller: Theorie elektrischer Maschinen, VHC, Weinheim, 1995

[2.53] A. E. Fitzgerald, Ch. Kingsley Jr., S. D. Umans: Electric Machinery, McGraw-Hill, Inc., New York, 1990

[2.54] G. R. Slemon: Electric Machines and Drives, Addison-Wesley Publishing Company, Inc., Reading, 1992

[2.55] Papst-Motoren, St. Georgen, Germany: Papst DC Motion, 1997

[2.56] P. Vas: Electrical Machines and Drives, Clarendon Press, Oxford, 1992
[2.57] W. Amrhein: Elektronische Korrekturstromspeisung, Elektrotechnik und Informationstechnik (e&i), 2/1997
[2.58] M. Schröder: Einfach anzuwendendes Verfahren zur Unterdrückung der Pendelmomente dauermagneterregter Synchronmaschinen, etz-Archiv, 10/1988
[2.59] R. Carlson, M. Lajoie-Mazenc, J. Fagundes: Analysis of Torque Ripple due to Phase Commutation in Brushless DC Machines, IEEE Transactions on Industrial Electronics 41, April 2/1994
[2.60] E. Favre, L. Cardoletti, M. Jufer: Permanent-Magnet Synchronous Motors, A Comprehensive Approach to Cogging Torque Suppression, IEEE Transactions on Industry Applications 29 Nov./Dec. 6/1993
[2.61] D. C. Hanselman: Minimum Torque Ripple, Maximum Efficiency Excitation of Brushless Permanent-Magnet Motors, IEEE Transactions on Industrial Electronics 41, June 3/1994
[2.62] W. Amrhein: Mechanisch-elektronische Systemlösungen im Bereich der Kleinantriebe, 16. Int. Kolloquium der Feinwerktechnik, Budapest, Oct. 1-3, 1997
[2.63] H. Weh: Ten Years of Research in the Field of High Density-Transverse Flux Machines, Symposium on Power Electronics, Industr. Drives Power Quantity, Traction Systems, 1996
[2.64] H. Weh: Permanentmagneterregte Synchronmaschinen hoher Kraftdichte nach dem Transversalflusskonzept, etz-Archiv, Vol.10, 1988
[2.65] G. Kastinger: Performance and Design of a Toroid-Coil Motor with Permanent Magnets, Symposium on Power Electronics, Electrical Drives, Advanced Machines, Power Quality, Sorrento, June 1998
[2.66] M. Linke, R. Kennel, J. Holtz: Sensorless Speed and Position Control of Synchronous Machines using Alternating Carrier Injection, IEEE International Electric Machines and Drives Conference IEMDC'03, Madison, USA, 2003
[2.67] K. Hyunbae, R.D. Lorenz: Carrier signal injection based sensorless control methods for IPM synchronous machine drives, IEEE Industry Applications Conference, 3-7 Oct. 2004
[2.68] M. Schrödl: Sensorless control of AC machines at low speed and standstill based on the "INFORM" method, IEEE Industry Applications Conference, 6-10 Oct. 1996
[2.69] S. Shinnaka: New Sensorless Vector Control Using Minimum-Order Flux State Observer in a Stationary Reference Frame for Permanent-Magnet Synchronous Motors, IEEE Transactions on Industrial Electronics, Vol.53, No.2, April 2006
[2.70] P. Vas: Sensorless Vector and Direct Torque Control, Oxford University Press, 1998
[2.71] K. Rajashekara, A. Kawamura, K. Matsuse: Sensorless Control of AC Motor Drives: Speed and Position Sensorless Operation, IEEE Press, 1996
[2.72] M. Schrödl: Sensorless control of permanent magnet synchronous machines - An overview, Proceedings of the 11th International Conference on Power Electronics and Motion Control, EPE-PEMC 2004, September 2004, Riga, Latvia
[2.73] J. Holtz: Developments in Sensorless AC Drive Technology, Proceedings of the 6th International Conference on Power Electronics and Drive Systems, PEDS 2005. November 2005, Kuala Lumpur, Malaysia

[2.74] R.H. Engelmann, W.H. Middendorf: Handbook of Electric Motors, Marcel Dekker, Inc., New York, Basel, Hong Kong, 1995
[2.75] C.-M. Ong: Dynamic Simulation of Electric Machinery, Prentice Hall PTR, New Jersey, 1998
[2.76] W. Leonhard: Regelung elektrischer Antriebe, Springer Verlag, Berlin, Heidelberg, New York, 2000
[2.77] Dirk Schröder: Elektrische Antriebe Grundlagen, Springer Verlag, Berlin, Heidelberg, New York, 2006
[2.78] Dirk Schröder: Elektrische Antriebe Regelung von Antriebssystemen, Springer Verlag, Berlin, Heidelberg, New York, 2001
[2.79] J. Boivie: "Iron loss model and measurements of the losses in a switched reluctance motor", IEE 6th International Conference on Electrical Machines and Drives, 1993
[2.80] A. Brösse: "Sensorloser Betrieb eines geschalteten Reluktanzmotors mittels Kalman Filter" ISBN 3-8265-3949-4, Aachen, 1998
[2.81] G. Henneberger, B. Fahimi, "Predicting the Transient Performance of a SRM Drive System Using Improved Magnetic Equivalent Circuit Method", PCIM Proceedings, Nürnberg, 1995
[2.82] D. Gospodaric: "Parameterbestimmung von Schrittmotoren und SR-Motoren unter Zuhilfenahme der Komponenten- und Systemsimulation", VDI Berichte NR. 1269, 1996, ISBN 3-18-091269-3
[2.83] J.R. Hendershot: "Comparision of AC, Brushless & Switched Reluctance Motors", Motion Control, April 1991
[2.84] E. Hoppach: "Sensorless control of an 8/6 Switched Reluctance drive", Power Conversion & Intelligent Motion, 8–96, Lafayette, 1996
[2.85] H. Huovila, O. Karasti: "A Sensorless SR Motor Position Measurement Method", ICMA'94, Tampere, Finland 1994
[2.86] Kaiserseder, Markus e.a.: Reduction of Torque Ripple in a Switched Reluctance Drive by Current Shaping, Proceedings Symposium on Power Electronics Electrical Drives Automation & Motion (Speedam), Ravello, Italy, 2002
[2.87] TJE Miller: "Switched Reluctance Motors and Their control" Magna Physics Publishing and Clarendon Press, Oxford, 1993
[2.88] TJE Miller: Optimal design of switched reluctance motors, Special Issue of IEEE Transactions on Industrial Electronics, Vol. 49, No. 1, February 2002, pp. 15–27
[2.89] T. Schenke: "Drehmomentglättung von geschalteten Reluktanzmotoren durch eine angepaßte Blechschnittgestaltung", ISBN 3-932633-01-6, 1997
[2.90] H.-J. Wehner: "Betriebseigenschaften, Ausnutzung und Schwingungsverhalten bei geschalteten Reluktanzmotoren", Erlangen, 1997
[2.91] A. Weller, P. Travinski: "Design and control of low power switched reluctance motors" EPE Firenze, 1991
[2.92] Stepina, J.: Die Einphasenmotoren. Berlin/Heidelberg: Springer-Verlag 1982 H. 5, S. 154–155
[2.93] Makowski, K.: Selection of the Running Capacitor Capacitance of a Single-Phase Capacitor Motor. ICEM München 1986, S. 924–927
[2.94] Stölting, H.-D.: Ungleichmäßige Wicklungsverteilung zweisträngiger Wechselstrom-Asynchronmotoren. etz-Archiv 8 (1985), H. 2, S. 61–65

[2.95] Konrad, P., Stölting, H.-D.: Berechnung des dynamischen Verhaltens von Wechselstrom-Asynchronmotoren. Archiv für Elektrotechnik 75 (1992), S. 109–119

[2.96] Brosch, P.F., Tiebe, J., Schudsdziarra, W.: Erwärmung kleiner Asynchronmaschinen bei Betrieb mit Frequenzumrichtern. etz-Archiv 7 (1985), H. 11, S. 351–355

[2.97] Tillner, S.: Auslegung und Betriebsverhalten moderner Spaltpolmotoren. F&M 83 (1975) H. 8, S. 372

[2.98] Stiebler, M.: Ein Modell für Spaltpolmotoren. Elektrie 48 (1994), H. 8/9; S. 299–304

[2.99] Oesingmann, D., Usbeck, S.: Spaltpolmotoren mit einteiligem asymmetrischem Ständerblechpaket. Elektrie 45 (1991), H. 4, S. 141–144

[2.100] Miller, T.J.E., Gliemann, J.H., Rasmussen, C.B., Ionel, D.M.: Analysis of Tapped Winding Capacitor Motor. Record of ICEM-1998, Istanbul, Vol. 2, pp. 581–585; Boldea, I., Dumitrescu, T., Nasar, S.A.: Steady State Unified Treatment of Capacitor AC Motor. IEEE Trans. Vol. EC-14, no. 3, 1999, pp. 557–582

[2.101] Williamson, S., Smith, A.C.: A Unified Approach to the Analysis of Single Phase Induction Motors. IEEE-Trans. Vol. IA-35, no. 4, 1999, pp. 837–843

[2.102] Rasmussen, C.B., Miller, T.J.E.: Revolving-field Polygon Technique for Performance Prediction of Single Phase Induction Motors. Record of IEEE-IAS-2000, Annual meeting

[2.103] Oesingmann, D., Schuder,R.: Zweisträngige Synchronmotoren kleiner Leistung. Innovative Kleinantriebe. VDI Berichte 1269. S. 161–172. Düsseldorf: VDI Verlag 1996

[2.104] Altenbernd, G., Wähner, L.: Kleine, permanenterregte Synchronmotoren mit und ohne elektrische Anlaufhilfe. Innovative Kleinantriebe. VDI Berichte 1269. S. 151–159. Düsseldorf: VDI Verlag 1996

[2.105] Hagemann, B.: Entwurf eines Mikromotors mit Permanentmagneterregung. Diss. U Hannover 1997

[2.106] Honds, L., Meyer, K.H.: Zweipoliger Spaltpol-Synchronmotor mit Hystereseläufer. F&M 86 (1978), H. 4, S. 168–171

[2.107] Gutt, H.J.: Reluktanzmotoren kleiner Leistung. etz-Archiv 10 (1988) H. 11, S. 345–354

[2.108] Brandes, J.: Das Betriebsverhalten eines permanenterregten Synchronmotors mit anisotropem Läufer. Fortschritt-Berichte VDI Reihe 21 Nr. 44. Düsseldorf: VDI-Verlag 1989

[2.109] Binns, K.J.: Permanentmagnet Drives: the State of Art. Proceedings Speedam, Taormina 1994, pp. 109

3
Electromagnetic Stepping Drives

Christian Richter, Thomas Roschke

3.1 Overview

Stepping motors are characterised by a stepwise movement of the rotor with a defined step angle θ_S. Each turn of the rotor can be divided into a defined number z of steps [3.1]. The common range of step angles θ_S lies between 0.36–180° or vice versa between 1000–2 steps per revolution. Step angles $\theta_S > 45°$ can be only found in special motors for watches, which are driven in micro stepping mode. This type of stepping motors cannot be used for positioning tasks.

The wide range of applications shall be shown with some examples:

- In the printing industry for typewriters, printers, XY-plotters, fax machines;
- In computer technology and peripherals for floppy disks, hard disks, CD-ROM-drives;
- In heating, ventilation and air conditioning (HVAC) for the control of flaps, valves and nozzles;
- In the automotive industry for mirror, headlight and seat adjustment, air conditioners and trip recorders;
- In the photo camera and cinema industry for standard and special cartoon projectors, aperture control, shutters, spotlight panning;
- In medical and laboratory equipment for infusion pumps, respiration measurement devices, dialysis machines and sample changers;
- In scientific devices and special purpose machines for x–y–φ-tables, bonding machines, manipulators, small robots, pick-and-place machines for printed circuit boards;
- In measuring instruments and control engineering for card and tape readers, sensor positioning, texture goniometers, sequencers and time delay relays;
- In entertainment for gaming machines, laser games and spotlight panning.

The structure of a stepping drive [1.13] is shown in Fig. 3.1. A characteristic feature of such a stepping motor is the operation in open loop based on its peculiar operating behaviour [3.28].

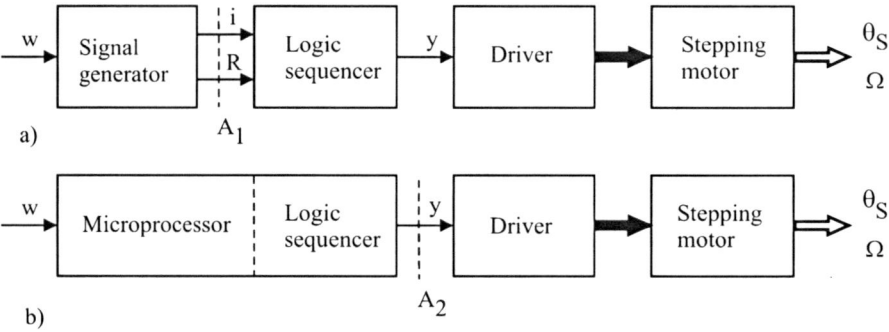

Fig. 3.1. Structure of a stepping drive. **a** With separate logic sequencer. **b** With microprocessor control

The signal generator in Fig. 3.1a can be built either with a simple optoelectronic transmitter or with a counter that stores the programmed number of steps and an oscillator that generates the clock signal i. The stepping motor executes a rotation of size θ_S with each pulse of the clock signal i. Additionally a signal for direction R must be given to control clockwise or counter clockwise rotation. Both signals R and i are supplied via interface A_1 to the "stepping motor specific" control electronics. The logic sequencer is an arithmetic circuit, which derives the control sequence of the motor phases from these both input signals. The driver amplifies the signals and supplies the motor phases. Finally the stepping motor converts electrical energy into mechanical and performs steps with the angle θ_S and angular velocity Ω in the desired direction according to the input signals R and i.

The way to use a microprocessor for the complete generation of pulses up to the direct control signals y for the driver (interface A_2) is shown in Fig. 3.1b. However, the programming of the logic sequencer inside the microprocessor needs memory and is usually slower and afflicted with jitter compared to an arithmetic circuit. Several component suppliers offer complete logic sequencers in commercially obtainable integrated circuits [3.2, 3.3]. Combinations of logic sequencers and drivers inside one IC are available for small power, too [3.15].

Three different types of stepping motors are largely in use:

- Claw-poled permanent magnet (PM) motor (permanent magnetically exited stepping motor, PM stepping motor, tin can motor)
- Variable-reluctance motor (motor with variable reluctance, VR motor)
- Hybrid stepping motor (hybrid derives from the use of both principles, that of the permanent magnet and the variable-reluctance motors)

Characteristic parameter ranges of these three types of construction are given in Table 3.1. Annually a review is listed in [1.18], showing suppliers with their produced motor types including parameter ranges.

Table 3.1. Characteristic parameter ranges of basic stepping motor types

Parameters	Claw-poled PM motor	Variable-reluctance motor	Hybrid motor
Step angle θ_S (°)	6 – 45	1.8 – 30	0.36 – 15
Holding torque T_H (cNm)	0.5 – 25	1.0 – 50	3 – 1000
Max. pull-in rate/start rate $f_{A0_{max}}$ (kHz)	up to 0.5	up to 1.0	up to 3.0
Max. pull-out rate/slew rate $f_{B0_{max}}$ (kHz)	up to 5.0	up to 20.0	up to 40.0

Generally claw-poled PM motors are used as a technically relative simple solution in small power applications with low or medium dynamic requirements. Its use in the automotive industry and building automation for air conditioning and heating equipment has caused two-digit growth rates during the last few years. The annual production worldwide is estimated at 120–150 million pieces.

The most efficient stepping motor with the highest performance is the hybrid motor. Its main field of application is characterised by a torque range up to approximately 100 cNm. It is rarely possible to improve the mass-performance ratio above this value. Motors of low power dominate in computer peripherals. Here the annual production is about 100 million pieces. Motors of higher power are typically used for top-quality positioning tasks, e. g. as required in robotics. Such applications are characterised by a small number of pieces, which leads to higher prices.

Today the variable-reluctance motor is less important, although it can be produced in a cost-effective manner. However, the VR motor shows only a comparatively low efficiency and provides no detent torque, i. e. without voltage supply no torque is provided. Moreover, it tends more than PM stepping motors to generate mechanical vibrations.

Stepping motors are usually produced with two monofilar winding phases for bipolar control or with four phases of bifilar windings for unipolar control. Bipolar control means that the current inside the winding can flow in a positive or negative direction. Meanwhile unipolar control is characterised by only one direction of current flow. These issues are addressed in detail in Sect. 3.3.

During the last few years hybrid motors with three and five winding phases were developed especially for small step angles $\theta_S < 1.8°$ [3.16]. They are characterised by a very low-vibration run. The significant disadvantage is the clearly higher requirement for control electronics. Therefore, such motors are used only in top-quality applications, where costs play a minor role.

3.2 Classification of Stepping Motors

3.2.1 Claw-Poled Stepping Motor

The claw-poled permanent magnet stepping motor (cp. also Sect. 2.4.2.2) is composed of at least two stators [1.13]. Usually the stator parts are made from sheet metal by punching, bending and/or deep drawing. Therefore, it is called in English-speaking countries a "tin-can motor". The windings of each stator are simple ring coils with one phase for bipolar control or two in the opposite direction wound windings with one common tap for unipolar control (Fig. 3.2). The magnetic flux encloses the ring coil. Hence north poles are formed at the claws on one stator side and south poles on the opposite stator side (Fig. 3.3). The magnetic flux path is closed across the magnetic rotor.

Both stators are shifted to each other by a half pole width β_s to achieve a well-defined sense of rotation. The rotor consists of a ring magnet, which is heteropolar magnetised at the circumference. The magnetisation of the rotor is constant in axial direction under both stators. The allocation between stator and rotor poles is shown in Fig. 3.3.

Main parameters of stepping motors using the claw-pole principle can be found in Table 3.1. Versions with four stators lead to higher torque and a smaller step angle down to $\theta_S = 3.75°$. Small torque corresponds to large step angles and *vice versa* due to the working principle. The claw-poled stepping motors offer a very advantageous price-performance-ratio. This is the main reason for their high volume production.

The special design of the claw-poled motor as a linear stepping motor [3.4, 3.5] became more and more important during the last few years (Fig. 3.4). Usually inside the rotor a lead-screw-nut-system is integrated for conversion from rotary into translational motion. A lead screw substitutes for the shaft and the nut is integrated into the rotor. The lead screw must be secured against rotation inside or outside the motor. When the stepping motor performs a step with the value of angle θ_S, the shaft will move according to its

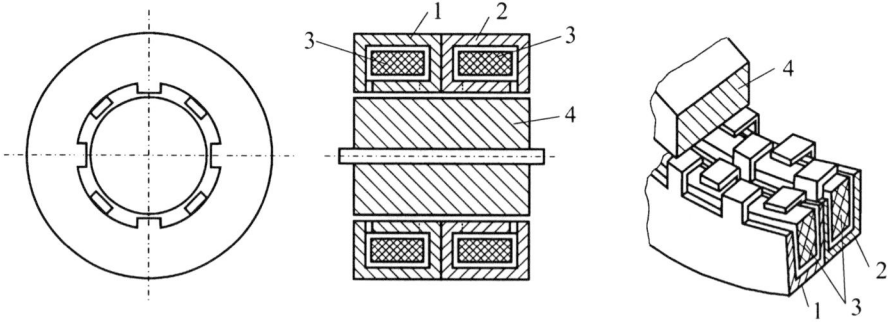

Fig. 3.2. Structure of a claw-poled stepping motor: *1* stator 1 with claw poles; *2* stator 2 with claw poles; *3* winding; *4* permanent magnet

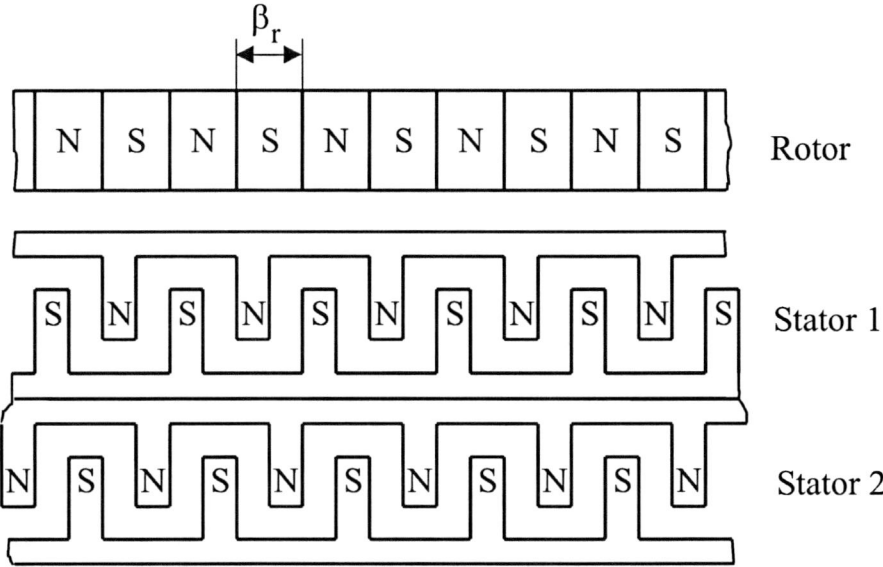

Fig. 3.3. Split-and-unrolled model of a claw-poled stepping motor

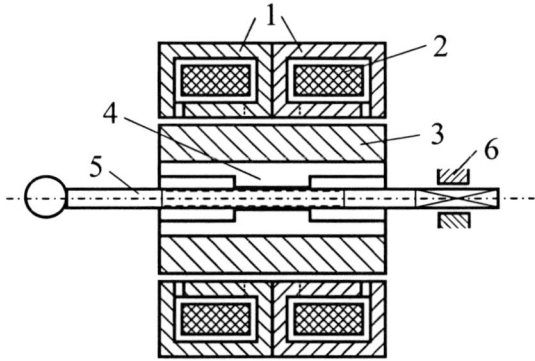

Fig. 3.4. Structure of a linear drive, based on a rotary claw-poled stepping motor and a converter from rotation into translation; *1* stator; *2* winding; *3* permanent magnet; *4* hub with internal thread; *5* shaft with thread and carrier/coupling element; *6* anti-rotation lock

pitch with a linear displacement x_S. Characteristic values of such linear steps are $x_S = 0.015$–0.050 mm. Benefits of such a motor design can be found in the possible use of the main parts of a rotary motor, the simple variation of linear step width x_S with constant step angle θ_S because of different pitch of the thread and the self-locking feature of such a lead-screw system. Therefore, energy is only needed during actuation. This leads towards a good overall-efficiency and low temperature rise. The wide use of such linear motors in the automotive industry for headlight adjustment and in building automation for miscellaneous valve actuation units is based on a good price-performance ratio, the sensitive positioning behaviour in open-loop control and the low energy consumption. Moreover, it is well suited for the actuation of small piston pumps, which are used to dose very small volumes of fluids.

3.2.2 Disk Rotor Stepping Motor

An interesting engineering design is the disc-type stepping motor [3.6, 3.7, 3.17] (Fig. 3.5). Its special feature is the extremely low inertia, because the rotor is made from a very small permanent magnetic ring, which is fixed with a support plate at the shaft. This design was only possible with the availability of permanent magnetic material with high energy density (e.g. based on NdFeB). Its high coercive field strength assures a short axial length.

C-shaped stacked sheets enclose the rotor. This rotating disk is axially magnetized with a high number of poles. North and south poles alternate at the rotor surface. The stacked sheets are very narrow and show only the width

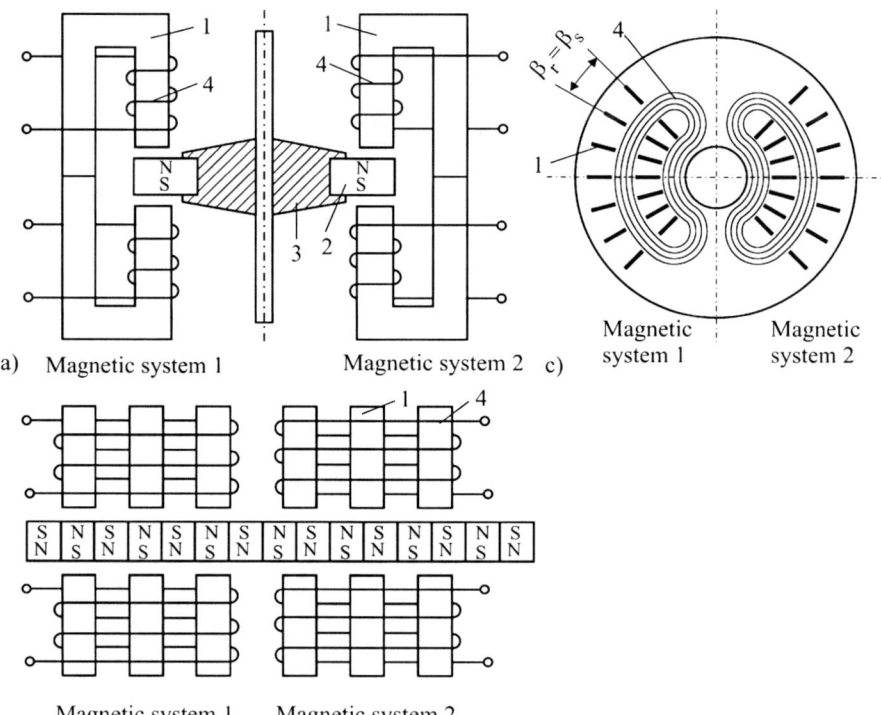

Fig. 3.5. Structure of a disk rotor stepping motor with permanent magnetic excitation. **a** Longitudinal section. **b** Developed view. **c** Top view on a stator from the rotor position. *1* C-shaped stacked sheets; *2* permanent magnetic ring; *3* support plate; *4* winding

of a rotor pole. Several laminated cores are in each half of a stator to take best advantage of the motor volume and generate a high torque. Figure 3.5 illustrates that the laminated cores of the magnetic system have the distance of a pole pitch from each other. The laminated cores of the second half of the stator are shifted half a pole pitch comparable to the claw-pole stepping motor.

The electrical excitation is generated by the windings, which could be placed at each single core (see Fig. 3.5a). However, the design allows exciting electrically each half of a stator with only one coil. Therefore, only two sickle-shaped coils are used for each half of a stator (Fig. 3.5c). Coils and lamination are cast with a sealing compound for fixation.

The significant advantages of this design are based on the extremely low inertia, the easily manufacturability of the coils and the small step angle θ_S of 1.8° comparable to hybrid stepping motors. The high manufacturing costs are unfavourable.

3.2.3 Reluctance Stepping Motor

Only reluctance stepping motors in single-stack design achieved practical importance. These motors are based on the reductor principle and show a similarity in design to switched reluctance motors (cp. Sect. 2.3). The pole pitch of the stator and the tooth pitch of the rotor are different (Fig. 3.6). The reductor principle says that the angular velocity of the magnetic field of the stator is larger than the mechanical angular velocity of the rotor. Between both angular velocities exists a fixed and invariant relation. This will be explained in detail when the operating behaviour is discussed in Sect. 3.4.1.

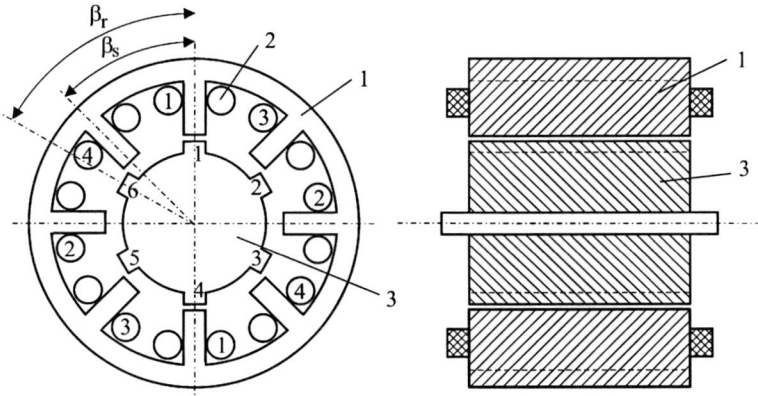

Fig. 3.6. Structure of a single-stack reluctance stepping motor: *1* laminated core with eight salient poles; *2* unipolar winding (the numbers designate the winding phases); *3* reluctance rotor with six salient poles

The stator of such a reluctance stepping motor is similar to that of a synchronous motor with continuously rotating field. However, the single winding phases are only wound around one pole or tooth, respectively. In synchronous motors the windings always enclose several teeth. Usually the control of these windings is unipolar. Nevertheless bipolar control is possible. The rotor is a simple reluctance armature made only from iron. Its impact is based on the reluctance differences in the regions of the slots and teeth. Consequently this motor is called variable-reluctance motor (VR-motor). For higher stepping rates the rotors must be laminated, at lower stepping rates massive rotors are feasible. At least three winding phases are necessary for an operation with a defined direction of rotation. Main parameters can be found in Table 3.1.

3.2.4 Hybrid Stepping Motor

The stator of a hybrid stepping motor (Fig. 3.7) corresponds to that of a reluctance stepping motor (Fig. 3.6).

The rotor consists of an axially magnetized permanent magnet, which is enclosed with two toothed pole caps or is positioned between two toothed disks [3.18]. The positioning of several equal magnetic systems behind each other in the axial direction is well known, too. Generally the teeth of one pole cap hold the same polarity. The two pole caps are shifted from each other by half a tooth pitch. Similar to the rotors of reluctance machines the pole caps are laminated for high stepping rates or made in solid iron for low frequencies.

The windings of one phase are connected in the circumferential direction in such a way that alternating north and south poles are created. Assuming that the top winding of phase 1 in Fig. 3.7 is generating a north pole, the opposite tooth of the pole cap in the foreground must display together with all other

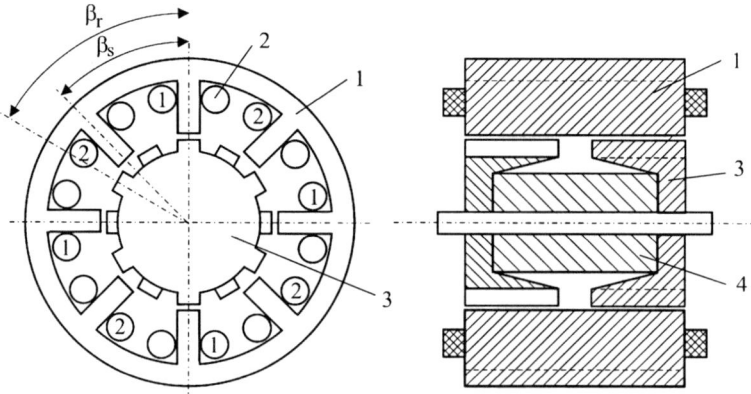

Fig. 3.7. Structure of a hybrid stepping motor: *1* laminated core with salient poles; *2* bipolar winding (the numbers designate the winding phases); *3* toothed pole cap; *4* permanent magnet

teeth of this pole cap a south pole. The second winding of phase 1 (on the right in Fig. 3.7) generates according to the above statement a south pole and is positioned across from one of the teeth of the rear pole cap, which exhibits at all teeth a north pole. In analogy the third tooth of phase 1 (bottom in Fig. 3.7) will generate a north pole again and corresponds to a south pole tooth of the front pole cap and so on. Due to that principle it is possible that, despite the axially aligned stator teeth, both rotor parts with different magnetic polarity will generate torque. The tooth pitch in the stator and in each pole cap of the rotor is identical with that of a reluctance motor (cp. Fig. 3.6).

Because of a required minimum width of slots and teeth it is not possible to manufacture step angles $\theta_S < 7°$ with a design according to Fig. 3.7. To reach smaller step angles it is necessary to arm the hybrid stepping motor at the air gap with a finer tooth pitch (Fig. 3.8). This subtle tooth pitch must be identical on stator and rotor poles. The angular distance between

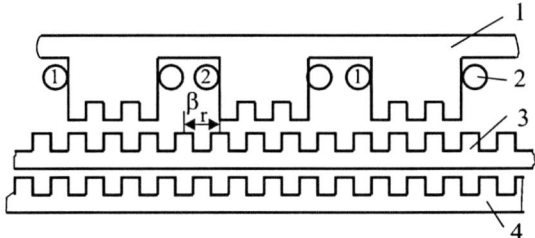

Fig. 3.8. Split-and-unrolled model of a hybrid stepping motor with a step angle $\theta_S < 7°$: *1* developed view of the stator; *2* winding (the numbers designate the winding phases); *3* developed view of the first pole cap; *4* developed view of the second pole cap

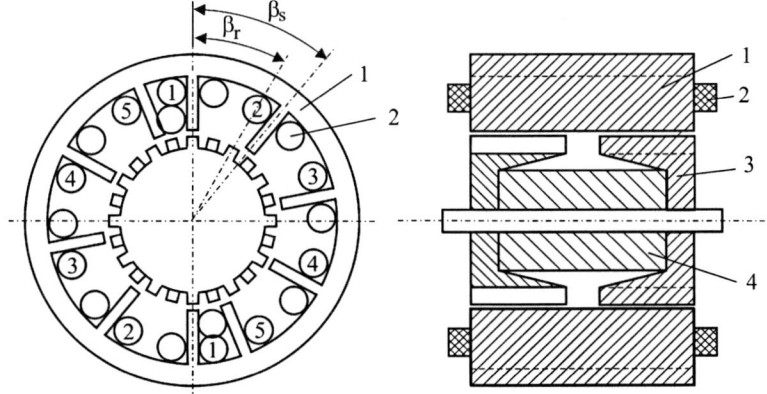

Fig. 3.9. Structure of a 5-phase hybrid stepping motor: *1* laminated core with salient poles; *2* bipolar winding (the numbers designate the winding phases); *3* toothed pole cap; *4* permanent magnet

the stator poles belonging to one phase is to arrange, that an alternating interaction is formed with the two pole caps which are twisted to each other. From the comparison of Figs. 3.7 and 3.8 it can be easily found that switching the magnetic field from one phase to the second causes a remarkably smaller angular movement of the rotor according to the smaller tooth pitch, although the stator pole distance is approximately the same.

Table 3.1 shows the range of parameters with the focus on hybrid stepping motors with step angles $\theta_S \leqq 1.8°$, holding torque $T_H \leqq 100\,\mathrm{cNm}$ and two phases in bipolar control.

With a five-phase hybrid stepping motor a very low-vibration run, high efficiency and high torque can be reached. The basic structure of such a motor is shown in Fig. 3.9. With 50 teeth per pole cap it offers a step angle of $\theta_S = 0.72°$ in full step operation, i. e. it performs 500 steps per turn. The control is generally bipolar, which needs an appropriate high level of electronics.

3.3 Control of Stepping Motors

The stepping motor is part of an electrical drive system (Fig. 3.1), which could only be reasonably carried out by means of electronic control. The breakthrough of this motor type was linked to its use in computer peripherals starting at the end of the 1970s. The two input signals clock i and direction R must be treated electronically and amplified (cp. Sect. 6.2.3) [3.8, 3.9, 3.26, 3.29] to supply the single winding phases with electrical power in such a way that the stepping motor works as an electromechanical energy converter and generates the required torque as well as makes steps in the desired number and direction.

Most of the stepping drives have two electromagnetic systems or winding phases, respectively, and therefore need a logic sequencer, which generates from the two input signals the four output signals $y_1 - y_4$ for the drivers (Fig. 3.10).

The full step mode is characterized by simultaneous operation of the same number of winding phases during each step and input signal, respectively. In common motors with two electromagnetic systems, both phases are energised at any given time. This is demonstrated in the sequence of output signals y_1–y_4 by the simultaneous change of the associated signals y_1 and y_2 or y_3 and y_4, respectively (Fig. 3.11). In wave mode only one phase is powered

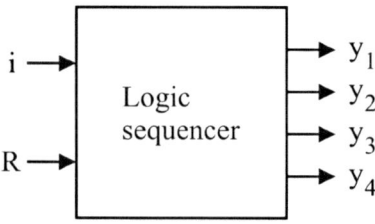

Fig. 3.10. Block diagram of a logic sequencer: i clock signal (input signal); R direction signal (input signal); y_1–y_4 output signals for control of the drivers

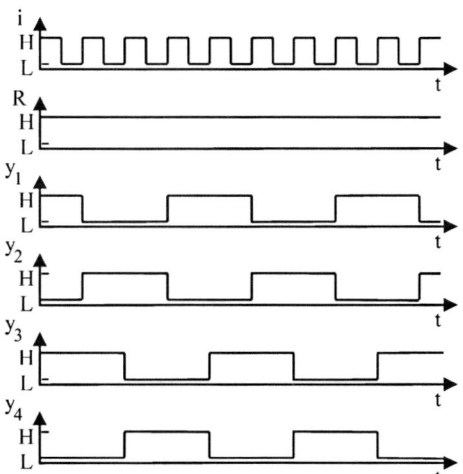

Fig. 3.11. Allocation of input and output signals in full step mode with continuous direction of rotation ($R =$ constant)

at any given time. The motor performs a full step as well, but only 50% of the copper is used and approximately 0.6–0.75 of the torque is generated in bipolar control with the same winding parameters.

In half step mode the number of switched winding phases changes from step to step by one. In common motors with two electromagnetic systems alternating one or two phases are energised. Hence always only one signal is switched at one time (Fig. 3.12).

Each electromagnetic system is carried out with one phase for bipolar control or with two inversely wound phases (bifilar winding) for unipolar control.

Unipolar control of winding phases (Fig. 3.13) stands for always one and the same direction of current flow in the coil. The current is alternately conducted inside the two in opposite direction wound phases for positive and negative directions of the magnetic flux (Figs. 3.15 and 3.16). The logic sequencer has to prevent a simultaneous current flow through both coils of a unipolar electromagnetic system, because in this case both components would compensate each other.

During bipolar control of winding phases (Fig. 3.14) the electrical current is alternately conducted in both directions through the coil. Therefore, magnetic fluxes of both polarities are established (Figs. 3.15 and 3.16). For bipolar supply a winding phase has to be connected inside a full bridge or H-bridge with four transistors.

More comments about sequences of control signals and current flow for alternating signal of direction R and micro step mode as well as control circuits for three-phase and five-phase stepping motors can be found in Sect. 6.2.3 and [1.13, 3.8], respectively.

The time-dependent behaviour of the current should be paid some attention, because stepping drives operate up to frequencies of some kHz. Each winding phase exhibits an ohmic resistance R and an inductance L. The tran-

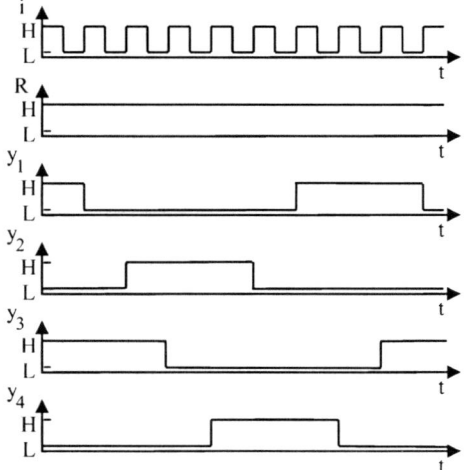

Fig. 3.12. Allocation of input and output signals in half step mode with continuous direction of rotation ($R =$ constant)

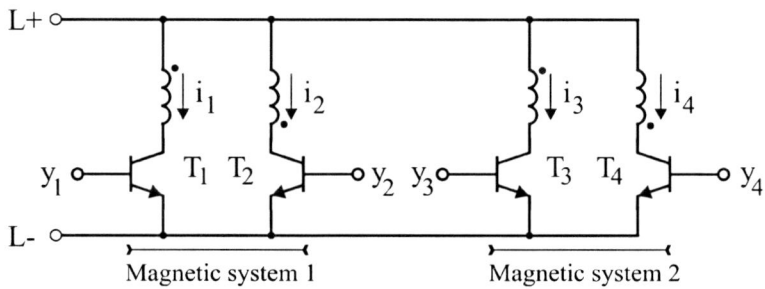

Fig. 3.13. Unipolar control of a winding phase

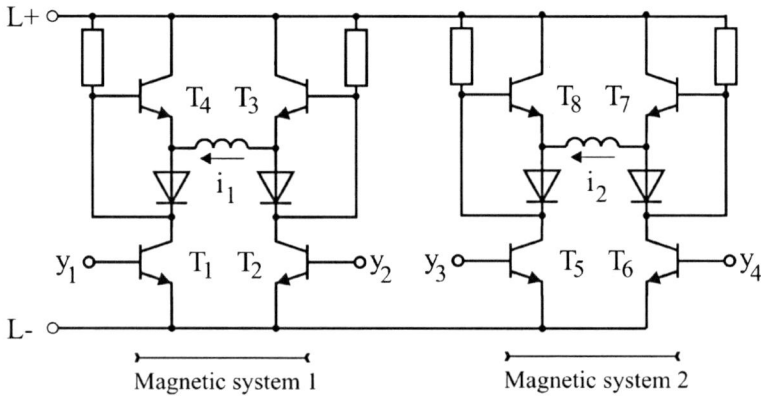

Fig. 3.14. Bipolar control of a winding phase

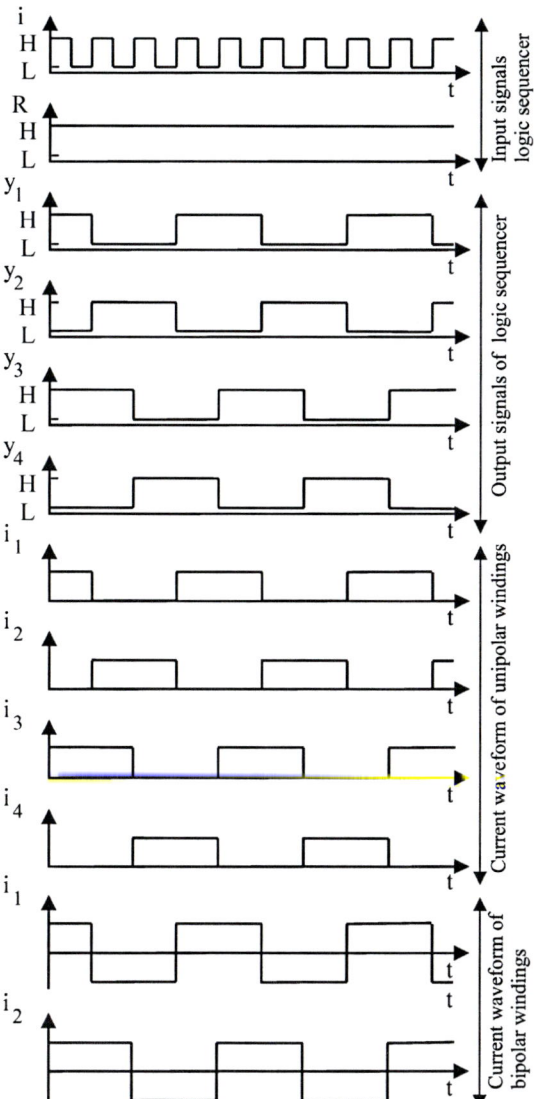

Fig. 3.15. Currents in unipolar and bipolar excited windings in full step mode with continuous direction of rotation

sistors will be considered in a first approximation as ideal switches, i.e. no voltage drop and no switching delay. If a voltage step of size U_0 is applied, the known time domain behaviour of the current $I(t)$ is given by

$$I(t) = I_0(1 - e^{-t/\tau_{el}}) \tag{3.1}$$

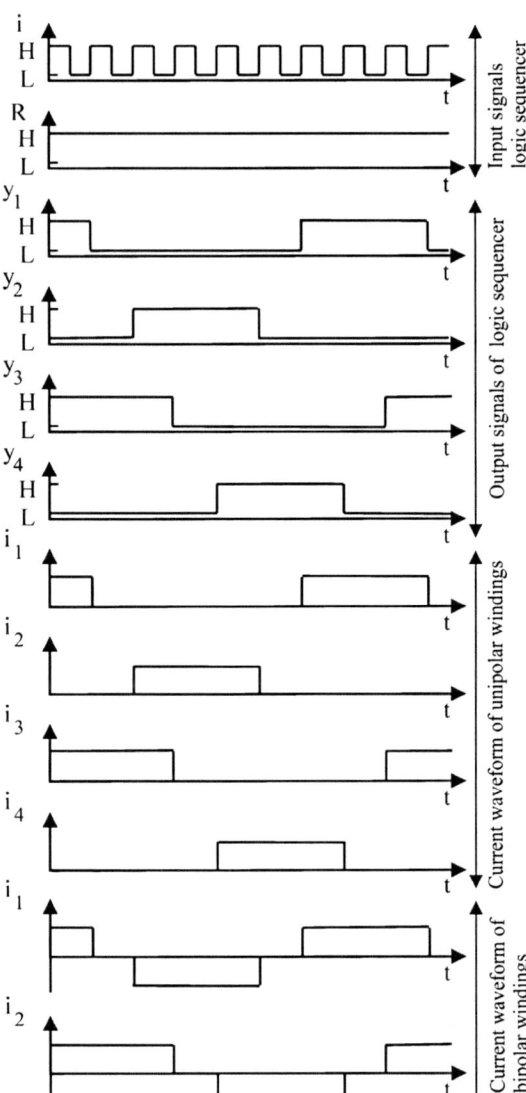

Fig. 3.16. Currents in unipolar and bipolar excited windings in half step mode with continuous direction of rotation

with the final value of the current $I_0 = U_0/R_0$ and the electrical time constant $\tau_{el} = L/R$. This current behaviour is valid only in the ideal case. Feedback effects caused by the rotor movement and eddy currents are neglected for these basic considerations. The current slope at time $t = 0$

$$\left.\frac{dI}{dt}\right|_{t=0} = \frac{I_0}{\tau_{el}} = \frac{U_0}{L} \qquad (3.2)$$

is determined by the applied voltage U_0 and the inductance L of the winding.

When the feedback effects described above are neglected the magnetic flux rises proportionally to the current and generates the torque. Precondition for high stepping rates is a strong slope of the current. Figures 3.17–3.20 present some options of circuits for unipolar control and the corresponding current vs time characteristics. Table 3.2 summarizes benefits and drawbacks.

A high supply voltage is, according to Eq. (3.2), the basis of a fast current slope. At the same time it should be noted that the maximum allowable thermal current I_0 is not exceeded to avoid damage of the motor by improper heating (cf. Sect. 9.3.2). A resistance R_v connected in series to the winding is used to limit the current (Fig. 3.17):

$$I_0 = \frac{U_0}{R_0 + R_v}. \tag{3.3}$$

In Fig. 3.18 a capacitor is connected in parallel to the coil. The capacitor will load to supply voltage level while the transistor is open. The stored energy of the capacitor is discharged across the winding when the transistor is switched on. This energy leads to a very fast current slope and can cause a short-term overcurrent. After discharging of the capacitor, the current will settle according to Eq. (3.3).

With the bi-level control according to Fig. 3.19 the voltage is excessively enhanced. First, a very high voltage is applied to terminal L1+ of the winding. After reaching the permissible current I_0, only voltage L2+ is supplied to the winding.

All these circuits for the control of unipolar windings are used because of its simple structure. However, one essential point has not been taken into consideration until now. The magnetic flux arises from the resulting magne-

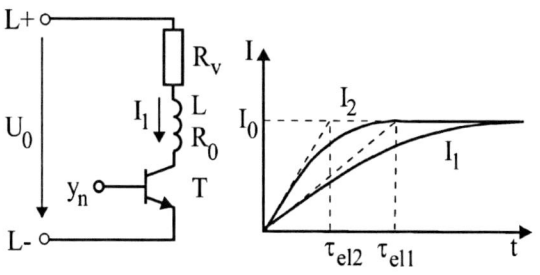

Fig. 3.17. Circuit of a unipolar controlled winding phase and its current characteristic: *Index 1:* current and time constant without series resistor; *Index 2:* current and time constant with series resistor R_v

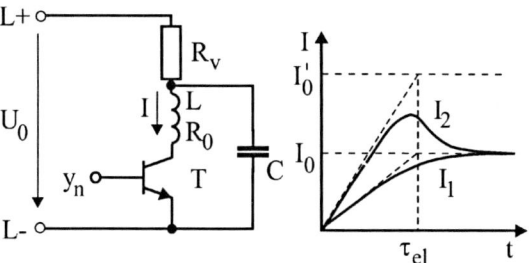

Fig. 3.18. Circuit of a unipolar controlled winding phase with shunt capacitor and series resistor and its current characteristic

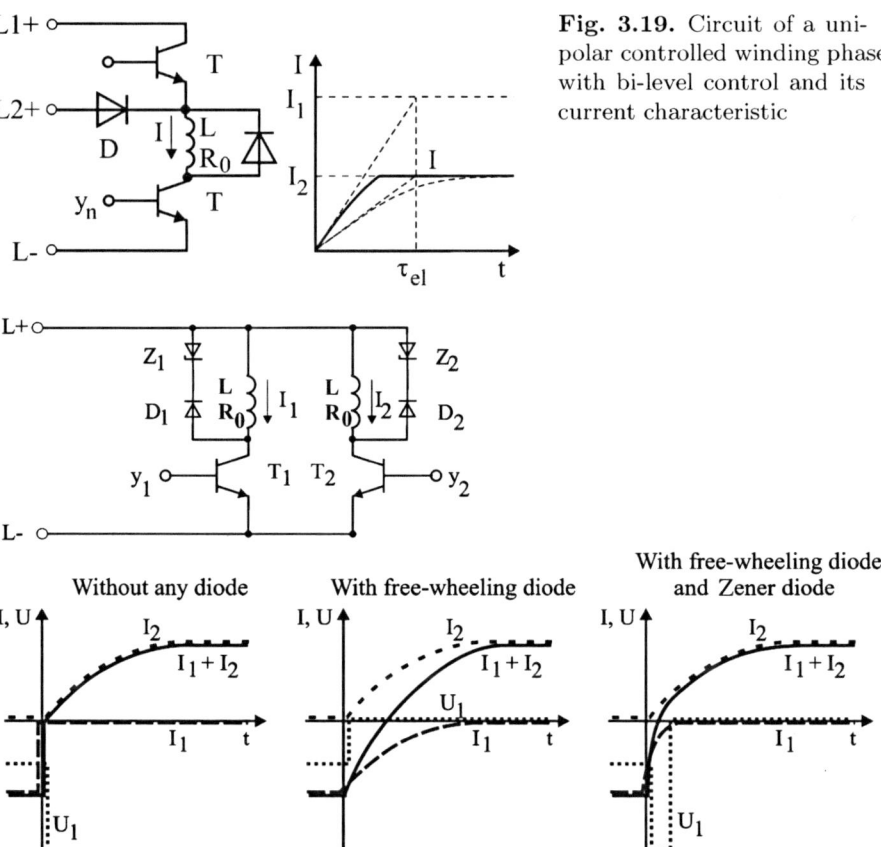

Fig. 3.19. Circuit of a unipolar controlled winding phase with bi-level control and its current characteristic

Fig. 3.20. Circuit of a unipolar controlled phase with freewheeling and Zener diode and its current and voltage characteristics

tomotive force, which is generated by the two phases of the unipolar winding. Both winding phases are wound in the opposite direction (bifilar winding) so that in the moment of change over between the phases both current components have to be considered. These coherences are shown in a simplified way in Fig. 3.20. Assuming that there is no snubber circuit with a freewheeling diode and the transistor T_1 would be an ideal switch, the current I_1 would go down abruptly to zero and only current I_2 in the newly connected phase could build up the magnetic flux. Simultaneously the voltage in the disconnected phase would have to go to an infinite value because of the sudden current decay. To avoid this impermissible voltage stress for the electric components, freewheeling diodes are used. The current will decay with time constant $\tau_{el} = L/R_0$ which is determined only by the winding parameters if ideal freewheeling diodes are assumed. The current vs time characteristics in Fig. 3.20 shows that only after a certain delay the sum of currents $I_1 + I_2$ changes the algebraic sign. For this reason the desired

Table 3.2. Features of unipolar snubber circuits for acceleration of current slope

Circuit	With series resistor R_v	With series resistor R_v and shunt capacitor C	Bi-level circuit	With freewheeling diode and Zener diode
Fig.	3.17	3.18	3.19	3.20
Benefits	Simple circuit, low costs	Simple circuit, low costs, increase of maximum stepping rate	Low losses, strong increase of maximum stepping rate	Low losses, low costs, strong increase of maximum stepping rate
Drawbacks	Joule's loss of R_v, only usable for low power, small increase of maximum stepping rate	Joule's loss of R_v, only usable for low power	Two supply voltages, complexity of circuit, high costs	Joule's loss of Zener diode

change of the direction of magnetic flux follows only after a sizable delay. There is the option to connect in series to the freewheeling diode a Zener diode. When the first winding phase is switched off the voltage increases up to the breakdown voltage of the Zener diode. Subsequently the current I_1 decays much faster than in the previous case, because the energy of the circuit is converted to heat inside the Zener diode. Consequently the time until the change of sign of the total current sum $I_1 + I_2$ can be remarkably reduced. The chosen breakdown voltage of the Zener diode is a measure of this effect. Moreover, it is not necessary to attach to each winding branch a single Zener diode. One Zener diode for all four winding phases of a unipolar stepping motor is enough, since all Zener diodes would be connected to the potential $L+$. However, the power rating of the Zener diode should receive attention.

Usually chopper circuits are used for current control of bipolar windings (see Fig. 3.21). Very seldom are choppers for current control used in a modified shape in unipolar control circuits. A high voltage leads towards a fast current slope. After reaching the upper limit or threshold of the current I_{og}, the winding phase is disconnected from the supply voltage. The current is conducted through the freewheeling diode and decays towards the lower limit I_{ug}. Thereafter the supply voltage is connected again. Thus the current oscillates around the averaged value of I_{og} and I_{ug}. The chopper frequencies are usually chosen higher than 13 kHz to exceed the human audibility limit. The rated voltage of the winding is chosen at a ratio of 1:5 to 1:10 to the supply voltage of the chopper circuit. The essential advantages of this circuit are the very high switching frequencies of up to 40 kHz, the small additional loss and that

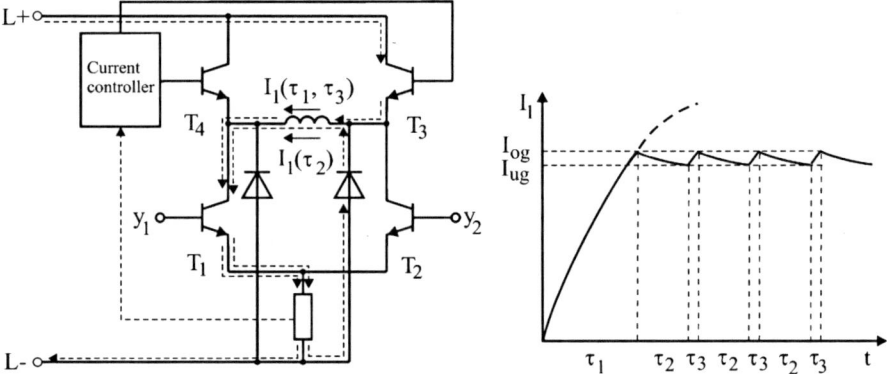

Fig. 3.21. Circuit and current characteristic of a bipolar controlled winding phase with current control (chopper)

Table 3.3. Comparison of different control types of windings

Type	Unipolar control	Bipolar control
Application	Small power, small frequencies	Small to high power, higher frequencies
Benefits	Simple structure, low costs	All winding phases simultaneously carry current, no additional power components for chopper control necessary, small power losses, with chopper control the current does not depend on the temperature, i.e. no influence on the torque
Drawbacks	Only 50% of the winding phases are simultaneously carrying current, additional loss caused by series resistors, current is reduced by heating, i.e. torque is reduced, too. Electromagnetic coupling between phases reduces torque and adds losses	Complexity of the circuit, higher costs

both upper power transistors work at the same time as controllers. Therefore, no additional power components are needed.

Table 3.3 summarizes the most important points for the application of unipolar and bipolar control circuits.

3.4 Operating Behaviour of Stepping Motors

3.4.1 Step Angle

At first the full step mode is considered, which is characterized by a simultaneous current flow inside the same number of winding phases. This means that all phases are energised at any given time.

For a claw-poled stepping motor (Fig. 3.2) a determined distribution of magnetic poles is shown in Fig. 3.3. The direction of the magnetic field of stator 1 has to change for a rotor movement towards the right-hand side with the next input signal i. This means for unipolar control that phase 1 will be de-energized and the in opposite direction wound phase 2 of stator 1 will conduct current. In bipolar control the direction of current flow inside phase 1 must be reversed. Starting from the original position a rotor movement towards the left-hand side is performed, when at first the direction of the magnetic field of stator 2 is reversed. Only the adequate currents inside the phases of stator 2 can cause this. The rotor with its polarity aligns itself towards the new allocation of the stator polarities due to the magnetic forces, i.e. it turns with an angle of $\beta_r/2$ to the left. Hence the step angle θ_S can be calculated from the rotor pitch β_r and the number of winding phases m:

$$\theta_S = \frac{\beta_r}{m} \text{ (bipolar control)} \qquad \theta_S = \frac{2\beta_r}{m} \text{ (unipolar control)}. \tag{3.4}$$

For the reluctance stepping motor of Fig. 3.6 it should be initially assumed that the stator poles 1 and 3 are electrically excited. The rotor teeth 1 and 2 or 4 and 5, respectively, will align towards the centre between the mentioned stator poles due to the magnetic attraction forces. That means the rotor will turn with an angle of 7.5° towards the left starting from the drawn position. Next the current is switched from pole 1 to pole 2. The rotor teeth 2 and 3 or 5 and 6, respectively, will align towards the stator poles excited from phase 2 and 3 and will conduct the magnetic flux. The step angle results from the difference of stator and rotor pole pitch:

$$\theta_S = \beta_s - \beta_r. \tag{3.5}$$

For the motor of Fig. 3.6 a step angle of 15° can be calculated.

The hybrid stepping motor (Fig. 3.7) can be considered in the same way as the reluctance motor. The winding sense of the coils of consecutive poles of one phase is always contrary. Thus magnetic fluxes of alternating polarity will be generated from these poles when the current is conducted through the phase. The following case is considered as a starting point. Both vertical teeth of phase 1 carry north poles, whereas the horizontal teeth generate south poles. Starting from the top tooth of phase 1 the adjacent pole of phase 2 in a clockwise direction and its *vis-à-vis* tooth generate, in turn, north poles. Thus both the thereto perpendicularly arranged teeth of phase 2 have to establish south poles. If the front pole cap of the rotor forms south poles, the rotor will turn with an angle of 7.5° compared to the drawn position in a mathematically positive direction; similary the rotor of the reluctance motor of Fig. 3.6. According to the assumption made, the teeth of the rear pole cap will show north poles. In this position the north poles of the stator will be aligned with the slots of the rear pole cap. When the matching of the stator and rotor teeth is tracked along the perimeter, it is discovered that rotor and stator teeth are always opposite to each other when different magnetic poles exist. A stator tooth will be lined up with a rotor slot if the same polarity exists. For this reason it is understandable that both pole caps of the rotor must be shifted against each other half a tooth pitch. From the above explanations it can be concluded that, comparable to the reluctance motor, the step angle is given by Eq. (3.5).

It is not possible to manufacture stator geometries of hybrid stepping motors with step angles smaller than 7° at reasonable costs. Therefore, fine teeth are introduced on the stator poles (Fig. 3.8). The tooth pitch of stator and rotor have to be identical to reach the maximum force. In order that the rotor is performing exactly one step after changing over of winding phases, adjacent poles have to be shifted to each other in analogy to Eq. (3.5) with the sum of step angle and rotor tooth pitch. The active principle is not destroyed when $(n-1)$ rotor tooth pitches are missed out. Hence the step angle yields

$$\theta_S = \beta_s - n \cdot \beta_r . \tag{3.6}$$

Alternately one or two winding phases are conducted in half step mode. In the full step mode described above where two winding phases always carry current, the rotor is directed to the centre position between adjacent poles. If only one winding phase is conducted, the rotor will align directly with the appropriate pole of the stator. Therefore, it performs only half a step from the position with two energized phases to the position where only one phase is energized. Advantages and disadvantages of half step mode are explained below.

With the geometrical layout the step angle θ_S of a stepping motor is fixed and can be calculated from Eqs. (3.4)–(3.6). This step angle is also named, in many cases, basic step angle. By a simple change of the electronic control in the logic sequencer this basic step angle can be divided in half.

The step angle of all kinds of stepping motors can be calculated using the following relation:

$$\theta_S = \frac{360°}{2p \cdot m \cdot k} \tag{3.7}$$

with

$2p$ number of poles of an electromagnetic system of a PM stepping motor, or

p number of rotor teeth of a reluctance stepping motor or number of teeth of one pole cap of a hybrid stepping motor,

m number of electromagnetic systems or winding phases,
$k = 1$ for full step mode; $= 2$ for half step mode.

3.4.2 Torque

The positions described in Sect. 3.4.1 are stable operating points of the stepping motor (Fig. 3.22). In this situation the appropriate magnetic poles of stator and rotor are aligned. Thus no torque is generated. If the rotor is moved from this equilibrium position to the left or right-hand side a torque is produced, which tries to turn back the rotor towards its rest position (Fig. 3.22). The preconditions for a stable operating point, $\Sigma T = 0$ and $dT/d\theta < 0$, are fulfilled. In the literature the torque vs angle characteristic is often described as a sinus function. However, the typical and – from the designer point of view-intended characteristic is more a trapezoidal function. The steep slope of the torque characteristic during zero crossing reduces the torsion from the zero point under the influence of external torque. After a rotation of $\pm 2\theta_S$ (generally valid is $\pm m\theta_S$ for bipolar and $\pm m\theta_S/2$ for unipolar control), an unstable equilibrium is reached ($dT/d\theta > 0$). In this case equal poles of stator and rotor are opposite to each other and the system achieves its maximum energy. If this unstable angular position is passed the torque changes its effective direction. Consequently the rotor is accelerated and moves towards the next stable operating point at an angle of $\pm 4\theta_S$. This means that if the maximum torque of the stepping motor, the holding torque T_H, is exceeded by external torque, the stepping motor will keep turning without any signal from the logic sequencer until it cannot go back to its initial position. Therefore, a step error arises. It is easy to see that step errors of one or two steps can never happen with a current flow through the winding phases according to a given signal pattern of the logic sequencer. The step error will be always a multiple of four steps.

Due to manufacturing tolerances not all crest values of the torque function along the perimeter are equal. Correspondingly the smallest amplitude will be defined as the holding torque T_H; i. e. at least this torque can be generated with electrical excitation at a standstill. Usually the holding torque is given for full step mode. Figure 3.22 is valid for full step mode, too, i. e. all winding phases carry current at any time.

In half step mode the number of excited phases alternates between all and half. For common designs with two phases in each second step only one phase is excited. Accordingly the holding torque T_{H1} of this state will be smaller. The ratio of torque between half and full step mode is in the range between $T_{H1}/T_H = 0.6$–0.75 because the function of torque has no ideal sinus shape and according to the actual angular positions the maximum of the algebraic superposition of both torque components is not always achieved in full step mode. Consequently the torque in half step mode varies between a full and a reduced value. This situation can be improved by current control of the winding phases. Thereby the current is increased in the case when only one

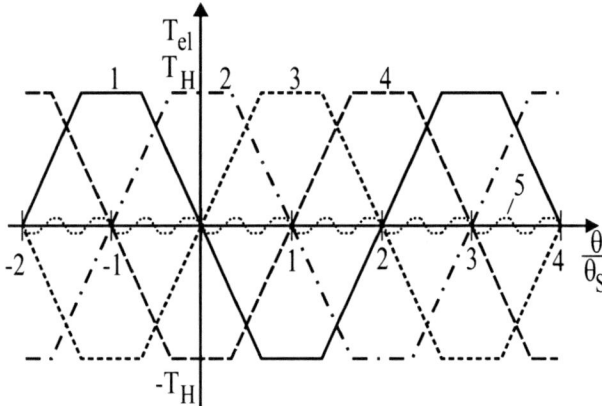

Fig. 3.22. Static torque as a function of the rotor angle for different control states: *1–4* torque with electrical excitation (the numbers designate the different excitation sequences of the winding phases); *5* torque without electric excitation (detent torque without friction)

phase is switched on in such a way that the ratio $T_{H1}/T_H = 1$ is reached. Most often this is referred to as compensated half step mode. For five-phase stepping motors the angular proportions are more favourable. In this case the torque difference between half and full step mode is only about 5% even without current control.

Permanently excited stepping motors (Figs. 3.2, 3.5 and 3.7) generate, similar to synchronous motors, a reluctance torque even without voltage supply. With stepping motors this torque is called detent torque T_s and is composed of a magnetic and a friction component. It typically ranges from $T_s = (0.05–0.30)T_H$.

3.4.3 Positioning Accuracy

For an ideal and unloaded stepping motor the zero crossings of the torque function are uniformly distributed along the perimeter. For this reason the motor has to perform with each input signal exactly one angular step θ_S. The real stepping motor possesses a friction torque T_f. If the electrically generated torque is smaller than the friction torque, the stepping motor cannot continue moving towards the ideal zero point (Fig. 3.23). Moreover, the zero crossings of torque are not ideal distributed along the perimeter due to magnetic and electric asymmetries. The real stepping motor will stop inside a range of $-\Delta\theta_S \leqq \theta \leqq +\Delta\theta_S$. The maximum deviation from the ideal position of an unloaded stepping motor is called positioning accuracy $\Delta\theta_S$.

Figure 3.24 shows possible angular positions in which an unloaded stepping motor could stop. The actual deviation from the desired position is a statistical value. Most important is that this error is not accumulated when several steps are performed. This inaccuracy is completely independent of the number and the direction of the performed steps and gives the maximum permissible final deviation. Common values amount to $\Delta\theta_S = (0.02\text{--}0.10)\theta_S$.

3.4.4 Step vs Time Characteristic

Assume that in a passive state an unloaded stepping motor is electrically excited due to curve 1 according to Fig. 3.22 and stands at the angular position $\theta/\theta_S = 0$. If now an input pulse is given to this stepping motor the next pattern of electrical excitation will be effective, i.e. for a clockwise rotation the curve 2 becomes valid. The positive torque of this curve at angular position $\theta/\theta_S = 0$ accelerates the stepping motor towards the angle $\theta/\theta_S = 1$. If no new pulse arrives the stepping motor will oscillate around this point until damping effects bring it to a halt.

The motion equation of this system is given by

$$T_{el} = T_d + J \cdot \frac{d^2\theta}{dt^2} \tag{3.8}$$

whereas the damping torque is $T_d = T_L + T_f + k_d \frac{d\theta}{dt}$

T_L static part of the load torque of the driven mechanism
T_f friction torque of the stepping motor
$k_d\, d\theta/dt$ speed proportional damping of the system

The described differential equation is nonlinear and cannot be solved in a closed shape. Common approaches for the solution are numerical methods or the phase portrait [3.10]. These solutions are not general, i.e. they are only valid for a specific case with fixed parameters. For the transient behaviour around the zero point the generated electromagnetic torque T_{el} can

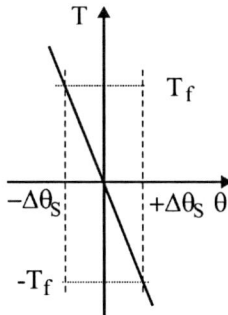

Fig. 3.23. Positioning accuracy as a result of friction torque

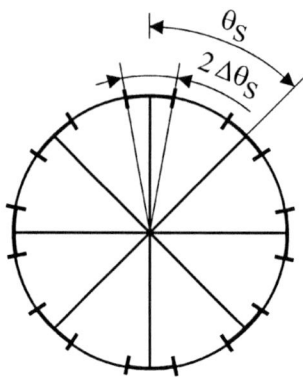

Fig. 3.24. Step angle taking into account the positioning accuracy

be linearised to

$$T_{el} = -k \cdot \theta. \tag{3.9}$$

Presumption for this linearisation is a steady-state approach, i.e. no feedback effect is caused by the rotor movement on currents and *via* the magnetic fields on the torque. Furthermore a straight line replaces the real static torque characteristic. Assuming for a basic consideration that the damping torque is initially given only by $T_d = k_d \, d\theta/dt$ and the stepping motor is rigid coupled with an inertia, i.e.

$$J = J_r + J_L = FI J_r \tag{3.10}$$

FI factor of inertia – the factor of inertia is the total
inertia of the drive related to the motor inertia

an oscillation equation can be obtained similar to a synchronous machine [3.11] or a mathematical pendulum [3.10]. The homogeneous solution of the differential Eq. (3.8) is known [1.13]. It reads as follows for a single step with the initial conditions $\theta(0) = 0$ and $d\theta/dt(0) = 0$:

$$\theta = \theta_S \cdot (1 - e^{-t/\tau_d} \cos(\omega_0 t)) \tag{3.11}$$

with

$$\omega_0 = \sqrt{\frac{k}{J_r FI}} \quad \text{and} \quad \tau_d = \frac{2 J_r FI}{k_d}.$$

The time domain behaviour of the rotor angle at a single step according to Eq. (3.11) is shown in Fig. 3.25. Figure 3.26 demonstrates the time domain behaviour at different stepping rates. In case a) the distance between the pulses is so high that the stepping motor will settle its oscillation at a final position after each single motion. Practically it is performing single steps one after the other. In case b) the pulse distance is reduced as far as the angular position can grow almost linear. That means that the stepping motor is moving with a nearly constant rotational speed. The pulse distance in case c) is too small. When the second pulse is fed the difference between desired and actual angular position is too large so that the motor cannot follow the input signal. It will stall in an undefined position.

3.4.5 Operating Characteristics

That the stepping motor is an oscillating system is the most important conclusion from the considerations of Sect. 3.4.4. It is described by different operation characteristics (Fig. 3.27), which separate ranges with different characteristic behaviour from each other. In these characteristic curves the specific conditions are quantitatively captured, e.g. including non-linearities, tolerances, etc.

It can be realised from a review of the time domain behaviour of steps (Fig. 3.26) that the stepping motor can follow the clock signal i without step error from standstill up to a maximum constant stepping rate f_A under specified load conditions (load torque T_{L0} and inertia J_r, factor of inertia $FI = 1$, Eq. (3.10)). This stepping rate f_A is the maximum operating frequency for dynamic operation in the pull-in range or start-stop region and is called start rate [3.27]. In a close sense it represents the start process and is valid in approximation for the stop process. Above this value f_A the stepping motor can be further accelerated but only with small changes in the rate. The maximum rate in the pull-out range which the stepping motor at a load torque T_{L0} can respond to at all is the pull-out or slew rate f_B. The relations between pull-in rate f_A and pull-out rate f_B can be read from Table 3.1.

The characteristic curves of stepping motors shown in Fig. 3.27 are less significant without a statement about control mode and circuit. Therefore, the whole stepping drive consisting at least of motor and electronic circuit should always be considered. The holding torque T_H and the maximum running torque T_{max} of the stepping drive depend only marginally on control mode and circuit, when it is designed for continuous operation (duty or ON duration is 100%). However, the further progression of the curve and especially the pull-out characteristics with its maximum pull-out rate $f_{B0_{max}}$ strongly

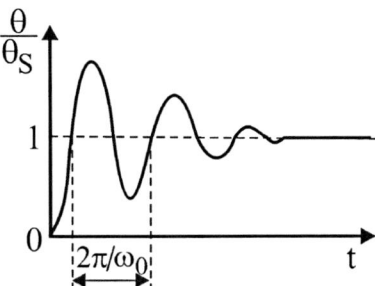

Fig. 3.25. Time domain behaviour of a single step

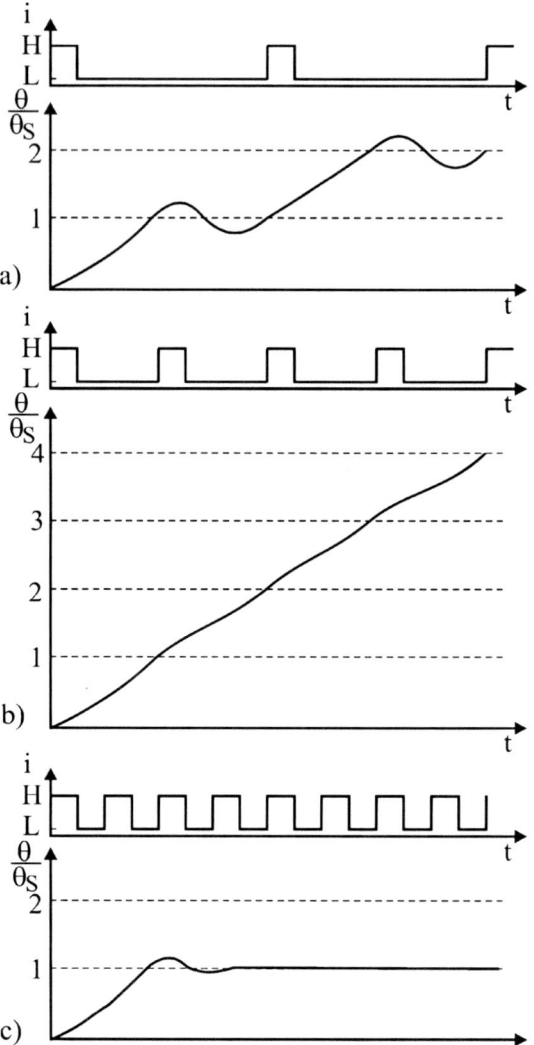

Fig. 3.26. Time domain behaviour of step sequences at different rates: **a** at low stepping rate; **b** at higher stepping rate; **c** at inadmissible stepping rate

depend on the control method. Thus up to five times higher values can be obtained with higher electrical excitation, thereby causing very steep current slopes (e. g. with bi-level or chopper control) like those described in Sect. 3.3. When parameters and operation characteristics are declared it is important whether the nominal voltage is given for the input of the control circuit or directly for the motor terminals. This is important especially for small rated voltages, e. g. 6 V. The characteristics will differ strongly if in such a case the voltage drop across the control circuit is 1 V.

The characteristic curves of a stepping drive have to be interpreted in a different way compared to other common electrical machines. They represent

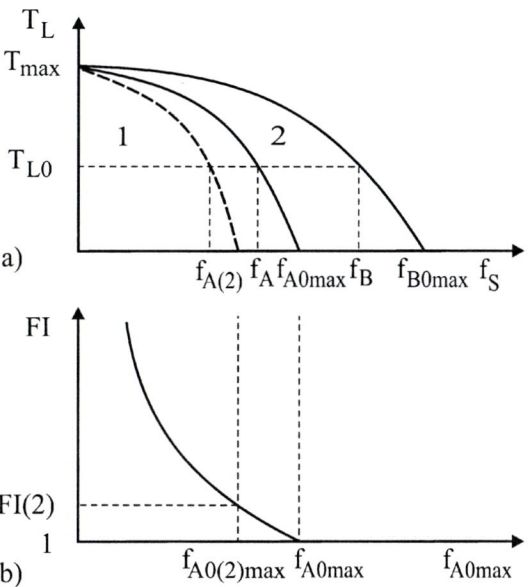

Fig. 3.27. Operation characteristics of stepping motors. a Operation characteristics for factor of inertia $FI = 1$, 1 pull-in curve; 2 pull-out curve. b Correction curve of the pull-in range for factor of inertia $FI > 1$

limiting curves or margins. In the whole area of the pull-in range a stepping motor can be started and stopped with any combination of load torque T_L and start stepping rate f_S without step errors. The pull-out range can only be accessed from the pull-in range while the stepping rate is increased during start or decreased during stopping, respectively [3.19, 3.20]. Usually rapid changes of the stepping rate from 0 to f_B and *vice versa* cause step errors.

The start-stop region or pull-in range is determined by the total inertia of the drive system. Without any additional remarks it is always given for the unloaded motor ($FI = 1$). Therefore, the characteristic curve $FI = f(f_{A0_{max}})$ (Fig. 3.27b) is complementary to describe the operational behaviour of a stepping motor. Its root point on the abscissa is given by the maximum pull-in stepping rate. The limiting curve of the pull-in characteristic can be converted with the following equation for a specific load situation with the factor of inertia $FI(2)$:

$$f_{A(2)} = f_A \frac{f_{A0_{max}(2)}}{f_{A0_{max}}} \quad \text{for} \quad T_L = \text{const}. \tag{3.12}$$

The converted curve is given as an example in Fig. 3.27a with a dotted line. If the conversion curve of Fig. 3.27b is unknown, it can be approximately determined with

$$f_{A0_{max}(2)} = \frac{f_{A0_{max}}}{\sqrt{FI(2)}}. \tag{3.13}$$

For the application of stepping drives it is very important to work within the limiting curves of the operation characteristics, because only in this case are the pulses of the clock signal and the mechanical steps synchronously performed. In this case no step errors arise. And only under this prerequisite can one of the essential advantages of stepping motors – the open-loop operation without feedback signals – be used.

3.4.6 Resonance Frequencies

That the stepping motor represents an oscillating system can be concluded from Eq. (3.8). Generally it possesses several natural or characteristic frequencies. The applied stepping rate f_S is the external excitation of the system. Resonant behaviour results from the interaction of these excitation frequencies and the natural frequencies. Unfortunately such ranges of resonances are characteristically for stepping drives.

So far no clear and detailed interpretation as well as no estimation of the whole resonant behaviour could be established. Both the nonlinearities of the stepping motor and the interaction with the electronic control circuit are theoretically difficult to handle. Additionally the constraints of the coupled load, e. g. gear backlash, complicate the situation. Therefore, only the following measures can be suggested for reduction or prevention of resonant behaviour in the practical application of stepping drives:

- Starting above the highest resonance frequency, if it is inside the pull-in range, $f_A > f_{Res}$
- Additional damping of the system by increasing the damping torque T_d, whereas the load torque T_L could help to do this, (cf. Eq. (3.8))
- Displace the resonance frequencies towards lower values by increasing the inertia of the load

$$f_{Res} = \frac{f_{Res0}}{\sqrt{FI}} . \tag{3.14}$$

- Faster passage through resonant ranges.

The five-phase stepping motor performs very small step angles of $\theta_S = 0.36°/0.72°$. Furthermore it shows only very small variations of the resulting torque, because the torque is generated simultaneously by four or five winding phases. These facts are significant reasons why five-phase stepping motors show in general almost no resonant behaviour.

3.4.7 Micro Stepping

The aim of the micro step mode is a further subdividing of the step angle θ_S given from motor design (Eqs. (3.4)–(3.7)) using an adequate electronic control circuit. Assuming that the torque *vs* angular position characteristic (Fig. 3.22) would be an ideal sinusoidal function a constant holding torque T_H for each position could be obtained by supplying the winding phases with currents, which are assigned likewise sinusoidals. The real stepping motor shows deviations from this ideal behaviour, because:

- The real torque characteristic shows a trapezoidal shape to achieve steep slopes at zero-crossings
- Caused by manufacturing tolerances of the motor and tolerances in the control circuit, the torque amplitudes of a half-wave can differ by up to 15%.

Usually no identical positions can be found when the single characteristic torque curves (Fig. 3.22) of a real stepping motor for different excitation patterns along the total perimeter can be compared with each other.

Different authors indicated approaches to measure the torque characteristic of real stepping motors and to reach with an appropriate current control based on that measurements a constant holding torque T_H = const. Apart from the high effort to do this for each single stepping motor, all authors came more or less to the conclusion that angular positions θ can be precisely approached, which are at least separated from each other by the step angle θ_S. Thereby the starting value plays a less important role. Generally the subdivision of the step angle θ_S in n partial steps of an angle θ_S/n will only be very inaccurately fulfilled. The absolute value of the positioning accuracy based on the full step mode (cp. Sect. 3.4.3) cannot be improved by the micro step mode.

Friction and load torque must not be neglected in the review of positioning accuracy. For instance although the stepping motor received input signals it will perform no movement if two positions in micro step operation are only so far from each other that the generated torque is smaller than the friction torque. In Fig. 3.28 the torque characteristic in full step mode is shown in a bold line (cp. Fig. 3.22). The thin lines represent the torque curves when the stepping motor is controlled with 12 micro steps. To clarify the point above, the friction torque T_f is chosen relatively large. It can be found that only after the third micro step does the generated torque become larger than the friction torque. Only then is the stepping motor starting its motion. The first two clock pulses are disregarded.

An explicit advantage of micro step operation is the change of the magnetic field of the stator with small angular positions and relatively high frequencies. In full step mode the magnetic field is jerkily moved from step to step with

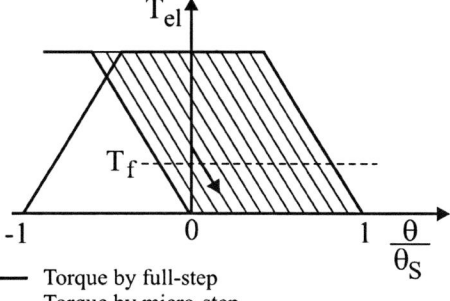

— Torque by full-step
— Torque by micro-step

Fig. 3.28. Motion sequence in micro step mode

relatively large angles. Consequently the micro step mode leads to a silent and almost vibration-free run.

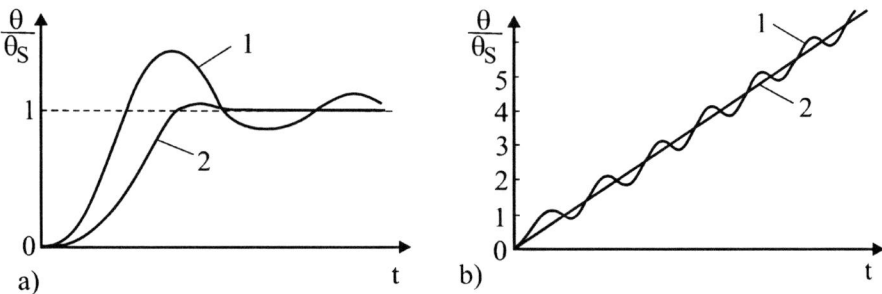

Fig. 3.29. Comparison of the step *vs* time characteristic of a stepping motor in full and micro step mode. **a** For a single step. **b** For a step sequence at a low stepping rate; *1* full step operation; *2* micro step operations

Figure 3.29 shows two cases whereby vibrations are reduced and possibly a desired position can be reached faster in micro step mode, especially at slow movements. Step fragmentations down to $n = 10\text{--}32$ are common and need only a justifiable effort [3.12]. Known peak values show a maximum number of steps $z = 50\,000$ steps per revolution.

Advantages of the micro step operation can be summarized as follows:

- Higher number of steps per revolution;
- Clearly lower vibrations;
- An almost constant torque of each step;
- Possibly gears can be omitted;
- Vibrations are reduced when approaching a desired position.

The following disadvantages must be traded off against the advantages:

- Higher effort for electronic control;
- Linked with that the clearly higher price;
- Bad positioning accuracy for a single step (the absolute value is not better than in full step mode).

3.4.8 Assigning Reference Position

Stepping drives work in open-loop control. No absolute position is defined, when a given pattern of electrical excitation is applied to a stepping motor. Stable operating points repeat along the perimeter with a distance of four steps θ_S (cp. Sect. 3.4.2). For instance, this can be observed in practise when manually moving the head of a disengaged printer. In this case the detent torque must be exceeded.

To ensure a constant print image it is necessary to approach a reference position before the real print process is started. For this purpose the stepping drive is controlled with a number of steps, at least equal to the maximum travel including possible tolerances. The move against a mechanical end stop is not critical for a stepping motor. Starting from this reference position the stepping drive works faultless as long as the operation parameters lie inside the permissible range.

A disadvantage of this method is evident in the case when the starting position is near to the reference position. The stepping motor will drive for a longer time against the mechanical end stop. The multiple impacts produce noise and possibly mechanical wear at the interconnected mechanical transfer elements. The same situation is seen in drives, which operate valves. With an increased number of steps safe closing of the valve should be ensured under different tolerances of the travel.

Therefore, different methods were developed to recognize the mechanical end stop of a stepping drive based on the evaluation of the time domain behaviour of the current [3.25]. An advantage of the referenced method is the definite stall detection of the motor within three or four steps. Furthermore, this evaluation is performed with little additional effort by the same microcontroller, which is already used for the functional control of the stepping drive. No additional external components, such as encoders are needed.

Advantageously the described method for end stop detection can be used with teach-in drive systems. Flap controls in heating, ventilation and air conditioning are a frequent application of stepping motors. Flaps with different angular travels can be easily realized with stepping motors. Thereto during setting-up operation a teach-in sequence is completed, where the two end positions are approached and detected by the stepping drive. From the number of steps performed between the two end stops the adjustment range is calculated and stored in the memory of the microcontroller as a maximum value. Thereby all existent tolerances of this specific drive system can be considered.

The electronic end stop or stall detection of stepping drives needs no additional hardware components. It leads to noise reduction and reduces the wear of mechanical components. Such stall detection methods extend the known applications of stepping drives with the described teach-in solution.

3.5 Motion Sequences

3.5.1 Classified Motion Sequences

Because of the multitude of specific cases with their particular parameters it is advisable to elaborate typical motion sequences of stepping drives [1.13] similar to duty types of electric machines [3.13]. These sequences are the basis of the analysis of individual driving tasks and their solution.

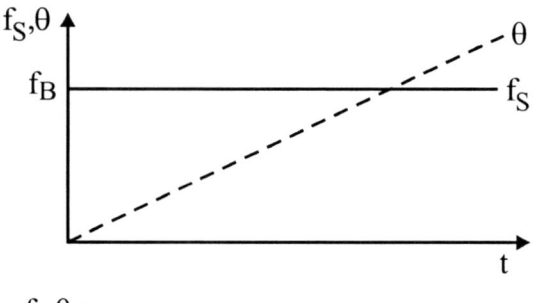

Fig. 3.30. Motion sequence at constant speed

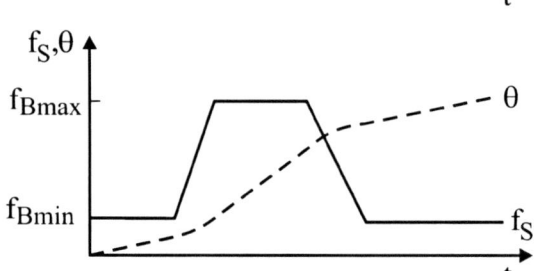

Fig. 3.31. Motion sequence at variable speed

Starting and stopping are without interest for the angular position *vs* time characteristic during motion sequences of long-term constant-speed drives (Fig. 3.30). The stepping rate f_S could take any value up to the maximum slew rate f_B. The long-term constancy is determined by the stability of the clock frequency, which is generated for high-quality drives from a quartz crystal oscillator. Common examples of such motion sequences are the form feed in recording instruments or infusion pumps of dialysis devices.

A second type of drive function needs operation at variable speed, i.e. between a maximum speed $n_{\max} \sim f_{B_{\max}}$ and a minimum speed $n_{\min} \sim f_{B_{\min}}$ (Fig. 3.31). If both stepping rates are situated in the pull-in range, the frequencies can be changed erratically. In the pull-out range the maximum permissible variation $|df_S/dt|_{zul}$ depending on the factor of inertia FI should be noticed. Thereby it is not necessary that the variation of the stepping rate is carried out continuously. This is especially interesting when a micro controller generates the frequency. In a first approximation sudden frequency changes in the range of the start rate are acceptable, taking into account the load torque and the factor of inertia.

Interesting applications in connection with such a motion sequence are so-called electronic gears, for instance if a rotation generated by any electric motor should be matched synchronously to a second motion. Therefore, the shaft that performs the first motion is coupled to a pulse encoder, which produces a number of pulses proportional to the angle of rotation. These pulses are given directly or *via* electronic pulse processing (reduction or multiplication) to a stepping drive. The proportional linkage between input pulse number and performed step number ensures that the first and second motions are synchronized. By means of an electronic frequency change the transformation ratio between both motions can be varied.

Moreover, it is conceivable that from an electronically generated frequency one or more control frequencies can be derived, which are supplied to particular stepping drives. In this case the transformation ratio is determined electronically, too. An example of the first case where a second motion is matched to the first is a trip recorder in a bus or a truck. Here the drive of the recording disk is linked directly to the revolution of the rear axle of the vehicle. For this reason a long and trouble-prone tachometer shaft between rear axle and trip recorder can be omitted. A pump system for analysis in chemical and medical laboratories should be given as an example of the assignment of different shafts together with presetting a fixed frequency and the change of the driving speed between the shafts.

Essential **advantages of these electronic gears** are:

- Absolute synchronism without mechanical coupling between the shafts;
- Arrangement of the driven shafts in any possible position in space;
- Electronic setting of the transmission ratio;
- Overload of a shaft does not cause mechanical damage;
- Wearless operation.

With such described electronic gear applications it is especially important to take care of the resonance frequencies, the maximum permissible frequency change and the maximum load torque.

One more drive function is the group step operation, particularly with positioning tasks. Thereby specific positions with small distances have to be approached at medium speed. A group of pulses is fed to a stepping drive during the operation time T_B (Fig. 3.32). Subsequently it is stopped until the end of the cycle time T. Then the next positioning occurs, i.e. the next pulse group starts. Usually the operation is performed in the pull-in range. An example of group step operation is the writing of single signs with printers. To implement proportional fonts each sign can be assigned to several groups of steps with different numbers. This is valid for the line feed as well. In each case a break is inserted during printing of a sign or a line. Afterwards the drive is fed, the next group of steps to move towards the next position.

Versatile types of oscillating drives are well known. Special possibilities are offered by a stepping drive, because miscellaneous timetables of the stepping rate can be programmed, e.g. see Fig. 3.33. By relatively simple means, different assignments of the position vs time characteristic can be generated. Conventional oscillating drives work at a fixed frequency (resonance frequency) and fixed amplitude. Samples of oscillating drives on the basis of stepping drives with programmable adjustment angle are the head of a photoplotter (cf. Sect. 9.7.4) and the feed across motion of the needle of a sewing machine for pattern generation.

For the implementation of motion functions with larger angular ranges (e.g. resetting to zero point or origin) an operation according to Fig. 3.32 is too time-consuming. The possibility of stepping drives to increase considerably the stepping rate towards the pull-out range is used for positioning operations with optimum time. After starting with the start rate f_A, the stepping rate is increased to the maximum pull-out rate f_B (Fig. 3.34). An exponential frequency change represents a technical optimum. However, linear or erratic frequency variations like those described for the motion function "variable speed" are possible, too. Shortly before achieving the target position the stepping rate is reduced from the slew rate f_B to the stop rate f_{Br} and then switched off. In general the stop rate f_{Br} can be chosen higher than the start rate f_A, because the load torque T_L supports braking and stopping. The time optimal positioning operation is used practically for filling and flushing of piston pumps and for the carriage return of printers, which are characterised by long strokes with high numbers of steps. Therewith the required time can be reduced remarkably compared to dosing or printing operations, which are carried out only in the pull-in range.

3.5.2 Closer Examination of Dynamic Operation

The motion sequences of Sect. 3.5.1 are based on the quasi-stationary operation characteristics of stepping drives. That kind of consideration is not sufficient for highly dynamic drives. For instance extremely fast positioning

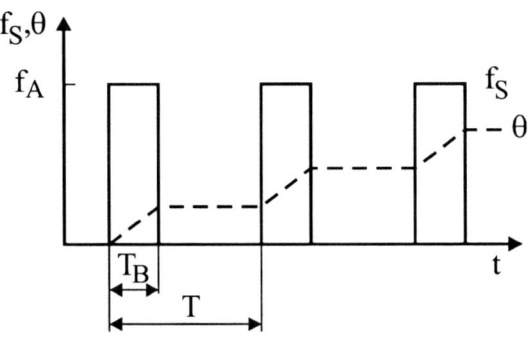

Fig. 3.32. Motion sequence of a group step operation

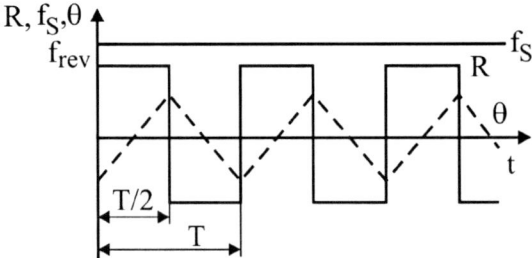

Fig. 3.33. Motion sequence of a reversing operation

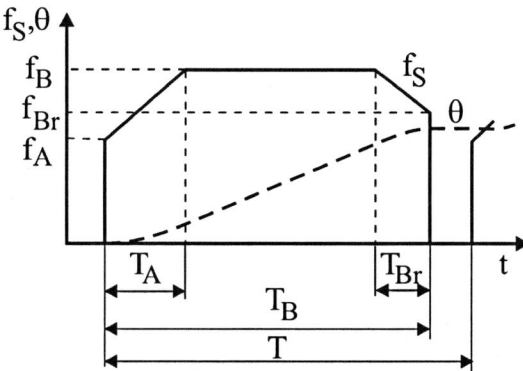

Fig. 3.34. Motion sequence of a positioning operation with time optimum

tasks demand varying pulse width of the clock signal in order to use the predisposition of the system for oscillation [3.12, 3.14, 3.21].

Numerical simulations are a feasible method to solve such complicated driving functions. Therewith it is possible to model the nonlinearities of stepping drives as well as sophisticated mechanisms. Critical operation states as well as optimal pulse sequences can be found without extremely high experimental effort by parameter studies and comparative simulations of different versions.

A clear representation of the computation results is given by the phase portrait, a kind of circular locus diagram [3.10]. In the following the main features of the work with phase portraits are shown using a free-oscillating stepping motor ($T_d = 0$ in Eq. (3.8)) as an example. Subsequently the discussion of the influence of damping will follow.

According to the equation of motion (Eq. 3.8) the torque T_{el} generated by the stepping motor is exclusively used for acceleration and braking of the complete drive system if the damping components inclusive of friction are neglected. The total inertia J contains all components of the rigidly coupled load. Basically the characteristic of the torque generated by the stepping motor is shown in Fig. 3.22. The detailed characteristic of supplying excitation pattern 1 to the windings (Fig. 3.35) is extracted there.

The trapezoidal function of the torque is given for distinctive regions:

$$\text{I}: \quad -\theta_1' = -\frac{\theta_1}{\theta_S} \leq \frac{\theta}{\theta_S} \leq \frac{\theta_1}{\theta_S} = \theta_1' \qquad T_{el} = -k \cdot \theta ,$$

$$\text{II}: \quad \theta_1' = \frac{\theta_1}{\theta_S} \leq \frac{\theta}{\theta_S} \leq \frac{\theta_2}{\theta_S} = \theta_2' \qquad T_{el} = -k \cdot \theta_1 ,$$

$$\text{III}: \quad \theta_2' = \frac{\theta_2}{\theta_S} \leq \frac{\theta}{\theta_S} \leq \frac{4\theta_S - \theta_2}{\theta_S} \qquad T_{el} = -\frac{k \cdot \theta_1}{(2\theta_S - \theta_2)}(2\theta_S - \theta) .$$

(3.15)

The equation of motion (Eq. 3.8) in region I is linear and shows the following solution:

$$\theta(t) = -\theta_0 \cdot \cos \omega_0 t \,, \tag{3.16}$$

$$\dot{\theta}(t) = \omega_0 \theta_0 \cdot \sin \omega_0 t \tag{3.17}$$

for the initial conditions $\theta(0) = -\theta_0$, $\mathrm{d}\theta/\mathrm{d}t|_{t=0} = 0$. The angular frequency ω_0 is given by Eq. (3.11).

The phase diagrams are displayed in the plane $\mathrm{d}\theta/\mathrm{d}t = f(\theta)$, whereas the time t is a parameter of the phase curves. The total sum of all phase curves is called the phase portrait. By elimination of time t in Equations (3.16) and (3.17) the following equation can be obtained:

$$\left(\frac{\theta}{\theta_0}\right)^2 + \left(\frac{\dot{\theta}}{\omega_0 \theta_0}\right)^2 = 1 \,. \tag{3.18}$$

The phase curve will become a circle if an adequate scale is chosen. The margin of the linear region I is the phase curve 1 in Fig. 3.36. When the stepping motor is compared to a pendulum the points on the abscissa axis will correspond to rest positions, where the pendulum or the stepping motor,

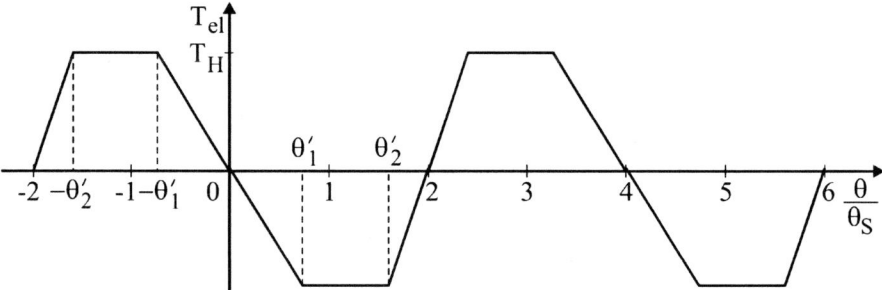

Fig. 3.35. Characteristic of the static torque as a function of the rotor angle

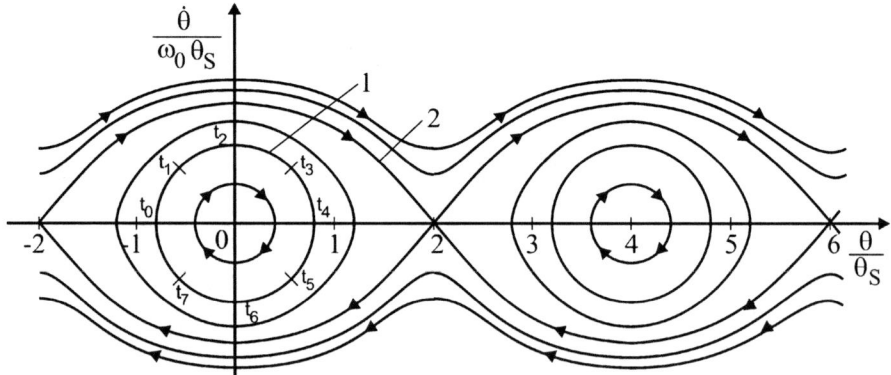

Fig. 3.36. Phase portrait of a free-oscillating stepping motor: *1* boundary curve of the linear region; *2* separatrix

respectively, are moved from this stable rest position by angle θ_0. Therewith the system exhibits at this position a potential energy of

$$W_{\text{pot}} = \int_{\theta_0}^{0} T_{\text{el}} \, d\theta = \frac{k \cdot \theta_0^2}{2} \, . \tag{3.19}$$

This energy will be completely converted into kinetic energy when the ordinate axis ($\theta = 0$) is passed. During energy conversion no losses will occur in a sustained oscillation system. Hence the equation

$$W_{\text{kin}} = W_{\text{pot}} \tag{3.20}$$

can be applied. Therefore, the maximum angular velocity can be obtained by

$$W_{\text{kin}} = \frac{J \cdot \dot{\theta}^2}{2} \tag{3.21}$$

$$\dot{\theta}_{\max} = \theta_0 \sqrt{\frac{k}{J}} = \theta_0 \omega_0 \, . \tag{3.22}$$

As a result of the periodicity of the torque characteristic phase curves will repeat with an interval of $\theta/\theta_S = \pm 4n$ ($n = 0, 1, 2, \ldots$).

The differential equation will become nonlinear in the regions II and III. Thus the solution cannot be given in a closed form. It is obtained with the aid of numerical simulations by determining single points of the curves [9.8]. Especially interesting is the boundary curve 2 in Fig. 3.36. This curve will be crossed when the energy at the point of $t = 0$ is equal to the maximum of the potential energy of the system:

$$W_{\text{pot}}(0) + W_{\text{kin}}(0) = W_{\text{pot max}} = \int_{2\theta_S}^{0} T_{\text{el}} \, d\theta \, . \tag{3.23}$$

Again the model of the pendulum can be used for interpretation. Maximum of potential energy means that the pendulum will be positioned in the upper unstable rest position. This corresponds to an angle of $2\theta_S$ of the stepping motor. A deflection out of this unstable position leads to a stable oscillation in the range $-2\theta_S \leq \theta \leq 2\theta_S$ without any overshoot. Such a boundary curve is named separatrix.

While the energy of the system at the moment $t = 0$ is larger than the maximum of the potential energy $W_{\text{pot max}}$, no stable oscillations around a defined rest position will be caused. Additional energy from outside is supplied to the system pendulum or stepping motor, respectively, so that the motor will continue to move even without supplied input pulses.

Considering the ever present damping in real stepping motors the phase curves will descend from closed to spiral curves (Fig. 3.37). The stepping motor moves from an initial deflection ($t = t_0$) towards a rest position under oscillations. The same behaviour in the time domain is shown in Fig. 3.25. Such characteristics can be quantified only for specific parameters.

Applying phase portraits to the solution of a positioning task the needed steps are shown in the following. The task is given that a stepping drive should approach with eight steps to a target position in the shortest possible time and without a significant overshoot. Load torque and friction can be neglected. Figure 3.38 shows the phase portrait of this stepping motor. Only the first and second quadrants are displayed, because oscillations should be suppressed according to the problem definition. Backward movements would be visible in the third and fourth quadrants. The time parameters are not shown in the diagram, since the sequence is treated in a qualitative manner.

Starting point is the rest position of the stepping motor, hence the origin of the coordinate system. While the first clock signal i is supplied to the stepping drive, the phase curves will be shifted left or right to a value of $\theta/\theta_S = \pm 1$. The stepping drive would start its movement on a phase curve, which derives from the origin. With each further pulse all phase curves must again be shifted to a value of $\theta/\theta_S = \pm 1$ left or right.

Therewith the method becomes complicated and confusing. Practically the approach is as follows:

- For each partial motion after a clock signal i the time measurement is started again with 0;
- Instead of shifting the phase curves to a value of $\theta/\theta_S = \pm 1$ after each clock signal i, the obtained angular position will be shifted to this amount in the opposite direction, i.e.

$$\theta_{i+1}(t_{i+1} = 0) = \theta_i(t_{ie}) \mp \theta_S ; \qquad (3.24)$$

- The obtained angular velocity is not changed, which means

$$\dot{\theta}_{i+1}(t_{i+1} = 0) = \dot{\theta}_i(t_{ie}) . \qquad (3.25)$$

In Fig. 3.38 each instant is marked with an encircled number, which is supplying the clock signal i to the stepping drive. The first pulse leads according to Eq. (3.24) to a shift of the coordinate system to $-\theta_S$ and the speed will remain zero due to Eq. (3.25).

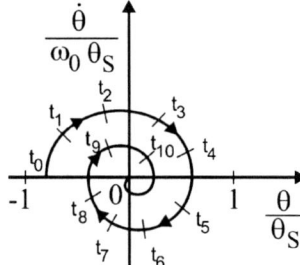

Fig. 3.37. Phase portrait of a damped stepping motor

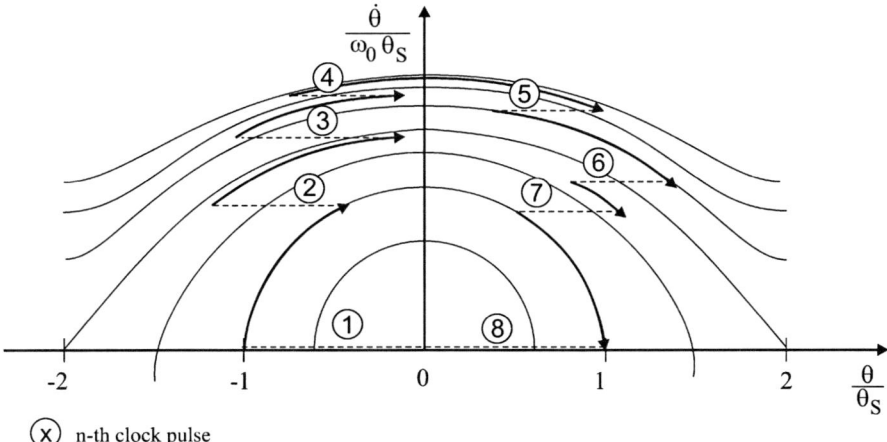

Fig. 3.38. Phase portrait of a stepping motor with a motion sequence consisting of eight partial steps

$$\theta_1(t_1 = 0) = 0 - \theta_S = -\theta_S \qquad \dot{\theta}_1(t_1 = 0) = 0 \ . \tag{3.26}$$

Thereafter the motion follows the phase curve until the second clock pulse is supplied. For instant t_2 is obtained:

$$\begin{aligned} \theta_2(t_2 = 0) &= \theta_1(t_{1e}) - \theta_S \\ \dot{\theta}_2(t_2 - 0) &- \dot{\theta}_1(t_{1e}) \ . \end{aligned} \tag{3.27}$$

With the third pulse the separatrix will be passed. That is possible and allowed during dynamic operation. The braking process starts when the ordinate axis is crossed. The seventh pulse will exactly hit again that phase curve, which goes across the point $\theta/\theta_S = 1$. When this point is touched with the last step, the stepping drive has performed $8 \times \theta_S$ angular steps and has reached zero speed again. The position is ensured without vibrations by feeding the 8. clock pulse towards the stepping drive. The target point will not be touched in the optimal time and oscillations will be unavoidable, when this clock pulse would be given earlier or later or the 7. clock pulse would be applied later respectively.

In Fig. 3.39 the torque vs angle characteristic is shown analogous to Fig. 3.22. Strictly speaking this picture is only valid for the steady-state or quasi-stationary operations. Deviations of the time behaviour of the current during the commutation of the winding phases are not considered. Nevertheless it can be used for orientation of how the sequence shown in the phase portrait is transferred into the torque characteristic. The points of switching over while a clock signal arrives are identical in both figures.

Figure 3.40 is formed while the individual sections of the motion sequence from Fig. 3.38 are composed in the phase plane like they really belong together. The time axis in the bottom part is nonlinear. The *phase portrait is a descriptive graphic tool for the solution of positioning tasks* with a few angular steps.

The composed phase curve of Fig. 3.40 will be basically idealised for the assessment if the chosen stepping drive is able to fulfil a required positioning function. It is divided into two equal motion sequences, one for acceleration and the second for deceleration with a constant torque (Fig. 3.41). Starting point of this consideration is the equation of motion (Eq. 3.8), which is converted to

$$\frac{d^2\theta}{dt^2} = \frac{T_{el} - T_d}{J_r \cdot FI} = \frac{T_0 f(t)}{J_r \cdot FI} = k \cdot f(t) \ . \tag{3.28}$$

The function $f(t)$ is positive during the first half of the motion sequence $t = 0 - T/2$, whilst it is negative in the second half $t = T/2 - T$. The torque T_0 is the torque which is effectively available for acceleration. It should be estimated as an averaged value out of Fig. 3.22.

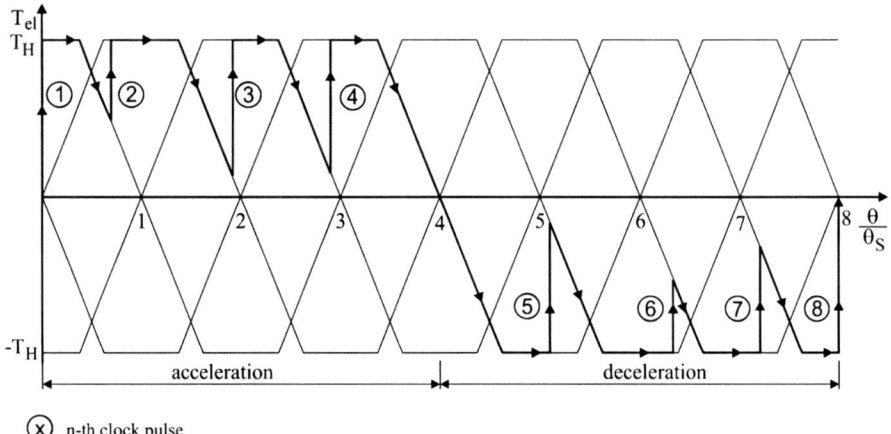

Fig. 3.39. Torque as a function of the rotor angle with the displayed motion sequence due to Fig. 3.38

3 Electromagnetic Stepping Drives   201

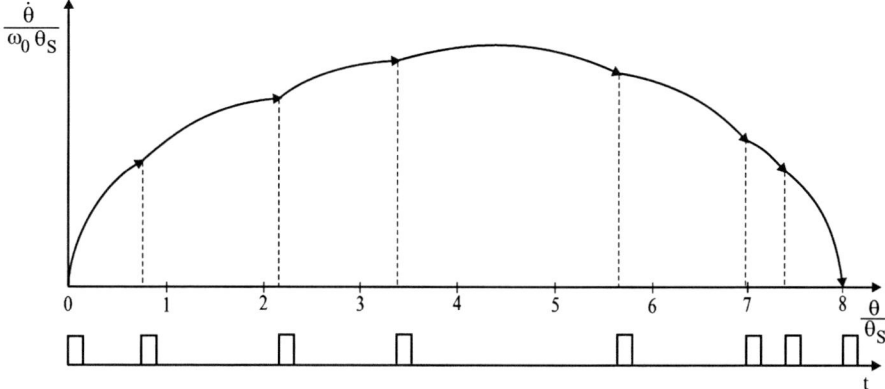

Fig. 3.40. Phase curve of the motion sequence of Fig. 3.38

The initial conditions are given for $t = 0$:

$$\theta|_{t=0} = 0 \qquad \frac{d\theta}{dt}\bigg|_{t=0} = 0 \qquad (3.29)$$

and for $t = T$:

$$\theta|_{t=T} = \theta_e \qquad \frac{d\theta}{dt}\bigg|_{t=T} = 0 \ . \qquad (3.30)$$

The minimum of time $T_{\min}$ needed for completion of the positioning sequence can be calculated from

$$T_{\min} = \sqrt{\frac{4 \cdot \theta_e}{k}} \ . \qquad (3.31)$$

In Fig. 3.41 $T_1 = T_{\min}$ is shown. No motion sequence exists that allows, with the given torque T_0, a faster positioning. Thus all real motion sequences

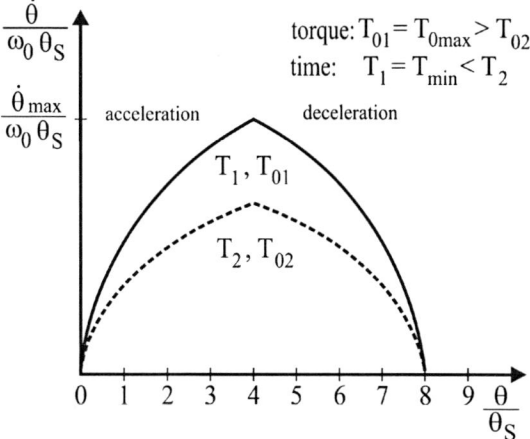

Fig. 3.41. Phase curve of optimal motion sequences

have to be situated below this curve. Therefore, it can be estimated from the time $T_{\min}$ and the number of steps θ_e/θ_S, if the demanded positioning time could be achieved or at which stepping rates the stepping motor should work. When these presumptions can be fulfilled, the time intervals of the single clock pulses i can be determined in the phase portrait.

The phase portrait is the total sum of all phase curves. The following statements can be derived from it:

- Turbulence points ($\theta/\theta_S = \pm 4n$; $n = 0, 1, 2, \ldots$) will always be enclosed in closed phase curves. They are the characteristic positions of equilibrium of a free-oscillating system.
- Further positions of equilibrium (singular points) appear at the intersection points of the phase curves with the abscissa axis ($d\theta/dt = 0$).
- Inside of the separatrix the total energy of a system will be at all times smaller then the maximum of potential energy $W_{\text{pot max}}$. Consequently the oscillations will be stable and the following equation is always true: $|\theta| < 2\theta_S$.
- The intersection point of the separatrix and the abscissa axis is an unstable point of equilibrium. This point is called the saddle point.
- Phase curves external from the separatrix reveal that the total energy of the system is larger than the maximum of potential energy. Consequently no stable oscillations with a constant value θ exist. The angle increases monotonously without any supplied pulse sequence on the clock input.
- In the upper phase plane the phase curves will always be directed from left to right; in the bottom plane *vice versa*.
- The time t appears only as a parameter of the phase curves.

The phase portrait can be used for the assessment or the dimensioning of a stepping drive in the following form:

- For a given stepping drive one general valid phase portrait can be established, when damping is neglected. The interrelation according to Fig. 3.36 has to be investigated by experiment in case it is not given.
- Starting from the initial conditions $\theta(0) = \theta_0$ and $d\theta/dt|_{t=0} = d\theta_0/dt$ the motion sequence could be tracked. Regions of stable and unstable oscillations can be identified.
- The motion follows the actual phase curve until the time instant t_{ie}, when the following input pulse should trigger the next step. *The final values of the just reviewed time period i will become the initial conditions of period $i + 1$.* Thereby the abscissa axis is to shift to a value of $\theta/\theta_S = \pm 1$. Advantageously the time will be counted in each period starting again from zero.
- The directives are given by Eqs. (3.24) and (3.25).
- After a first rough examination and the choice of the varying pulse distances (or the stepping rates, respectively, which must not be constant too), the calculation of the total sequence with individual parameters follows under consideration of the transient behaviour of the electric circuit as well as the damping torque.

References

[3.1] DIN 42021-2 Schrittmotoren; Begriffe, Formelzeichen, Einheiten und Kennlinien, 10/1976
[3.2] Rudolph, G.; Kluge, R. Step by Step, elektronik industrie 1998, Heft 7, S. 46–50
[3.3] Oberwallner, C., Stürzer, M. Mit dem 2-Phasen-Schrittmotortreiber TCA 3727 weniger externe Bauelemente und bessere Eigenschaften, Siemens Components 1991, S. 242–246
[3.4] Richter, C. Linearantriebe mit Schrittmotoren, VDI-Berichte 1269 "Innovative Kleinantriebe", S. 217ff., VDI-Verlag 1996
[3.5] Richter, C. Neuer Linearschrittmotor, TR Transfer, Nr. 49, 1994, S. 38–40
[3.6] Firmenschrift Portescap Deutschland GmbH, D-75179 Pforzheim
[3.7] API Portescap Scheibe contra Trägheit, KEM 1998, September S2, S. 62–64
[3.8] Schörlin, F. Mit Schrittmotoren steuern, regeln und antreiben, Francis' 1996
[3.9] Förstl, S. Schrittmacher, ELEKTRONIKPRAXIS Nr. 17, 1998, S. 116–118
[3.10] Fischer, U. u.a. Mechanische Schwingungen, 2. Auflage, Fachbuchverlag Leipzig, 1984
[3.11] Müller, G. Theorie elektrischer Maschinen, VCH Verlag Weinheim New York Basel Cambridge 1995
[3.12] Löwe, B. u.a. Auf den Punkt gebracht, F&M, Nr. 12; 1998, S. 931–934
[3.13] EN 60034-1, Drehende elektrische Maschinen – Teil 1: Bemessung und Betriebsverhalten, 11/1995
[3.14] Kuntz, W. Mehrachsige, schnelle Schrittmotorsteuerungen mit beliebigem Frequenz-Zeit-Profil, Elektronik 1982, Heft 18, S. 35–38
[3.15] Förstl, S. Schrittmacher, Elektronikpraxis, Nr. 17, 1998, S. 116–118
[3.16] Yukio Miura u.a. Development of Three-phase Stepping Motor, SANYO DENKI Technical Report, Nr. 5, 1008, S. 31 37
[3.17] Scheibe contra Trägheit, KEM, Nr. S2, 1998, S. 62–64
[3.18] Gollhardt, E. Optimierte Magnete für Hybridschrittmotoren, F&M, Nr. 7–8, 1998, S. 503–506
[3.19] Eissfeldt, H. Regelung von Hybridschrittmotoren durch Ausnutzung sensorischer Motoreigenschaften, Dissertation TU München, 1991
[3.20] Schatter, G. Mikrorechnereinsatz zur Beschleunigungssteuerung von Schrittmotoren, Feingerätetechnik, Nr. 3, 1979, S. 103–105
[3.21] Maas, S. u.a. Auslegung eines Schrittmotorantriebs mit einem Modell hoher Ordnung, antriebstechnik, Nr. 7, 1996, S. 52–54, Nr. 8, 1996, S. 57–60
[3.22] Kreuth, H.P.: Schrittmotoren. München/Wien: Oldenburg Verlag 1988
[3.23] Kenjo, T., Sugawara, A.: Stepping Motors and their Microprocessor Controls. Reprint 2. Edition (with corrections) Oxford: Clarendon Press 1995
[3.24] Rummich, et al.: Elektrische Schrittmotoren und -antriebe. 2. Auflage. Renningen-Malmsheim: expert-Verlag 1995
[3.25] Ritschel, S.: Anschlagerkennung von Schrittmotoren, Tagungsband SPS/IPC/DRIVES 2001, Hüthig Verlag Heidelberg, S. 487–494
[3.26] H. Sax: Stepper motor driving. Application note ST Microelectronics 1995
[3.27] IEC TS 60034-20-1 Technical Specification: Rotating electrical machines – Part 20–1: Control motors – Stepping motors. 01/2002
[3.28] Kuo, B.C.: Theory and applications of step motors. St. Paul: West Publishing Co. 1974
[3.29] Kuo, B.C.: Step motors and control systems. Champaign, Illinois: SRL Publishing Co. 1979

4

Drives with Limited Motion

4.1 Electromagnets

Eberhard Kallenbach

4.1.1 Electromagnets as Drive Sections

Electromagnets are simple constructed, robust drive sections that are used in mechanical, automotive and automation engineering applications with a growing scale and large quantities. Their field of application is especially the generation of small linear (0.1 100 mm) or angular (0.1 90°) strokes. Normally they are not stand-alone solutions, but an integrated part of complex functional modules (e. g. electromagnetic valves, clutches or relays).

The main parts of electromagnets are the armature as moving element, the stator (yoke) as closing element for the magnetic iron circuit and the exciting coil (Fig. 4.1).

The function principle of electromagnets is based on the generated forces onto interfaces between two different materials in inhomogeneous magnetic fields. Classified by energy conversion, we deal with electro-magneto-mechanical energy converters that transform the electrical energy input *via* an interstage form of magnetic energy into mechanical energy used for motion generation.

Of vital importance for the mode of operation are the constructive design of the magnetic circuit, the air gap between armature and stator and

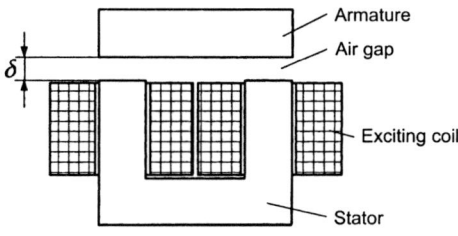

Fig. 4.1. Basic structure of an electromagnet with U-shaped magnetic circuit

guidance and/or bearings (depending on the designated movement pattern), which avoid magnetic forces orthogonally to the operation direction, especially in electromagnets with manipulated magnetic force characteristics. The components of the magnetic force are generated at the bounding surfaces between iron and air around the operating air gaps. Their vectorial sum represents the effective summary magnet force F_m. The dependence of the stroke on the magnetic force at constant exciting current is called magnetic force characteristic. It is pivotal for the application of electromagnets and should be above the characteristic of the load $F_L = f(\delta)$ (Fig. 4.2).

The operation principle of electromagnets is of cyclic nature. The armature travels under excitation from a starting position (δ_{max}) to a final position (δ_{min}) and is moved back to the initial position after switching off the excitation, which leads to a reversal of the motion direction by external forces (spring forces, gravity forces, magnetic forces of a second electromagnet) (Fig. 4.3).

In most cases, the armature moves the end effector directly. Magnet drives are preferably direct drives (Fig. 4.14). For an assured and effective way of operation of these direct drives, it is necessary to align magnetic and reset force characteristics.

Because of manifold applications and the necessity of the adaptation of electromagnets for special motion tasks or for the functional integration of electromagnets with other functional elements (e. g. the contacts of relays), different types of magnets with their specific advantages and disadvantages have been developed.

Electromagnets can be divided by:

- Their way of excitation into DC, AC and impulse magnets
- Force action into push, pull, rotary, oscillating and retaining magnets
- The magnetic circuit's design into U-shaped, E-shaped and tubular magnets
- Length of the stroke in proportion to the armature's diameter into long, medium and short-stroke magnets
- Their field of application into hydraulic, air gap switch, pressure, locking magnets and so on

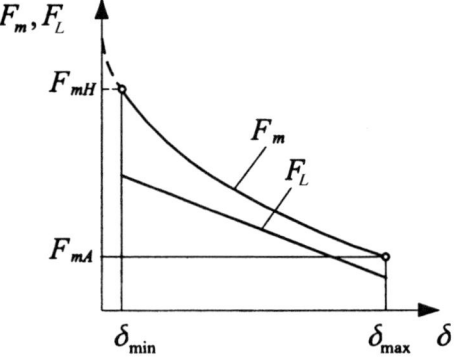

Fig. 4.2. Operation diagram of an electromagnet for steady state, F_m magnetic force, F_{mA} pull-in force, F_{mH} retention force, F_L counter force (force of the load F_L)

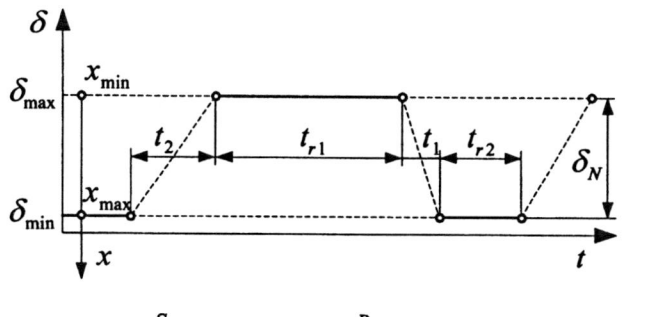

Fig. 4.3. Motion sequence of electromagnets, t_{r1}, t_{r2} pause times, t_1 pull-in time, t_2 drop-out time

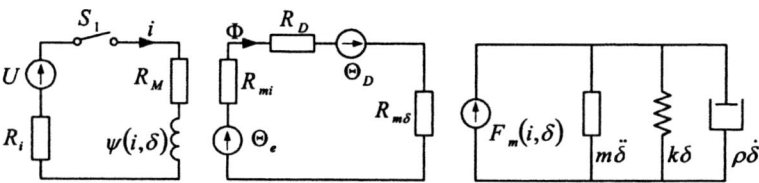

Fig. 4.4. Basic structure of electromagnets; neutral DC magnet: $\theta_D = 0$, $R_D = 0$, AC magnet $\theta_D = 0$, $R_D = 0$, $U = \hat{U} \sin \omega t$, polarized electromagnet $\theta_D \neq 0$, $R_D \neq 0$

Normally, electromagnets are used as drive sections with limited linear or angular stroke with an open-loop control. Exceptions to this are proportional magnets for hydraulic or pneumatic valves with requirements for higher accuracy that are part of closed-loop controlled drives.

Exact computation of electromagnets requires knowledge about the electromagnetic field, transient processes during switching on and off and heating effects. Numeric field computation is very extensive in many cases. That's why, field computation is often reduced to the evaluation of the integral parameters (Φ, ψ, R_m, θ) of the magnetic field with network calculations (Fig. 4.33) [4.16, 4.17].

The simplest equivalent network, which accounts for electrical, magnetic and mechanical effects, is the basic structure.

Figure 4.4 shows the complete basic structure of electromagnets that with modifications allows describing DC, AC, and polarized electromagnets.

4.1.2 Direct Current Magnets

4.1.2.1 Specifics

An electromagnet whose exciting coil is operated with direct current is called a direct current magnet. Nowadays they are the most common type of magnets with a multitude of different shapes and fields of application.

Advantages: simple construction, high reliability, good adaptation to the load by manipulating the magnetic force characteristic or with circuitry methods, constant actuation times.

4.1.2.2 Steady-State Behaviour of DC Electromagnets

Energetic Conditions in Non-Linear Electromagnets

In order to obtain a good ration between mass and power, the magnetic circuit has to consist of preferably small reluctances. Therefore, a non-linear behaviour dependent on i and δ has to be considered for the energetic analysis of the electromagnet (Fig. 4.5). When the electromagnet is switched on, an electro-magneto-mechanical transient process is taking place. This is caused by the mutual induction voltage (consisting of the two parts $u_1(t)$ and $u_2(t)$, which represent the voltages induced because of the changing current and the motion) created in the coil.

$$\begin{aligned} U &= iR + \frac{\partial \psi(\delta, i)}{\partial i} \frac{di}{dt} + \frac{\partial \psi(\delta, i)}{\partial \delta} \frac{d\delta}{dt} \, . \\ &= iR + u_1(t) + u_2(t). \end{aligned} \qquad (4.1)$$

In the initial stage ($\delta = \delta_{\max}$) with constant armature position, we have $\frac{\partial \psi(\delta, i)}{\partial \delta} \frac{d\delta}{dt} = 0$; only the magnetic energy stored in the coil changes.

$$W_{\mathrm{m}} = \int_0^{t_{11}} i u_1 \, dt = \int_0^{t_{11}} i \frac{\partial \psi}{\partial i} \frac{di}{dt} \, dt = \int_0^{\psi_1} i \, d\psi \, . \qquad (4.2)$$

The magnetic co-energy dependent on the stroke is represented by the area $0I_010$ (linear case) or by the area $0I_020$ (non-linear case) in the ψ–i-curve family (Fig. 4.5a). During the travel of the armature from $\delta_{\max}$ to $\delta_{\min}$ the energy $\Delta \psi I_0$ (area $\psi_1 1 2 \psi_2$) is added to the electromagnet's energy content and mechanical work represented by the area 0120 is set free.

The area $\psi_0 I_0 - \int_0^{\psi_0} i \, d\psi = W_{\mathrm{m}}^*$ represents the magnetic co-energy.

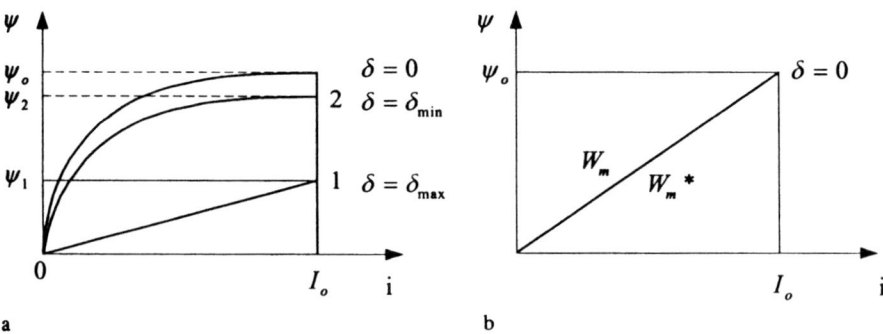

Fig. 4.5. Energy conversion of electromagnets; **a** non-linear case $W_{\mathrm{m}} \neq W_{\mathrm{m}}^*$; **b** linear case $W_{\mathrm{m}} = W_{\mathrm{m}}^*$

For linear magnetic circuits the flux linkage can be calculated as $\psi = L(\delta)i$ (Fig. 4.5b).

Because of

$$\int_0^{\psi_0} i\,d\psi = \int_0^{I_0} \psi\,di = \frac{I^2}{2}L(\delta)$$

the magnetic energy W_m is equal to the magnetic co-energy W_m^*.

Magnetic Force of the Non-Linear DC Electromagnet

Due to the law of energy conservation, the applied electrical energy W_{el} is converted into thermal losses W_{therm}, magnetic field energy W_m and mechanical energy W_{mech}.

For small variations, we have

$$dW_{el} = dW_{therm} + dW_m + dW_{mech} . \tag{4.3}$$

For the stored energy content of the magnetic field as a function of δ and ψ, we obtain from Eq. (4.2)

$$W_m = W_m(\delta, \psi) = \int_0^{\psi_0} i(\delta, \psi)\,d\psi . \tag{4.4}$$

The total differential can be calculated by

$$dW_m = \frac{\partial W_m}{\partial \delta}\,d\delta + \frac{\partial W_m}{\partial \psi}\,d\psi . \tag{4.5}$$

We substitute Eq. (4.4) into (4.5) and obtain

$$dW_m = \frac{\partial}{\partial \delta}\int_0^{\psi_0} i(\psi,\delta)\,d\psi\,d\delta + i(\psi,\delta)\,d\psi . \tag{4.6}$$

By comparing Eqs. (4.6) and (4.3) and with respect to $W_{mech} = F_m\,d\delta$ and $W_{therm} = 0$. We can compute the magnetic force as follows:

$$F_m = -\frac{\partial}{\partial \delta}\int_0^{\psi_0} i(\psi,\delta)\,d\psi . \tag{4.7}$$

If we emanate from the magnetic co-energy for calculating the magnetic force, we obtain:

$$F_m = \frac{\partial}{\partial \delta}\int_0^{I_0} \psi(i,\delta)\,di . \tag{4.8}$$

Equations (4.7) and (4.8) are universally valid and equivalent. Both consider flux leakage of the magnetic circuit and saturation effects. They show for DC electromagnets the proportionality of the generated magnetic force and the change of the energy within the magnetic field (4.7) or the magnetic co-energy (Eq. 4.8).

The aim of every electromagnet dimensioning is the optimization of the relation between magnetic co-energy and the volume of the magnet. For the ideal case (no flux leakage with dropped out armature and rectangular ψ–i-characteristic of the closed magnetic circuit) the energy to be converted can be obtained from $W^*_{\text{mech}} = \psi_0 I_0$. A measure for the closeness of a certain magnet design to the ideal case is the magnetic effectiveness:

$$\kappa_m = \frac{W_{\text{mech}}}{\psi_0 I_0} \,. \tag{4.9}$$

For DC magnets, this value varies within a wide range: $0.1 \leq \kappa_m \leq 0.75$ (Figs. 4.7b and 4.8b).

κ_m is dependent on the magnet's design the stroke (change of the operating air gap), the parasitic air gaps with pulled in armature, the flux leakage with the armature at its starting position, the used magnetic material and the operation mode.

For electromagnets with linear ψ–i-characteristics for the nominal stroke range, κ_m is always smaller than 0.5. Electromagnets with non-linear ψ–i-characteristics normally have a larger magnetic effectiveness (Fig. 4.8).

Especially for large operating air gaps the ψ–i-characteristics of most electromagnets are linear. Therefore we have

$$\psi(\delta, i) = i L(\delta) \,. \tag{4.10}$$

We substitute Eq. (4.10) into (4.8) and obtain

$$F_m = \frac{\partial}{\partial \delta} \int_0^i i L(\delta) \, \mathrm{d}i = \frac{i^2}{2} \frac{\mathrm{d}L(\delta)}{\mathrm{d}\delta} \,. \tag{4.11}$$

With respect to $L(\delta) = w^2 G_m(\delta)$, we obtain

$$F_m = \frac{i^2 w^2}{2} \frac{\mathrm{d}G_m(\delta)}{\mathrm{d}\delta} = \frac{\Theta^2}{2} \frac{\mathrm{d}G_m(\delta)}{\mathrm{d}\delta} \,. \tag{4.12}$$

$G_m(\delta)$ represents approximately the magnetic permeance of the operating air gap.

By assuming that the magnetic force is only generated at the bounding surfaces of the operating air gap and with neglecting the magnetic potential difference within the iron circuit, Eq. (4.12) can be written as follows:

$$F_m = \frac{B^2 A_{\text{Fe}}}{2\mu_0} \,. \tag{4.13}$$

This equation is known as Maxwell's tensile force law. In comparison to Eqs. (4.12) and (4.13), this formula has a higher generalisation level for linear magnet systems. The geometry of the magnetic circuit and flux leakage are already considered within the inductivity. That's why, we get advantageous relations for the synthesis of magnets.

In relevant literature [4.1], appropriate mathematical descriptions for the inductivity of usual magnet circuit geometries for approximate calculation of force-vs-stroke characteristics can be found.

Linked with the improvement of mass-power relations of electromagnets and thereby improved utilisation of the magnetic material and the increasing application of special electronic driver circuits (over-excitation and impulse-excitation), a rising number of DC magnets with non-linear ψ–i-characteristic can be found in industrial applications. Electromagnets with non-linear ψ–i-curve families even at large operating air gaps, can be created by adequate design of the armature/armature-complement system (manipulated force characteristic with design modifications of the armature's complement and unaltered magnet dimensions) (Fig. 4.6). For the approximate calculation of the operating reluctance of the armature/armature-complement system, an equivalent network with a non-linear ferromagnetic parallel reluctance $R_{\mathrm{mFe}}(\delta, i)$ can be established (Fig. 4.6b).

If we know the characteristic curve families – $\psi(\delta, i)$ or $\Phi(\delta, \theta)$ – the magnetic force characteristic can be computed, the following ways:

- Approximation of the ψ–i- or Φ–θ-characteristics with adequate mathematic functions. Computation of the analytic formulas for obtaining the magnetic force of non-linear DC magnets with respect to Eqs. (4.7) and (4.8) [4.5].
- Approximate computation of the magnetic force in consideration of saturation and flux leakage aspects with magnetic networks [4.5, 4.16, 4.17].

Force-Stroke Characteristics of DC Magnets

They show the dependency of magnetic force from stroke with the magnetomotive force (mmf) or, respectively, the excitation current as parameter and are the basis for selection and application of electromagnets. Without special requests for actuation times, a high static utilisation

$$k = \frac{W_{\mathrm{N}}}{W_{\mathrm{mech}}} \qquad (4.14)$$

is aspired in order to get small magnet dimensions.

Thereby $W_{\mathrm{mech}} = \int\limits_0^\infty F_{\mathrm{m}}(\delta, i)\,\mathrm{d}\delta$ represents the mechanical energy generated by the electromagnet and $W_{\mathrm{N}} = \int\limits_{\delta_{\min}}^{\delta_{\max}} F_{\mathrm{geg}}(\delta)\,\mathrm{d}\delta$ represents the effective energy generated by the electromagnet (Fig. 4.2).

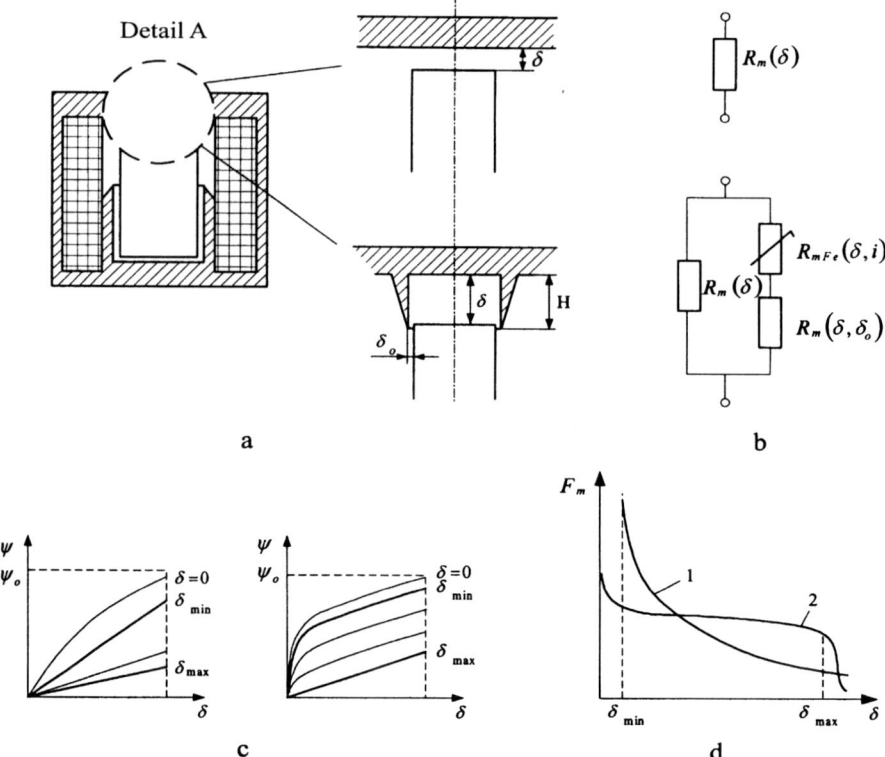

Fig. 4.6. DC electromagnet with and without manipulated magnetic force characteristic. **a** Geometrical design. **b** Equivalent magnetic network for armature/armarture-complement systems. **c** ψ–i-characteristic curve families. **d** Typical force-stroke characteristics, *curve 1* without *curve 2* with manipulated magnetic force characteristic

It is advisable to select or to design magnets with characteristic curves that are adapted to the static load characteristics throughout a wide range. This can happen by means of design (selection of the optimal basic shape for the magnetic circuit, modification of the geometry of the armature's complement) or circuitry (control of the exciting current dependent on stroke).

DC Magnets without Manipulated Magnetic Force Characteristic

are electromagnets, whose force-stroke characteristic are only dependent on the geometry of the magnetic circuit. Figure 4.7 shows such an armature/armature-complement system. For the magnetic permeance of the operating air gap, approximative we have

$$G_{m\delta} = \frac{\mu_0 A_{\text{Fe}}}{\delta}. \tag{4.15}$$

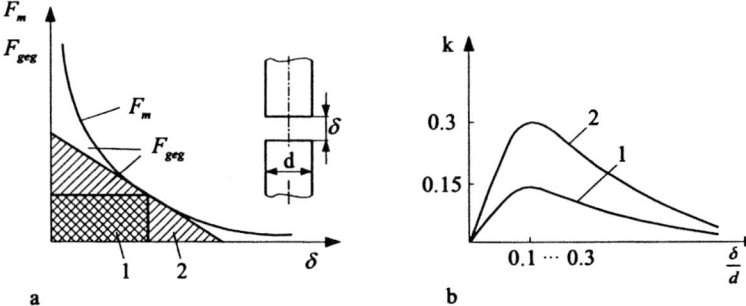

Fig. 4.7. Utilisation k of DC magnets without manipulated magnetic force characteristic. **a** Typical magnetic force and load curves, *area 1* mass load, *area 2* spring load. **b** Dependency of the utilisation k on mass load (*curve 1*) and spring load (*curve 2*)

From Eq. (4.12) we obtain for the magnetic force

$$F_\mathrm{m} = \frac{\Theta^2}{2} \frac{\mu_0 A_\mathrm{Fe}}{\delta^2} \; . \tag{4.16}$$

According to Eq. (4.16), the magnetic force decreases fast with larger operating air gap.

For most loads such a characteristic is disadvantageous. For spring or mass load we only obtain a very small utilisation k $(0.2\ldots 0.3)$, whose maximum is at very small operating air gaps (Fig. 4.7).

DC Magnets with Manipulated Magnetic Force Characteristic

A huge improvement of the utilisation k can be obtained, when the magnetic permeance of the operating air gap is changed by design modifications (at constant main geometric dimensions (height and diameter) and constant magnetomotive force Θ). This can be achieved by using specific armature/armature-complement systems (Fig. 4.6). Thereby the dependency from the stroke of the magnetic permeance of the operating air gap and parts of the adjacent sections of the armature's complement (Fig. 4.6b) is changed, which creates a completely different force characteristic for the same main geometric dimensions of the electromagnet.

Relevant literature describes different armature/armature-complement systems [4.1].

They allow a variation of the force-stroke characteristics in wide ranges and thereby an optimal adaptation to the specific load.

Figure 4.8 shows the possibility of increasing the utilisation k from 0.3 up to 0.75 and moving its optimum towards larger air gaps. With that improvement, volume reductions up to 50% can be achieved. Another typical application for manipulated magnetic force characteristics are for instance relays with enlarged pole faces of the core in comparison to the cross section of the iron circuit in order to increase the pull-in force.

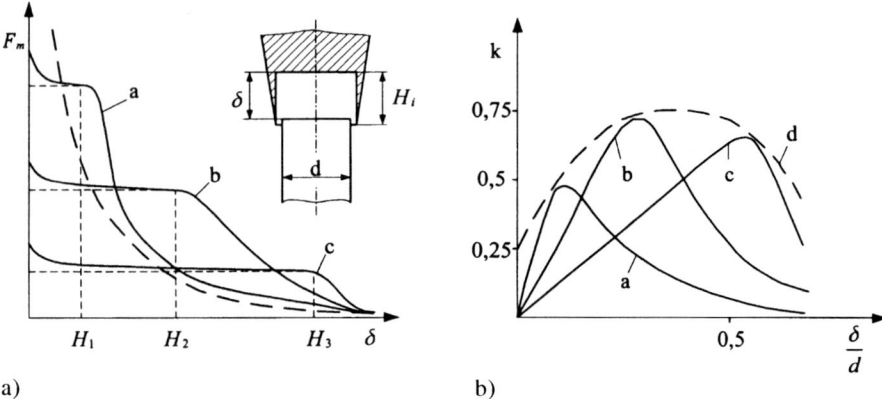

Fig. 4.8. Utilisation k of DC magnets with manipulated magnetic force characteristic. **a** Magnetic force characteristic with different geometric dimensions of the armature/armature-complement systems. **b** Dependency of the utilisation k on stroke

It has to be considered that every increasing of the pull-in force at the begin of the stroke results in a decreasing magnetic force at smaller air gaps and a smaller retention force, because the created mechanical energy stays the same. The areas underneath the different force-stroke graphs are identical:

$$\int_0^{\delta_{\max}} F_{\mathrm{m}}(\delta, i)\, \mathrm{d}\delta = \mathrm{const.} \text{ with constant magnetomotive force}. \qquad (4.17)$$

Besides the increasing of the utilisation k, manipulated magnetic force characteristics can improve the dynamic properties of the electromagnet (Fig. 4.16).

4.1.2.3 Dynamic Behaviour of DC Electromagnets

The requirements for automotive and mechanical engineering applications on the dynamic behaviour of switching and proportional electromagnets has increased over the last years. Therefore, the dynamic behaviour has to be considered during the design or selection process.

"Dynamic behaviour" of DC magnets shall denote the entirety of all electric, magnetic, mechanical and thermal transient processes during switching on and off.

Matters of particular interest are:

- Dependency of the armature's position on time ($x = f(t)$),
- Dependency of the velocity on position or time ($\dot{x} = f(x), \dot{x} = f(t)$),
- Dependency of the instantaneous value of the excitation current on time ($i = f(t)$),
- Dependency of the dynamic pull-in force ($F_{\mathrm{dyn}} = f(t), F_{\mathrm{dyn}} = f(x)$) at impressed voltage or current.

Great importance is attached to the overall dynamic behaviour for the application and synthesis of magnet drives. Each of the transient processes during switching on and off are separated into two stages with different couplings of the differential equations or, respectively, different circuitry structure (switches of the transistor stages, effects of the free-wheeling diodes) (Fig. 4.9).

The pull-in time t_1 is divided into the pull-in delay t_{11} – the time from switching on the excitation to the start of the motion – and the travel time t_{12} – the time from the beginning of the motion to its end in the stop position of the armature. In a similar way, the drop-out time t_2 is divided into the drop-out delay t_{21} – the time from switching off the excitation to the begin of the return movement – and the return travel time t_{22} – the time from the begin of the return motion to its end in the initial position of the armature.

The dynamic behaviour of DC magnets can be described with the following equations that can be derived from the basic structure (Fig. 4.4) by equating δ and x

$$U = iR + \frac{\mathrm{d}\psi(x,i)}{\mathrm{d}t} \tag{4.18}$$

$$F_\mathrm{m}(x,i) = m\ddot{x} + F_\mathrm{geg}(x) + F(\dot{x}) \tag{4.19}$$

$$F_\mathrm{m}(x,i) = \frac{\partial}{\partial \delta} \int \psi(i,x)\,\mathrm{d}i\ . \tag{4.20}$$

F_m magnetic force; $F_\mathrm{geg}(x)$ steady-state counter force; $F(\dot{x})$ friction force.

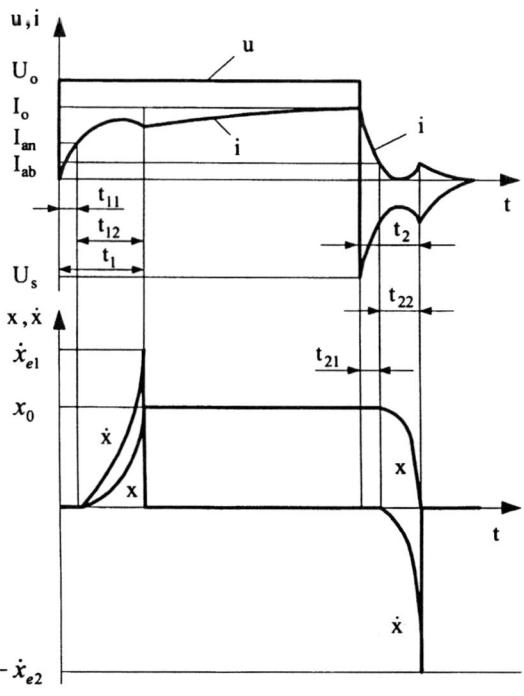

Fig. 4.9. Dynamic behaviour of DC magnets during one duty cycle with impressed voltage; t_1 pull-in time, t_{11} pull-in delay, t_{12} travel time, t_2 drop-out time, t_{21} drop-out delay, t_{22} return travel time, U_S interrupting voltage peak

Usually, even with linear ψ–i-characteristics, these equations lead to non-linear differential equations that cannot be solved exactly and therefore have to be calculated simplified with adequate approximation procedures or computer-aided to obtain the dynamic properties and the actuation times [4.2, 4.16].

Computation of the Pull-In Delay t_{11}

During the time interval of the pull-in delay, the magnetic field is built up with the applied electric energy. The armature is not moving. Only electromagnetic transient procedures take place. Equations (4.18) and (4.19) are decoupled. For computing the transient process, only Eq. (4.18) needs to be evaluated.

Initial conditions:

$$x = \dot{x} = \ddot{x} = 0 \quad \text{at} \quad t = 0$$
$$F_\text{m} \leq F_\text{geg} \quad \text{at} \quad 0 \leq t \leq t_{11} \ .$$

For many cases, due to the shear of the ψ–i-characteristic curve family at $\delta = \delta_\text{max}$ or, respectively, $x = 0$, the assumption of a linear ψ–i-characteristic is valid to compute the pull-in delay with dropped-out armature.

For the instantaneous value of the exciting current we have

$$i = \frac{U}{R}\left(1 - e^{-\frac{Rt}{L}}\right) \tag{4.21}$$

and thus

$$t_{11} = \frac{L}{R} \ln \frac{k}{k-1} \quad \text{with} \quad k = I_0/I_\text{an} \ , I_0 = U/R; I_\text{an} - \text{pull-in current} \ . \tag{4.22}$$

The solution of magnet systems with non-linear characteristics can be found with graphical methods or analytically with approximation procedures [4.12].

In general we obtain

$$t_{11} = \int_0^{\psi_\text{an}} \frac{\mathrm{d}\psi}{U - I_\text{an} R} \ . \tag{4.23}$$

Computation of the Travel Time t_{12}

With the start of the travel of the armature, an electro-magneto-mechanical interaction takes place, which requires a solution of the complete motion describing differential equation system. For the computation of transient processes during the pull-in time, many methods have been published, and they can be separated into three groups:

- Graphical analytical methods
- Analytical approximative methods
- Computer-aided simulation

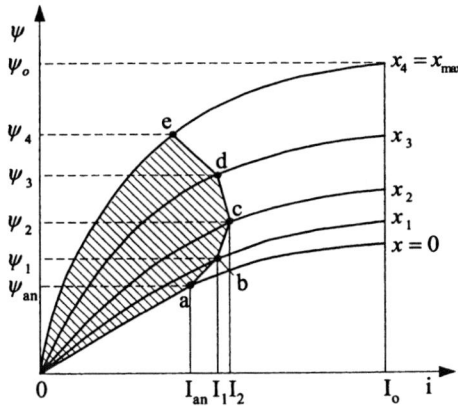

Fig. 4.10. ψ–i-characteristic for determining the travel time

The first two methods are problem-specific, but lead in special cases to useable solutions.

For the computation of the pull-in time with graphical analytical methods (e. g. method of Lysov [4.1]) knowledge about the ψ–i-characteristic $\psi = \psi(x, i)$ is needed.

From Eqs. (4.18) and (4.19), written in difference form, with neglecting friction forces, we obtain

$$U = iR + \frac{\Delta \psi}{\Delta t} \tag{4.24}$$

$$F_m \Delta x = \Delta \left(\frac{m\dot{x}^2}{2} \right) + f(x) \Delta x \ . \tag{4.25}$$

The pull-in time can be determined the following way: Selecting a point a (current I_{an} at the beginnig of the motion (Fig. 4.10)), we draw a straight line that crosses the adjacent curve in point b. The area 0–a–b represents the mechanical energy set free during the travel of the armature from x_0 to x_1. Computation of $\dot{x}$ from Eq. (4.25). Determination of the motion time for the first stage presuming a constant acceleration:

$$t_{12} = \frac{x}{\dot{x}_m}$$
$$\dot{x}_m = (\dot{x}_0 + \dot{x}_1)/2 \approx \dot{x}_1/2 \ .$$

The determined values for $\Delta \psi_1$, Δt and $i = I_0 + I_1/2$ are put into Eq. (4.24). If we obtain equity, the selection of the line segment a–b was correct; if not, the line segment a–b has to be newly chosen and the calculation has to be repeated until equity is reached. In the same manner, the motion time for all other stages can be determined. The travel time t_{12} is calculated as the sum of the motion times over all stages:

$$t_{12} = \sum_{i=1}^{n} t_{12i} \ .$$

Computer-aided calculation of the dynamic behaviour during the pull-in time. With the aid of digital computers, a solution of the motion-describing differential equation system is possible without too many simplifications or neglects. Figure 4.11 shows a block diagram for the computation with the tool SIMULINK. Effective computation methods for the dynamic behaviour can be derived from network models [4.16].

Transient Processes During the Drop-Out Delay t_{21}

The drop-out delay is the time period required for the degradation of the magnetic field, i. e. for the decrease of the magnetic force below the counter-force.

For this time period, the following conditions are valid:

$$\frac{d\psi}{dt} + iR = 0 \tag{4.26}$$

with the initial condition $I = I_0$ at $t = 0$.

Because the armature is not moving during that period, it is sufficient to compute the electro-magnetic transient procedures, i. e. to solve Eq. (4.18) at $U = 0$ with the initial conditions

$$F_m \geq F_{\text{geg}} \; ; \; x = x_0 \; ; \; \dot{x} = \ddot{x} = 0 \; ; \; \psi = \psi_3 \; ; \; i = I_0 \quad \text{at} \quad t = 0 \; .$$

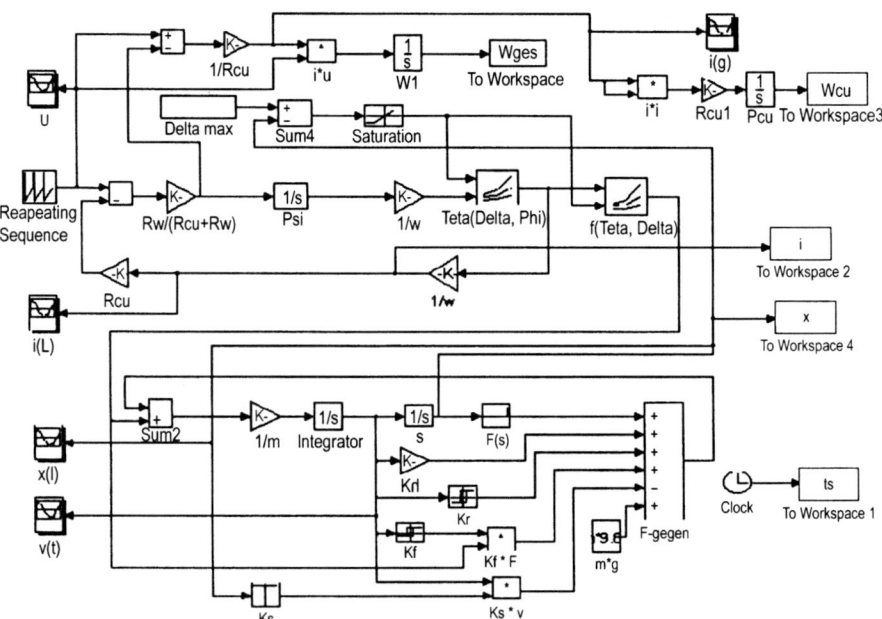

Fig. 4.11. Block diagram for the computation of the dynamic behaviour of DC electromagnets

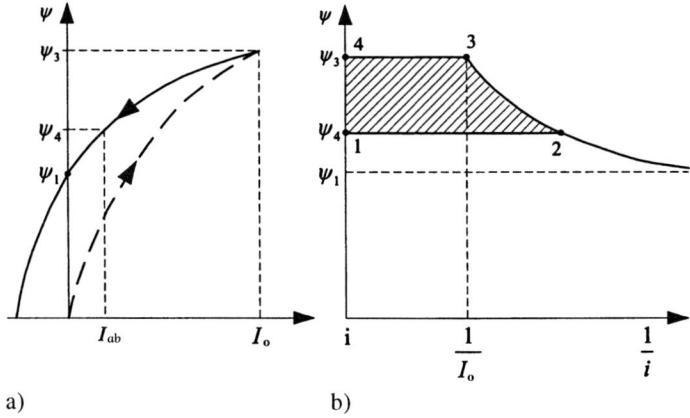

Fig. 4.12. ψ–i-characteristic for determination of the drop-out delay

The magnetic circuit is closed during the drop-out delay. That's why, a nonlinear relation between ψ and i often has to be considered. Because of hysteresis effects, the ψ–i-characteristics at rising and falling currents are different in good magnetic circuits (Fig. 4.12).

By solving Eq. (4.26) for t, we obtain for the drop-out delay:

$$t_{21} = R \int_{\psi_4}^{\psi_3} \frac{\mathrm{d}\psi}{i}, \qquad (4.27)$$

where ψ_3 is the flux linkage at I_0 and ψ_4 denotes the flux linkage at the drop-off current I_{ab}.

After drawing $\psi = f\left(\frac{1}{i}\right)$ (see Fig. 4.12b), the hatched area represents the drop-out delay.

If we deal with a linear ψ–i-characteristic even at pulled-in armature, with $L\frac{\mathrm{d}i}{\mathrm{d}t} + iR = 0$, we obtain for the drop-out delay

$$t_{21} = \frac{L}{R} \ln \frac{I_0}{I_{\mathrm{ab}}}, \qquad (4.28)$$

where I_0 symbolizes the continuous current at steady state and I_{ab} represents the drop-out current.

R can be affected heavily by measures for damping the interrupting voltage peak U_S (Fig. 4.15).

Transient Processes During the Return Travel Time t_{22}

The return travel time is the time period the armature needs to move back from the pulled-in position to its initial position due to reset forces. For magnets without manipulated magnetic force characteristic (very high retention force F_H), it can be assumed that the return motion does not start until $i = 0$ and thereby no electro-mechanical interaction has to be considered.

In this case, the return motion is a solely mechanical process and we have

$$t_{22} = \sqrt{\frac{2mx_0}{F_{\text{geg m}}}}, \qquad (4.29)$$

where m is the mass of all moving parts, $F_{\text{geg m}}$ is the average value of the counter-force and x_0 is the travel of the armature.

For magnets with manipulated magnetic force characteristic, normally an electro-mechanical interaction is taking place, which delays the motion and increases the return travel time t_{22} [4.1].

4.1.2.4 Influence of Eddy Currents on the Dynamic Behaviour of DC Electromagnets

In every magnetic body with an electric conductivity $\kappa > 0$, changes of the magnetic flux cause eddy currents, which reduce the rate of build up and degradation of the magnetic field and thereby affect the dynamic behaviour (e. g. increase the actuation times t_1 and t_2). This effect is occurring in every operation mode, but is highly noticeable in the operation mode of extremely rapid action (Fig. 4.18). For electromagnets with high switching frequency, e. g. magnets for injection valves, fast switching magnets for hydraulic applications, this influence has to be considered during the design process from the outset of the synthesis (selection of the most advantageous basic shape of the magnetic circuit and the optimal magnetic material). The computation of the influence of eddy currents requires powerful computers even for simple shapes of the magnetic circuit. The examination of the field's build-up and degradation in infinitely long cylindrical ferromagnetic bodies leads to mathematically well manageable analytic solutions. This case is especially interesting for DC tubular magnets. By assuming the same permeation rate of the magnetic field into all parts of the magnetic circuit, the eddy currents in the armature are the main contributor to the delay of the field's build-up or degradation.

By assuming a step function for the exciting current, an equal distribution of the magnetic flux at steady state and neglecting flux leakage and bulging of the magnetic flux lines in the operating air gap, we obtain from Maxwell's equations the following approach for the magnetic field:

$$\frac{\partial B_z(r,t)}{\partial t} = \frac{1}{\mu_R \kappa} \left[\frac{\partial^2 B_z(r,t)}{\partial r^2} + \frac{1}{r} \frac{\partial B_z(r,t)}{\partial r} \right]. \qquad (4.30)$$

With the initial and boundary conditions for the switch-on process

$$B(r,0) = 0 \quad r \in [0, r_0], \quad B(r_0, 0) = B_0, \quad \text{and} \quad \Phi(t) = \int_0^{r_0} B(r,t) 2\pi r \, dr$$

we find that

$$\Phi(t) = \Phi_0 \left(1 - 4\sum_{i=0}^{\infty} \frac{1}{x_i^2} e^{-x_i^2 \frac{t}{\mu_r \kappa r_0^2}}\right).$$

This equation can be approximated with a simple exponential function:

$$\Phi = \Phi_0 \left(1 - e^{-\frac{t}{\tau}}\right). \tag{4.31}$$

By analogy we obtain for the switch-off process:

$$\Phi = \Phi_0 e^{-\frac{t}{\tau}}, \tag{4.32}$$

with $\tau = \frac{\mu_r r_0^2}{\kappa}$.

With Eq. (4.13) and the initial conditions mentioned previously – the exciting current follows a step function – the following magnetic force *vs* time characteristic during switching on and off can be found:

$$F_\mathrm{m}(t) = \frac{\Phi(t)^2}{2\mu_0 A_\mathrm{Fe}} = \frac{\Phi^2}{2\mu_0 A_\mathrm{Fe}} \left(1 - e^{-\frac{t}{\tau}}\right)^2, \tag{4.33}$$

$$F_\mathrm{m}(t) = \frac{\Phi_0^2}{2\mu_0 A_\mathrm{Fe}} e^{-2\frac{t}{\tau}}. \tag{4.34}$$

The effects of eddy currents (Fig. 4.13) on the magnetic force *vs* time characteristic can be approximatively modelled with a resistance R_W parallel to

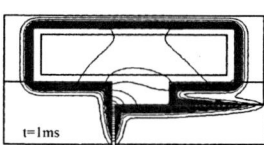
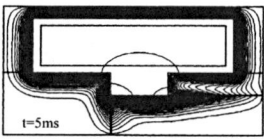
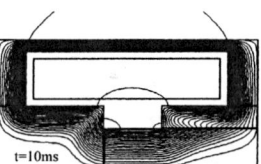

a

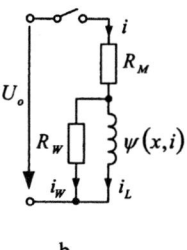
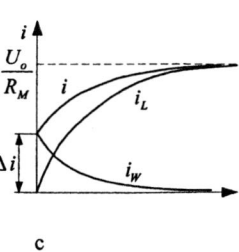

b   c

Fig. 4.13. Influence of eddy currents on the build-up of the field. **a** Penetration of the B-field into the magnetic circuit (current step). **b** Equivalent network. **c** Current characteristic (voltage step)

the coil's inductance (Fig. 4.13b). This resistance reduces the changing rate of the excitation current i_L that causes the magnetic force.

Effective measures for reducing eddy currents are:

- Assembly of the magnetic circuit from thin sheets (lamination)
- Slitting of rotationally symmetric parts
- Selecting magnetic materials that have low electrical conductivity
- Selection of advantageous designs for electromagnets

The faster the build-up or degradation of the magnetic field has to be, the more important is the reduction of the electrical conductivity of the magnetic material.

4.1.2.5 Measures for Manipulating the Dynamic Behaviour of DC Magnets

Electromagnetic drives with few exceptions (e. g. proportional magnets for hydraulic valves) are operated in open-loop controls (Fig. 4.14).

Thus, a targeted manipulation of the dynamic behaviour can be accomplished the following ways:

- Specification of an optimal input $w(t)$ (e. g. optimal current function for minimized thermal losses)
- Selection of an optimized driver circuit for the power controller
- Selection/design of the electromagnet with appropriate manipulated magnetic force characteristic (Fig. 4.8)

The last two measures have to be especially considered, if the control of the electromagnet is only possible with contact switches.

With the development of low-cost ICs, the specification of an optimal input function is used more and more often. With the help of driver circuits, the pull-in and drop-off processes can be manipulated within wide ranges.

The return-travel procedure over time is dependent on the degradation of the magnetic field after switching off the excitation and thereby on the electrical conducting paths in the system after breaking the excitation circuit.

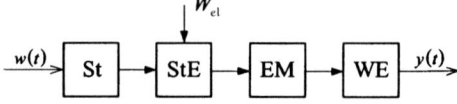

Fig. 4.14. Basic structure of an electromagnetic drive: *St* controller, *StE* driver, *EM* electromagnet, *WE* end effector. The dynamic output $y(t)$ is dependent on the input $w(t)$ and on the properties of the serial connection of all elements of the control path

Damping of Interrupting Voltages

Electromagnets possess high inductance values and thus act as sources of high overvoltages [4.2] after switching off the excitation that can jam adjacent electronic circuits. These overvoltages have to be reduced to permissible values (EMC) with appropriate measures of circuitry. A jam-protector circuit should possess a good damping and low power losses, enable a simple test of functional capabilities and have only minor influence on the actuation times of the electromagnet – combined with a space-saving design.

Real set-ups match these requirements up to different degrees. In [4.2] and [4.12] advices for reasonable protector circuits can be found. Figure 4.15 shows the schematic network for an electromagnet that can be retrofitted with different jam-protection elements (alternatives a–d):

- *Diode* (so-called free-wheeling diode or damping diode)
  While the pull-in time is not affected by the diode, the drop-out time increases eminently (factor 10 to 20). The reverse voltage of the diode should be at least $1.5U_N$ and the forward current should not be below $1.5I_N$. An interrupting overvoltage is not occurring. The energy losses due to the reverse current of the diode are very small and normally can be neglected.
- *Series connection of diode and resistor R*
  By choosing $R \approx R_M$, a maximum interrupting overvoltage of $U_{smax} \approx -U_N$ is occurring. The return travel time is reduced to about half of the value of alternative a).
- *Diode and Zener-diode*
  This alternative realizes adjustable interrupting overvoltages ($U_s \approx U_Z$) without mentionable delays of the return motion. The time until the current reaches 0 is $t_0 = \frac{L}{R_M} \ln\left(1 - \frac{1}{U_{s\,max}/U_N}\right)$.
  For a Zener voltage $U_Z = U_N$, we have $U_{smax} = -U_Z = -U_N$. Both diodes have to be dimensioned for the nominal current I_N.
- *RC-combination*
  Normally, the dimensioning is made in a way that allows a damped oscillation after switching off: $R = (0.2\ldots 1)R_M$, $C \approx L/4R_M^2$.

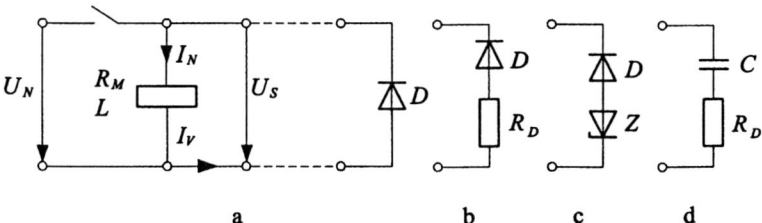

Fig. 4.15. Damping interrupting voltage spike. **a** Basic circuit with diode. **b** Diode and resistor. **c** Diode and Z-diode. **d** RC-combination

It is $U_{s\,\max} \approx -\frac{U_N}{R_M}\sqrt{\frac{L}{C}}$, $t_0 \approx \frac{\pi}{2}\frac{L}{R_M}\sqrt{\frac{L}{C}}$.

R and C must be dimensioned for a (short-term) peak value of I_N. The capacitor C (no electrolytic capacitor because of alternating polarities!) has to tolerate about two to three times of the nominal voltage.

Dynamic Behaviour of DC Electromagnets with Modified Magnetic Force Characteristic

DC magnets without modified magnetic force characteristic normally possess a high gradient of the force for the last stage of the stroke and thus, with mass load a disadvantageous dynamical behaviour. The speed of the armature increases eminently at the end of the stroke, which causes a high impact velocity. Figure 4.16 shows in curve 1 the velocity characteristic of such an electromagnet. If the magnetic force characteristic is modified by design modifications, the velocity characteristic is influenced as well. Curves 2 and 3 show the velocity characteristic of the same electromagnet – but with a horizontal force stroke characteristic – for different loads. Although the average value of the velocity is increased, the impact speed is much lower. The typical hard impact for electromagnets can be avoided this way.

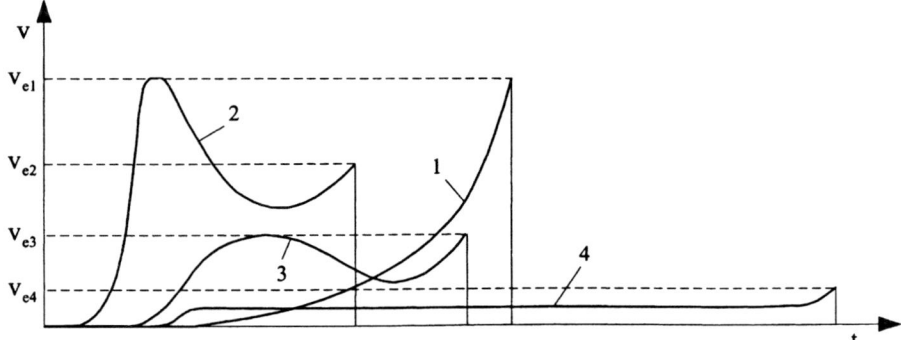

Fig. 4.16. Velocity characteristics of DC magnets with and without manipulated magnetic force characteristic (F_N – nominal load). *1* without manipulated magnetic force characteristic; *2* with manipulated magnetic force characteristic and without load; *3* with manipulated magnetic force characteristic, load $F_{\text{geg}} = F_N/2$; *4* with manipulated magnetic force characteristic, load $F_{\text{geg}} = F_N$

Dynamic Family of Characteristics for DC Electromagnets

Because of the steadily increasing requirements on the dynamic behaviour of electromagnetic actuators, the dynamic properties need to be considered in the design-development process from the very beginning and the electromagnetic drives have to be optimized under a dynamic point of view. Thus, the developer is obliged to evaluate the dynamic properties. Therefore a fundamental knowledge of the dynamic behaviour of existing magnets as well as possibilities and limits of targeted manipulation is essential. This is why criteria for the evaluation of the "total dynamic behaviour" of electromagnets are necessary. Because the dynamic properties are highly dependent on operation mode and drivers, an objective rating only can be achieved by knowing the dynamic behaviour over the whole operating range. The user of the electromagnet always wants to have information about the dynamic properties of the chosen system at different loads and operation modes. He also wants to know which possibilities exist to change the dynamic behaviour to desired values by measures of circuitry or change of the load. Both of these demands can be fulfilled with the dynamic family of characteristics (Fig. 4.17).

The first quadrant shows the pull-in time dependent on the normalised impressed voltage (t_{1u}) and the normalised impressed current (t_{1i}) with the load as parameter.

t_{1u} is the longest pull-in time that can occur during the switching-on process for a given case of operation because the voltage retroaction in Eq. (4.18) caused by $\frac{d\psi(x,i)}{dt} = \frac{\partial \psi}{\partial x}\frac{dx}{dt} + \frac{\partial \psi}{\partial i}\frac{di}{dt}$ is maximal.

t_{1i} is the shortest pull-in time, that can occur for the given case of operation, because the voltage retroaction does not take effect because of the impressed current.

At impressed exciting current, the pull-in delay t_{11} converges to zero, if no eddy currents appear. The mechanical motion and the travel time t_{12} can

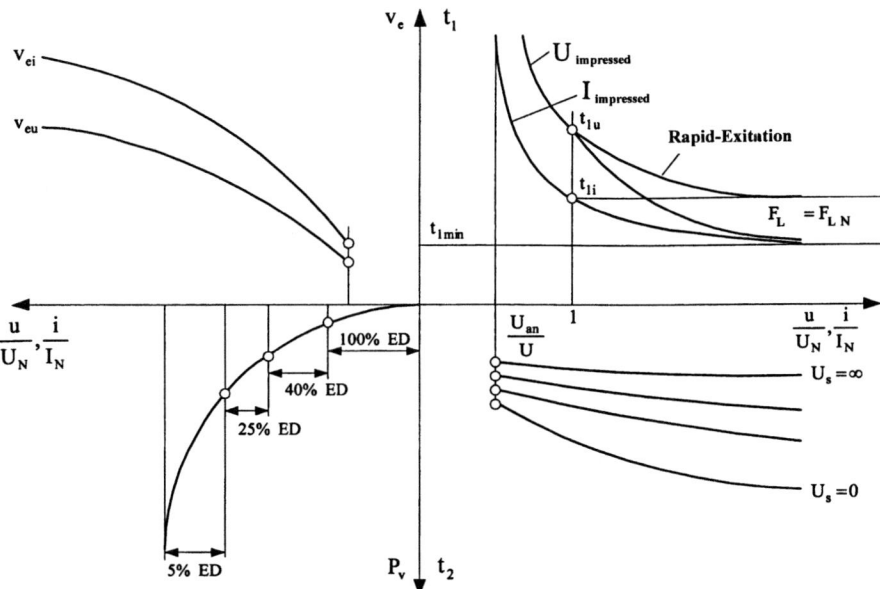

Fig. 4.17. Dynamic family of characteristics for DC electromagnets

be calculated solely with Eq. (4.19), where the magnetic force F_m can be set to its value at steady state at impressed current. The difference $t_{1u} - t_{1i} = \Delta t_1$ is a quantity for the possibility to reduce the pull-in time with a rapid-excitation circuit (changeover from the operation mode impressed voltage to the operation mode impressed current).

As we can see from the chart, the curves $t_{1u} = f(u/U_\mathrm{N})$ and $t_{1i} = f(i/I_\mathrm{N})$ converge dependent on load to the limiting value $t_{1\,\min}$, that cannot be underrun only with circuitry measures, i.e. the electric input signal. The shortest with a given electromagnet achievable pull-in time $t_{1\mathrm{mingr}}$ occurs without load.

The second quadrant of the dynamic family of characteristics shows the final velocity of the armature v_e dependent on the normalised impressed voltage (v_eu) and the normalised impressed current (v_ei).

v_e is a quantity for the kinematical energy $W_\mathrm{ü}$, that is set free at the impact of the armature into its complement. $W_\mathrm{ü}$ strongly influences impact chatter oscillations and abrasion and therefore should be limited, unless this energy is vital for the l operation (e.g. printing or stamping).

The third quadrant shows the characteristic of power loss and the ranges for different duty ratios (ED) and is necessary for the application of the electromagnet with respect to its thermal load. The tolerable increase of the electromagnet's power loss for decreasing duty ratios can be determined and by projecting the operating points into the first quadrant, the influence of this power increase onto the actuation times can be acquired.

Finally, the characteristic of $t_2 = f(u/U_\mathrm{N})$ with U_S (U_S – interrupting peak voltage, Fig. 4.9) as parameter shown in the fourth quadrant, allows to determine the change of the drop-out time t_2 in dependency on the normalised impressed voltage and the amplitude of the interrupting peak voltage influenced by damping elements.

These dependencies are of special interest for users of fast-acting electromagnets driven with electronic switches. An optimal dynamic behaviour always requires a compromise between damping of the interrupting voltage peak to get tolerable values for the electric circuit and the increase of the drop-out time.

Classification of electromagnets due to their dynamic properties: Electromagneto-mechanical drives can be separated into three operating ranges with respect to their dynamic properties on the basis of the dynamic family of characteristic curves (Fig. 4.18).

The three operating ranges are defined as follows:

Range I: $\frac{t_{1u}-t_{1i}}{t_{1i}} \geq 1$ drive with dominating electric inertia,

Range II: $1 > \frac{t_{1u}-t_{1i}}{t_{1i}} \geq 0.1$ drive with electro-mechanical inertia (fast action),

Range III: $\frac{t_{1u}-t_{1i}}{t_{1i}} < 0.1$ drive with dominating mechanical inertia (ultra-fast action).

The dynamic family of characteristic curves is also appropriate to show the influence of eddy currents onto the dynamic behaviour (Fig. 4.19) [4.13].

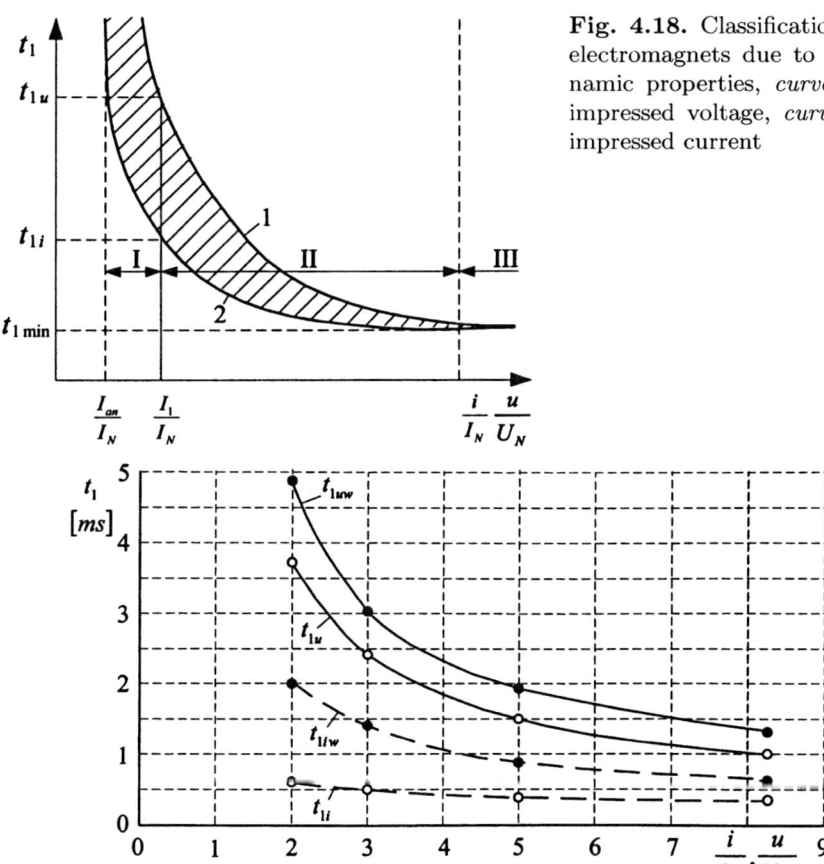

Fig. 4.18. Classification of electromagnets due to dynamic properties, *curve 1* impressed voltage, *curve 2* impressed current

Fig. 4.19. Influence of eddy currents on the pull-in time t_1, t_{1u} impressed voltage without eddy currents, t_{1uw} impressed voltage with eddy currents, t_{1i} impressed current without eddy currents, t_{1iw} impressed current with eddy currents

4.1.3 Alternating Current Magnets

4.1.3.1 General Issues

Alternating current magnets are electromagnets whose exciting coils are operated with alternating current and thereby create an alternating magnetic flux. Electromagnets with built-in rectifiers are DC magnets. AC magnets possess significantly different functional properties compared to DC magnets:

- The magnetic force is time-dependent.
- The root mean square current is dependent on the armature position, which leads to a smaller change of the magnetic force due to the travel of the armature, but can cause a thermal overload of the exciting coil, if the armature seizes in its initial position $\delta_{\max}$, because $I_{\mathrm{an}} \gg I_{\mathrm{h}}$ (Fig. 4.20).

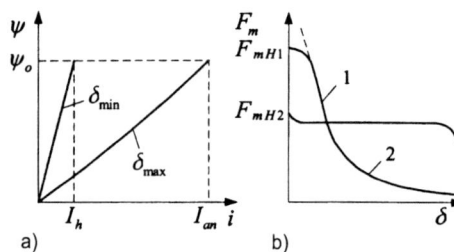

Fig. 4.20. Characteristic curves of the AC magnet ψ–i-characteristic of a linear AC magnet $\omega L \gg R$, qualitative comparison of force stroke characteristics of DC and AC magnets with identical geometric dimensions, *curve 1* DC magnet, *curve 2* AC magnet

- Appearance of eddy currents, eddy current losses and hysteresis losses within the magnetic circuit. Therefore, the magnetic circuit needs to be laminated with thin sheets or consist of materials with a low electrical conductivity.
- For AC magnets used for switching or positioning applications, measures for creating a constant component of the magnetic force in the pulled-in position have to be taken to prevent the magnet from humming.
- The dynamic properties are dependent on the switch-on phase angle α_1 and the switch-off phase angle α_2 (Figs. 4.25 and 4.27).

AC magnets can be used advantageously, when a drive element is needed that can be connected directly to the industrial AC electricity network or that shall generate mechanical oscillations with a frequency of 50 or 100 Hz [4.8].

For AC magnets with a small magnetic work ($W_{\mathrm{mech}} < 10 \, \mathrm{N \, cm}$) a lamination of the magnetic circuit can be omitted because of the small geometrical dimensions. For applications in this range, electromagnets can be designed, which can fulfil the same driving task for DC or AC excitation without changing the geometric dimensions of the magnetic circuit, but only the coil's parameters (e. g. relays, magnetic valves).

4.1.3.2 Computation of the Magnetic Force of Single-Phase AC Magnets

The mathematical relations for the magnetic force can be obtained from the law of energy conservation by analogy to the DC magnet (see Sect. 4.1.2.2).

Generally, the relations for AC and DC magnets are identical, but it has to be considered, that the current is time-dependent. Mostly, AC magnets are designed with a linear ψ–i-characteristic.

The magnetic force of a linear AC magnet can be computed with:

$$F_{\mathrm{m}} = \frac{1}{2} I^2 \frac{\mathrm{d}L(\delta)}{\mathrm{d}\delta} \, . \tag{4.35}$$

For impressed voltage, we obtain from Eq. (4.35)

$$F_{\mathrm{m}} = \frac{1}{2} \frac{U^2}{R^2 + [\omega L(\delta)]^2} \frac{\mathrm{d}L(\delta)}{\mathrm{d}\delta} \, . \tag{4.36}$$

Furthermore, we can derive from Eq. (4.35) by analogy to the DC magnet:

$$F_\mathrm{m} = \frac{B(t)^2 A_\mathrm{Fe}}{2\mu_0}, \tag{4.37}$$

$$F_\mathrm{m} = \frac{\Phi(t)^2}{2\mu_0 A_\mathrm{Fe}}. \tag{4.38}$$

Especially for large AC magnets, the ohmic voltage drop in the exciting coil is neglectably small compared to the inductive component.

Thus, the flux linkage is independent of the armature position for impressed voltage. Therefore, the ψ–i-characteristic shown in Fig. 4.20a is valid for AC magnets.

In comparison to pulled-in armature (retention current I_h), the AC magnet consumes a much higher current for dropped-out armature (pull-in current I_an):

$$I(\delta) = \frac{U}{\sqrt{R^2 + |\omega L(\delta)|^2}} \approx \frac{U}{\omega L(\delta)}. \tag{4.39}$$

For technically realised magnets, we have

$$\begin{aligned} I_\mathrm{an} &= 3\ldots 10 I_\mathrm{h}\,; \quad I_\mathrm{an} = I(\delta_\mathrm{max}) \\ I_\mathrm{h} &= I(\delta_\mathrm{min})\,. \end{aligned} \tag{4.40}$$

Thus, a complete pulling in of the armature must be guaranteed to avoid thermal overload of the exciting coil.

4.1.3.3 Time Dependency of the Magnetic Force of Single-Phase AC Magnets

Because of the time dependency of the magnetic flux, the magnetic force is also time-dependent. For a linear magnetic circuit, a sinusoidal exciting voltage causes a sinusoidal exciting current and a sinusoidal magnetic flux.

$$\Phi = \hat{\Phi} \sin \omega t\,. \tag{4.41}$$

With Eqs. (4.41) and (4.38), we obtain:

$$\begin{aligned} F_\mathrm{m}(t) &= \frac{1}{2\mu_0 A_\mathrm{Fe}} \hat{\Phi}^2 \sin^2 \omega t \\ &= \frac{1}{4\mu_0 A_\mathrm{Fe}} \hat{\Phi}^2 (1 - \cos 2\omega t)\,. \end{aligned} \tag{4.42}$$

The magnetic force changes with the double frequency of the exciting current between zero and $\frac{\hat{\Phi}^2}{4\mu_0 A_\mathrm{Fe}}$.

This magnetic force vs time characteristic (Fig. 4.21) is very unfavourable for the use of AC magnets in switching or positioning applications. If counterforces F_geg appear at the end of the stroke, the force $F_\mathrm{m} - F_\mathrm{geg}$ is applied to

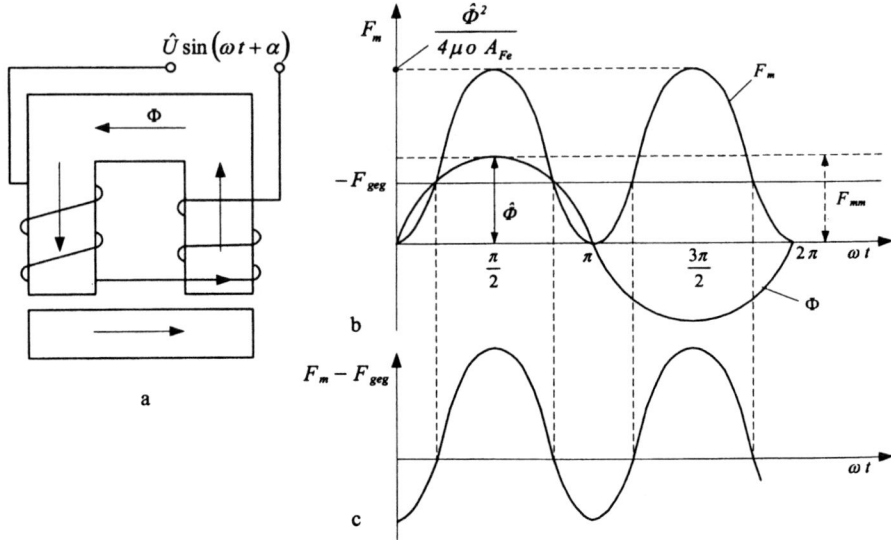

Fig. 4.21. AC magnet without short circuit ring. **a** Structure. **b** Magnetic force over the time. **c** Armature force over the time

the armature and can cause oscillations (the magnet is humming). Thus, the chatter oscillations cause a quick abrasion of the pole surfaces. That's why measures to eliminate oscillations have to be taken. The aim is, to obtain a high constant component (DC component) of the magnetic force at the end of the stroke to keep the resulting force $F_\mathrm{m} - F_\mathrm{geg}$ positive and to retain the armature at its end position.

Therefore, the following measures are appropriate:

- Attaching a short circuit ring at the magnetic poles at the operating air gap in single-phase AC magnets;
- Attaching multiple spatially distributed windings, that generate phase-shifted forces in different areas of the pole surfaces.

Both measures have to cause phase-shifted forces that accumulate at the armature.

Magnetic Force of the Single-Phase AC Magnet with Short Circuit Ring

Figure 4.22 shows the magnetic pole of a single-phase AC magnet with short circuit ring. The magnetic flux Φ_z caused by the short circuit ring leads by summation with the partial fluxes Φ_1 and Φ_2 to two phase-shifted fluxes through the air gap:

$$\Phi_\mathrm{f} = \hat{\Phi}_\mathrm{f} \cos(\omega t + \varphi_\mathrm{fk}) \tag{4.43}$$

$$\Phi_\mathrm{k} = \hat{\Phi}_\mathrm{k} \cos \omega t \ . \tag{4.44}$$

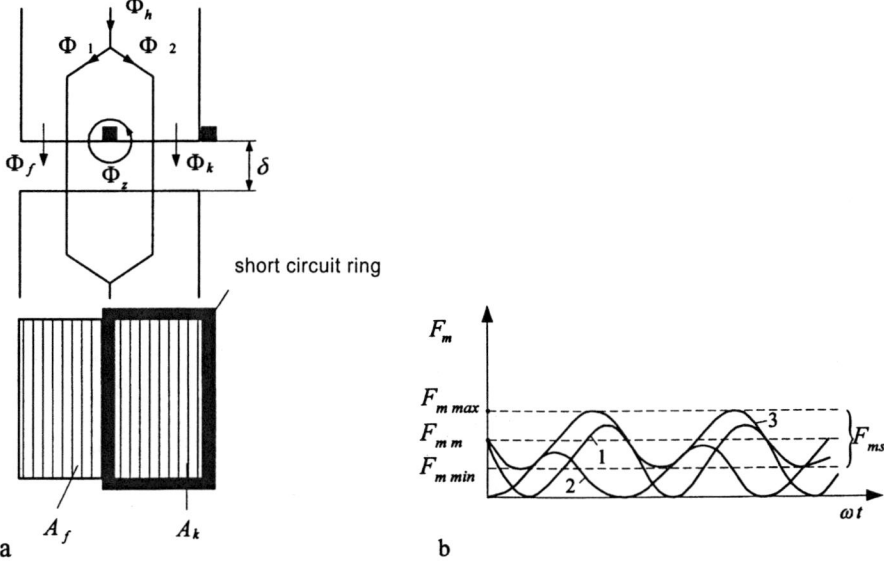

Fig. 4.22. Magnetic fluxes of a single-phase AC magnet with short circuit ring; flux distribution and surface separation at the magnetic pole; time-dependency of the magnetic force at $\delta = 0$; curves *1, 2* magnetic forces of the surface areas $A_{\rm f}$, $A_{\rm k}$, *curve 3* is the resulting force

By adding the phase-shifted forces at the surfaces $A_{\rm f}$ and $A_{\rm k}$, we obtain the following relations for the resulting magnetic force $F_{\rm mg}$.

$$F_{\rm mg} = \frac{1}{4\mu_0}\left[\frac{\hat{\Phi}_{\rm f}^2}{A_{\rm f}} + \frac{\hat{\Phi}_{\rm k}^2}{A_{\rm k}} + \frac{\hat{\Phi}_{\rm f}^2}{A_{\rm f}}\cos(2\omega t + \varphi_{\rm fk}) + \frac{\hat{\Phi}_{\rm k}^2}{A_{\rm k}}\cos 2\omega t\right] . \qquad (4.45)$$

For example, the magnetic force consists of a constant component

$$F_{\rm mm} = \frac{1}{4\mu_0}\left[\frac{\hat{\Phi}_{\rm f}^2}{A_{\rm f}} + \frac{\hat{\Phi}_{\rm k}^2}{A_{\rm k}}\right] \qquad (4.46)$$

and an alternating component

$$F_{\rm mS} = \frac{1}{4\mu_0}\left[\frac{\hat{\Phi}_{\rm f}^2}{A_{\rm f}}\cos(2\omega t + \varphi_{\rm fk}) + \frac{\hat{\Phi}_{\rm k}^2}{A_{\rm k}}\cos 2\omega t\right] . \qquad (4.47)$$

The time dependency of the resulting force $F_{\rm mg}$ is shown in Fig. 4.23.

A complete suppression of the alternating component $F_{\rm mS}$ according to Eq. (4.47) would be optimal for the application, i.e.:

$$\frac{\hat{\Phi}_k^2}{A_k} = \frac{\hat{\Phi}_f^2}{A_f} \quad \text{and} \quad \varphi_{\rm fk} = 90° .$$

These two conditions cannot be fulfilled with a short circuit ring all at once under real circumstances.

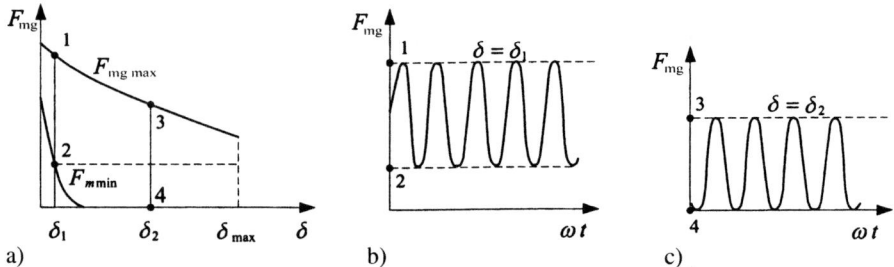

Fig. 4.23. Effects of the short circuit ring for a shaded pole magnet. **a** *Curve 1* $F_{mmax} = f(\delta)$, *curve 2* $F_{mmin} = f(\delta)$. **b** Time-dependent magnetic force at operating air gap length δ_1. **c** Time-dependent magnetic force at operating air gap length δ_2

With increasing δ, φ_{fk} converges to zero. The short circuit ring only takes the desired effect for small operating air gaps, i.e. for pulled-in armature (Fig. 4.23). That's why the constant component of the magnetic force decreases fast with increasing air gap δ. Soiled pole surfaces or abrasion can lead to humming effects.

4.1.3.4 Magnetic Force of a Three-Phase Alternating Current Magnet

Figure 4.24 shows the general design of a three-phase AC magnet.

When this magnet is connected to a symmetrical three-phase electricity network, the magnetic fluxes through the branches of the magnetic circuit are

$$\Phi_1 = \hat{\Phi}^2 \sin \omega t \,, \quad \Phi_2 = \hat{\Phi}^2 \sin\left(\omega t + \frac{2}{3}\pi\right) \,, \quad \Phi_3 = \hat{\Phi}^2 \sin\left(\omega t + \frac{4}{3}\pi\right) \,.$$

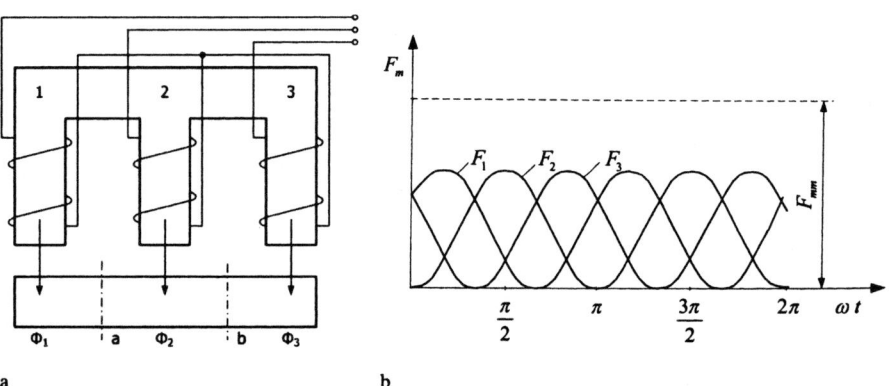

Fig. 4.24. Three-phase alternating current magnet. **a** Basic configuration. **b** Magnetic force *vs* time characteristic

With Eq. (4.38) and summation of the forces generated at the different pole surfaces, we obtain:

$$F_{\text{mg}} = \frac{\hat{\Phi}^2}{2\mu_0 A_{\text{Fe}}} \left[3 - \cos 2\omega t + \cos\left(2\omega t + \frac{4\pi}{3}\right) + \cos\left(2\omega t + \frac{8}{3}\pi\right) \right] \quad (4.48)$$

$$F_{\text{mg}} = \frac{3\hat{\Phi}^2}{2\mu_0 A_{\text{Fe}}} = F_{\text{mm}} . \quad (4.49)$$

The phase-shifted fluxes cause sinusoidal forces at different areas. With the typical conditions of a symmetrical three-phase network a constant magnetic force with a ripple independent of the length of the operating air gap δ is generated.

4.1.3.5 Dynamic Behaviour of AC Magnets

The computation of the dynamic behaviour of AC magnets can be done by evaluating Eqs. (4.18) and (4.19) with $U = \hat{U} \sin(\omega t + \alpha)$ and considering the magnetic force's time-dependency.

$$\hat{U} \sin(\omega t + \alpha) = iR + \frac{d\psi}{dt} \quad (4.50)$$

$$m\ddot{x} + \rho\dot{x} + F_{\text{geg}}(x) = F_{\text{m}}(t) . \quad (4.51)$$

Where $\alpha = \alpha_1$ represents the switch-on phase angle of the exciting voltage for examining the switch-on process or, respectively, $\alpha = \alpha_2$ is the switch-off phase angle for the switch-off process.

To compute the pull-in delay, the integration of Eq. (4.50) is sufficient.

For a linear magnetic circuit, we obtain with the initial conditions $u = \hat{U} \sin \alpha_1$ and $i = 0$ for $t = \vartheta$

$$\frac{i}{\hat{I}} = \sin(\omega t + \alpha_1 - \varphi) - e^{-\frac{t_{11}}{\tau}} \sin(\alpha_1 - \varphi) \quad (4.52)$$

with $\hat{I} = \frac{\hat{U}}{\sqrt{R^2 + [\omega L(\delta)]^2}}$; $\varphi = \arctan \frac{\omega L_0}{R}$.

The dependency of the exciting current on α_1 leads to huge variations for t_{11}.

At the time t_{11}, we have

$$\frac{I_{\text{an}}}{\hat{I}} = \sin(\omega t_{11} + \alpha_1 - \varphi) - e^{-\frac{t_{11}}{\tau}} \sin(\alpha_1 - \varphi). \quad (4.53)$$

As we can clearly see, t_{11} is a function of α_1 (Fig. 4.25). This equation cannot be solved offhand for t_{11}. A generalized solution is only possible for a few special cases. For $\alpha_1 - \varphi = 0$, we obtain

$$t_{11} = \frac{\arcsin \frac{I_{\text{an}}}{\hat{I}}}{2\pi f} . \quad (4.54)$$

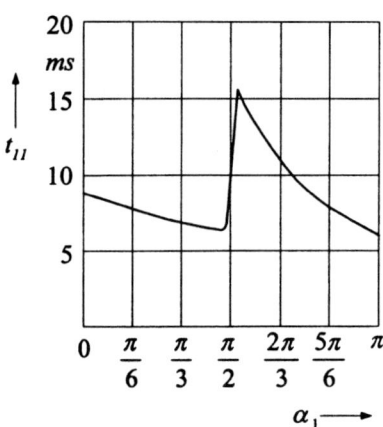

Fig. 4.25. Dependency of the pull-in delay t_{11} on α_1

This shows, that depending on the load and thus depending on I_{an}, t_{11} can vary within the range of $0 \leq t_{11} \leq \frac{1}{4f}$.

For $\omega L \gg R$, we have

$$\frac{I_{\text{an}}}{\hat{I}} = \sin(\omega t_{11} + \alpha_1 - \frac{\pi}{2}) \tag{4.55}$$

and thus

$$t_{11} = \frac{\arcsin \frac{I_{\text{an}}}{\hat{I}} - \alpha_1 + \frac{\pi}{2}}{2\pi f}. \tag{4.56}$$

In [4.10], the influence of the geometry of the magnetic circuit on the pull-in delay t_{11} is investigated.

Computer-Aided Calculation of the Travel-Time t_{12}

Exact calculations of the dynamic behaviour of AC magnets can be done with digital computers excellently. For AC magnets with a linear ψ–i-characteristic, computer programs and computation results are shown in [4.11].

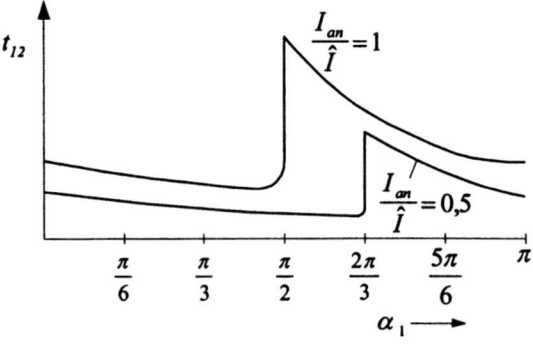

Fig. 4.26. Travel time t_{12} as a function of α_1

The time dependency of the magnetic force leads to much more complicated transient processes for each variable during the switching procedures compared to DC magnets.

Especially the strong dependency of the travel time t_{12} on the switch on phase angle α_1 has to be considered (Fig. 4.26).

Computation of the Drop-Out Delay t_{21}

The drop-out delay depends strongly on the value of the magnetic flux through the magnetic circuit at the turn-off moment, i. e. the value of α_2. If conductive paths with the resistance R for the current through the coil are existing after switching off the exciting voltage, we have

$$\Phi = \Phi_0 \, e^{-\frac{t}{\tau}} \quad \text{with} \quad \tau = \frac{L}{R} . \tag{4.57}$$

With $\Phi_0 = \hat{\Phi} \sin \alpha_2$, we obtain

$$\Phi = \hat{\Phi} \sin \alpha_2 \, e^{-\frac{t}{\tau}} \tag{4.58}$$

or, respectively,

$$t_{21} = \ln \frac{\hat{\Phi} \sin \alpha_2}{\Phi_R} . \tag{4.59}$$

Φ_R represents the reset flux.

According to Eq. (4.59), the drop-out delay falls within the range

$$0 \leq t_{21} \leq \tau \ln \frac{\hat{\Phi}}{\Phi_R} . \tag{4.60}$$

After disconnecting the AC magnet from the electricity network, AC and DC magnets show the same behaviour.

Computation of the Return Travel Time t_{22}

For the computation of the return travel time, the same relations compared to the DC magnet are valid. The principle of manipulated force characteristics is not applied for AC magnets. Therefore it can be assumed, that the return motion starts, when $i = 0$.

Dynamic Family of Characteristics for AC Magnets

For application and evaluation of the properties of AC magnets as driving element, by analogy to DC magnets, the most important dynamic properties can be outlined in a dynamic family of characteristic curves (Fig. 4.27). Because of the strong influence of the switch-on phase angle α_1 and the switch-off phase angle α_2 on the dynamic behaviour, an adaptation of the dynamic family of characteristic curves of the DC magnet has to be made [4.9].

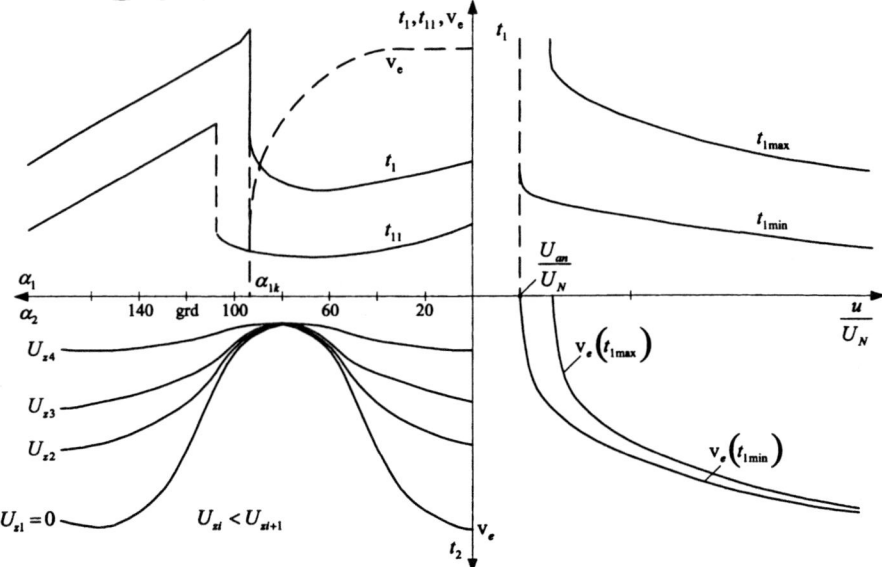

Fig. 4.27. Dynamic family of characteristics of AC magnets

4.1.4 Polarized Electromagnets

4.1.4.1 Specifics

Besides one or more exciting coils and magnetically soft iron parts for the guidance of the magnetic flux, the magnetic circuit of polarized electromagnets contains elements that can store magnetic field energy and can act as magnetic energy sources in specific operation modes and thus generate magnetic forces. Functional and structural properties of polarized electromagnets are primarily characterised by the interaction of the magnetic flux of the coil and of the permanent magnet.

With the rapid development of permanent magnetic materials over the last decades (increase of the energy content by two powers of ten), much better preconditions arose to improve the specific properties of polarized magnets, to reduce their size and to widen their area of application. Besides the classical classification criteria like the basic shape of the magnetic circuit, position of the operating air gap, movement pattern of the armature and area of application, polarized electromagnets can be classified by their guidance of the magnetic flux and the play of the operating point.

By means of flux guidance, they can be divided into polarized electromagnets with

- Polarized series circuit (Fig. 4.28a)
- Polarized parallel circuit (Fig. 4.28b)
- Polarized bridge circuit (Fig. 4.28c) [4.20]

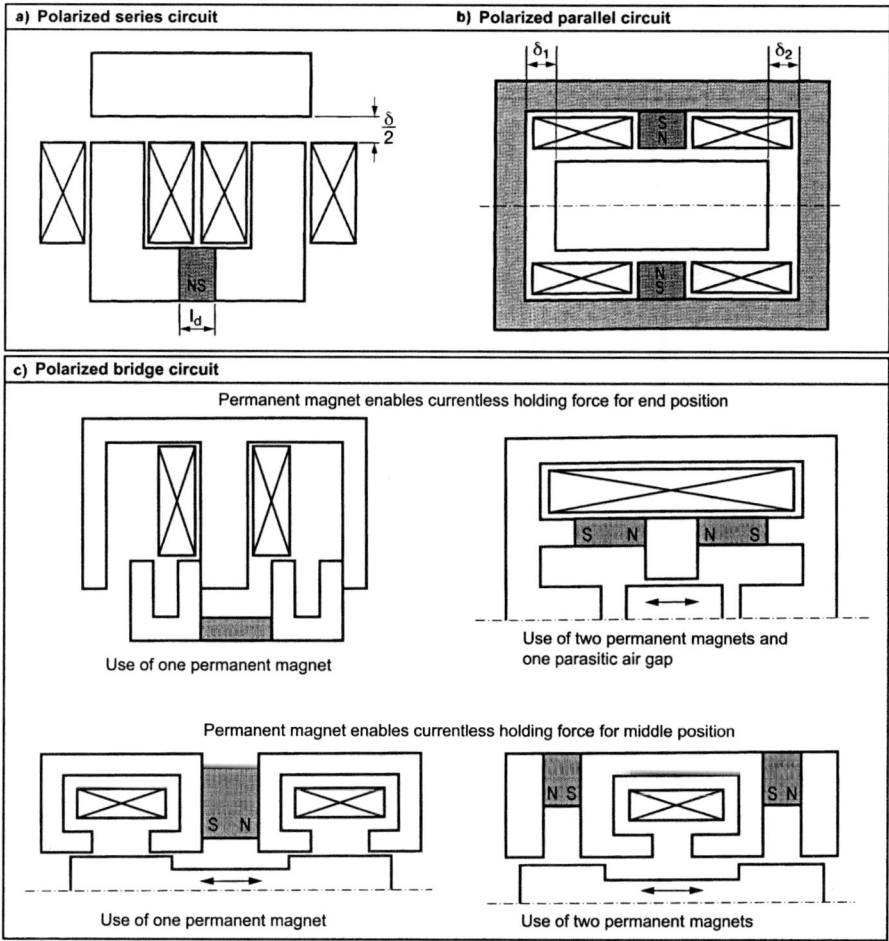

Fig. 4.28. Basic shapes for polarized electromagnets. **a** Polarized series circuit. **b** Polarized parallel circuit. **c** Polarized bridge circuit

By means of the operating point play, they can be divided into polarized electromagnets with

- Remanent magnetic circuit
- Dynamic permanent magnetic circuit [4.14]

4.1.4.2 Magnetic Force of Polarized Electromagnets with Series Circuit

The computation of the magnetic force of polarized electromagnets can be done similarly to the computation of the magnetic force of neutral electromagnets, if all permanent magnetic parts in the network representation are replaced by active two-terminal networks – consisting of the equivalent permanent magnetic source Θ_D and its internal reluctance R_D.

For the polarized series circuit, we obtain a simplified equivalent magnetic network (Fig. 4.29).

To determine Θ_D and R_D of the respective permanent magnetic part, the following way is advisable: Starting from the material-specific demagnetization curve of the selected permanent magnetic material, the circuit-specific Φ–Θ-characteristic is calculated with consideration of the geometric dimensions of the permanent magnetic circuit and is depicted as an active two-terminal network.

From the equivalent magnetic network, we obtain the magnetic flux as follows:

$$\Theta_D + \Theta_e = \Phi \left(R_{mi} + R_D + \frac{1}{G_{m\delta} + G_{m\sigma}} \right) \tag{4.61}$$

$$\Phi = \frac{\Theta_D + \Theta_e}{R_{mi} + R_D + R_{m\delta}^*}, \tag{4.62}$$

with $R_{m\delta}^* = \frac{1}{G_{m\delta} + G_{m\sigma}}$.

With Eqs. (4.62) and (4.8), we obtain the magnetic force of the polarized electromagnet:

$$F_m = \frac{(\Theta_D + \Theta_e)^2}{2} \frac{1}{(R_{mi} + R_D + R_{m\delta}^*)^2} \frac{dR_{m\delta}^*}{d\delta}. \tag{4.63}$$

Equation (4.63) reveals the following interesting properties in comparison to neutral electromagnets:

- Powerless generation of magnetic forces, especially retention forces.
- Reduction of power losses for constant retention forces, because power losses are only generated by the exciting current, or respectively increase of the magnetic forces for constant power losses.
- To compensate the permanent magnetic field during the turn-off process a inversion of the direction of the exciting current is necessary (bipolar drive, Fig. 4.30b).
- For a specific magnetic circuit (bridge circuits) it is possible to generate a dependency of armature position on current direction and decrease the needed activation power. Decreasing the needed activation power is especially important for the construction of sensitive relays.

Figure 4.30 shows the magnetic force-stroke characteristic and the exciting current vs time characteristic for a magnetic clamp. With polarized DC magnets, shorter pull-in times compared to neutral magnets can be realized [4.13].

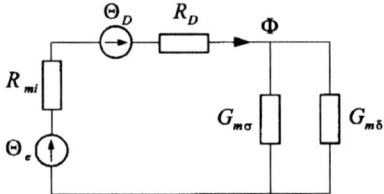

Fig. 4.29. Equivalent magnetic network of a polarized series circuit, $G_{m\sigma}$ leakage permeance

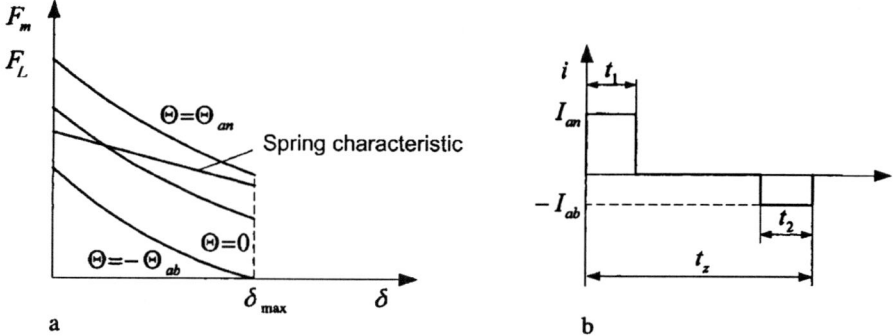

Fig. 4.30. Polarized series circuit. **a** Force-stroke characteristics. **b** Time characteristic of the excitation impulses

4.1.4.3 Application of Polarized Electromagnets

Well-known fields of application for polarized electromagnets are:
- Polarized magnetic drives with powerless lock.
- Polarized magnetic drives as impulse actuators with small volume (Fig. 4.31) and safety switching by retaining the current position in case of voltage drop.
- Polarized relay drive for realizing extremely sensitive relays or relays for indicating the direction of the exciting current. Normally, the magnetic circuit of these relays consists of a parallel connection of two magnetic circuits.
- Drives for trigger or lock switches with extremely small turn-off times.

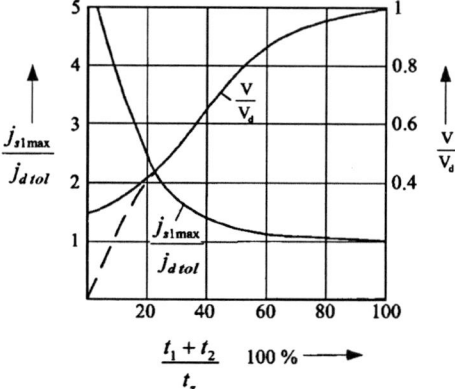

Fig. 4.31. Dependency of the magnet's volume V and the tolerable current density dependent on the duty ratio for constant magnet work. V volume of the magnet, V_d magnet volume for 100% ED, j_{dzul} tolerable current density for continuous duty, j_{s1max} tolerable current density for short time duty, t_1 length of the turn-on current impulse, t_2 length of the turn-off current impulse, t_z cycle time (Fig. 4.30b)

4.1.5 Project Planning of Electromagnets and Magnetic Drives

Electromagnets and magnetic drives have a wide range of technical applications. Because of the large quantities and the important driving tasks some of them are produced for, an optimal design has a vital role in the development process. The steady upward tendency towards the development of highly specialized magnets causes the urgent need for development methods and tools including the involvement of digital computers to calculate optimal main geometric dimensions and to simulate the dynamic behaviour. Therefore, the electromagnet has to be regarded as technical system (Fig. 4.14) that is optimized in its entirety.

Numerous advises for a systematic project planning of electromagnets can be found in relevant literature, while hints for CAD of electromagnets are given rarely, yet [4.15].

Of vital role for the process of project planning for electromagnets are:

- Selection of the shape of the magnetic circuit (U-shaped magnet, tubular magnet), position of the operating air gap in relation to the exciting coil (flat face or plunger armature), guidance or bearing of the armature, choice of the excitation (AC, DC or impulse excitation).
- Length of the nominal stroke δ_N is defined by the static and dynamic operation circumstances (relays $0.3\ldots 5$ mm, contactors $10\ldots 20$ mm, braking magnets $30\ldots 150$ mm). Electromagnets are usually used as direct drives. Thus, the force-stroke characteristic and the load characteristic have to be adapted adequately. For DC magnets, an adaptation to the load characteristic with geometric elements to manipulate the magnetic force characteristic is highly efficient. By means of miniaturization, solutions with unification of drive element and end effector in terms of functional integration are targeted increasingly (e. g. magnetic valves, magnetic clutches, reed relays).
- Selection of adequate electronic driver circuits. The progress of electronics/micro electronics increases the possibilities to adapt the static and dynamic properties of electromagnets to the user's application with circuitry measures for economically justifiable costs. Thus, a huge reduction of the magnet's size can be possible under certain circumstances (Fig. 4.31). DC and AC magnets generally vary in their dynamic behaviour (force-stroke characteristics, dynamics, life cycle length, heating). The retention force of DC magnets normally is much higher than its pull-in force (depending on the size of the operating air gap about 10 to 100 times), so that measures for increasing the pull-in force are favourable. In contrast to that, the constant component of the retention force of AC magnets is often too small.
- The selection of advantageous magnet circuit shapes is made easier with the help of data bases, that contain the relation between function and structure (function-structure database). In [4.15], a program system is presented that allows a complete computer-aided development process of tubular magnets.

- Heating computations based on network methods [4.18] and numeric field computation methods.

4.1.6 Design of Magnetic Drives

4.1.6.1 Magnetic Drives as Mechatronic Systems

The driving properties of magnetic drives are dependent both on the electromechanical energy converter and on the other drive components, control unit, electronic driver, transmission element and end effector (Fig. 1.2).

The development of magnetic drives is characterized by the following attributes:

- Within drive systems material flows, energy flows and information flows appear, that interfere with each other
- The number of elements and their functional linkage increase, which causes an exponential growth of the solution set
- Components can be assigned to different domains, which leads to an increasing heterogeneity
- In contrast to the previously common optimization of the single components, drive systems have to be optimized in their entirety. That results in the creation of mechatronic systems with a high innovation potential

Today there is no generally valid computer-aided design tool for the development of magnetic drives that matches the requirements on an entire optimization and leads to a fast solution. With the guideline VDI 2206 [4.23] since 2004 a general development methodology for mechatronic systems exists, which establishes a basis for a cross-domain design methodology as an advice for the development procedure.

VDI 2206 suggests executing of the development process for mechatronic systems in the following three stages:

- System design as conceptual design, during which the functional structure of a drive section is split into ϑ subfunctions F_i to which adequate solution elements as substructures S_i are assigned. These assignations should be done under consideration of knowledge-based function-structure databases with respect to a possible later integration (volume integration, functional integration).
- Domain-specific design, during which the substructures are designed with domain-specific tools. Normally, existing design tools for the single domains can be used efficiently.
- System integration of the domain-specific substructures into an optimized entire system with respect to functional as well as technological aspects.

Because of the high complexity of mechatronic systems, synthesis-friendly design tools are necessary, that can be cross-domain coupled and are able to consider effects that have been neglected in the past, but are influencing the drive section's properties.

For instance, these can be the consideration of the non-linearity of the magnetic circuit (that limits the force density of magnetic actuators), electromagnetic hysteresis (that leads to a hysteresis of the force of proportional magnets and magnetic bearings) and eddy currents (that affect actuation times and fast action properties).

4.1.6.2 Computation Methods for Magnetic Drives

Analytic computation methods are still important nowadays in this age of powerful digital computer and program systems. Valuable models with only small modelling depth that are based on idealized assumptions allow to determine limits, e. g. of the dynamic properties.

Field computation methods for the solution of boundary condition tasks of the magnetic field are especially useful for the domain-specific design and for analysis tasks, e. g. for a given structure, functional properties can be calculated. Often their usage is limited by long computation times. Thus, they are only of restricted suitability for synthesis tasks or optimization procedures.

Network methods possess the advantage, that – with respect to the physical laws of the magnetic fields – equivalent network models can be set up, which require only short computation times and allow optimization tasks. Besides that, physical phenomena like saturation, hysteresis and eddy currents can be considered easily. Network computation methods are more efficient, but less accurate than field computation methods. Figure 4.32 shows advantages and disadvantages of each computation method.

4.1.6.3 Network Method for Magnetic Drives

For the application of magnetic drives in order to generate alternating motions (Fig. 4.3), the static magnetic force characteristic, actuation times in both motion directions and velocity characteristics are relevant. Ströhla [4.16] established the mathematical basis for the computer-aided simplification of the network method, so that even challenging design tasks (optimization of static and dynamic behaviour with respect to hysteresis and eddy currents) can be solved efficiently.

Computation of Static Magnetic Force Characteristics – Theoretical Basis

Starting point for the computation of the magnetic force characteristic is the change of the magnetic co-energy according to Eq. (4.8) when the armature is moving. Electromagnets, whose magnetic circuit is excited far into saturation during the duty cycle, e. g. for over-excitation (Fig. 4.31), single-mesh magnetic networks are not sufficient because of the arising flux leakage. For

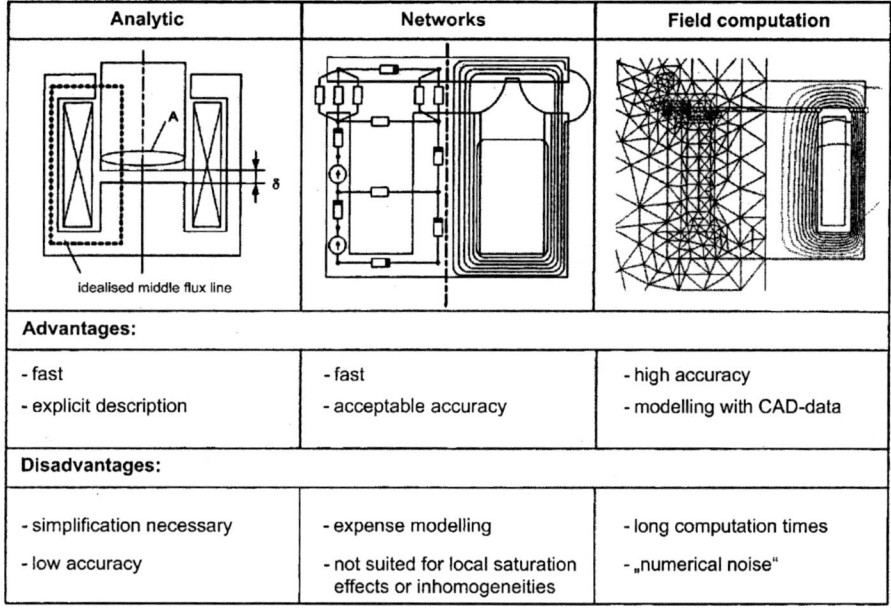

Fig. 4.32. Advantages and disadvantages of different computation methods [4.19]

this case, adequately accurate solutions can only be obtained with multi-mesh networks and distributed sources of the magnetomotive force Θ_v (Fig. 4.32).
The respective magnetic co-energy can be calculated with Eq. (4.8):

$$W_m^* = \int_0^{\Theta_0} \Phi(\Theta, \delta) \, d\Theta \qquad (4.64)$$

$$= \int_0^{\Theta_0} \sum_{\Theta_1}^{\Theta_v} \Phi_v(\Theta_v, \delta) \, d\Theta_v \ ,$$

where Θ_v represents the magnetomotive force sources of the single meshes. Within a network the sum of the energies the sources feed in the network equals the sum of the energies that are accepted by the reluctances as load

$$W_m^* = \sum_{\lambda=1}^N \int_0^{V_\lambda} \Phi_\lambda \, dV_\lambda \ . \qquad (4.65)$$

V_λ represent the magnetic voltage drop over the reluctances. If a network contains N_l linear and N_{nl} non-linear reluctances, the co-energies can be written as two sums:

$$W_m^* = \sum_{\lambda=1}^{N_l} \int_0^{V_\lambda} \Phi_\lambda \, dV_\lambda + \sum_{\lambda=1}^{N_{nl}} \int_0^{V_\lambda} \Phi_\lambda \, dV_\lambda \qquad (4.66)$$

with $N = N_l + N_{nl}$. With the application of Hopkinson's law $V = R_m \Phi$, we obtain from Eq. (4.66)

$$W_m^* = \sum_{\lambda=1}^{N_l} \frac{V_\lambda^2}{2R_{m\lambda}} + \sum_{\lambda=1}^{N_{nl}} A_\lambda l_\lambda \int_0^{V_\lambda} B_\lambda \, dH_\lambda \ . \qquad (4.67)$$

Equation (4.67) contains only one left integral, if the magnetic circuit is made from one material with one B–H-characteristic. This integral represents the co-energy density of the magnetic material with

$$w_m^* = \int_0^H B \, dH \ . \qquad (4.68)$$

From Eq. (4.67) we obtain with Eq. (4.68);

$$W_m^* = \sum_{\lambda=1}^{N_l} \frac{1}{2} V_\lambda \Phi_\lambda + \sum_{\lambda=1}^{N_{nl}} A_\lambda l_\lambda w_m^* \left(\frac{V_\lambda}{l_\lambda} \right) \ . \qquad (4.69)$$

Thus, the magnetic co-energy only is dependent on the magnetic branch voltages V_λ and the partial fluxes Φ_λ. Because the co-energy density function

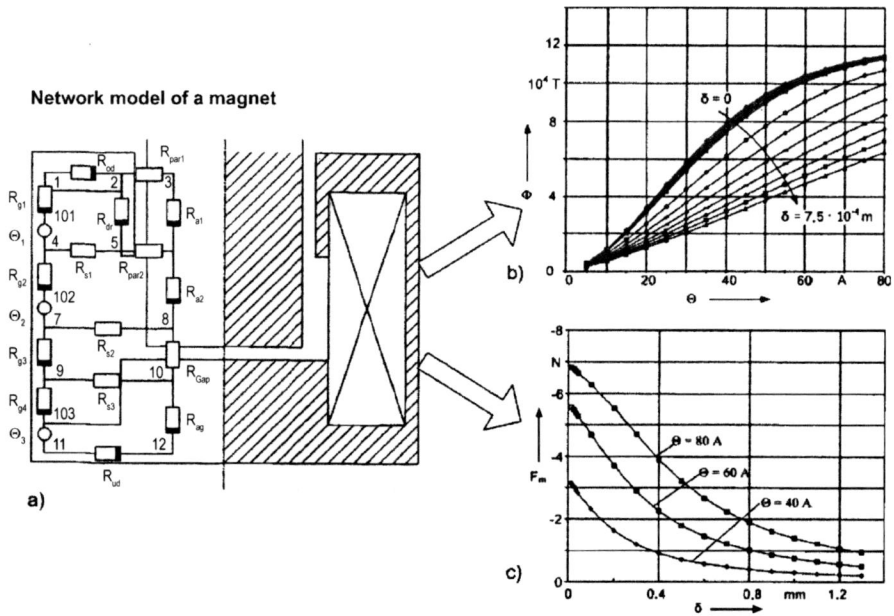

Fig. 4.33. Computation of the static magnetic force characteristic of a tubular magnet with the network method. **a** Network model of a tubular magnet. **b** $\Phi(\Theta, \delta)$-family of characteristics of a tubular magnet. **c** $F_m(\Theta, \delta)$-family of characteristics

describes the non-linearity of the magnetization characteristic, the co-energy of the system can be determined by a single call of this function, which is advantageous for computing. Derivation of Eq. (4.69) with respect to the operating air gap, after some transformations, we obtain:

$$F_\mathrm{m} = \sum_{\lambda=1}^{N_\mathrm{nl}} \Phi_\lambda \frac{\mathrm{d}V_\lambda}{\mathrm{d}\delta} - \frac{1}{2}\sum_{\lambda=1}^{N_1} \Phi_\lambda^2 \frac{\mathrm{d}R_{\mathrm{m}\lambda}}{\mathrm{d}\delta}. \qquad (4.70)$$

With help of Telegen's law it can be shown, that the first summand in Eq. (4.70) equals zero and we obtain a surprisingly simple expression for the magnetic force:

$$F_\mathrm{m} = -\frac{1}{2}\sum_{\lambda=1}^{N_1} \Phi_\lambda^2 \frac{R_{\mathrm{m}\lambda}}{\mathrm{d}\delta}. \qquad (4.71)$$

For example, for the computation of the magnetic force in a non-linear circuit, the magnetic flux through the operating air gaps has to be calculated and the reluctances have to be differentiated. The implementation of Eq. (4.71) leads to a drastically reduced computation effort. Figure 4.33 shows exemplarily the computation of the force-stroke characteristic of a tubular magnet with magnetic networks.

Computation of the Static Magnetic Force Characteristic with Respect to the Magnetic Material's Hysteresis Effects

For some technical applications, e. g. proportional magnets or linear magnetic guidance, the computation of the influence of hysteresis effects onto the static magnetic force characteristic or, respectively, the existence of a model to compensate the influence of the hysteresis onto the force-stroke characteristics are of special interest.

Relevant literature describes different hysteresis models [4.21], that are more or less appropriate for different applications.

In the following paragraphs, the modelling of the hysteresis with the Preisach model will be explained. The Preisach model does not use the physical phenomena to describe the hysteresis, but bases on a phenomenological mathematical approach that uses the superposition of elementary rectangular hysteresis loops with the amplitude 1. The raising branch crosses the abscissa at $u = \alpha$ and the falling branch at $u = \beta$ (Fig. 4.34a). In case of magnetic hysteresis u represents the magnetic field strength H. We assume that α is always larger than β and above a value H_0 the outer loops coincide. Within the α–β-plane (the Preisach plane), we obtain a circumscribed triangle (Fig. 4.34b) with a separation line $L(t)$. The steps outline the memory of the Preisach model. Thereby it can save information about previous higher excitation states.

The hysteresis behaviour of specific materials can be modelled by specific distribution functions. The determination of the Preisach function with

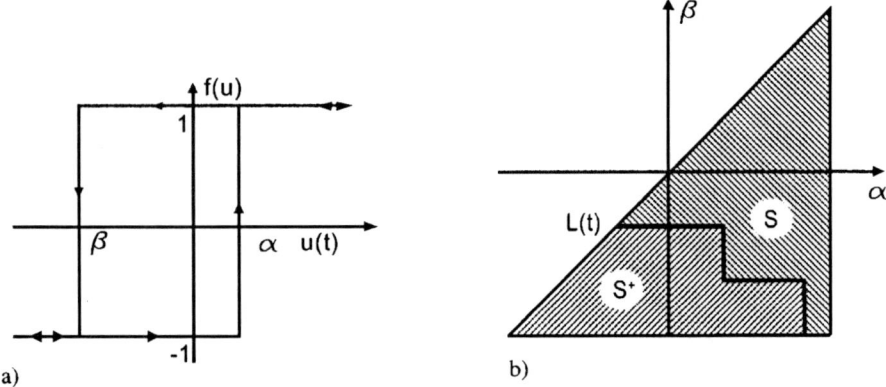

Fig. 4.34. Preisach model. **a** Elementary loop. **b** Preisach plane with circumscribing triangle

adequate accuracy for network method applications can be achieved from measurements of ring core samples (shown by Ströhla [4.16]).

Figure 4.35 shows the measurement date of the hysteresis curve of a ring core sample and the thereby determined Preisach function. With the help of this, the force-current characteristic of electromagnets with and without hysteresis can be computed and show significant hysteresis behaviour in the force-stroke characteristic at steady state (Fig. 4.36).

Computation of the Dynamic Behaviour of Electromagnet under Consideration of Eddy Currents

Fast acting electromagnets without a lamination of the metallic parts of the magnetic circuit require a consideration of eddy currents for an adequately accurate simulation of their dynamic behaviour (see Sect. 4.1.2.4). With the "magnetic inductance" introduced by Schweer [4.22], adequate models that can be implemented into magnetic networks can be derived, which solely require geometric and material parameters.

For the deriving of the modelling equations, we assume a magnetic cylindrical solid of iron with a magnetic flux Φ, in which by flux alteration an eddy current i_w is generated (Fig. 4.37).

For the magnetic cylindrical solid, we have

$$V = i_w + R_m \Phi . \qquad (4.72)$$

For a first approximation, the eddy current i_w can be calculated as follows:

$$i_w = \frac{u_{\text{ind}}}{R_{\text{el}}} = -1 \frac{1}{R_{\text{el}}} \frac{d\Phi}{dt} . \qquad (4.73)$$

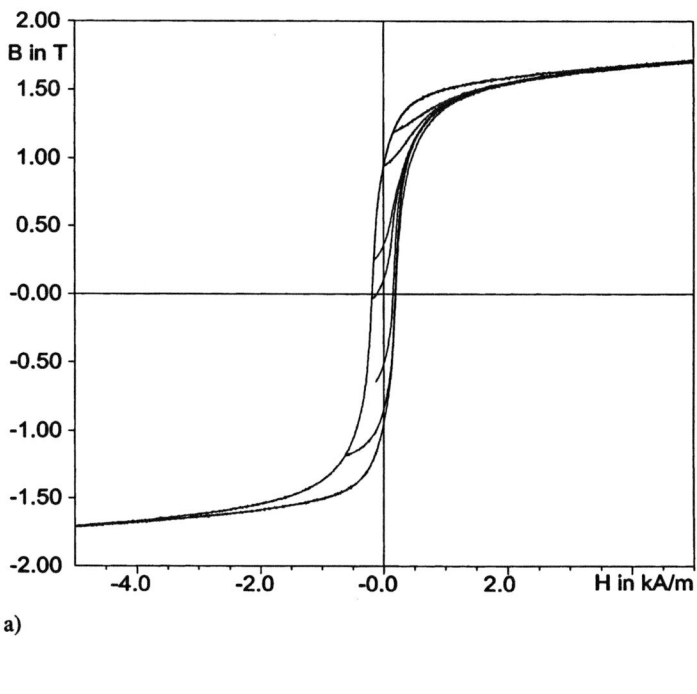

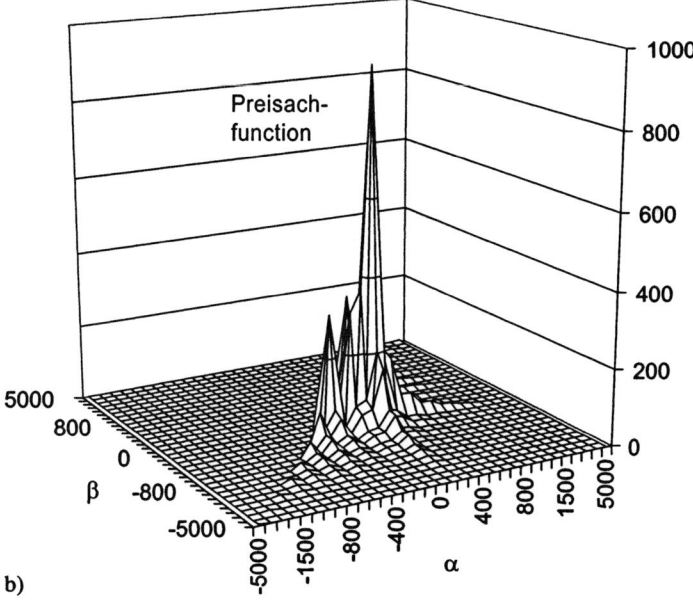

Fig. 4.35. Determination of the Preisach function [4.16]. **a** Measured hysteresis curve of a ring core sample. **b** Preisach function

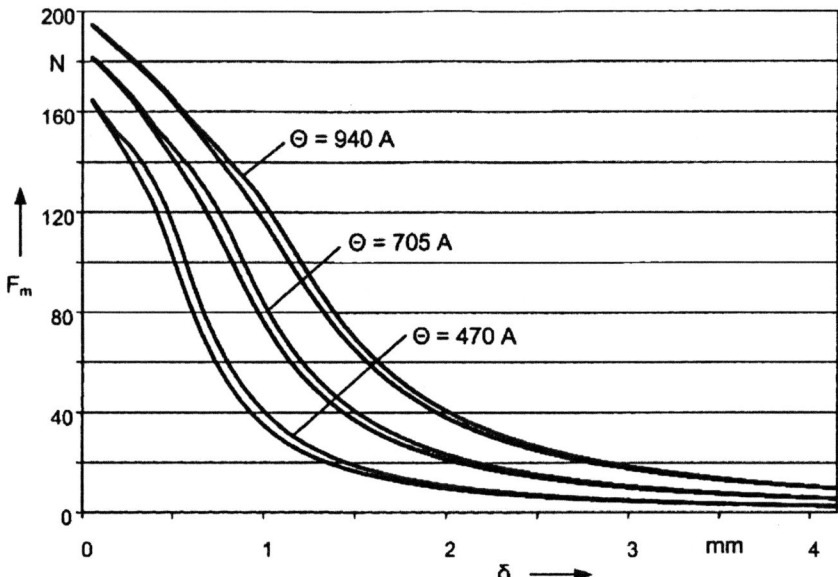

Fig. 4.36. Hysteresis of the computed force-stroke characteristics

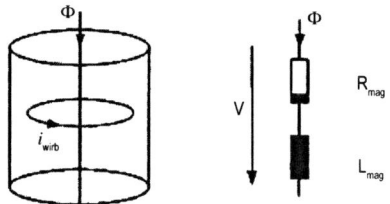

Fig. 4.37. Eddy current in a cylindrical solid

By analogy to the electrical circuit

$$L_\mathrm{m} = \frac{1}{R_\mathrm{el}} = \frac{\chi_\mathrm{Fe}}{l_\mathrm{w}} A_\mathrm{Fe} , \qquad (4.74)$$

we obtain from Eq. (4.72)

$$V = R_\mathrm{m} \Phi + L_\mathrm{m} \frac{\mathrm{d}\Phi}{\mathrm{d}t} . \qquad (4.75)$$

A significantly more accurate modelling can be achieved, if the cylinder is separated into several shells the magnetic flux passes through axially [4.16]. Figure 4.38 shows a magnetic network for a tubular magnet with magnetic inductances and its respective simulation (Fig. 4.39).

The influence of eddy currents on the dynamic behaviour can be recognized easily. The operation mode impressed current was chosen, because the influence of eddy currents is larger than for the operation mode impressed voltage.

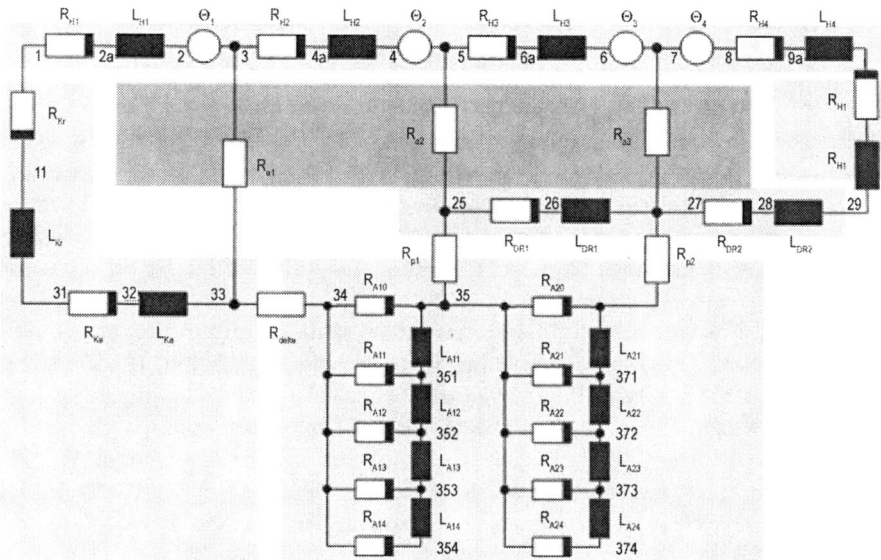

Fig. 4.38. Network model with magnetic inductances

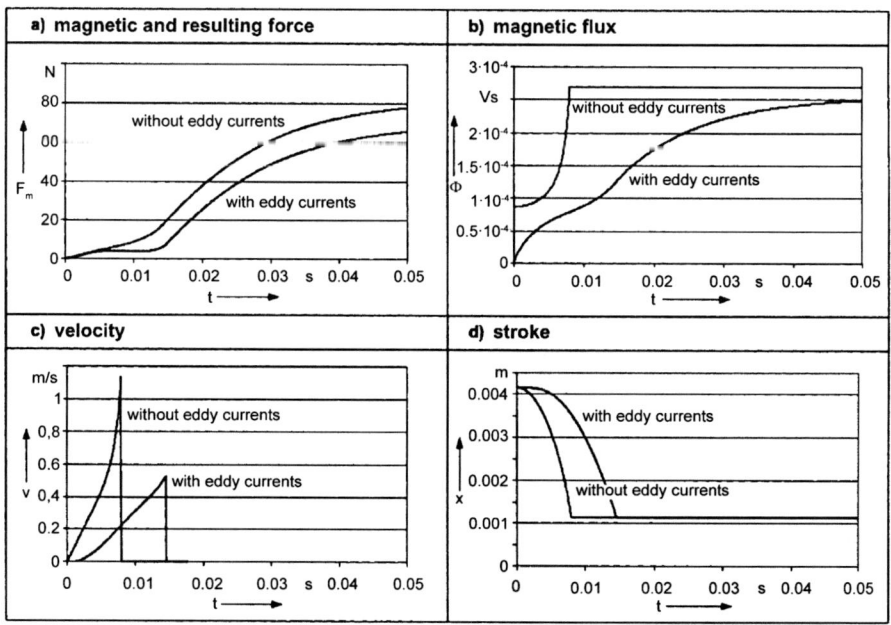

Fig. 4.39. Dynamic behaviour for impressed current with and without eddy current phenomena [4.16]. a Magnetic force $F_m = f(t)$. b Magnetic flux $\Phi = f(t)$. c Armature velocity $v = f(t)$. d Armature position $x = f(t)$

4.2 Electrodynamic Linear and Multi-Coordinate Drives

Wolfgang Schinköthe, Hans-Jürgen Furchert

Many of the technical movements to be realized are linear movements in one or several axes. It would have to be expected that drives for the direct generation of linear movements therefore are just as common as rotary drives. However, this applies only to electromagnets, with a travel range of some millimetres up to a few centimetres (see Sect. 4.1). In the generation of larger linear movements rotary motors with coupled gearings (rotation converters) to adjust torque as well as speed and following rotation translation movement converters to transform the rotary into a linear movement, Fig. 4.40a, predominate. This permits the use of common inexpensive rotary motors produced in high quantities, which are adapted to the movement task using application-specific gearings.

However, these movement converters have considerable disadvantages for the complete system. Large masses to be moved, friction, gear backlash as well as elasticity limit the achievable positioning accuracy and dynamic. The costs of the subsequent mechanics, its generation of noise or also a particle emission by wear cannot be neglected. On the other hand linear direct drives and multicoordinate drives have none of these mechanical movement converters (Fig. 4.40b). Out of this mechanical advantages, low wear as well as a high life time arise with regard to a quiet operation. From a control perspective these drives are outstanding due to low elasticity and friction; gear backlash in the motion transmission does not appear. Linear direct drives therefore have the conditions for highest dynamic and positioning accuracy. Therefore requirements of high positioning accuracy and dynamics and sometimes also questions of cost or even the operating conditions (clean room conditions) lead to the application of specially adapted linear direct drives.

Because of the necessary adaptation to the travel range linear and multicoordinate drives represent application specific drives, which are frequently also integrated into the complete design of a device especially in the workspace. Through this on the one hand very dynamic drives with a very high positioning accuracy can be realized. However, on the other hand common pro-

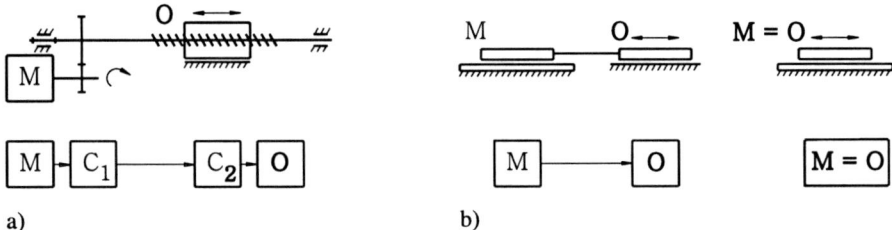

Fig. 4.40. Linear drive systems in comparison. **a** With rotary motors. **b** With linear direct drives. C_1, C_2 movement converter (gearbox); M motor; O output

duction series and with that an effective production in large quantities is still rare. The bad mass performance relation has an adverse effect since such motors have generally to be built as long as the travel range. An integration of the direct drive into the complete design can compensate this to a certain degree. At present a quick increase in use of linear motors can be recorded.

4.2.1 Working Principle and Basic Structure

From the rotary small motors working continuously and discontinuously linear motors can be derived by a linear unwinding of stator and rotor [4.25]. Analogous relations and connections are then valid, so that it can be referred to these chapters (see Chaps. 2 and 3). Asynchronous or synchronous motors or also reluctance or hybrid stepping motors are comparatively rarely found as linear motors in small performance ranges, however, at higher power (i. e. in machine tools [4.39]) they are used increasingly.

Here only linear and multicoordinate drives working in accordance with the classic electrodynamic action principle, preferably direct current linear motors, shall be represented in greater detail [4.26, 4.27, 4.34, 4.35, 4.42]. They often differ considerably from the small rotary motors as there is usually no commutation, a single-phase air gap winding and an adapted permanent magnetic excitation. Also pivot drives as used in hard disk drives are part of it. They could be characterized as direct drives with a bended path of motion.

Working Principle

Electrodynamic drives use the force on moving charges in the magnetic field (see Sect. 2.1). Reluctance forces arise in some drives in addition to the electrodynamic force. Maxwell's equations form the base for the calculation.

Electrodynamic force of action appears if current-carrying conductors are arranged in the magnetic field. The force on moving charges (Lorentz force) in the quasi stationary magnetic field causes a force on current-carrying conductors with a current i and an active winding length l arranged in the

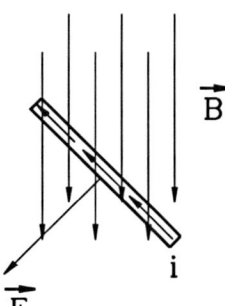

Fig. 4.41. Force on current-carrying conductors in the magnetic field

magnetic field with the magnetic flux density B if the current direction is perpendicular to the field direction (Fig. 4.41) in form of

$$F = Bli = k_F i \,. \tag{4.76}$$

In linear drives Bl or k_F will also be called force constant or motor constant. In rotary motors, however, a motor or torque constant k_T will be defined by the torque equation

$$T = Fr = k_T i \,. \tag{4.77}$$

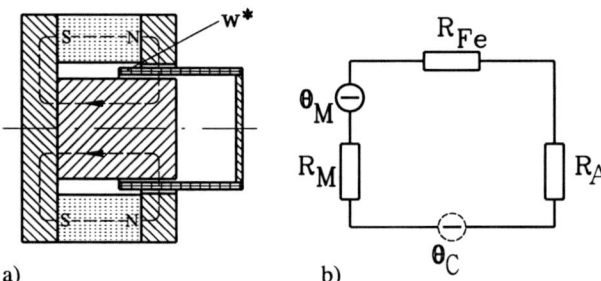

Fig. 4.42. Magnetic circuit of a voice coil motor (**a**) and equivalent network (**b**)

The working principle shall be clarified exemplarily at a voice coil motor according to Fig. 4.42 [4.41]. The operating air gap takes up the moving air gap winding (voice coil), which forms the output. Parts of the coil with the number of windings w^* (the number of turns of the coil part within the magnetic circuit) in Fig. 4.42a can involve the magnetic circuit and work as additional electrical excitation iw^* in which the number of the effective windings for this so-called armature reaction permanently changes at movement in this example. The magnetic field is generally generated by permanent magnetic excitation at small motors [4.32]. This avoids heat losses, which otherwise appear in the excitation winding in addition to the heating of the moving coil (see Sect. 2.1) and should have to be led away. Electrical excitations with series or shunt characteristics are, however, in principle also useable.

If leakage fields are neglected, we have a non-branched magnetic series circuit according to Fig. 4.42b. The following applies to the flux in the air gap

$$\Phi = \frac{\Theta_{\text{sum}}}{R_{\text{msum}}} = \frac{\Theta_M \pm \Theta_C}{R_{\text{msum}}} \quad \text{with} \quad R_{\text{msum}} = \sum R_i \,. \tag{4.78}$$

For the air gap flux density B_A considering only simplified magnetic reluctances follows

$$B_L = \frac{\frac{B_r l_M}{\mu_0 \mu_p} \pm iw^*}{A_A \left[\frac{l_M}{\mu_0 \mu_p A_M} + \frac{l_A}{\mu_0 A_A} + \frac{l_{Fe}}{\mu_0 \mu_r A_{Fe}} \right]} \tag{4.79}$$

and therefore for the force on the coil

$$F = B_A \cdot l \cdot i \,. \tag{4.80}$$

Equation (4.79), however, applies only to magnet materials with linear demagnetization characteristics or with nonlinear characteristics in the area above the break in the characteristic (see Sect. 2.1.2.2). The operating point also should lie in this area.

However, the calculation is strongly simplified because of the neglect particularly of the lateral leakage fields parallel to the air gap. These leakage fluxes also contribute to the action force. If they are not symmetrical or if the coil does not lie in them symmetrically, this leads to nonlinearities in the force-path characteristic [4.30]. Nonlinearities of the force-current characteristic result from the armature reaction according to (4.79) or more precisely out of the additional electromagnetic excitation iw^* (self-excitation), which leads to a reinforcement or a reduction in the air gap flux density and therefore using (4.80) to a component for the force that depends on the square of the current i.

Reluctance forces or forces on boundary surfaces appear if the magnetic circuit is not rigid in itself, if the air gap extension or the field distribution changes by moving or if a steel core or magnetic armature is brought into action [4.76]. Then additional forces add to the electrodynamic force due to reluctance changes or boundary surface forces, which can be calculated by the Maxwell stress tensor. For forces of action on iron parts with a magnetic field $B_\perp$, which leaks perpendicular into the air space, the Maxwell force equation follows

$$F = \frac{\mu_r - 1}{2\mu_r \mu_0} A B_\perp^2 \quad \text{resp. with} \quad \mu_r \gg 1 \quad F \approx \frac{A}{2\mu_0} B_\perp^2 \,. \tag{4.81}$$

However, one often also uses the energy theorem for the calculation which led to the naming reluctance force. The reluctance force then arises from the path-dependent energy changes and with it reluctance changes in the magnetic circuit or in the winding [1.7]. It is valid

$$F_R = \frac{dW}{dx} = \frac{1}{2}\frac{d(Li^2)}{dx} = \frac{1}{2}w^2 i^2 \frac{d}{dx}\left(\frac{1}{R_m}\right) \quad \text{with the inductance} \quad L = \frac{w^2}{R_m}\,. \tag{4.82}$$

Basic Structure of Electrodynamic Drives

Electrodynamic drives can be divided into four partial systems, Fig. 4.43 [1.7]:

1. an electrical partial system, consisting of windings, accompanying power supply, control devices and required measuring systems,
2. a magnetic partial system, consisting of permanent magnetic or/and electrical excitation, flux-guide parts as well as operating air gaps,

3. a mechanical partial system, consisting of support and guiding systems as well as moving masses, counteracting forces, friction forces and other influences,
4. a thermal partial system, consisting of electrical, magnetic and mechanical components and their heat conduction, convection and thermal radiation characteristics as well as the surrounding environmental conditions.

At first, Fig. 4.43 shows the relations between the single partial systems. The interactions and limitations in electrodynamic systems as well as the coupling equations are represented in Fig. 4.44.

The magnetic field is excited permanent magnetically. The electric loss appears only in the armature winding then. The heating of this winding therefore limits the possible current and thus the force in continuous operation. P_{Cu} corresponds here to the electric losses at an ohmic resistance and causes a heating on which the thermal layout of the motor is based.

In short time or intermitted operation, where thermal limitations have no effects due to the high thermal inertia, the field weakening by the above-mentioned armature reaction iw^* can be a limitation. The main problem at the layout of electrodynamic drives remains, however, in leading the appearing power dissipation away.

The generated motor force serves the movement of the mass and thus the overcoming of an inertia force mostly $(m\ddot{x})$. Counteracting forces can appear as constant and velocity-dependent friction forces ($k_1\dot{x}$ and F_F), as load forces (F_L), and under circumstances also as spring or elastic forces (k_2x),

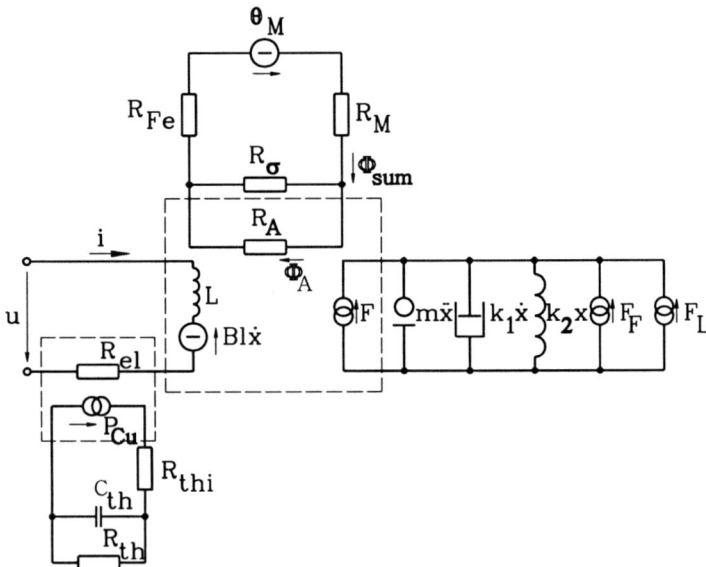

Fig. 4.43. Basic structure of electrodynamic drives, according to [1.7]

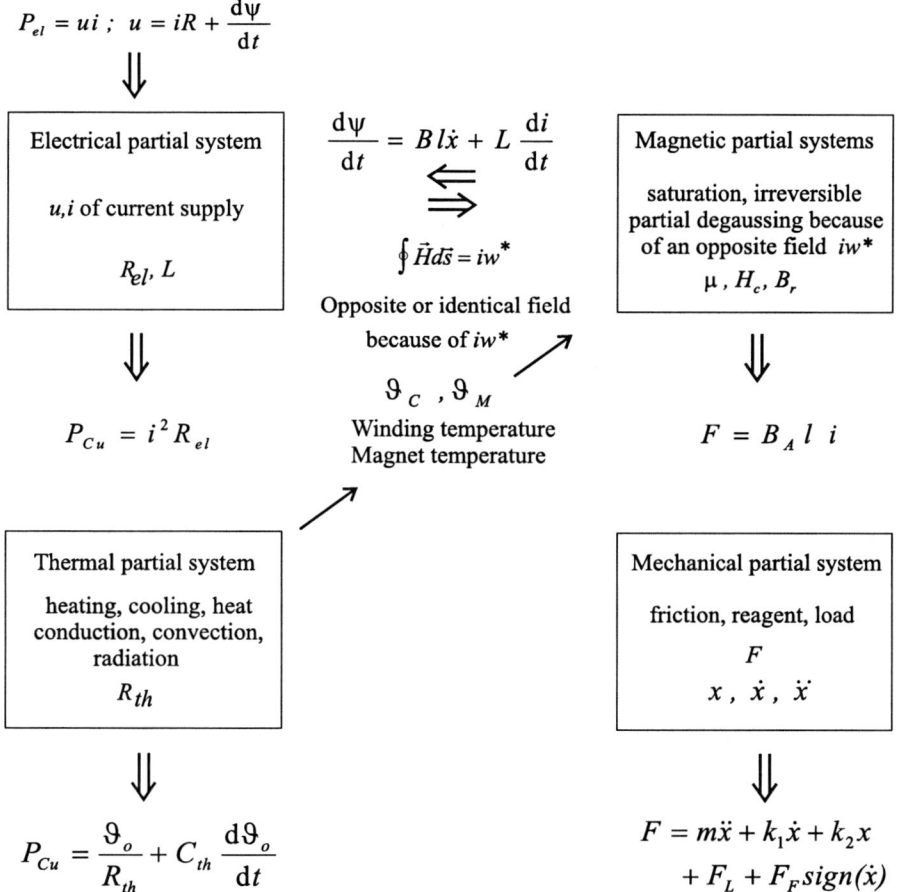

Fig. 4.44. Interactions of the four partial systems of an electrodynamic drive

for instance for the compensation of weight forces in vertical arrangements. Reluctance forces are additionally available if necessary.

4.2.2 Configurations of Electrodynamic Linear Motors

For a comprehensive systematization of electrodynamic linear motors we have to look at the partial systems mentioned above one by one at first, to vary them in their components and then put all approaches for the individual partial systems together again for a complete solution.

Essential distinction criteria are:

1. in the mechanical partial system
    the type of the moving component (moving coil, moving magnet, moving permanent magnetic circuit resp. moving yoke),

the geometry of the design (rotational-symmetric, prismatic etc.),
2. in the magnetic partial system
the type of the excitation (permanent magnetic or electrical), where the electrical excitation will be insignificant in the area of small performances, the polarity of the used fields (homopolar or heteropolar layout),
3. in the electrical partial system
the existence of a commutation,
the type of the commutation (electronic or mechanical), where the commutation has to be seen in relation to the magnetic polarity.

The thermal partial system shall stay largely unconsidered. Also the classic mechanical components such as the guide of the armature, the appearing friction forces, the realization of the force output, the effecting load forces and similar questions, are not explained in detail. On the one hand linear motors are often offered only as separate armatures and stators and external guides are used in the device or a translation stage. On the other hand, the sliding guides, rolling guides, aerostatic or magnetostatic guides, well known from the device technology, can be applied between the moving parts (see Chap. 8).

4.2.2.1 Configurations with Moving Coils

The arrangement of the field-generating permanent magnets and accompanying flux-guides or yokes in the fixed stator as well as the current-carrying windings in the armature represent the standard solution first of all like at rotary motors. The complete magnetic circuit, which consists of permanent magnets and flux-guide parts (yokes), is therefore fixed in the stator. One or more air gaps of the magnetic circuit contain moving current-carrying coils (air gap windings). The moving coils are generally non-ferrous and therefore in a low-mass design, which enables high dynamic drives, though a current supply to the moving parts (trailing cable, slider) will be required and has the resulting disadvantages. Iron yokes can be massive in these cases since almost permanent magnetically excited fields are used.

Moving flux-guides or magnetic yokes that are tightly connected with the winding are also possible, however, are only common at greater forces and rarely at small drives. Configurations where the winding is found on a linearly moving iron armature or connected with a magnetic yoke plate single-sided as at flat coils would have here to be mentioned. This enables very high current densities and thus also great forces due to the good heat removal, however, this increases the moving mass considerably. Boundary surface and reluctance forces as well as hysteresis and eddy-current losses in the iron also then have to be taken into account.

Homopolar Configurations without Commutation

Homopolar types without commutation are configurations with moving coils that are geometrically most simply built and are therefore the most common configurations. Homopolar type means that the coils or the partial coils are always in a magnetic field of constant polarity at the movement or being in interaction with such one. A self-supporting coil or a moving coil that is wrapped on a bobbin is in an air gap of a permanent magnetically excited magnetic circuit with a massive magnetic yoke and does not leave the field area at movement, at least not complete. A commutation therefore is not required. Differences arise mainly from the geometric design of the mechanical and electrical circuit. Figures 4.45 to 4.47 show basic configurations [4.28,4.33,4.36,4.40,4.41,4.49,4.51].

Rotational-symmetrically configurations with cylindrical coils (Fig. 4.45) usually have only a cylindrical air gap with one coil in it. Axial (a) and radial (b) magnetized toric magnets or as well several magnetic segments instead of radially magnetized toric magnets, which then can be realized anisotropically, are used in magnets arranged outside the coil. Thereby outer leakage fluxes appear. Core magnetic systems (c) avoid this.

Box coils (Fig. 4.46) are needed at a prismatic basic construction, however, they lead to similar drive configurations. However, because of constructive and assembly reasons, one frequently does not make use of the box coils on all sides in prismatic systems but arranges them only in one or two parallel levels in air gaps. Not all coil segments can then contribute to the force generation, but heat dissipation loss appears also in the unused segments. The magnets, particularly the high-energy ones, are mostly arranged directly at the air gap; concentrated, flux expanding arrangements not directly placed at the air gap (c) are, however, also possible. The magnetic yokes with constant thicknesses could also be designed according to the flux load with different cross sections in order to minimise mass (see Fig. 4.45b with reduced cross sections at the outer yoke).

Forms with plane flat coils (Fig. 4.47) require a polarity change of the magnetic field within the air gap prolongation because of the different current directions between outgoing and returning conductors. In the magnetic circuit two air gaps with different field directions are required, in the equivalent network the air gaps are in series. Unused coil segments also exist here, com-

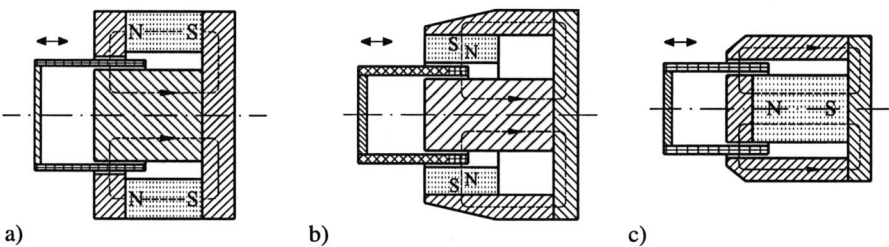

Fig. 4.45. Linear motors in homopolar configurations with moved rotational-symmetrically cylindrical coil (voice coil motor). a Long-coil system. b Short-coil system. c Short-coil system with magnet core. Typical parameters: $s \leq 100\,\mathrm{mm}$; $F \leq 100\,\mathrm{N}$; Applications: loudspeaker, short-stroke positioning systems

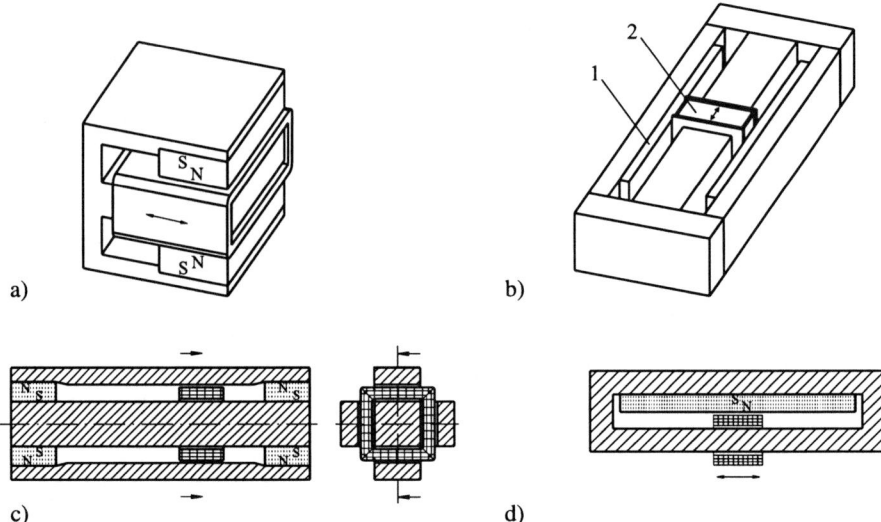

Fig. 4.46. Linear motors in homopolar configurations with moved box coil. **a** One-side opened circuit with a long-coil system. **b** Symmetrical configuration with a short-coil system. **c** Flux-expanding magnet arrangement. **d** Asymmetrical magnetic circuit. *1* magnet; *2* moved coil. Typical parameters: $s \leq 200$ mm; $F \leq 50$ N; Applications: positioning systems, writing measuring instruments

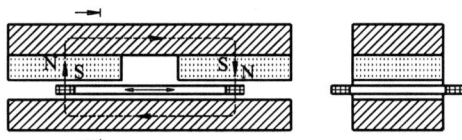

Fig. 4.47. Linear motor in homopolar configuration with moved flat coil. Typical parameters: $s \leq 50$ mm; $F \leq 100$ N; Applications: short-stroke positioning systems

parable with end windings of rotary motors. A commutation is not required, as the outgoing and returning conductors stay in a respective field segment of the same polarity at movement. The name homopolar type is in this respect correct, too.

Figure 4.48 shows similar forms for pivot drives. At present, these are the standard drive solutions for head positioning at hard disk drives, particular configurations according to a or b. They represent linear motors arranged on a circular arc and are considerably closer to the linear direct drives in their configurations than to rotary direct drives. At such high dynamic drives also the coil inductance is of special interest. If the moving coil does not contain any iron cores, the inductance remains small, and consequently the time constant as well (Fig. 4.48a and b). In the other case (c), additional fixed short-circuit rings or shading coils on the iron core, which consist of a layer of foil, are used in order to ensure short response times.

Another essential distinction criterion represents the relation between air gap and coil dimensions into the direction of motion. Short-coil systems are mostly found. The coil stays in movement completely within the air gap, so the travel range arises from the air-gap extension into direction of motion less the coil length, and perhaps additional reductions depending on linearity demands. The supplied electrical energy is optimally used for the force generation, but not the magnetic one. The magnetic circuit (and with it the stator) then have a comparatively large volume, since the air gap extension must be greater in the direction of motion than the travel range.

Long-coil systems in contrast show coil dimensions exceeding the air gap extensions in the direction of motion. The coils shall always still completely cover the air gap at movement here. The travel range arises from the coil length less the air gap extension into the direction of motion and if necessary further reductions. The greater the travel ranges, the larger are the parts of the coil outside the air gap, which do not contribute to the force generation but to the heat dissipation loss. However, often outside the air gap, but nearby, are considerable magnetic leakage fluxes which are used to a certain extent also for force generation. In comparison with short-coil systems, a considerably stronger force is available. Here the magnetic energy, the construction volume of the motor and particularly the volume of the permanent magnets are completely used, however, the electrical energy only incompletely.

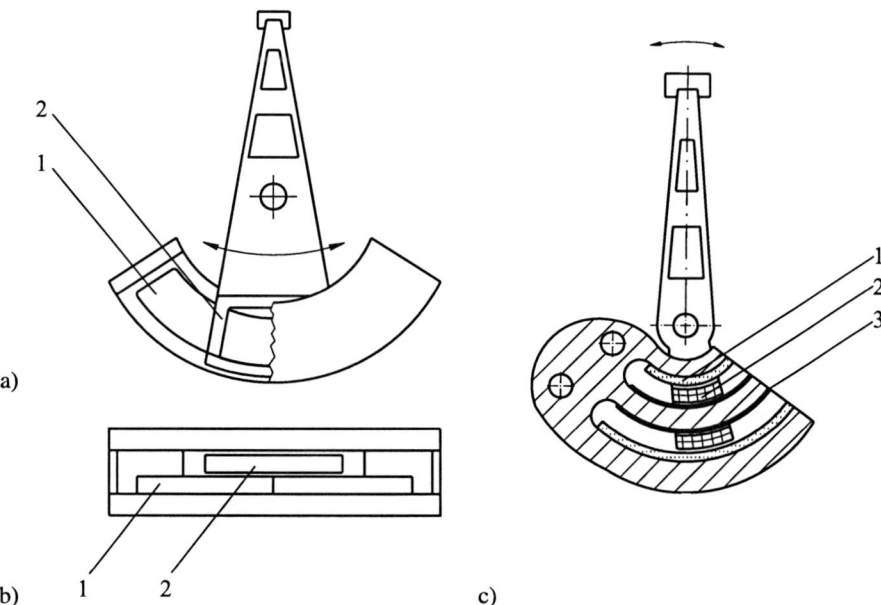

Fig. 4.48. Pivot drives. **a** With flat coil. **b** Side view (view on backside). **c** With box coil. *1* magnet; *2* moved coil; *3* shading coil. Typical parameters: $\varphi \leq 30°$; $F \leq 10$ N; Applications: positioning systems at hard disk or CD-ROM drives

A substantially higher heat dissipation loss appears. Long-coil systems can be recommended at high thrust demands. On the other hand, configurations for high dynamic applications with few moving mass, which are electrically well used, are mostly designed as a short-coil system. Nonlinearities, however, characterise both variants. A specific air gap design (constrictions, extensions) enables a linearization of the induction characteristic in the air gap and thus for the force-path characteristic.

Homopolar Configurations with Commutation

Motors with long coil arrangement are in principle conceivable also with mechanical commutation by sliders to carry current to the coil segments only within the air gap and to avoid additional heat dissipation loss. However, the expensive mechanical sliders and the life time problems caused by it are generally not accepted, so that not-commutated arrangements dominate. Comparable configurations are practicable also with electrical excitation.

Figure 4.49 shows configurations with electrical excitation with and without mechanical commutation. Their distribution is, however, low because of the additional excitation losses.

A variety of configurations modified or optimized after certain criteria can be derived from the basic models now. For all, a variety of optimization suggestions may be found in patent literature but practical applications are mostly close to the represented basic configurations. Figure 4.50 shows some application examples.

Advantages of the homopolar configurations arise primarily from the very simple complete construction of the magnetic, mechanical and the electrical part. Disadvantages appear, however, at larger travel ranges. So travel ranges above 100 mm lead to comparatively large motors. Since the stator is at least as long as the travel range, the air gap dimensions grow proportionally in the

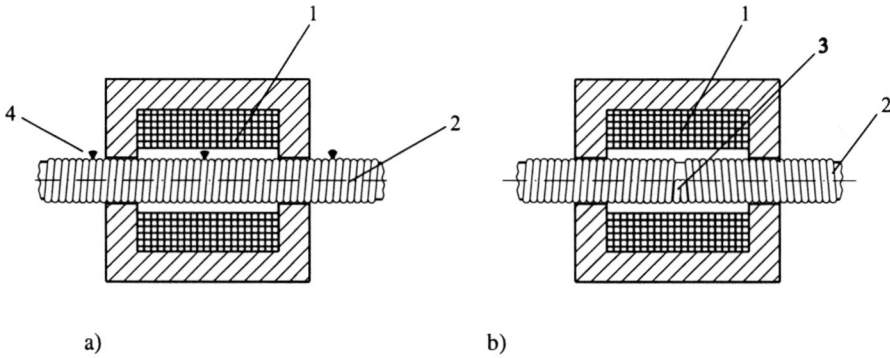

Fig. 4.49. Electrical excited linear motors in homopolar configuration. **a** With and **b** without mechanical commutation. *1* excitation winding; *2* armatures winding with iron core; *3* returning of the winding direction; *4* brushes

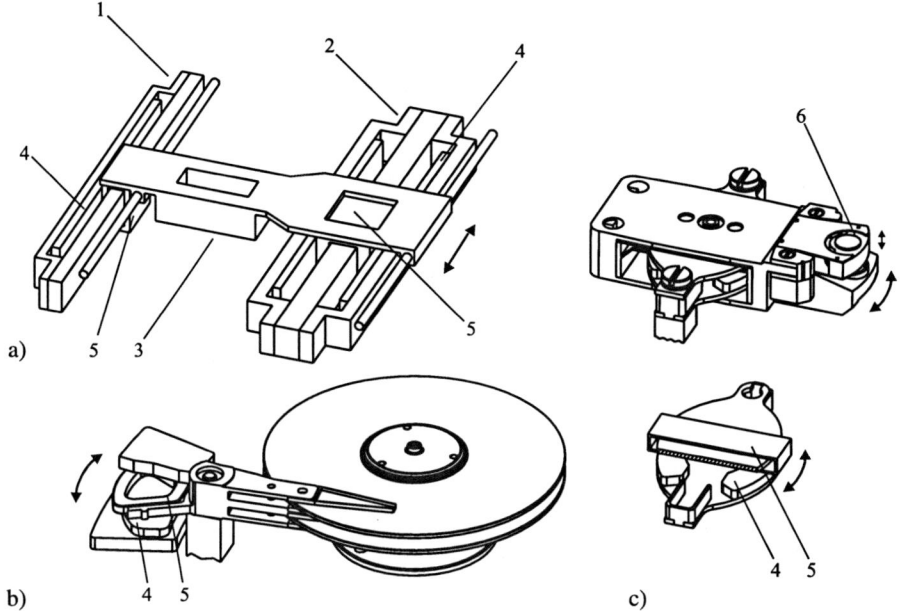

Fig. 4.50. Typical applications of electrodynamic linear motors (simplified). **a** Linear motor and a second linear motor used as tacho-generator in a CD player. **b** Pivot drive in a hard disk drive. **c** Pivot drive in a CD-ROM drive, down pivot drive separated. *1* tacho-generator; *2* linear motor; *3* read head slot; *4* magnet; *5* moved coils; *6* leaf spring guided voice coil motor integrated for the z-movement of the optical components

direction of motion, the air gap space and with that the magnetic fluxes expand themselves. The flux guide cross sections in the magnetic yoke also have to increase proportionally since the flux uses only the same magnetic path.

Heteropolar Configurations with Commutation

The realization of larger travel ranges therefore requires the transition to the alternating polar or heteropolar configuration and thus also the use of a position-dependent phase-winding commutation, preferably electronically controlled. By the alternating arrangement of magnetic sections with changed field directions the magnetic flux closes in each case only *via* the adjacent segments and not *via* the complete motor length, the yoke and with that the motor cross sections remain small.

The basic configurations of magnetic circuits from Figs. 4.45 to 4.47 can principally also be lined up as a heteropolar configuration. The winding system then consists of at least two mechanically connected short coils or two-phase windings. Figure 4.51 shows the principal design. Figure 4.52 shows practical used configurations with non-ferrous winding (non-ferrous core) and electronic commutation (brushless DC linear motor) [4.48]. Only in some parts of the figures can you see decided distances between the magnets. These gaps represent an optimization problem and a question of the optimization criterion. Too small distances between the magnets lead to a weakening of the adjacent magnets; too large distances enlarge the construction length.

In this case again a constant magnetic field generated by permanent magnets is present in the stator and thus a massive stator yoke can be used. However, each partial-coil or, respectively, phase winding now has to leave air gaps of one field direction and travel into areas with opposite field direction. This requires an inversion of the current direction in every phase winding by a position-dependent mechanical (brush) or preferably electronic (brushless) commutation. In addition, several phase windings displaced to each other and to the pole partitioning (overlaid commutation) are usually required to enable the transition of a phase winding into a new air gap segment. If a phase winding leaves a field limitation, at least a second phase winding still must be completely engaged in an air-gap section.

Two or three-phase windings are common. The phase windings can be inserted into one another and therefore have distinctive end windings because the previous phase winding is always led over the next phase winding. This permits to line discrete phase windings without blank spaces. However, a non inserted design can be used, too, with the phase windings placed next to each other and therefore without distinctive end windings to simplify the production. But this requires blank spaces within a phase winding. These blank spaces are then used most for bobbins. The current has to be switched over with fluent transmissions if possible. One usually uses a position-dependent sinusoidal or trapezoidal commutation (see Sect. 2.2.2.1). The commutation is controlled by a path-measurement signal or by a Hall-effect sensor. Due to the sinusoidal commutation some manufacturers call such configurations also electronically commutated AC synchronous linear motors, operated in the DC-servo mode.

The fixed stator can be adapted to the demanded travel range in the length while the moving coil system is remaining comparatively small and low in masses regarding the travel range. However, due to the larger travel range, the current supply of the coils using flexible lines or even sliders is more complex.

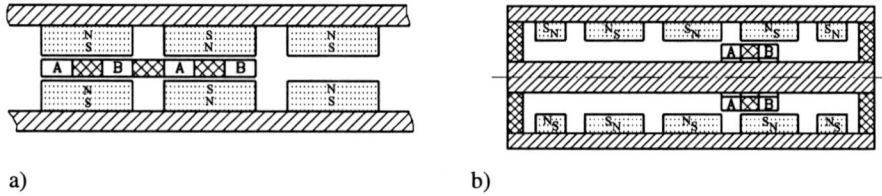

Fig. 4.51. Principal design of linear motors with moved coil in a heteropolar configuration. **a** With flat coil. **b** With box coil. A, B phase windings. Typical parameters: s accordingly to the movement requirements; $F \leq 200$ N. Applications: general positioning systems

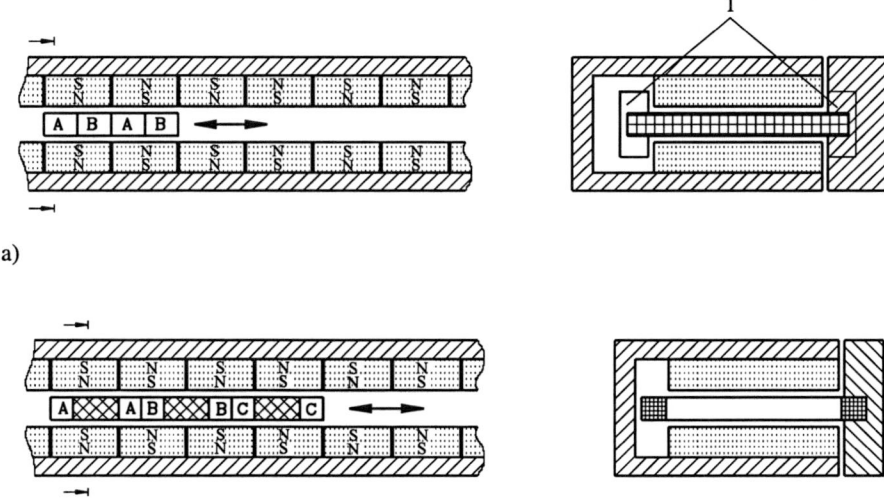

Fig. 4.52. Practical used designs of linear motors in heteropolar configurations with moved non-ferrous coils (flat coils) [4.48]. **a** With two-phase winding and distinctive end windings. **b** With three-phase winding without distinctive end windings, but with blank spaces inside the coils. *1* distinctive end windings; A, B, C phase windings. Typical parameters: $s \leq 1000$ mm (also further); $F \leq 1000$ N. Applications: general positioning systems, also at the mechanical engineering

At greater forces, ferrous cores in mostly unsymmetrical design are usually found beside the non-ferrous cores, Fig. 4.53 [4.48,4.51–4.53]. However, considerable boundary surface forces as attraction or normal forces between armature and stator arise from it, which have to be absorbed in the guide and thus generate strong friction forces at gliding or rolling guides. The magnitude of boundary surface forces can be ten times higher than the thrust. The yoke in the armature is mostly laminated to counter eddy currents. When crossing magnetic pole transmissions, the armature experiences significant magnetic cogging forces in the range from 5–10% of the thrusts. The asymmetrical, single-sided arrangement also leads to strong leakage fluxes in the environment. In addition, reluctance forces appear especially at the ends of the travel ranges and at unevenly distributed ferromagnetic masses in the surrounding of the armature. In the case of very strong forces, such motors are also applied with air, water or oil cooling. They are offered as mounting solution, most without a guide of their own. The transmission of small drives to comparatively large electrical engines is fluent here.

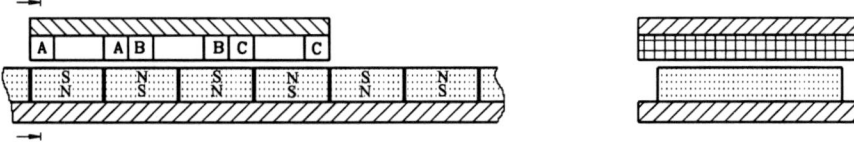

a)

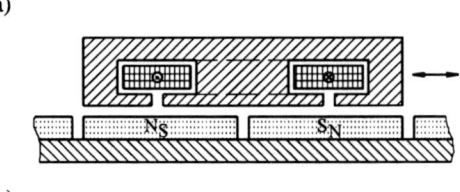

b)

Fig. 4.53. Linear motors in heteropolar configuration with ferrous core armature. **a** Unsymmetrical design with a three-phase winding on a ferrous plate. **b** Principle of a slotted winding (other phase windings unembodied). A, B, C phase windings. Typical parameters: $s \leq 1000$ mm (also more); $F \leq 1000$ N. Applications: general positioning systems, especially at the mechanical engineering

4.2.2.2 Configurations with Moving Magnets

The electrodynamic force of action appears between an electrical and a magnetic partial system; therefore, also parts of the magnetic circuit can be arranged as mobile and the coil system fixed, though the magnetic circuit itself is no longer rigid. By relative motions within the magnetic circuit reluctance forces or forces on boundary surfaces as well as hysteresis losses appear in the iron yoke and with that considerable damping under circumstances.

Advantages:

1. No current supply to moving parts (also the scale of the measuring system should then be arranged at the armature and not the receiver)
2. Improves heat removal since the windings then lie at or on yoke parts
3. Under circumstances a defined position of the current-free armature by reluctance forces exists

Disadvantages:

1. Great dissipation loss, especially at the most common long-coil device
2. Additional radial reluctance forces on guides, if no compensating rotational-symmetric or double designs are used
3. Additional axial reluctance forces generate back-driving forces and increase nonlinearities
4. Damping by hysteresis losses

Homopolar Configurations without Commutation

By exchanging the moving and fixed parts, solutions with moving magnets can be realized with the above-mentioned motors [4.31, 4.37, 4.38]. In practice, the multitude and complexity of possible designs, however, are restricted to some comparatively simple arrangements, Fig. 4.54. Basically the design considerations from moving coil motors are valid. As magnets are dipoles, systems open on both sides and no single-sided systems like voice coil motors, are applied. A reversal of the winding sense in the middle of the coil may be the required. By use of high-energy magnet materials with a small volume or plastic-bound magnets, the moving mass is kept low and strongly miniaturized motors can be fabricated (using conventional technologies down to 2 mm of outside diameter with configuration given in Fig. 4.54a). Since the magnets move within the coil, the coil extension in the direction of motion is larger than the length of the moving magnets. These designs are referred to as long-coil systems with comparatively high dissipation loss. Configurations with concentrated windings were also suggested in [4.27, 4.37].

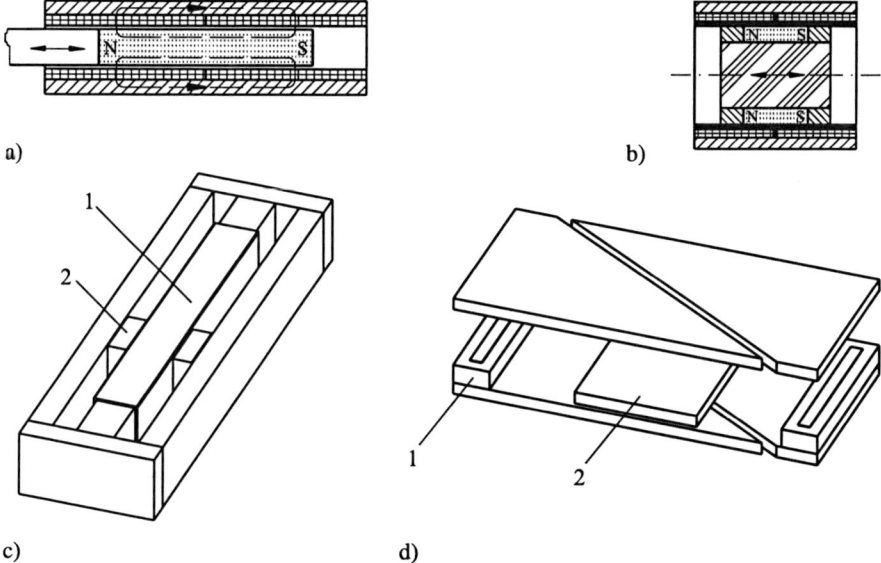

Fig. 4.54. Linear motors with moved magnets in homopolar configuration (without commutation). **a** Basic principle [4.43]; **b** design with open channel in the middle (here for an optical component) [4.43]; **c** prismatic design [4.27, 4.31]; **d** design with concentrated windings [4.27, 4.37]. *1* winding; *2* magnet. Typical parameters: $s \leq 100$ mm; $F \leq 50$ N. Applications: servo and positioning systems, also at the mechanical engineering

Heteropolar Configurations with Commutation

The realization of larger travel ranges requires the transmission to heteropolar configurations and thus the application of a preferably electronic phase winding commutation, Fig. 4.55 [4.48, 4.49, 4.50]. In comparison to motors with moving coils, the construction and technology effort as well as the use of copper for the realization of a multi-phase linear stator winding along the whole travel range is, however, substantially higher than the effort for an alternating magnet arrangement in the stator. This applies particularly to the design of the end windings [4.47], to be seen in Fig. 4.55a and b in the cross section. The statements about commutated motors with moving coils apply analogously. In principle designs with concentrated windings (Fig. 4.55d) can also be applied. The attraction forces that appear particularly in asymmetrical designs cause strong normal forces and again result in high friction forces.

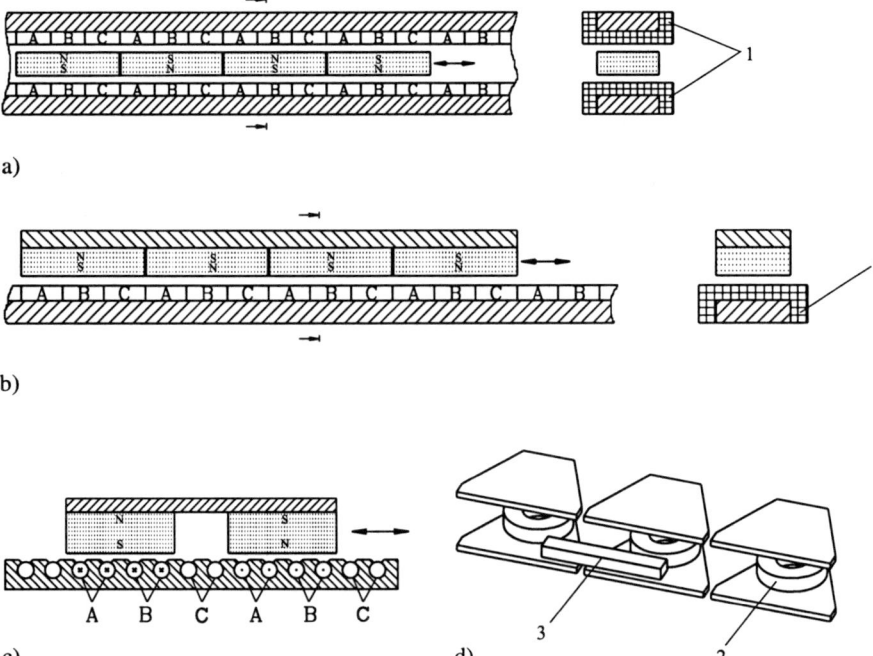

Fig. 4.55. Linear motors with moved magnets in heteropolar configuration (with commutation). **a** Three-phase linear motor. **b** Three-phase linear motor with single-side asymmetrical design and moved yoke. **c** Three-phase linear motor with fixed slotted windings and moved yoke. **d** Linear motor with concentrated windings and only one moved magnet [4.37]. *1* distinctive end windings, *2* winding, *3* magnet, A, B, C phase windings. Typical parameters: $s \leq 200$ mm; $F \leq 500$ N. Applications: general positioning systems, also at the mechanical engineering

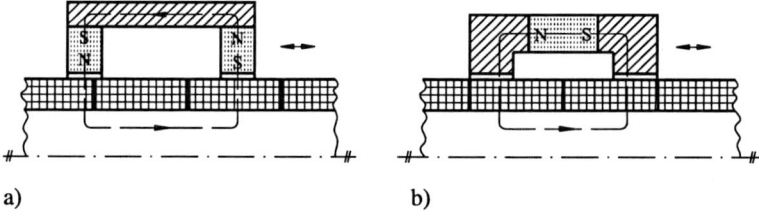

Fig. 4.56. Linear motors with moved magnets and moved yoke, electronically commutated (mechanical commutation also possible). **a** With radial magnetisation of the permanent magnets. **b** With axial magnetisation. Typical parameters: $s \leq 100$ mm; $F \leq 200$ N. Applications: general positioning systems

Motors with fixed electronically commutated cylindrical coil systems attached to a ferromagnetic cylinder and moving permanent magnets in the armature are also possible, Fig. 4.56. In this case, no ineffective end windings and no blank spaces occur in the coils but the current supply of the middle axis is more difficult to design. The guide of the armature on the coils is also more complex. In the armature, yoke parts are also moved, the arising problems were already mentioned. The represented rotational-symmetric design, however, is balanced as far as attraction forces due to forces on boundary surfaces are concerned. Hysteresis losses appear, though.

4.2.2.3 Configurations with Moving Permanent Magnetic Circuits

In principle, all types of permanent magnet circuit configuration designed for use with moving coils can also be executed in mobile mode opposite coils which are a fixed part of the stator. However, only a few of these configurations are serviceable for movement purposes, being magnetic circuit structures which are light in mass, and it will be only these which are dealt with here.

Configurations which have an armature with a moving permanent magnetic circuit (as opposed to coils fixed firmly in the stator) have the advantage over moving permanent magnets in that any boundary surface forces will appear only as an attraction between surfaces which are not moving in relation to each other. An exception arises when the iron yoke is partitioned into one immobile and one moving part. This will be treated last.

The advantage of the absence of trailing cable for the armature can also be obtained in the case of moving permanent magnetic circuits (as opposed to coils fixed firmly in the stator), but the armature mass will be rather large in relation to coils or permanent magnets which are normally the moving parts, particularly where the travel is over larger distances. It is now possible to configure drives either without or with commutated coils, depending on the length of travel required.

Homopolar Designs without Commutation

Moving permanent magnetic circuits with a homopolar arrangement and non-commutated coils are only feasible for short distances of travel (up to approx. 50 mm) because the mass to be moved is otherwise too great. This mass comprises the iron yoke and the permanent magnet mass (relatively large in this case) and has to be moved the whole distance above the stationary coil. In this situation, the cost of high-energy permanent magnets is considerable.

So that there is full iron usage at the same magnetic flux density B along the whole length, special adaptations of the circular iron cross section can permit thinning of the iron core cross section, for example, at the open ends of the iron yokes, considerably reducing the mass. Figure 4.57 demonstrates flat and box-coil arrangements with differing thickness of the iron yokes as relevant to the magnetic flux, achieved by cross-section adaptation (dotted lines marked on the cross-hatched thicknesses).

In the case of the flat windings, the thickness of the iron yoke (and thus the mass) decreases from a to d because of the different partitioning of the magnetic fluxes. With box coils, there is an increase from e to h in the iron yoke mass because of the variation in the usage and effectiveness of the coil vis à vis the amount of magnetic flux in the different permanent magnet circuits available. However, the configuration is of far higher importance in the case of moving heteropolar magnetic circuit designs with commutation.

Heteropolar Designs with Commutation

An armature with little mass can be achieved by using interleaved flat windings in the stator, two coils being shaped and arranged each time in such a way there is never a gap between the windings. This also guarantees optimal use

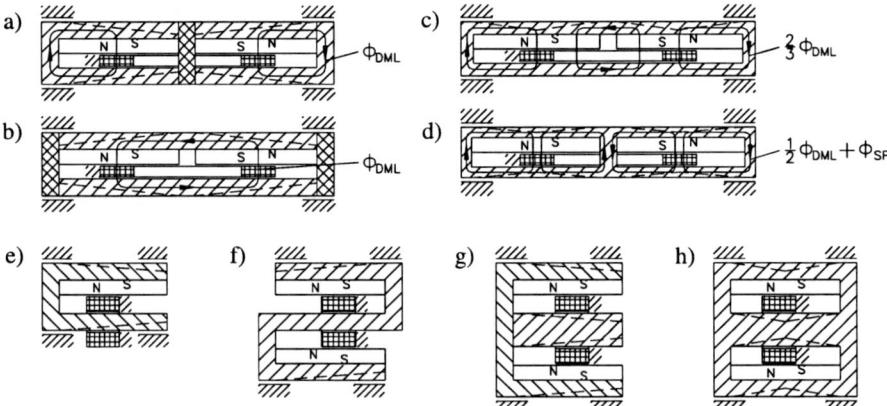

Fig. 4.57. Flat and box-coil arrangements with differing thickness of the iron yokes, as a function of the structure of the windings and magnetic circuit

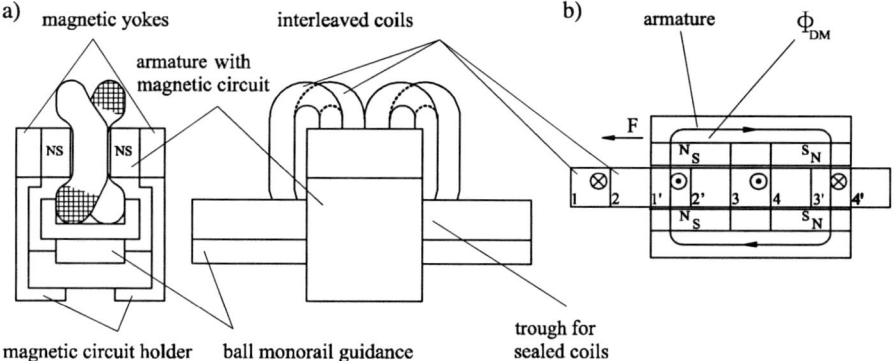

Fig. 4.58. Electrodynamic linear motor with moving permanent magnetic circuit, interleaved, commutated flat windings (a), and diagram showing development of force (b) when current flows in coils *1* and *3*

of space, obviation of a trailing cable for the armature and favourable heat dissipation from the coils to the surrounding air, as the coils are embedded on one side only in epoxy resin with the end windings in, for example, a small, shallow trough. However, special devices will be necessary for the manufacture of the coils to enable the end windings to be fed round each other. Figure 4.58 shows a motor of this kind, with the windings plus aluminium trough attached to the guide rail of a ball monorail guidance system and the permanent magnetic circuit in the armature. It is also possible to use the other familiar types of guide (glide, roll, aerostatic and magnetic guide).

Another advantage of this variant is that it lends itself to modular design so that different movement paths and forces can be provided for, depending on the number and length of the coils in relation to the length of guide rail employed and the accessibility of the slide for load coupling purposes. The commutation regime will be derived from whatever measuring system is necessary to regulate the drive. The physical measure will be attached to the armature and the sensor to the stator, which will mean that the armature is free of any trailing cable.

Overlapped commutation is useful to preclude major losses of force or force peaks in the force-path ratio.

4.2.3 Types of Construction of Integrated Electrodynamic Multi-Coordinate Drives With Single-Mass Armatures for xy-, $x\varphi$-, $xy\varphi$- and $xy\varphi z$-Travel

In contrast to driven cross-tables that have rotating and lifting mechanisms involving the movement relative to each other of a number of masses coupled together by bearings or guide systems, multi-coordinate drives will achieve motion with a compact (or less than compact) single-mass armature in several coordinates without mechanical converters being necessary for the movement. Many and multifarious constructions of electronic multi-coordinate motors have been developed since 1979 [4.72, 4.76, 4.77, 4.79, 4.82, 4.86, 4.96, 4.97]. Figure 4.59 shows the structures of multi-coordinate movement equipment as demonstrated in [4.83]. Following the development that has taken place since the mid-1980s [4.95] (which were advances of the idea contained in a US patent of 1971 [4.85]), an integrated $x\varphi$ motor (described in [4.66, 4.67]) can be employed as a rotating and lifting mechanism.

Very complex electrical magneto-mechanical structures, which are essentially the basic structures of linear motors, can be produced by means of the integration of motor assemblies for several coordinates. Travel of up to 200 mm, and more, is realisable by means of motors working on either two or three axes without commutation: examples of possibilities are $xy, xy\varphi, yz\varphi$.

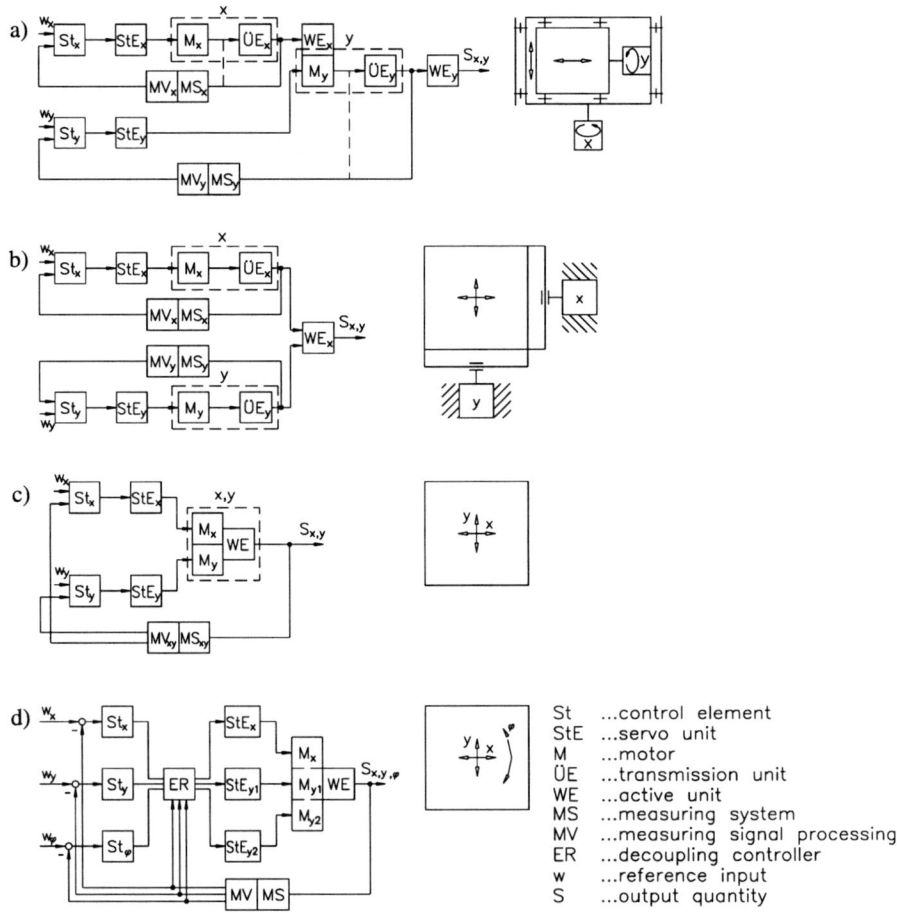

Fig. 4.59. The structures of multi-coordinate movement equipment (as in [4.83]) with **a** serial, **b** parallel, **c** integrated xy-, **d** integrated $xy\varphi$-drive

z, the fourth coordinate, can only be integrated into the travel in a minute area of a few μm to a few mm. This can be done, for instance, by the change of pressure of an aerostatic guide system, by the (controlled) change of attraction in a magnetic guide system, or, in the case of serially linked linear motors, by swinging a functional surface with a spring suspension. Then, the last-mentioned variant is not really a genuine, integrated, four-coordinate motor.

The basic concept of the two-coordinate and multi-coordinate motors is that of coils with long end windings or very large magnetic poles, to enable, for instance, a tabletop to be moved on two axes. The coils with the magnetic poles must be guaranteed not to go outside the area of influence of the Lorentz force generated. The coil geometry used, which will considerably affect the further construction of the apparatus, is itself a means of recognising the different motors.

4.2.3.1 Electrodynamic Multi-Coordinate Motors with Flat Coils

A very flat type of construction is obtained by flat coils without an iron core, a method which means the complete apparatus can be shallow, and will avoid interference in the form of reluctance forces and of same-direction Lorentz forces in consequence of self-excitation. To serve their purpose, flat coils are arranged symmetrically, either singly or in groups requiring commutation: to be mobile, they will be in the armature; to be stationary, they will be in the stator (in this case, with mobile permanent magnets or parts of magnetic circuits). The actual, physical measurement system of a multi-coordinate measurement system is usually integrated within the armature, so that the spatial data and velocity and acceleration information for position and movement control can be obtained without need of a trailing cable.

Also the guide elements for whatever type of guidance system is used (sliding, rolling, aerostatic or magnetic [4.88]) are normally placed in the armature, because they are less massive than the surface guide elements required for larger distances of travel, although in aerostatic and possibly also in magnetic guidance systems, trailing hoses or cables will be present which always generate interfering forces.

Configurations with Moving Coils

Multi-coordinate motors with moving flat coils can be made for any type of movement, from xy to $xy\varphi z$. The abbreviation *MCM* (MKM in German) is used henceforward in this chapter for multi-coordinate motor and *MCD* (MKA in German) for multi-coordinate drive.

The principle of an electrodynamic MCM giving $xy\varphi$-movement is shown in Fig. 4.60. It has integrated flat coils in the armature [4.72].

In this MCM, there are two widely spaced flat coils for each axis above the sets of twin permanent magnets, which have opposite magnetisation and are firmly attached to the ferromagnetic stator. By this means, each long side of the winding in the rectangular flat coil contributes to the production of Lorentz force and the large end windings, which produce no force, enable orthogonal movement along the twin permanent magnets along the two axes. Ferromagnetic yoke elements are so arranged above the coil pairs as to complete the magnetic circuits. The magnetic circuits comprise the ferromagnetic baseplate, the permanent magnet pairs, the yoke, and the air-gap.

If the two flat coils, each intended for one axis, are supplied with different current, different amounts of force are generated in the individual coils and, with that, the torque to achieve a tiny rotary movement $\Delta\varphi$. This rotation must be registered simultaneously with the xy-movement of the armature by a high-resolution measuring system and communicated after very rapid signal processing with the transputers or the signal processors in an associated multi-variable control. Unintentional rotations of the armature must be systematically excluded using $\varphi = 0$ and the desired tiny angles of rotation must be systematically regulated in relation to predefined values for the armature position. It is useful if the measuring system to capture the $xy\Delta\varphi$-movement is an optical, physical grid situated in or on the armature, which is scanned optoelectronically in the stator, to avoid the use of trailing cables [4.92, 4.93]. The y_1 and y_2 positions will be recorded as base distance b, thus capturing the rotation of the armature. In addition, the linear shift is calculated from x_1 and y_1. The signal processing must take into account that the sensor signal

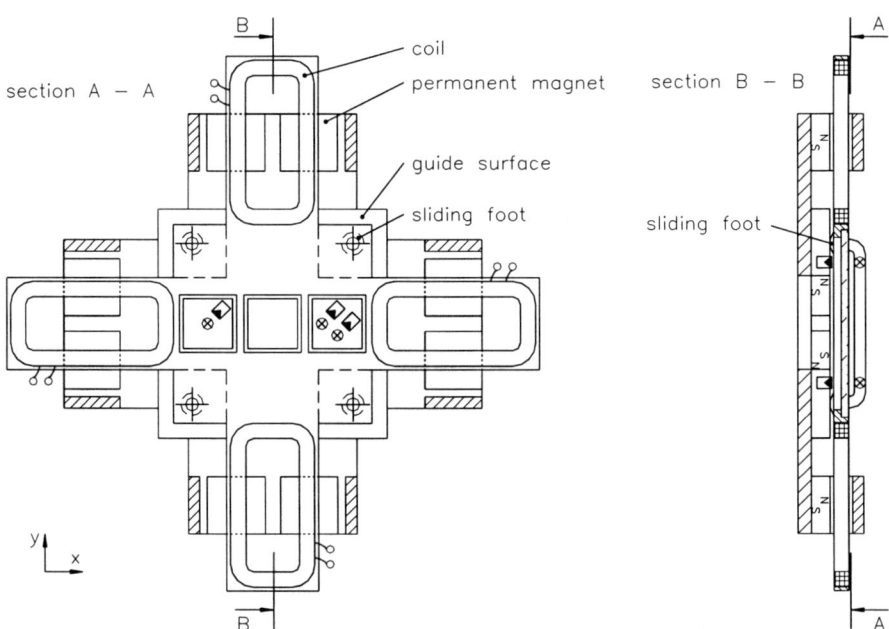

Fig. 4.60. Principle of an electrodynamic MCM with flat coils for $xy\varphi$-movement

will reduce (in an almost linear ratio) until rotation of approx. 3° is reached, if a certain geometry is adhered to for the scanner in relation to the geometry of the physical grid [4.93]. The scanner heads will use gridlines as their scanning geometry. Figure 4.61 demonstrates the principle of the multi-coordinate regulation to be carried out on the basis of the three measurements, x, y_1 and y_2.

It is absolutely necessary that the system should be able to eliminate $\Delta\varphi$-rotation, not only because of the torque arising from trailing cables (in the case of moving coils) but also because of the forces associated with the eccentric position of the armature in relation to the stator with its fixed arrangement of permanent magnets, which thus do not operate upon the centre of gravity of the armature and will thus generate torque. This source of error can only be removed in the case of moving coils by appropriate curving of the operative parts of the winding around the centre of gravity of the armature or by using large magnetic pole surfaces so that the operative parts of the winding never change their position in relation to the centre of gravity of the armature. Figure 4.62 demonstrates a combination of both possibilities.

Movement on two axes can, however, also be produced with a single coil in the armature and at least two controllable magnetic poles in the stator. A variant is shown in Fig. 4.63.

This last configuration will only achieve a movement on two axes if there is a protective mechanism against twisting. For all the principles described so far, pantographs or cross-push guides can serve as anti-twisting mechanism. It must be admitted that the advantages of the pure movement of the single mass without mechanical friction in the anti-twisting mechanism are then lost. However, the control process will be simpler, because each axis of movement is mechanically independent of the other, as on a cross-table.

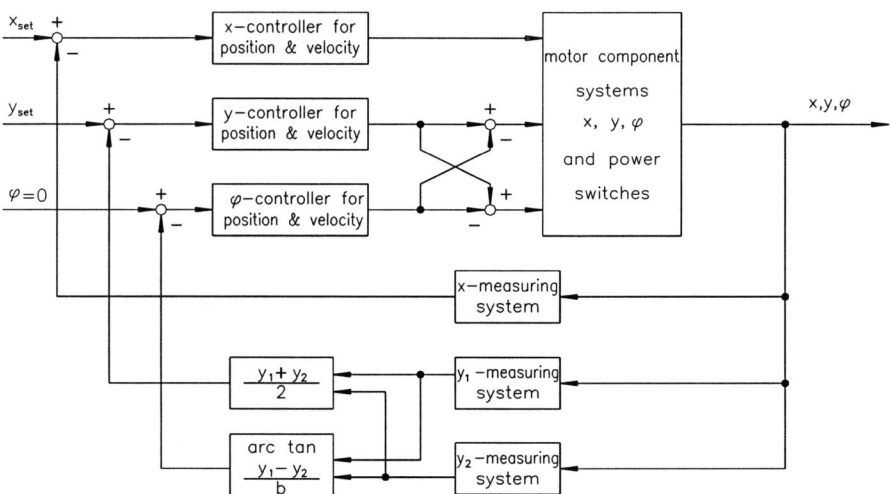

Fig. 4.61. Principle of control in the case of multi-coordinate capture

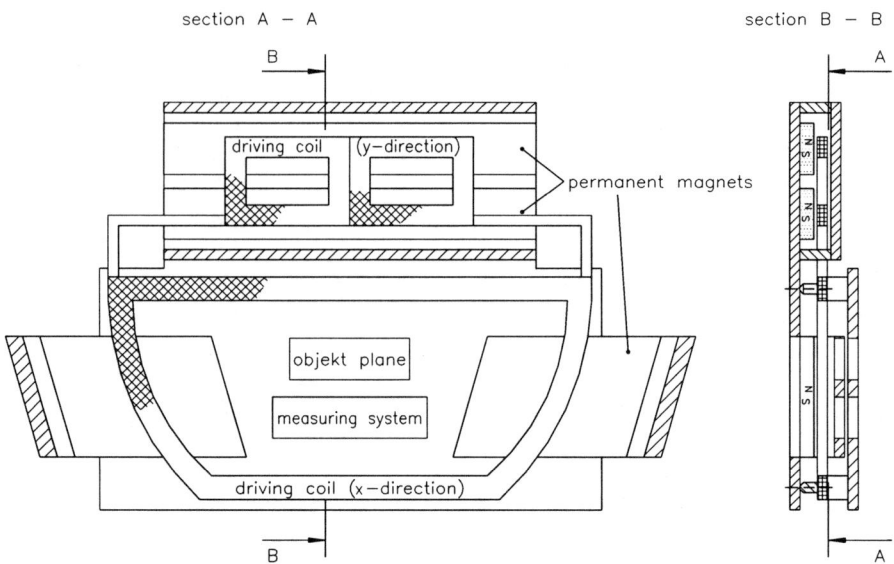

Fig. 4.62. Principle of an electrodynamic MCM with the combination of long magnetic poles for two y-axis coils and a single curved coil for the x-axis

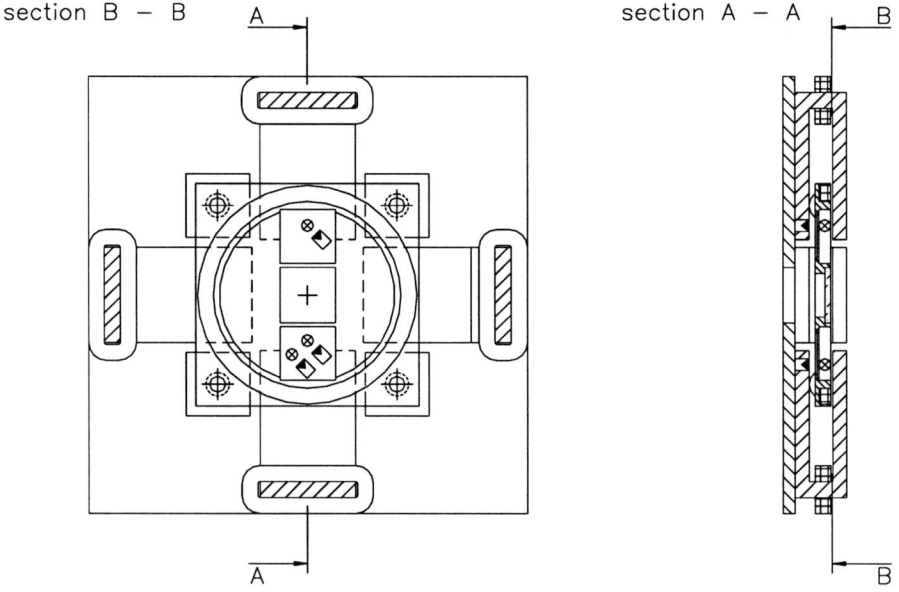

Fig. 4.63. Principle of an electrodynamic two-coordinate motor with one drive coil in the armature and four excitant windings in the stator

4 Drives with Limited Motion   275

For some movement tasks, large angles of rotation, up to 360°, are necessary. It is only with completely different configurations as regards coils and measurement system that such large angles of rotation are practicable [4.73]. A very shallow configuration with moving coils and a measuring system is shown in Fig. 4.64.

Fixed into a socket in the armature there are two layers of coils shaped as a 90°-sector, the coils displaced in relation to each other by 45°, above a permanent magnet ring installed in heteropolar arrangement on the ferromagnetic stator plate.

A special measuring system [4.63, 4.100] is installed in the empty space in the middle of the armature that consists of a disc with a radial grid on the rim, a hole in the centre, and three CCD lines separated by less than 90° to permit scanning in the stator. The measuring system is a three-coordinate system for the simultaneous measuring of x, y and φ (φ up to 360°). Very fast signal processors are essential here to capture and communicate in real time the data necessary for control of position and velocity.

The armatures in the $xy\varphi$-level can generally be carried at a constant distance from the stator by sliding, rolling, aerostatic or magnet guide systems in such a way that the measured values given by the incremental [4.71, 4.93], or absolute [4.94] sensors are sufficiently constant.

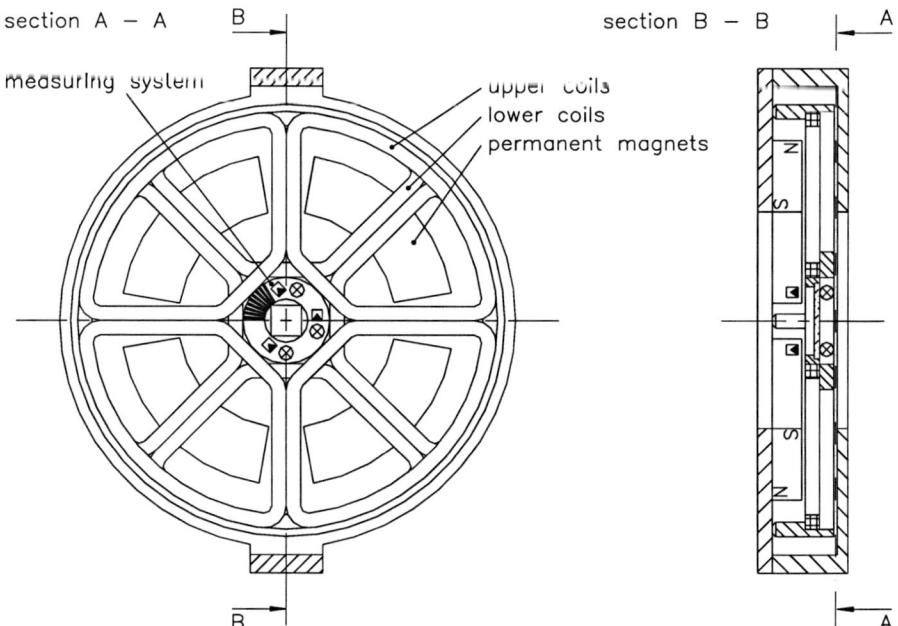

Fig. 4.64. Principle of an electrodynamic MCM with flat coils for $xy\varphi$-movement (rotating up to 360°)

For the pair of materials in the case of sliding friction, PTFE steel with a relatively high, velocity-related coefficient of friction is suitable, making damping of vibrations and good positioning accuracy achievable. The wear of the PTFE is, it must be said, relatively great. There are great hopes of lubricant made of fullerenes for the pair of sliding friction materials [4.65]. This material is reputed to encourage very low friction values constant over very long periods.

Roller-guidance systems with special spherical rollers [4.76], or, even better, with simple materials paired at ball level without a retaining cage for certain load conditions are also known [4.75].

Aerostatic guides can be easily produced, since the air guide elements and carrying air elements are available commercially [4.88], and the compressed air is produced very quietly. The interference forces from the trailing hose may still be seen as a disadvantage if the air is supplied to the armature.

Magnetic guides for low-mass loads are more trouble to design.

For longer distances of travel, the coils can be equipped with oversized end windings, commuted and arranged in series. Because it is an arrangement that involves less mass, it would, however, be preferable to use only two coils (requiring commutation), one behind the other, on each side, above a large number of permanent magnets in heteropolar order [4.93].

As the mass of the armature is relatively great in the case of longer areas of travel because of the large end windings which are then necessary but relatively ineffective, the possibility of using high-energy permanent magnets with less mass in the armature should be explored to improve the dynamics.

Configurations with Moving Magnets

Configurations with moving permanent magnets are only useful if equipped with aerostatic or magnetic guidance systems, since, in the case of glider or roller guides, the great attraction would generate too much friction. The principle of an electrodynamic MCM for $xy\varphi$-movement is demonstrated in Fig. 4.65. The fixed arrangement of the coils in the stator and the permanent magnets in the armature is, in effect, just the reverse of the pattern shown in Fig. 4.60.

The air guide elements or, as the case may be, the carrying air elements must be attached to the upper plate on the yoke since the very tiny airgap is what produces the direction of the attraction, and the top side of the armature must have smooth, flat areas to match the required range of movement.

As to the way the measuring system is arranged, the same applies as in the case of moving coils: the physical measuring equipment is in the armature and the scanner in the stator. This will give the absolutely optimum solution of arrangement a trailing hose and cable, which is here present.

For areas involving long distances of travel, it is again possible to commutate a large number of coils fixed in the stator one behind the other (this

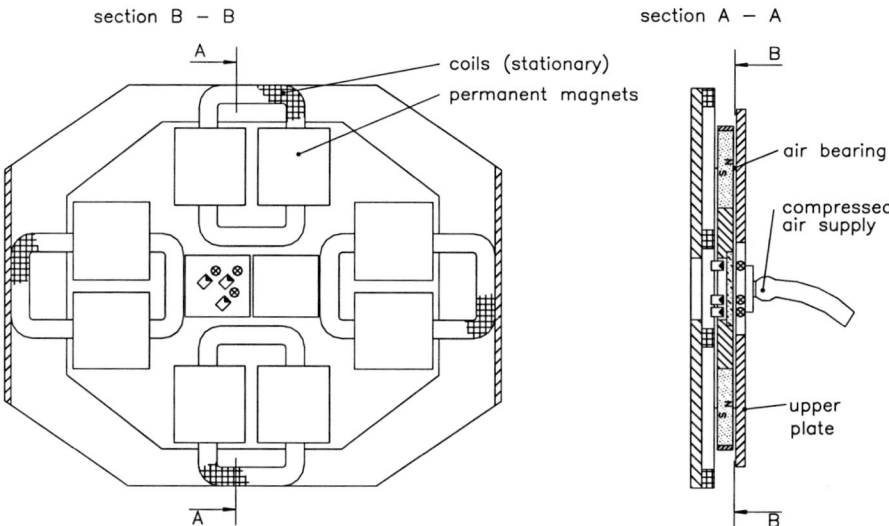

Fig. 4.65. Principle of an electrodynamic MCM with moving permanent magnets for $xy\varphi$-movement

represents the reverse of the case with moving coils), if the two oppositely polarized permanent magnets attached to each side of the armature are at the appropriate spot. This configuration is then optimal for moving a mass even a long distance and keeps down the cost in terms of high-energy permanent magnets.

The only disadvantage of this type of construction arises if the armature is unevenly supported as it changes position over the large distances of travel, the effect being that of an erratic load. A guide system over a more extensive area is made available if instead of nozzles and distribution channels, perforated air supply elements are used.

Configurations with Moving Permanent Magnetic Circuits

The interrupted permanent magnetic circuit consists of the permanent magnet, the iron yoke and the operative air-gap for the winding arrangement.

The yoke can be divided into the immobile baseplate (stator) and the moving yoke with its two permanent, oppositely magnetized magnets. The moving parts are relatively low in mass because the length of the yoke only bridges the magnets and the distance between them, and the thickness of the yoke depends on the magnetic induction of the permanent magnets (Fig. 4.66). Both dimensions (length and thickness) are dependent only apart from the material of which the permanent magnet is made and its magnetization, on the height of the air-gap ensuing on the coil thickness and the guide tolerance. The breadth of the permanent magnet will depend on the selected breadth of

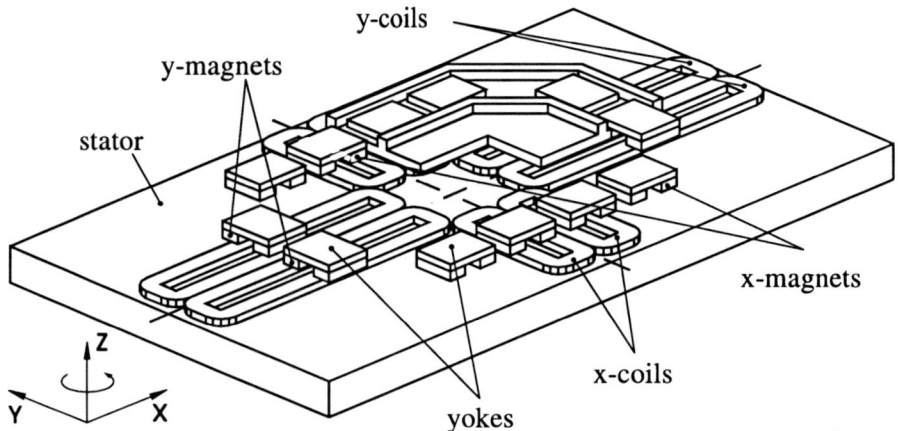

Fig. 4.66. Configuration of an electrodynamic MCM with moving section of magnetic circuit as referred to in [4.93]

winding. The thickness of the yoke is given by

$$d_{\text{Fe}} = \frac{B_{\text{L}}}{B_{\text{zulFe}}} b_{\text{DM}} \ . \qquad (4.83)$$

At $B_{\text{L}} = 0.6T$ and permitted magnetic induction in the iron $B_{\text{zulFe}} = 1.8T$ and $b_{\text{DM}} = 20$ mm, the minimum thickness of the iron is $d_{\text{Femin}} = 6.67$ mm.

The structure shown in the last figure has thoroughly proven itself in association with the previously mentioned two-coordinate, longitudinal, illuminated measurement system as described in [4.93], an aerostatic guide and the appropriate control system for multiple coordinates. The angle of rotation is restricted to approx. 3°, the resolution of the travel is down to nanometres and positioning accuracy is better than 1 μm. The force-path characteristic has been optimized to a force fluctuation of only approx. 3%, by trapezoidal commutation derived from the coordinates registered by the position sensor in the measuring system.

The only disadvantage of this configuration is the effect of the trailing hose. Using an active magnetic guide will obviate this problem [4.87]. At hovering height, the armature is regulated by means of three cylindrical magnets and three iron plates fastened to the armature, which together form an adjustable magnetic circuit. The distance of the iron plates acting as armature to the cylindrical magnet yokes is registered by three eddy current sensors. This information is sent to the control circuits for the z-axis. A completely freely floating armature with six degrees of freedom has thereby been created.

The task can also be managed by simultaneous production of xz-forces for the forward shunt and the magnetic guide using a HALBACH array and corresponding stator coils as described in [4.80, 4.101] or an AUER arrangement as in [4.90]. Figure 4.67 shows the arrangements described in [4.81, 4.86],

4 Drives with Limited Motion   279

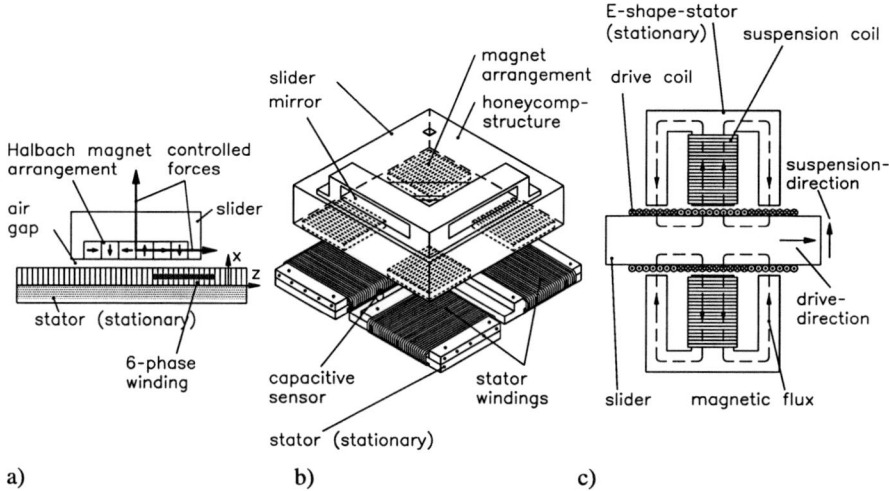

Fig. 4.67. Arrangements for simultaneous production of forward-moving and lifting force

and [4.90], with the mirrors for measurement by interferometry shown at the armature in part (b) of the picture.

4.2.3.2 Electrodynamic Multi-Coordinate Motors with Box Coils

Box coil is the name here given to the rectangular form of a frame-type winding. Box coils are suitable for MCMs because long end windings can be realized for the two-coordinate movement; this is not possible with circular windings (see also Fig. 4.57).

With box coils, the force is not produced at right angles to the central coil axis as in the case of flat coils, but to or in alignment with it, and there is always an iron core inside the coils. The iron circuits are either U-shaped or E-shaped.

If the winding is being used on two sides, E-shaped iron circuits are often used to keep the iron yoke material to a minimum. By the shear of the iron circuit, the thickness of the yoke is less because the magnetic resistance is increased and the magnetic flux correspondingly decreased, but reluctance forces appear [4.76]. Besides these reluctance forces, which are always directed to the closed end of the iron circuit, Lorentz force arises because of the magnetic flux generated by the winding itself, and this is always directed towards the closed end of the iron circuit [4.76]. These forces can be described as interference forces because their direction is always independent of the direction of the current in the coil and always the same [4.78]. Both interfering forces vary with the square of the current and therefore in one direction they can considerably weaken the Lorentz force produced by the flux from the permanent magnet, even cancel it out, and, if the current is further increased, can produce force in the opposite direction [4.78], and cf. Sect. 4.2.4.

Figure 4.68a shows the alteration to the distribution of the additional magnetic flux in the winding as a function of the box coil's position. When the current is in the opposite direction, the opposing magnetic flux will appear. This is true if the opposing field changes accordingly to the field of the permanent magnet. The force-path and force-current characteristic curves that arise are shown in parts (b) and (c) of the figure.

The reluctance force attendant on the change of field distribution can be expressed by Eq. (4.82) for the alteration of the energy of the magnetic field in relation to the distance travelled. The Lorentz force (in the same direction as the reluctance force) as a result of self-excitation can be calculated by means of Eq. (4.79) inserted into Eq. (4.80) or as an approximation by substituting ($\mu_{rFe} \to \infty$, $\mu_{rDM} \to 1$) with the equation

$$F_L^* = \frac{\mu_0 i^2 w^2 b_{Fe}}{2\delta} \qquad (4.84)$$

in which δ is the thickness of the air-gap and the permanent magnet.

The reluctance force F_R is nonlinearly dependent on the distance travelled and the Lorentz force in the same direction because self-excitation is almost constant. Both forms of interference always appear together in such sheared-circuit arrangements. If not taken into account, these forces can lead

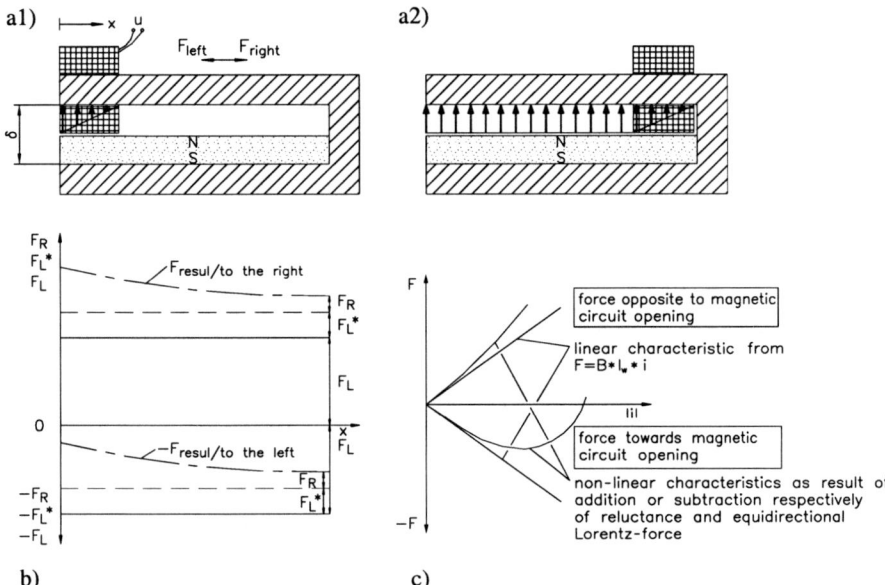

Fig. 4.68. Changes (a_1 and a_2) in the distribution of the additional magnetic field in relation to the position of the box coil; force-path and force-current characteristics (b and c) reflecting the changes

to significant deviation from the force-path characteristic curve, as is shown schematically in Fig. 4.68b.

The two forms of interference can be minimized by using double or Z-type arrangements and can be compensated for or optimized to the desired force-path characteristics by further combination with similar structures and/or items with stored force (such as springs) [4.76].

If it is desirable to eliminate these two forms of interference completely and if one takes the thickness of the iron into account when there is not too large a number of turns of the winding – for small forces – and the distances are above 50 mm, one can use closed yoke circuits in the permanent magnetic circuit (Fig. 4.57h) for almost pure permanent magnetic excitation.

Configurations with Moving Coils

A mature type of construction with moving box coils and mechanical block against twisting has already established itself on the market because of its simple control associated with mechanically decoupled systems. There is a sketch of this system in Fig. 4.69.

For bilateral usage of the coil, E-shaped iron yokes are used here, together with a cross-push guide to block torsion mechanically. For a movement area of $100 \times 100 \, \text{mm}^2$, four box coils with large end coils are connected to a tabletop and arranged symmetrically. The windings are self-supporting and embedded in epoxide resin (or are made of self-bonding wire) with-

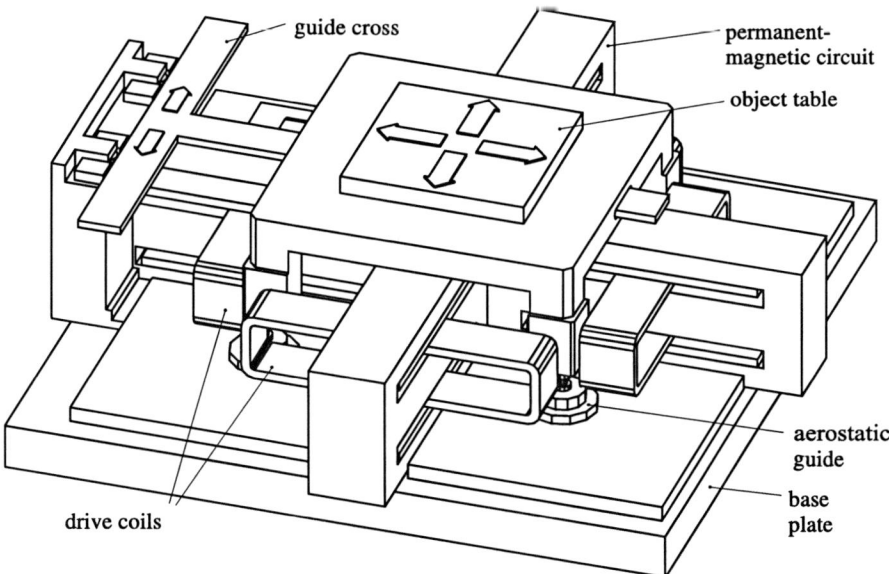

Fig. 4.69. Multi-coordinate motor with box coils, E-shaped iron yokes and mechanical torsion blocking as described in [4.84]

out bobbins. They can also be made from copper or aluminium foil. The current is supplied *via* trailing cables and the tabletop is guided aerostatically on guide surfaces in stators; the air enters through a trailing hose. Because of the mechanical block against twisting, however, the trailing cable and hose are not a critical problem. It must, however, be said that the friction in the mechanical twisting guard and the fact that the motive forces work eccentrically when the tabletop is in off-centre generate interfering forces that result in torque and must be dealt with by the anti-twist mechanism.

The measuring system used is a simple two-coordinate system with a cross grid under the tabletop, scanned by a two-coordinate incident light scanner in the stator. Here the commercially available two-coordinate system suggested in [4.69] could, in principle, also be used.

The control system consists of two discrete systems [4.68]: "The transputer system takes over the table's path data from a PC via a RS interface. The transputer (a CISC, or complex instruction set computer) regulates the movement impulses for the two drive axes, synchronizes the axes and processes the monitoring information and error messages. It also controls the communication with the operator, who will enter his information from the keyboard.

The drive-specific system consists of two servo regulator modules. A configurable cascade control without scanning time, which governs the position and velocity of the table, is integrated into each axis."

The technical data for the table are given in [4.68].

Taken as a whole, the two-coordinate motors take up rather a lot of space by having the box coils in an E-shaped iron yoke. However, with an U-shaped iron yoke one can achieve also flatter forms by sacrificing the more efficient coil usage.

To enable the φ-coordinate to be incorporated in addition, a circuit can be formed (see Fig. 4.70) from a curved three-part U-shaped iron yoke [4.74]. With this motor, rotation can be achieved through an angle up to $\varphi \approx 100°$.

To be able to execute the z movement, important for microscope focusing, for instance, even when the guidance system is of the gliding or rolling type, the functional area to be moved is suspended (by diaphragm spring) and thus can be deflected a few millimetres by means of a linear motor assembly in the armature.

The measuring system can be that of a disc with a radial grid on the rim and CCD scanning already described, which will genuinely measure $xy\varphi$-movements. To regulate the z position, a linear measuring system can be used, or, in the case of microscope focusing, a microscope image clarity analyser which permits automatic adjustment of object under examination into the necessary correct position.

A multivariable control must be used for the $xy\varphi$-movement and a simple analogue or digital servo regulator for the z movement.

There will be trailing cables and, in the case of an air guide, also trailing hoses as potential causes of interference.

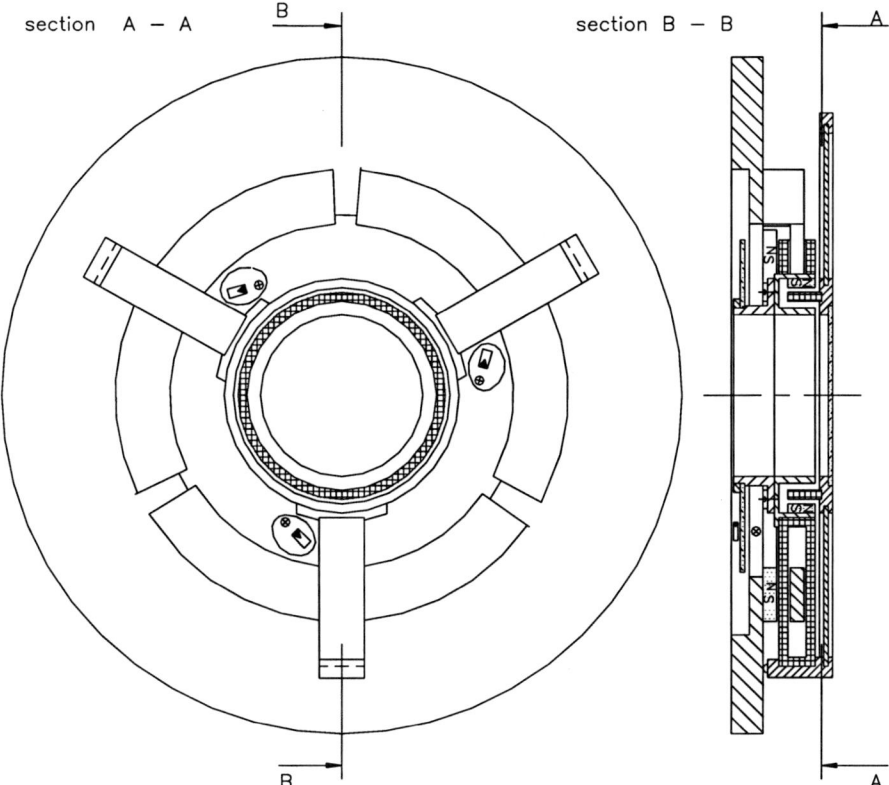

Fig. 4.70. Principle of an electrodynamic MCM with moving box coils for $xy\varphi z$-movement

Configurations with Moving Magnets

Motor types with moving permanent magnets require a relatively large space for installation and considerable static iron mass for larger distances of travel. In order that the iron yokes for large distances do not get too big, commuted coils in a state of counter-excitation must be employed, so that the magnetic flux in the iron of the coil excitation is compensated as well as possible.

Configurations with Moving Permanent Magnetic Circuits

The total mass of the permanent magnetic circuits to be moved would, even in the case of box coils, be very large if the movement areas were large, more than 50 mm, and the forces in excess of 5 N. However, this variant is definitely practicable in the case of smaller paths and forces [4.76]. Therefore, the same mass reduction as in the case of the flat spooling forms in Sect. 4.2.3.1 – by partitioning of the iron yoke into one motionless and one moving part, which also contains the permanent magnets – must be applied here also.

4.2.3.3 Electrodynamic Multi-Coordinate Motors with Cylindrical Coils and Curved Box Coils for an $x\varphi$-type Movement

The $x\varphi$-motors (see [4.67, 4.95]) already mentioned in the introductory Sect. 4.2.3 can be derived from the classic voice-coil motor for movements combining lift and rotation. The armature has a cylindrical coil for the x-movement and two curved box coils for the φ-movement. These are arranged in the relevant operative air gaps in permanent magnetic circuits. The box coils must be commuted according to the rotation: the commutation signals will be derived from an integrated incremental xy-measuring system. The current supplies to the windings are made *via* trailing cable. The bearing of the armature can be of the gliding, rolling or aerostatic type. Figure 4.71 shows a section through $x\varphi$-motors as described in [4.66, 4.67, 4.95], with either separate or identical air gaps for the x- and the φ-winding.

The measuring system required is not yet commercially available. A transmitting and reflecting cross grid on a glass cylinder serves as the physical measure [4.71]. Optoelectronic scanning can be used, in incident or transmitted light, as in the case of the flat xy-measuring systems. The scanning grid must be curved to match the cylinder bearing the physical measure and it must have orthogonal line grids for the x- and φ-coordinates.

The control system should be a multivariable control even though only two values are coupled together in this instance. Actually, the movements can also be made in controlled fashion one after the other, in which case reference marks are used as zero or base marks [4.95]. The performance parameters of the $x\varphi$-drive are indicated in [4.95].

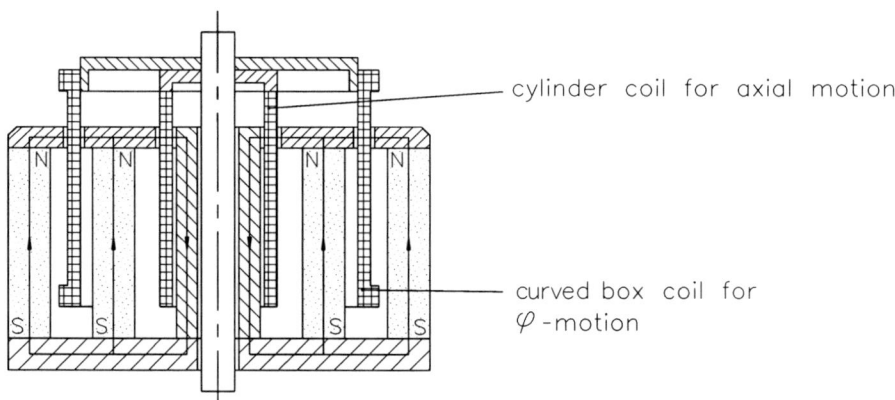

Fig. 4.71. $x\varphi$-motor as described in [4.66] without measuring system and with two separate air-gaps for the rotating and push winding

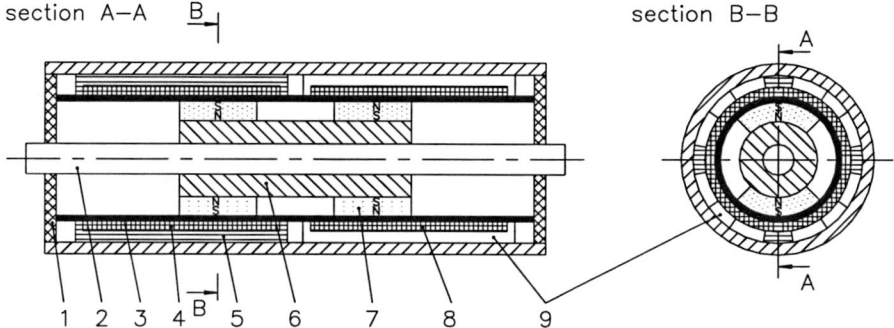

Fig. 4.72. $x\varphi$-motor as described in [4.95] without measuring system and with one common air gap for the two drive windings. *1* cover; *2* drive shaft; *3* bush of gliding-bearing; *4*, *8* push windings; *5*, *9* rotating windings; *6* inner magnetic yoke; *7* magnets

4.2.4 Behaviour of Electrodynamic Linear and Multi-Coordinate Motors

Operating performance of linear and multicoordinate motors will be considered only for permanent magnetically excited systems as this is the predominant implementation in motor design. For static and quasistatic operation a restriction to basic models of non-commutated homopolar configurations suffices.

4.2.4.1 Static and Quasistatic Operation of Permanent Excited Motors

The permanent magnetic excitation field is almost constant at the basic models. Distinct air gap and with thus air gap field variations appear only in systems with a variable air gap or moving magnets. Great air gap variations with the danger of an irreversible partial degaussing have to be expected during the assembly and disassembly, i. e. at the removal of the permanent magnet armature. Temperature influences on the permanent magnets and with that on the air gap flux density B_A can be determined by the temperature coefficients of the remanence $T_C(B_r)$. With hard ferrites the temperature coefficient of the coercive field strength $T_C({}_BH_c)$ is in the opposite direction to the temperature coefficient of the remanence $T_C(B_r)$. The danger of an irreversible partial degaussing then also exists at a temporary cooling at comparatively strongly sheared magnetic circuits (see Sect. 2.1.2.2). The electrical partial system shows stronger temperature influences by the temperature coefficient of the conductor materials (e. g. $T_{CCu} = 0.39\%/K$). This is generally compensated for using a current control. Armature reactions lead to the weakening or arise of the operating point according to Eq. (4.79). There is also the danger of an irreversible partial degaussing at nonlinear degaussing characteristics.

The amplitude of the coil current is thus respectively limited in the short-time or intermitted operation.

The operating air gap flux density is approximately constant only in the middle. A clear decrease has to be expected towards the edges of the air gaps (Fig. 4.73a). The force-path and the force-current characteristics show corresponding nonlinearities (Fig. 4.73b, c). Generally, the force-path characteristics decline towards the edges of the air gaps due to the flux density decrease. The integrating effect of the coil, however, causes a less strong decline than the one of the flux density. The force-current characteristics show an opposite override (identical field and opposite field influence depending on the current direction) in the two directions of motion. This is the result of a force component (Eqs. (4.79) and (4.80)) which is proportional to the square of the current. In short time operation with high currents a reversal of the direction of force may appear, however, it barely appears.

Reluctance forces or forces on boundary surfaces in magnetic fields appear additionally in systems with variable air gaps or moving magnets. Figure 4.74 clarifies qualitatively the effect of the reluctance forces at a motor with moving magnets [4.43] according to the basic concept introduced in Fig. 4.54a. A deflection from the middle position will result in restoring forces on the magnet armature. As these do not depend on the direction of the

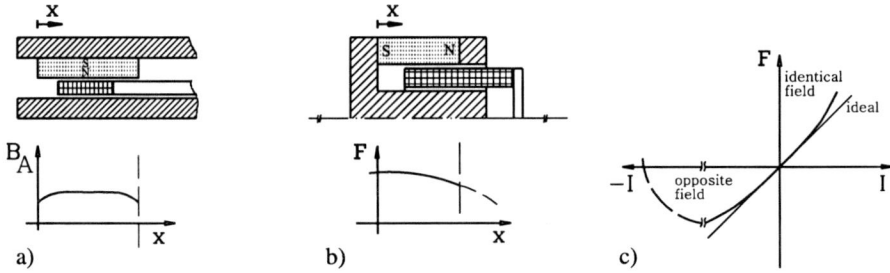

Fig. 4.73. Characteristic non-linearities at electrodynamic linear motors (examples, qualitative) of the air gap flux density of a flat-coil motor (**a**), of the force-path characteristic of a voice coil motor by a non-constant air gap flux density (**b**), and of the related force-current characteristic of the voice coil motor by field overlay (**c**)

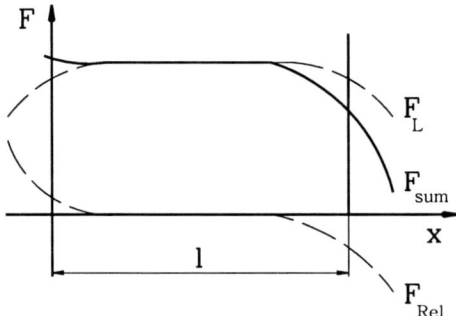

Fig. 4.74. Forces at a motor with moving magnets (qualitatively) F_L Lorentz force; F_{Rel} reluctance force (force on boundary surfaces); F_{sum} resulting force effect

current but only on the moving direction, this practically leads to a force reduction leaving the middle-position and to reinforcement returning to the middle position. They appear in addition to the nonlinearities shown above.

4.2.4.2 Dynamical Operation

Characteristic motor and dynamic equations which interconnect the four partial systems apply to permanent magnetically excited linear motors as well as to rotational permanent magnetically excited DC motors. These equations are already contained in Fig. 4.44.

Motors in homopolar configurations realize only low strokes. Stationary operating states are achieved only for a short time. Positioning processes with starting and braking phases as well as the holding of achieved positions dominate. Stationary characteristics, comparable to the speed, torque and power or efficiency characteristics respectively at rotary motors therefore are rarely used. Such characteristics can be represented at heteropolar configurations with commutation, when there is a sufficiently large travel range analogue to the permanent excited rotary DC motors or when there is a quite rare electric excitation analogue to series or shunt-wound electric motors. At heteropolar configurations, there are also influences by the commutation or by corresponding wavy force characteristics. These are already displayed at the rotary motors (see Sect. 2.2).

If the force-path and the force-current characteristics are linearised and if the nonlinear adhesion and sliding friction forces F_F, stationary load forces F_L, reluctance forces F_{Rel} as well as the commutation at first are neglected, the transfer functions of the drives can be found by Laplace transformations with the help of the above-mentioned equations and the signal-flow diagram can be prepared to clarify the reaction effects [4.41]. Table 4.1 shows selected path-transfer functions. Figure 4.75 shows the appropriate signal-flow diagram

Table 4.1. Selected path-transfer functions $G(s)$ of linear motors (without influences of friction, load and reluctance forces, simplified for $T_{el} \ll T_{mech}$)

---

1. Motor with spring force

a) at impressed voltage

$$G_u(p) = \frac{\frac{Bl}{Rk_2}}{(1+sT_1)(1+2DTs+s^2T^2)}$$

$$T_1 = \frac{L}{R}; \quad T = \sqrt{\frac{m}{k_2}}; \quad D = \frac{1}{2}\frac{k_1+\frac{(Bl)^2}{R}}{\sqrt{k_2 m}}$$

b) at impressed current

$$G_i(s) = \frac{\frac{Bl}{k_2}}{1+2DTs+s^2T^2}$$

$$T = \sqrt{\frac{m}{k_2}}; \quad D = \frac{1}{2}\frac{k_1}{\sqrt{k_2 m}}$$

2. Motor without spring force

a) at impressed voltage

$$G_u(s) = \frac{\frac{Bl}{R(k_1+(Bl)^2/R)}}{s(1+sT_1)(1+sT_2)}$$

$$T_1 = T_{el} = \frac{L}{R}; \quad T_2 = T_{mech} = \frac{m}{k_1+\frac{(Bl)^2}{R}}$$

b) at impressed current

$$G_i(s) = \frac{\frac{Bl}{k_1}}{s(1+sT_1)}$$

$$T_1 = \frac{m}{k_1}$$

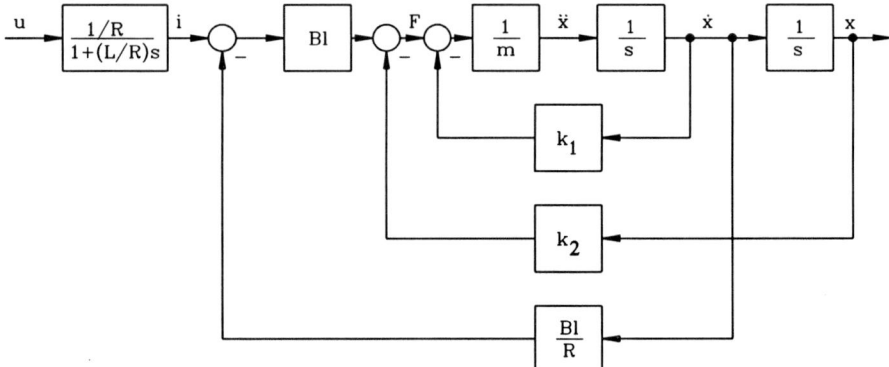

Fig. 4.75. Signal-flow diagram of an electrodynamic linear motor with spring counter-force at impressed voltage (simplified for $T_{el} \ll T_{mech}$)

of a motor with spring counter-force at impressed voltage. By corresponding simplifications signal-flow diagrams are derivable for all other possible load constellations.

At impressed voltage, both the electrical time constant and the mechanical time constant have a damping effect on the movement due to the induced counter voltage at movement of the coil. At impressed current there are no effects of the coil inductance and the movement-induced back-e.m.f. $Bl\dot{x}$. Also oscillating elements appear at motors with spring counter-force with the known transfer behaviour [4.41]. In addition to the above-mentioned equations and transfer characteristics, there are systems with moving magnets or such with moving flux-guide with further damping effects such as remagnetization or hysteresis losses.

4.2.5 Control of Electrodynamic Linear and Multi-Coordinate Motors

From their movement behaviour, electrodynamic linear and multi-coordinate motors are continuous drives, and correspond to rotary permanent magnetically excited DC motors. A closed loop is required for positioning processes since electrodynamic drives have no internal measure. It should be referred to Chap. 6 for this. Only selected examples of linear direct drives and the special features of multi-coordinate drives shall be elaborated. Positioning applications are the most frequent applications of electrodynamic linear motors, only these are explained in the following [4.24].

4.2.5.1 Control of Electrodynamic Linear Motors

Closed loops for electrodynamic linear motors are built up most as cascaded closed loops as known of rotary motors. A typical closed-loop structure is shown in Fig. 4.76, as it is often applied at positioning systems with electrodynamic linear motors.

A subordinated current control serves the improvement of the response times by compensation of the movement-induced back-e.m.f. and of the electrical time constant as well if necessary the suppression of nonlinearities of the forces due to the temperature-dependent variation of the coil resistance.

An overlaid velocity control is applied for highly dynamic systems, provided a velocity signal is derivable with acceptable effort. The outer closed-loop represents a position control loop. Here, mostly PI or PID controllers are applied, for the subsidiary control often P-controllers are applied. In principle, controllers can be digital or analogue, as indicated in Fig. 4.76. For reasons of the necessary computing speed at present, the current controls are usually analogue and the position controls digital, while for an additional velocity control both usually depend on the computing power. A setpoint setting and calculation resulting from the application is made primarily by a host computer.

As a final control element for the current and thus the motor force, only a power amplifier which is used particularly for small mechanical output power and the demand for smooth operation is required. At higher mechanical output, power pulse-width modulated power converters are applied. Since these output stages work in switching operation, dissipation losses are small. The usual pulse-width modulation operates at 20...25 kHz and contains high harmonic content in a wide frequency spectrum that possibly creates EMC problems. Also at motor standstill ($I = 0$) there is dissipation caused by the use of pulse-width modulated amplifiers as the current mean average becomes zero but there still is a permanent periodical current change, carried out around zero with the pulse frequency. This periodical current change can not be neglected ($P_{el} \sim i^2$), especially with the frequently used drive windings with very small inductances, e. g. with flat coils. The final control element often takes over additional tasks, like current control, excess temperature and operation voltage monitoring, at commutated systems also the commutation, mostly controlled by the path-signal or hall-effect sensors.

Linear measuring systems for movement tasks to register the position and maybe the velocity have to be attached as close as possible to the output. Preferably relative or absolute incremental encoders and increasingly also in-

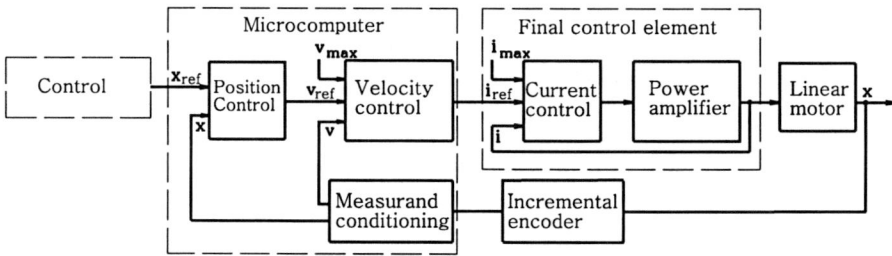

Fig. 4.76. Cascaded closed-loop control with microcontroller

terferometric position encoders are applicable. Information about velocity can be generated by the counting rate. The achievable resolutions then increase to the nm area at high absolute precision; the costs are, however, still considerable. Analogue position encoders with the disadvantages resulting from the analogous operating principle are more economic but appropriate only for small travel ranges, for example inductive position encoders according to the differential transformer or differential throttle principle. Resolutions of 1 µm and better can be obtained with small measuring lengths and restricted absolute precisions.

The positioning accuracies of linear direct drives mainly depend on the measuring system and the friction conditions and reach at interferometric path measurement and aerostatic guide up to the nm area, for example. A comprehensive general representation of the control of linear direct drives is not possible in this place (see Chap. 6), solution trials shall, however, be clarified for problem areas represented above at two examples.

Example 1: Analogous Final Control Element for Linear Motors (Voice Coil Motor Driver)

Analogous power-control elements for not commutated small power motors are available as a so-called Voice Coil Motor Driver in the form of an IC (e. g. Vishay Siliconix Si 9961 [4.54]). Such final control elements are preferred for pivot drives in hard disks or CD-ROM drives. A series of important auxiliary functions are integrated in the circuit, such as a current control, an operation voltage monitoring, a head retraction in case of average with adjustable retraction current (using the induced back-e.m.f. in the rotary disk motor when the operating voltage breaks down), a short-circuit monitoring as well as a commutation of the current control between rail seek and tracking mode. With the position information recorded on the hard disk as a path signal a highly dynamic, highly accurate and also economical positioning system then can be realized at use of a micro computer controller with only few components.

Example 2: Control of a Motor with Moving Magnets Under Use of an Integrated Path-Signal Generation

A further example shall clarify that at medium precision and dynamic demands also position information, in other cases also velocity information can be won from the motor itself. This reduces the total effort and the costs considerably due to the abolition of additional measuring systems and thus opens new attempts at miniaturization. The central idea is to record the inductance variation in two partial coils of the motor as a path proportional signal when moving the armature [4.29,4.44–4.46,4.56]. Figure 4.77 shows the principle circuit of a closed loop with integrated path-signal generation.

Motor configurations with moving magnets are used in this case, for example according to Fig. 4.54a or b. The coil and the magnetic yoke of such a motor are stationary. The field directions at the two ends of the moving

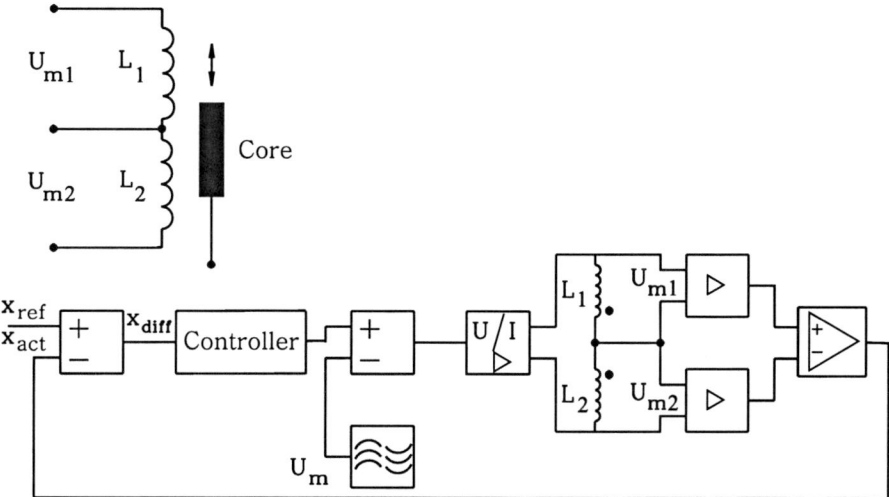

Fig. 4.77. Closed-loop control at a motor with moving magnets with integrated path-signal generation by impressing a measuring alternating voltage frequency into the analogous final control element [4.45, 4.46]

magnets and thus the directions of the air gap field are opposite, therefore a reversal of the direction of the current between the two magnetic poles by a reversal of the winding direction of the coil or at the same winding direction by separation into two partial coils with opposite current feed is required. Each of the two partial coils extends over half of the length of the yoke.

These two partial coils can be used as a path measuring system, if a high frequency signal is overlain on the motor current and measured in a bridge circuit. This strategy can be found in differential throttle-measurement systems using a carrier frequency. An analogue power amplifier is used here as a final control element. The functional principle is based on an impedance decrease when the armature changes its position and leaves a partial coil within the coil system. To the same degree the armature is entering another partial coil and a increasing impedance can be measured there. For this to work the armature needs to have a relative permeability that is decisively higher than one. The difference of the impedances of the two partial coils (AC voltage divider) will be recorded and conditioned as a path proportional signal. Thus a multi-purpose coil system for measuring and driving is established. To achieve a separate detection of the measuring voltages *via* the partial coils, a tapping in the middle of the coil at the connection of the two partial coils will be led through the motor. The alternating voltages are filtered, amplified and then subtracted electronically in the measuring circuit in order to remove offset disturbances. The signal that is received after the subtraction of the two partial voltages represents a path signal for the absolute position of the armature. With this path signal a closed loop can be realized.

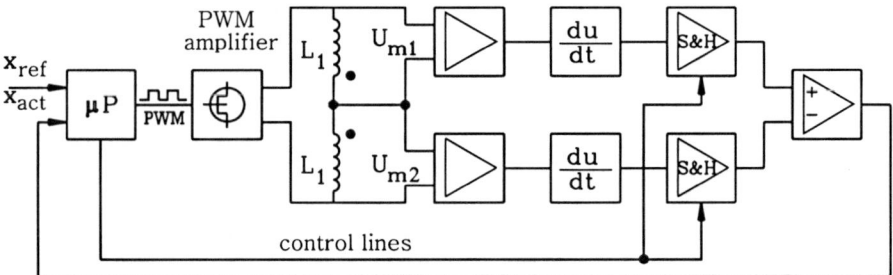

Fig. 4.78. Closed-loop control of a motor with moving magnets with integrated path-signal generation by use of a chopper amplifier [4.29]

Figure 4.78 shows a block diagram of a similar solution, however with the application of a chopper amplifier. In this case, the high switching frequency of the converter can be used and non additional measuring frequency is modulated. The inductance difference is computed based on measurements of the voltage rise of the middle voltage between the two coils at certain intervals [4.29]. The position resolution is typically $0.1\ldots0.5\%$ of the measurement range. The absolute accuracy depends on the production precision of the coils and reaches at quasistatic operation approx. 1%.

4.2.5.2 Control of Electrodynamic MCMs

Multi-coordinate motors without a mechanical torsion block are systems with multiple variables coupled together, and, as such, require a multivariable control system. If there is a mechanical torsion block, the position and the speed of each separate coordinate can be regulated either in turn or simultaneously, using the regulators already mentioned for electrodynamic linear motors.

Multivariable controls make high dynamic demands on electronic assemblies to be used for the information processing. Fast signal processors or transputers are therefore used which are able to process in real time the information gathered by the multi-coordinate measuring systems from the xy-, $xy\Delta\varphi$- and $x\varphi$-measuring systems. These multi-coordinate measuring systems are the basis of the multivariable control and it is, therefore, proper to treat them again at this point in addition to the explanations in Sect. 4.2.3, where their function is described.

Multi-Coordinate Measuring Systems for Movement Control

For multi-coordinate drives with areas of movement $>10\times10\,\mathrm{mm}^2$, the measuring systems most often used are those with relative or absolute physical measures for two coordinates; also those with a multi-coordinate measuring system based on interferometry. The most common types are incremental cross-grid measures with reference tracks for two coordinates on a special

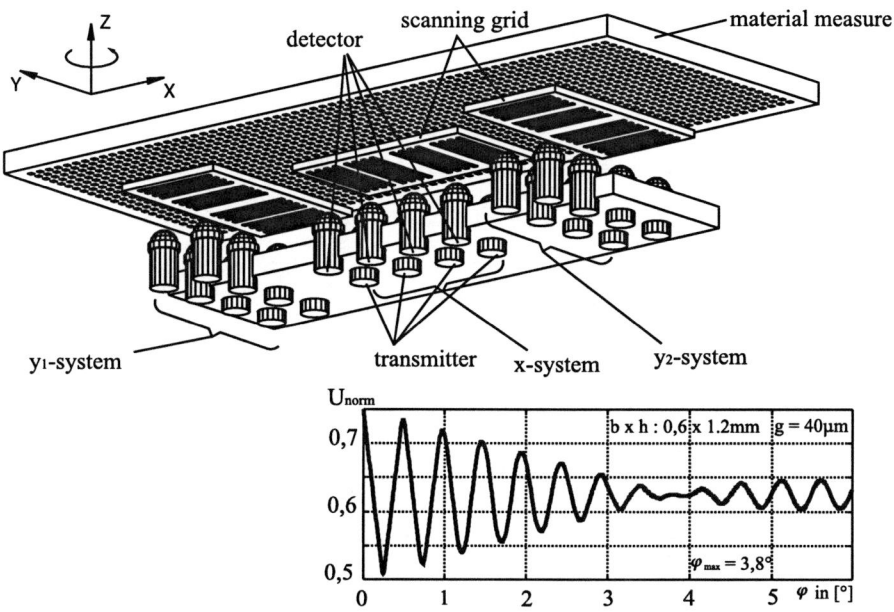

Fig. 4.79. Principle of the arrangement of an $xy\Delta\phi$ sensor movement as described in [4.93] and reduction of the measurement amplitude when the armature undergoes torsion

silicate glass substrate scanned with incident or transmitted light by means of optoelectronic illumination sources (LED) and sensors such as photodiodes or phototransistors [4.89].

One offered commercially is an incremental two-coordinate measuring device for a measurement range of 68×68 mm with a resolution of 0.01 µm and accuracy to 2 µm, see [4.69]. A two-coordinate phase grating with the structure running at 45° to the axes, made of DIADUR, on glass, with a graduation interval of 8 µm, a signal interval of 4 µm and a reference mark for each coordinate, made 3 mm after each measurement cycle begins, constitutes the actual physical measure.

Newer developments in x-, xy- and $xy\Delta\varphi$-measuring systems have recently become well known from [4.64, 4.70, 4.98] and [4.99]. An interesting laser-interferometric $xy\Delta\varphi$-measuring system with positioning accuracies of 30 nm (for 120 mm) and a tilt angle of at most 10' across the area of movement of up to 300×300 mm is also described in [4.98]. Figure 4.80 shows the principles of this measuring system.

A measuring system with reference mark capture has been specially developed for $xy\Delta\varphi$-data acquisition from three-coordinate drives which is called by [4.70] and [4.64]) a 2-D Epiflex module and works with incident light. With the photodiode array employed, multiple scanning with error reduction is achieved, ensuring the highest resolution by means of the interpolation of

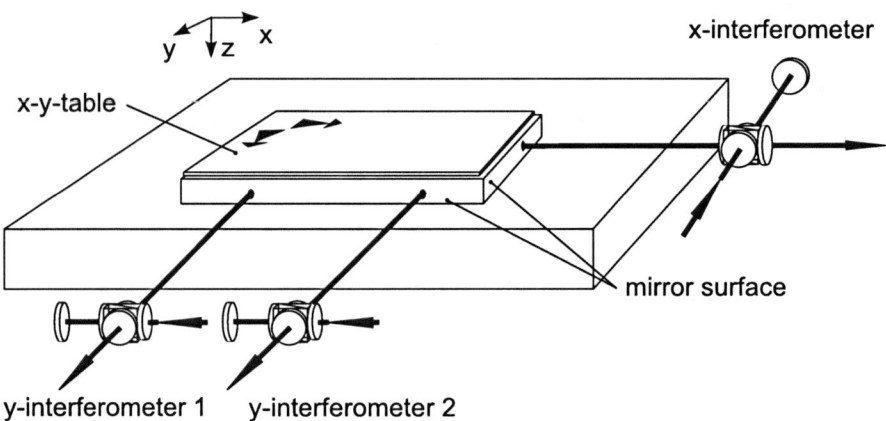

Fig. 4.80. Laser-interferometric $xy\Delta\varphi$-measuring system as described in [4.98]

the analogue measuring signals. The reference detectors consist of individual photodiodes which scan particular individual lines as reference marks. Interpolation factors are given up to 4000 for the various versions of the three individual sensors, so that resolutions of 0.01 µm and 0.068″ (depending on base distance) are reached for a grid constant of 40 µm.

In the case of working measuring systems operating incrementally, external influences due to signal processing errors may cause errors in counting. The measuring system described in [4.93] can be used for the direct extraction of $xy\Delta\varphi$-signals encoded absolutely. Figure 4.81 shows a coding field arrangement with overlaps and one possible pattern of CCD lines for acquisition of absolute location data in respect of x, y and $\Delta\varphi$ (a). Coding fields which have

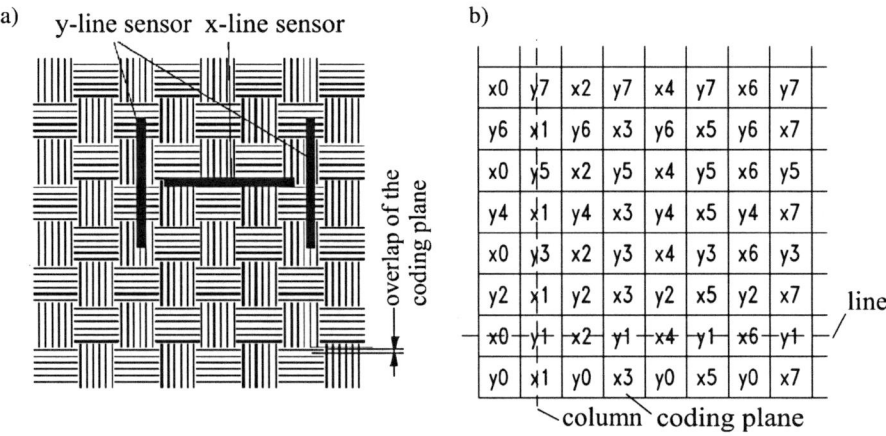

Fig. 4.81. Area scale coded absolutely with **a** sensor lines, and **b** coding fields, arranged according to [4.93]

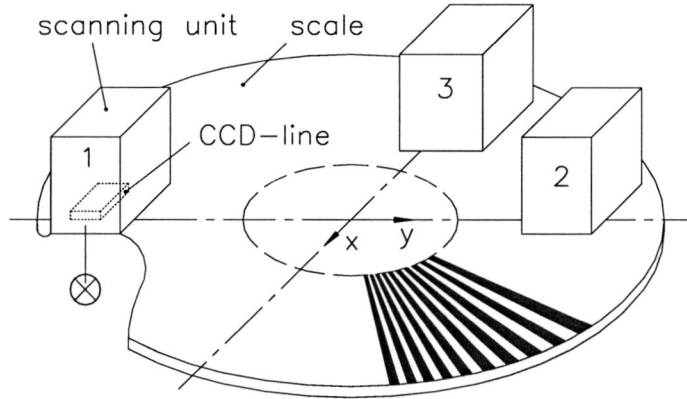

Fig. 4.82. Measuring system for simultaneous capture of the coordinates $xy\phi$ as described in [4.100]

a linear code are so arranged that in each row of the matrix and in every second field the y-coordinate information is available and in each column and in every second field the x-coordinate information is available (b). Here, the angle of rotation is also calculated from the difference between the measured values of two CCD-lines and the distance between them. With this measurement system, no reference marks are necessary.

Absolute coding is also found in [4.99] among the so-called transformation measuring systems (with CCD line scanning) considered

For wide angles of rotation, up to 360°, as they are realized using the principle represented in Fig. 4.64, measuring systems with a disc with a radial grid on the rim and three CCD lines arranged by less than 90° to permit scanning can be used as the physical measure for the simultaneous capture of the $xy\varphi$-coordinates, see [4.63, 4.100]. There is a sketch of this measuring system in Fig. 4.82.

Multivariable Control for Movement on Multiple Axes

The performance parameters of MCAs are basically determined by the control method. Therefore, besides the control systems mentioned in the explanation of the principles, it is necessary to note here the most recently published developments. The hardware structure of the system of control is shown in Fig. 4.83, see also [4.91].

The large number of analogue outputs allows the separate control of at least eight coils, as it is necessary for certain of the principles detailed, where a torque is generated for torsion control. Commercially obtained switched output stage systems with output currents up to ±20 A and with reduced loss of power can be employed as output stages for the current.

"The control concept is based on dedicated time-discrete linear-and-square control with estimated state, following an incremental model which eliminates

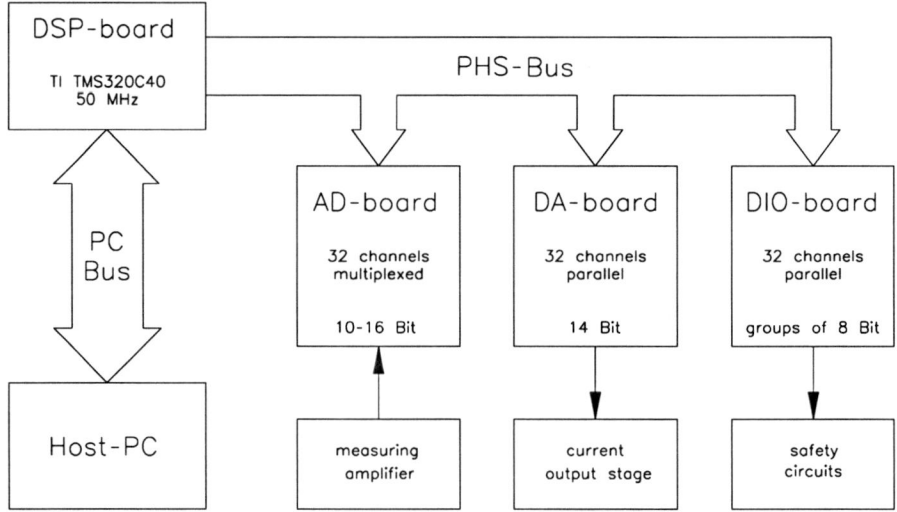

Fig. 4.83. Hardware structure of a multi-coordinate system of control as described in [4.91]

any remaining deviation with the help of the compensating interfering forces which arise. Restrictions of the controlled variables are also taken into account" [4.91].

Figure 4.84 shows the structure of the control of one coordinate, as described in [4.91].

The control concept combines the advantages of a state controller with the robustness and other properties of a classic PID-controller and tackles three

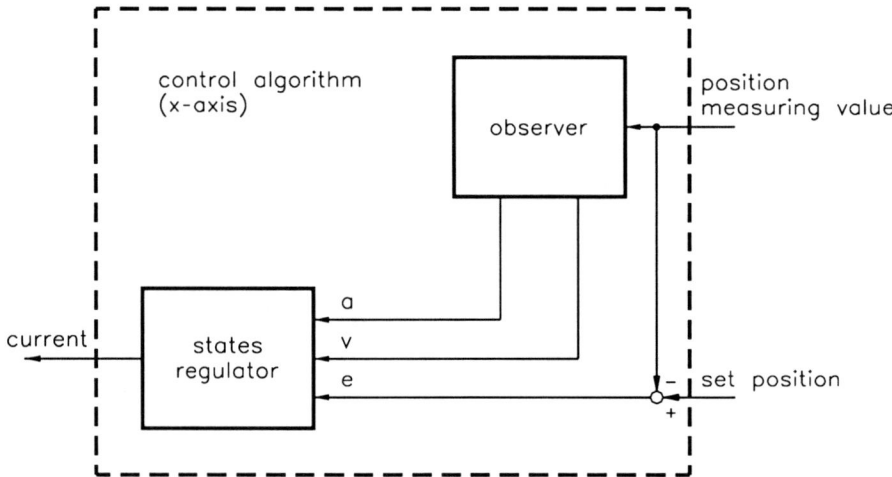

Fig. 4.84. Structure of the control of one coordinate, as described in [4.91]

decoupled coordinates. The decoupling concept includes different support levels, which transform the real drive into three individual virtual coordinates, visible by control algorithms. In Fig. 4.85, the coupling structure is shown as described in [4.91].

Besides high dynamic positioning, movements along a path also have to be performed with the multi-coordinate drives. A very effective path control concept has been drawn up for this, see [4.91]. It is the principle of vector control, and is based on the fact that every path movement along a required path can be analysed in a desired direction and in an error direction. This principle is shown for an orbit in Fig. 4.86 (cf. [4.91]).

Transformation is required into a mobile coordinate system. After control in the desired path's mobile coordinate system, the forces required are transformed back into the basic coordinate system of the MCA. The computation effort is relatively low in comparison with multi-axis path generators using inverse kinematics. At least the results are equal, since the velocity along the path and the deviation from the path are immediately corrected, as it has been experimentally confirmed, see [4.91].

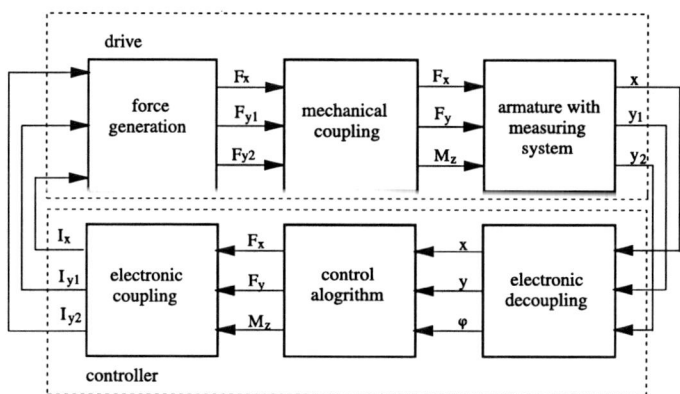

Fig. 4.85. Coupling structure of a three-coordinate drive as described in [4.91]

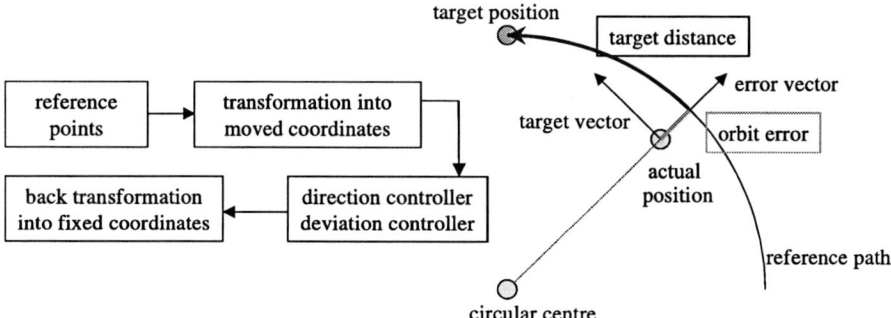

Fig. 4.86. Concept of vector control, exemplified in the orbit described in [4.91]

4.2.6 Commercially Offered Systems

In general, linear and multi-coordinate motors are realized as problem-specific drives. Because motion converters are missing, they must be designed in accordance with the scheduled travel range, so a problem-specific development up to a very extensive integration into the complete design of the device is obvious. Thus the commercial availability of many of the mentioned designs of linear and multi-coordinate motors is restricted or intensively tied to equipment or system manufacturers, e. g. hard disc drive manufacturers or CD-ROM drive manufacturers. Certain basic models, preferably commutated heteropolar systems, however, are offered increasingly in series. They can be easily adapted to other construction lengths by string together of further subsystems. Also non-commutated linear and special multi-coordinate motors are commercially available to a certain extent.

Again, a summary of the advantages and disadvantages of the described linear direct drives and multi-coordinate drives is given. Linear direct drives and multi-coordinate drives lack mechanical motion converters. This results in mechanical advantages with regard to a quiet operation, low wear as well as a potential high lifetime. From control aspects, these drives have no backlash as well as low elasticities and friction. The moving armatures are usually very lightweight. In addition, electrodynamic drives stand out due to a relatively simple motor design with comparatively low mechanical precision requests. The attainable positioning accuracy is depending only on the measuring system, mechanical friction will be ignored at first (aerostatic guide). Linear direct drives and multi-coordinate drives therefore have the prerequisite for highest dynamic and positioning accuracy. For positioning applications, additional measuring systems and a closed-loop control are required of course, which increase the effort, particularly the electronic one, significantly.

The bad weight-power relationship has an adverse effect, however, since such motors are as long as the travel range. An integration of the direct drive in the driven unit can compensate it to a certain extent. However, at the comparison with rotary drives the complete drive system has always to be taken into account that includes guides and required rotation-translation transformers.

4.3 Linear and Planar Hybrid Stepping Motors

Eberhard Kallenbach

4.3.1 Linear Hybrid Stepping Motors

Currently, a strong tendency towards linear direct drives can be observed in mechanical and automation engineeringapplications – linked with the development of increasingly powerful electronics. With these drives, larger traverse travels, improved dynamic properties, more accurate positioning and better integration ability compared to linear drives with a rotatory translatory motion converter can be achieved.

The operating principle of linear hybrid stepping motors according to Sawyer [4.105] is shown in Fig. 4.87.

If both exciting coils (phases) are turned off, the permanent magnetic flux Φ_p closes its circuit over the blades of the two magnetic circuits. Both exciting coils are designed for weakening the permanent magnetic flux through one pole, while the flux through the other pole is strengthened. The thereby generated driving forces (tangential magnetic forces) cause independency on the momentary armature position and the coil excitation a defined displacement of the armature in the travel direction x. If the exciting coils are excited individually with current impulses, a step sequence is generated that correlates for two-phase hybrid stepping motors with half of the tooth width or respectively a quarter of the pole pitch τ_z (Eq. 4.86), and causes a facing of the teeth of those poles, where the highest magnetic flux passes through. This shall be considered as the assumed starting point in Fig. 4.87 (upper schematic). If exciting coil 1 is switched off and coil 2 is switched on, the highest magnetic flux passes through the pole to the right of magnet 2 (Φ_p and the coil's flux Φ_{e2} sum up), while the fluxes through the left pole of magnet 2 compensate each other. The armature travels for the increment x_0 to the left (Fig. 4.87). Reversing the current direction in coil 1, the next step can be generated and

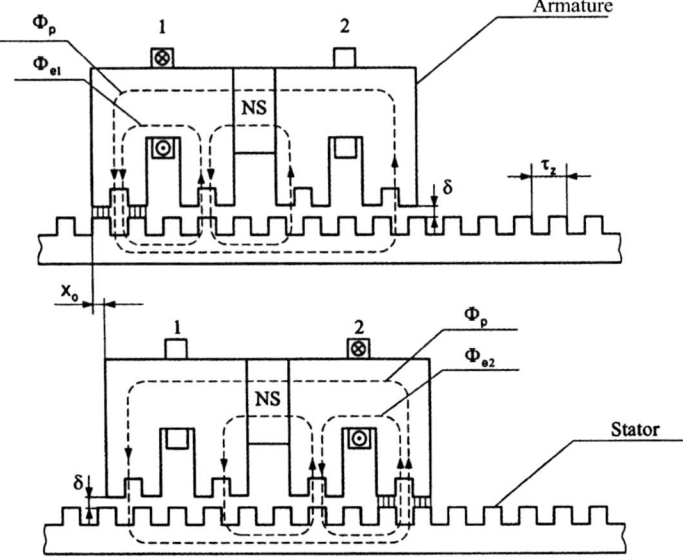

Fig. 4.87. Operating principle of linear hybrid stepping motor

so on. Besides the tangential forces F_t, also normal forces F_n are generated at the pole surfaces. These can be up to ten times higher than the tangential forces.

Tangential and normal forces increase the smaller the air gap δ between the armature's and the stator's poles is. Thus, the quality of the armature guidance is essential for the parameters of the linear drive. The relatively large normal forces require a high stiffness for mechanical guidances. The best results have been achieved with air guidances. The high magnetic normal forces are even advantageous for a high stiffness of the air guidance. Another advantage is the absence of mechanical friction and abrasion. On the other hand, the supply of pressured air requires an air tube to the armature and the finite stiffness of the air guidance leads to oscillations of the armature in the nanometer up to the micrometer range, if the normal forces vary. For normal technological manufacturing conditions, air gaps $\delta = 10$ up to $20\,\mu m$ occur, dependent on evenness and finish of the stator's and the armature's guiding surfaces.

For the dependency of the magnetic force F_m of linear stepping motors on the displacement x with an excitation only in the linear ranges of the magnetic circuit, a sinusoidal characteristic can be assumed as a first approximation

$$F_m = -F_k \sin \frac{2\pi x}{\tau_z} . \qquad (4.85)$$

F_k represents the amplitude of a tilting force (Fig. 4.94). We obtain the design dependent increment x_0 with a phase number $m_p = 2$ and a pole pitch τ_z for the stepping motor shown in Fig. 4.87 as follows:

$$x_0 = \frac{\tau_z}{2m_P} = \frac{\tau_z}{4} . \qquad (4.86)$$

Figure 4.88 shows a possible design layout of a linear hybrid stepping motor with air guidance.

The teeth grooves are filled with a non-magnetic material in order to obtain even surfaces for the air guidance. Technical parameters of commercially available linear hybrid stepping motors are listed in Table 4.2.

Fig. 4.88. Design layout of a linear hybrid stepping motor with air guidance

4 Drives with Limited Motion

Table 4.2. Technical parameters of linear hybrid stepping motors

Traverse travel	Up to 2 m (limited by stator length)
Tilting force F_k	20 ... 1000 N
Tilting force F_k per driven armature's mass m	90 ... 120 N/kg
Maximum velocity v_{max}	2.5 m/s
Guidance accuracy	5 μm/300 mm

Linear hybrid stepping motors have the following advantages compared to classic linear drives with rotatory stepping motors and rotatory to translatory motion converter [4.107]:

- The driving properties are independent on the traverse travel, which is limited by the armature's and the stator's geometric dimensions and can be enlarged almost endlessly.
- The acceleration (F/m) and the resonance frequency f_r are independent of the armature's length.
- The stationary positioning accuracy is independent of the number of steps (positioning travel) and the stepping frequency (positioning velocity). It is dependent on the accuracy of the pole pitch and the triggering (electronic step pitch).

Because of the advantageous driving properties of linear hybrid stepping motors, several manufacturers developed different designs consisting of flexibly usable modules with specific properties (Table 4.3).

Different from rotatory drives, linear motors as drive sections for direct drives always have to be selected problem-specific and adapted with an adequate electronic driver design to the application. This requirement can be matched excellently with mostly standardized drive modules. The developer thereby gets the opportunity to benefit from the following advantages [4.112]:

- With mechanical coupling (series or parallel connection) of modules, a huge variety of possible motions can be realized and thus a better adaptation to the specific technical task can be achieved.
- With the help of the drive modules, decentralized drive systems can be built. To each motion coordinate a drive module can be assigned.
- The decentralization allows by omission of force transmitting elements to reduce the driven masses of a machine drastically and thereby increase velocity and accuracy capabilities.
- The drive section allows developing application-specific cost-effective machines with improved technical parameters. However, according to a mechatronic design (see Sect. 4.1.5), the use of linear motors has to be considered from the very beginning with the design of the entire system in order to utilize their complete potential. The stator of respective machines for ex-

Table 4.3. Configurations of linear hybrid stepping motors

Nr.	Configuration	Comments
1		• Linear basic module • Flat design • By changing the armature position, the tilting force f_k can be increased while the acceleration stays constant
2		• Good linear guidance • Prevention against rotation • Duplication of the tilting force for the same armature length in comparison to 1
3		• Very good linear guidance • Prevention against rotation • Triplication of the tilting force for the same armature length in comparison to 1
4		• Very good linear guidance • Quadruplication of the tilting force for the same armature length in comparison to 1 • High acceleration • Especially appropriate for vertical applications

ample has to be designed stiffer due to the higher acceleration of linear stepping motors.

The decision whether a linear stepping motor or a rotatory stepping motor with a rotatory to translatory motion converter is the optimal drive solution is application dependent and needs to be examined thoroughly for each single case. A comparison of several properties for both driving methods is listed in Table 4.4. Today, validated simulation tools for the description of the dynamic behaviour of linear hybrid stepping motors are available [4.108, 4.111] and can be included into the simulation of a drive system in its entirety.

Table 4.4. Comparison of linear positioning systems with rotary and linear stepping motors

Assumption:
Rotatory stepping motor with the phase number m_p, whose rotatory motion is converted into a linear motion with an ideal rotatory to translatory motion converter with variable transmission ratio $\ddot{u}$ (drive I)
Linear stepping motor, whose armature has the same volume of iron and the same active armature mass as the rotatory stepping motor (drive II)

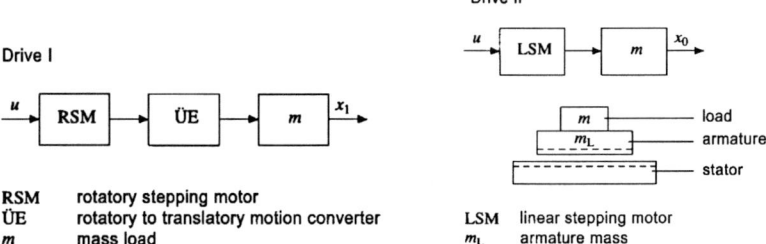

RSM	rotatory stepping motor
ÜE	rotatory to translatory motion converter
m	mass load

| LSM | linear stepping motor |
| m_L | armature mass |

- φ_0 stepping angle of the rotatory stepping motor
- $\ddot{u} = x_1/\varphi_0$ transmission ratio of the rotatory to translatory motion converter
- x_1 equivalent linear increment
- stiffness of the linear stepping motion

$$C^*_{\text{magn}} = \frac{C_{\text{magn}}}{\ddot{u}}$$

- tilting force increases with falling $\ddot{u}$

$$F^*_k = \frac{M_k}{\ddot{u}}$$

- x_0 increment of the linear stepping motor
- linear stepping motor is directly coupled to the load $\ddot{u} = 1$
- increment of the stepping motor can be reduced with electronic step pitch
- for a step pitch, the stiffness C_{magn} and F_k are approximately constant
- higher possible velocities, but also higher compensation times compared to drive I

Drive II is generally faster for small travels than drive I, but more sensitive towards load or disturbance fluctuations. Drive II has to be optimized problem-specific.

4.3.2 Multi-Coordinate Hybrid Stepping Motors

With the help of volume integration of two or more hybrid stepping motors with a motion degree of freedom DOF = 1, compact drive sections with a DOF > 1 can be built. Figure 4.89 shows such a two-coordinate module. The advantage of these compact multi-coordinate drive modules is the ability to generate desired moving patterns (x–y-movements or paths) flexibly with the control of the moving patterns of the integrated single modules.

The stator of the x–y-module consists of a ground plate (1), whose active layer is made of soft magnetic material with a toothed structure. The armature

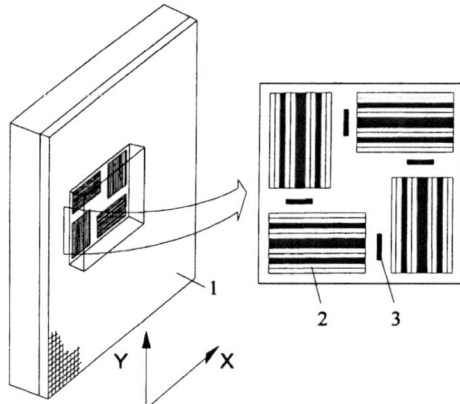

Fig. 4.89. Design layout of an x–y-hybrid stepping motor with air guidance. *1* Ground plate, *2* Armature module, *3* Air nozzle

consists of at least two orthogonal linear modules (2), which are arranged to avoid torques.

To compensate torques, normally four linear modules are arranged symmetrically to the armature's centre of gravity. Of course linear modules can be integrated together with rotary modules as well, which results in x–Φ-modules (Fig. 4.90). Finally, the combination of several compact modules allows to create drive units with a DOF = 4 up to 6, that can be adapted flexibly.

Figure 4.91 shows a respective drive section with DOF = 4.

The function of the integrated modules is highly dependent on the geometrical arrangement of the force-generating surfaces. In Table 4.5, different possible designs for planar stepping motors and their most important properties are listed.

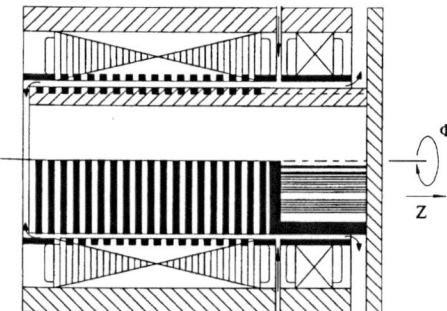

Fig. 4.90. Design layout of a z–Φ-hybrid stepping motor with air guidance

Table 4.5. Variants and technical parameters of planar stepping motors

Design layout	Properties	Comment
1 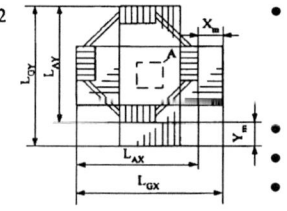	• L_{AX}, L_{AY} and the armature mass m_A are independent of X_m, Y_m • X_m, Y_m only affect the dimensions L_{GX}, L_{GY} of the toothed ground plate • Due to the toothed structure (intersection of the grooves), we have $$f_A = 4\ldots 6 \cdot 10^3 \text{ N} \cdot \text{m}^{-2}$$ $$a_m = 40\ldots 50 \text{ m} \cdot \text{s}^{-2}$$ • Stability against external counter-torques smaller compared to 2, 3	a_m maximum acceleration m_A armature mass f_A usable force per area X_m, Y_m maximum displacement in x- and y-direction L_{GX}, L_{GY} length in x- and y-direction of the toothed or grooved ground plate made of soft magnetic material L_{AX}, L_{AY} length in x- and y-direction of the armature A border of the operating area, relating to the centre of the armature
2 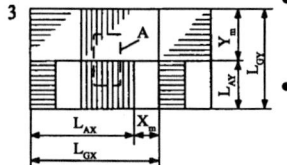	• L_{AX}, L_{AY} and the armature mass m_A are dependent on X_m, Y_m; thus the traverse area is limited • $f_A = 6\ldots 12 \cdot 10^3 \text{ N} \cdot \text{m}^{-2}$ • $a_m = 80\ldots 1000 \text{ m} \cdot \text{s}^{-2}$ • higher stability against external torques • operating area can be built as "window" • for larger X_m, Y_m the dynamic is worse compared to 1	
3	• L_{AX}, L_{AY} and the armature mass m_A are dependent on X_m, Y_m • Useable force per area $$f_A = 6\ldots 12 \cdot 10^3 \text{ N} \cdot \text{m}^{-2}$$ • higher stability against external torques • limited traverse area	

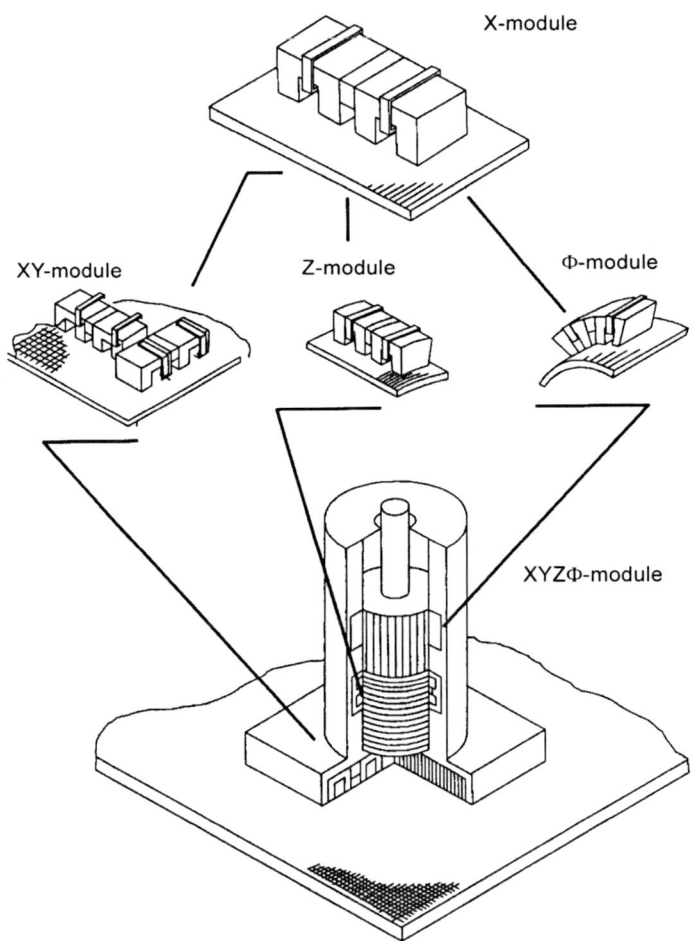

Fig. 4.91. Combination of single modules to an x–y–z–Φ-module (DOF = 4)

4.3.3 Dynamic Properties of Linear Hybrid Stepping Motors

For the application of stepping motors, the knowledge of the dynamic behaviour under concrete circumstances is relevant. Because of the direct coupling of linear stepping motors with the load, a stronger influence of the load onto the dynamic behaviour compared to rotary stepping motors with a high transmission ratio gear, that reduces load reaction occur.

In comparison to the computation of the dynamic behaviour of DC motors, open-loop controlled stepping motors require a verification of their errorless step execution – only in this case they are appropriate for positioning tasks. For linear and rotatory stepping motors, nearly the same relations are valid (Fig. 4.92). The operation has to be within the areas for the start-stop-operation or, respectively, within the operating frequency range. For the equivalent network shown in Fig. 4.93, the following differential equation system can be set up for the two-phase hybrid stepping motor to compute the dynamic behaviour:

$$U_1 = i_1 R_1 + \frac{d\psi_1(i_1, i_2, x)}{dt} \tag{4.87}$$

$$U_2 = i_2 R_2 + \frac{d\psi_2(i_1, i_2, x)}{dt} \tag{4.88}$$

$$F_{m1} + F_{m2} = m\ddot{x} + K\dot{x} + F_{geg}(x) \tag{4.89}$$

with

$$F_m = \frac{\partial}{\partial x} \int \psi(i, x)\, dx . \tag{4.90}$$

Because of the toothed structure of and/or because of the electromagnetic force, the ψ–i-characteristics are non-linear and magnetically coupled. De-

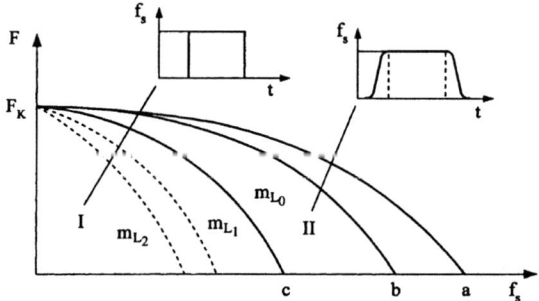

Fig. 4.92. Stepping motor characteristic of a linear hybrid stepping motor. *I* start-stop-frequency range, *II* operation frequency range, m_L armature mass, m_{Li} armature mass plus additional mass, $m_{L1} < m_{L2}$. **a** Boundary characteristic of the operating frequency range for impressed current; **b** boundary characteristic of the operating frequency range for impressed voltage; **c** boundary characteristic of the start-stop-frequency range

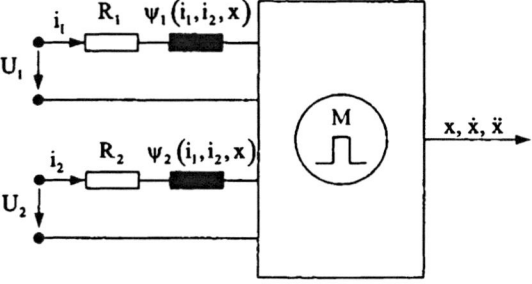

Fig. 4.93. Equivalent schematic of a linear two-phase hybrid stepping motor

pendent on the triggering sequence, often the single phases are excited consecutively with an overlap (e. g. for electronic step pitch). Thus, the coupling of the differential equations can change from one cycle to the next.

Often it is advisable, to aim for problem-specific approximative solutions dependent on the specific design of the stepping motor, the application and the control regime. In [4.105] for example, the solution is based on the stationary force-stroke characteristic of one excited phase and it is considered that the voltage induced through the armature motion $u(\mathrm{d}x/\mathrm{d}t)$ acts in the same way as if a velocity-dependent friction force.

If

$$u(\dot{x}) = K_\mathrm{m}\dot{x},$$

we obtain from Eq. (4.89) the following approximative equation:

$$F_\mathrm{m} = m\ddot{x} + (K + K_m)\dot{x} + F_\mathrm{geg}(x). \tag{4.91}$$

If we further assume a sinusoidal force characteristic of one phase

$$F_\mathrm{m} = -F_\mathrm{k} \sin\left(\frac{2\pi x}{\tau_\mathrm{z}} - x^*\right), \tag{4.92}$$

for small values we can approximatively describe the stationary force–stroke characteristic as a straight line with the slope of $(-C_\mathrm{magn})$.

With this simplification and the precondition that the setpoint setting for the position x^* is made by the reference input impulses $w(k)$, the transfer function of the linear stepping motor can be derived [4.114]. It is the transfer function of an oscillation element

$$G(p) = \frac{1}{1 + 2DT_\mathrm{M}p + T_\mathrm{M}^2 p^2}$$

with

$$T_\mathrm{M} = \sqrt{\frac{m}{C_\mathrm{magn}}} \quad \text{and} \quad D = \frac{K + K_\mathrm{M}}{\sqrt[2]{mC_\mathrm{magn}}}.$$

With the help of this transfer function, the step response for the start–stop-operation and for the operation frequency range can be calculated [4.105]. Further, we obtain the resonance frequency

$$f_\mathrm{r} = \frac{1}{2T_\mathrm{M}\pi}\sqrt{1 - D^2}.$$

For linear stepping motors with air guidances, the eigenfrequencies are placed at low stepping frequencies that have to be avoided during operation. For run-up controls, this frequency range has to be passed quickly.

4.3.4 Principle of Microstepping

The control of stepping motors in basic stepping mode (the increments the motor moves are identical with the pole pitch x_0) has the following disadvantages:

- The increment based on the pole pitch is too large for many applications and limits the positioning accuracy.
- For small velocities, the motor switches literally into single-step mode. Strong dampened oscillations occur that can act disturbing as well.

An effective way to eliminate these disadvantages is microstepping. The basic principle of the electronic step pitch can be described as follows: By simultaneous excitation of both coils, the roots of the resulting $F(x, i)$-characteristics can be shifted between the roots I and II that occur when only a single phase 1 or 2 is excited (Fig. 4.94). Because stepping motors in nominal rating normally work only slightly below the saturation of the magnetic circuits, an effective control can only be achieved by reducing the exciting currents. If the current were increased drastically above the nominal value, heavy distortions of the $F(x, i)$-characteristic away from its sinusoidal shape will occur due to saturation effects of the iron circuit. Under real conditions with the electronic step pitch, a higher resolution can be achieved for open-loop control applications, but no or only a slight improvement of the positioning accuracy can be accomplished.

An improvement of the positioning accuracy can be realized by calibrating the exciting currents [4.107] or, respectively, with closed loop control. For the last mentioned case, there are several possible solutions:

- A position sensor as an indiscrete part is built into the drive section control loop.
- A sensor is included into the magnetic circuit of the hybrid stepping motor that generates an output signal, which allows calculating a position signal [4.109].

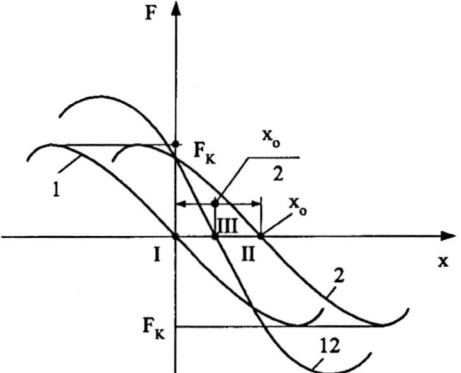

Fig. 4.94. Principle of microstepping: The resulting force–stroke characteristic. $F_{12} = k_1 F_1 + k_2 F_2$; F_1: $k_1 = 1$, $k_2 = 0$; F_2: $k_2 = 0$, $k_2 = 1$; F_{12}: $k_1 = 1$, $k_2 = 1$ (half stepping), microstepping mode: $0 < k_1 < 1$, $0 < k_2 < 1$

- Based on the current and voltage input signals, a position or velocity signal is determined [4.115].

4.3.5 Linear Hybrid Stepping Motors as Non-Linear Magnetic Drives

Hybrid stepping motors are usually excited to a high level in order to obtain high force densities, a linear model is often not sufficiently accurate to describe especially the dynamic behaviour of a stepping motor. Thus, non-linear models have to be used that include the non-linearities of the $F_m(x, i)$-characteristic and the non-linear $\psi(x, i)$-characteristic. In addition, it is desirable to determine eddy currents and hysteresis losses for higher stepping frequencies.

In [4.108] a similar model is presented and shows good simulation results. While the hysteresis losses can be emulated by Coulomb friction, the influence of eddy currents is modelled by a resistor R_W that is connected parallel to the inductance (Sect. 4.1, Fig. 4.13a).

The computation of the non-linear $F_m(x, i)$- and $\psi(x, i)$-characteristics can be achieved with numeric field computation tools (e.g. Maxwell, Profi) by varying i and x, if the geometric parameters of the stepping motor and material date of the magnetic materials is known. The results of the field computation are listed in matrices. Each matrix represents a look-up-table that describes the families of characteristics as single interpolation point. Figure 4.95 shows an example of the characteristic for $F_m(x, i)$ caused by one excitation coil with different amplitudes of the exciting currents [4.116]. Easily recognizable is the difference towards an ideal sinusoidal curve and the influence of saturation onto the amplitude of force.

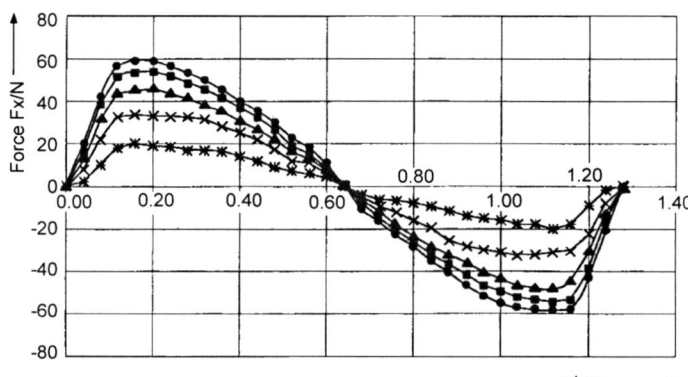

Fig. 4.95. Stationary force-stroke-characteristics with excitation of one phase of a linear hybrid stepping motor (parameter exciting current), determined with FEA, $I_{min} = 0.2A$, $I_{max} = 1A$ [4.116]

Starting from these field computations, non-linear state space models can be derived that can be integrated into the simulation system MATLAB/SIMULINK. Especially for higher excitation of the magnetic circuit of the hybrid stepping motors and larger stepping frequencies, significantly better simulation results for the dynamic behaviour can be achieved [4.110, 4.111, 4.117].

References

[4.1]   Kallenbach, E.: Der Gleichstrommagnet. Leipzig: Akademische Verlagsgesellschaft Geest & Portig KG 1969
[4.2]   Kallenbach, E.; Eick, R.; Quendt, P.: Elektromagnete. Stuttgart: Teubner 1994
[4.3]   Kallenbach, E.; Eick, R.; Quendt, P.; Ströhla, T.; Feindt, K.; Kallenbach, M.: Elektromagnetische Grundlagen, Berechnung, Entwurf und Anwendung. 2. überarbeitete und ergänzte Auflage. B. G. Teubner Stuttgart Leipzig Wiesbaden 2003
[4.4]   Ljubcik, M. A.: Optimal'noe projektirovanie silovych elektromagnitnych mechanizmov (Optimale Projektierung von elektromagnetischen Mechanismen). Moskau: Energija 1974
[4.5]   Kallenbach, E.: Untersuchungen zur systematischen Projektierung nichtlinearer gleichstromerregter elektromagneto-mechanischer Antriebselemente mit translatorischer Ankerbewegung. Diss. B TH Ilmenau 1978
[4.6]   Aldefeld, B.: Felddiffusion in Elektromagnete, Feinwerktechnik und Messtechnik, 90 (1982) 5, S. 222–226
[4.7]   Kallenbach, E.; Feindt, K.; Hermann, R.; Schneider, S.: Auslegung von schnellwirkenden Elektromagneten unter Berücksichtigung von Wirbelströmen bei bewegtem Anker. 3. Magdeburger Maschinenbautage, 1997, Tagungsband II, S. 59–68
[4.8]   Oesingmann, D.: Systematisierung der Schwinganker. Berechnung und experimentelle Überprüfung elektromagnetischer Schwinganker. Wissenschaftliche Zeitschrift, TH Ilmenau 20 (1974) 2, S. 51–62
[4.9]   Hermann, R.: Untersuchungen zur Dynamik von wechselstromerregten elektro-mechanischen Antrieben. Diss. A, TH Ilmenau 1983
[4.10] Nikitenko, A.G. u. a.: Matematiceskoj modelirovanije i avtomatisazija projektirovanija tjagovych elekticesky apperatov (Mathematische Modellierung und Automatisierung der Projektierung elektrischer Apparate). Moskau: Vyssaja skola 1996
[4.11] Hermann, R.: Zum dynamischen Verhalten von wechselstromerregten Magnetsystemen. Feingerätetechnik 28 (1979) 12, S. 547
[4.12] Habiger, E. u. a.: Elektromagnetische Verträglichkeit. Berlin/München:, Verlag Technik, 2. Auflage 1992
[4.13] Kallenbach, E.; Feindt, K.; Hermann, R.; Schneider, S.; Nikitenko, A.G.: Dynamische Leistungsgrenzen von Elektromagneten, DRIVES 97, Nürnberg 1997 Tagungsband, S. 462–471
[4.14] Kallenbach, E.; Bögelsack, G.: Gerätetechnische Antriebe. Berlin: Verlag Technik, 1991; München/Wien: Hanser Verlag 1991

References

[4.15] Kallenbach, E.; Birli, O.; Dronsz; F., Feindt; K., Spiller, S.; Walter, R.: STURGEON – an existing software system for the completely CAD of elektromagnets. IDED 97 Tampere Finnland Proceedings, Bd. I

[4.16] Ströhla, T.: Ein Beitrag zur Simulation und zum Entwurf von elektromagnetischen Systemen mit Hilfe der Netzwerkmethode. Dissertation 2002, TU Ilmenau, Fakultät für Maschinenbau

[4.17] Birli, O.; Kallenbach, E.: Grobdimensionierung magnetischer Antriebssysteme mit dem Programmsystem SESAM. Tagungsband des Statusseminars Simulationswerkzeuge für schnelle magnetische SEnSor- und Aktorelemente der Mikrosystemtechnik (SESAM), 24. 9. 2001 Ilmenau, Herausgeber: Technische Universität Ilmenau und VDI/VDE-Technologiezentrum Informationstechnik GmbH Teltow S. 35–42

[4.18] Spiller, S.: Untersuchungen zur Realisierung eines durchgängigen rechnergestützten Entwurfssystems für magnetische Aktoren unter Einbeziehung von thermischen Netzwerkmodellen. Dissertation 2001, TU Ilmenau, Fakultät für Maschinenbau

[4.19] Feindt, K.: Untersuchungen zum Entwurf von Elektromagneten unter Berücksichtigung dynamischer Kenngrößen, Dissertation TU Ilmenau 2002

[4.20] Riethmüller, J.: Eigenschaften polarisierter Elektromagnete und deren Dimensionierung anhand eines Entwurfsalgorithmus mit einem Optimierungsverfahren. Dissertation TU Ilmenau 2004

[4.21] Kleineberg, T.: Modellierung nichtlinearer induktiver Bauelemente der Leistungselektronik. Dissertation TU Chemnitz 1994, VDI Fortschrittsberichte Reihe 20 Rechnerunterstützte Verfahren, Nr. 161, VDI Verlag Düsseldorf 1995

[4.22] Schweer, J.B.: Berechnung kleiner Wechselstrom-Ventilmagnete mit massivem Eisenkreis. Dissertation Universität Hannover 1997

[4.23] VDI 2206 Beuth-Verlag 2004

[4.24] Blank, G.: Untersuchungen zur Steuerung inkremental geregelter linearer Ein- und Mehrkoordinatengleichstrommotoren für Positioniersysteme. Diss. TH Ilmenau 1982.

[4.25] Draeger, J.; Moczala, H.: Linearkleinmotoren. München: Franzis-Verlag 1985.

[4.26] Draeger, J.; Moczala, H.: Gleichstrom-Linearantriebe kleiner Leistung. Technische Rundschau Bern 77 (1985) 25, S. 80–88.

[4.27] Draeger, J.; Moczala, H.: Gleichstrom-Linearmotoren kleiner Leistung ohne Kommutator. Feinwerktechnik & Messtechnik 87 (1979) 4, S. 157–162.

[4.28] Glöß, R.: Schnelle Präzisionspositioniersysteme für magnetomotorische Speicher. Feingerätetechnik 39 (1990) 2, S. 61–63.

[4.29] Hartramph, R.: Integrierte Wegmessung in feinwerktechnischen elektrodynamischen Lineardirektantrieben. Diss. Universität Stuttgart, 2001.

[4.30] Honds, L.; Meyer, H.: Nichtlinearität der Kraft-Weg-Kurven von Tauchspullinearmotoren. Feinwerktechnik & Messtechnik 88 (1980) 4, S. 162–166.

[4.31] Honds, L.; Meyer, H.: Unipolar-Linearmotor mit bewegtem Magneten für verschleißarme Linearantriebe. Feinwerktechnik & Messtechnik 87 (1979) 4, S. 152–156.

[4.32] Joksch, C.; Oettinghaus, D.: Magnetsysteme für Gleichstromlinearmotoren. Thyssen Edelstahl, Technische Berichte 6, Bd. 1, 1980.

[4.33] Katterloher, R.; Menzel, K.: Linearmotor zum Einsatz bei kryogenen Temperaturen. Feinwerktechnik & Messtechnik 93 (1985) 4, S. 165–168.

[4.34] Krause, W.; Schinköthe, W.: Linearantriebe für die Feinwerktechnik. Feinwerktechnik & Messtechnik 98 (1990) 7–8, S. 303–306.
[4.35] Krause, W.; Schinköthe, W.: Gleichstromlinearmotoren in der Feinwerktechnik - Robust, schnell und genau. Technische Rundschau Bern 83 (1991) 28, S. 42–45.
[4.36] Kühnel, A.: Einsatzuntersuchungen zu Tauchspullinearantrieben in Magnetfolienspeichern. Diss. TU Karl-Marx-Stadt 1986.
[4.37] Moczala, H.: Bürstenlose Gleichstrom-Linearmotoren kleiner Leistung mit gegenüber dem Ständer kurzem Läufer. Feinwerktechnik & Messtechnik 88 (1980) 4, S.177–182.
[4.38] Moczala, H.: Ein Beitrag zur Gestaltung bürstenloser Gleichstrom-Linearmotoren für kurze Wegstrecken. VDI-Berichte Nr. 482 (1983), S. 43–47.
[4.39] Nasar, S. A.; Boldea, I.: Linear motion electric machines. New York, London, Sydney, Toronto: Jon Wiley & Sons 1976.
[4.40] Olbrich, O. E.: Aufbau und Kennwerte elektrodynamischer Linearmotoren als Positionierer für Plattenspeicher. Feinwerktechnik & Micronic 77 (1973) 4, S. 151–157.
[4.41] Schinköthe, W.: Dimensionierung permanenterregter Tauchspullinearantriebe für gerätetechnische Positioniersysteme. Diss. TU Dresden 1985.
[4.42] Schinköthe, W.: Gleichstromlinearmotoren für die Gerätetechnik. Feingerätetechnik 35 (1986) 5, S. 207–211.
[4.43] Voss, M.; Schinköthe, W.: Miniaturisierte Linearmotoren erschließen neue Anwendungen. Tagung Innovative Kleinantriebe, Mainz, 9.-10.05.1996. VDI-Berichte 1269, S. 105–119.
[4.44] Schinköthe, W.; Hartramph, R.: Miniaturlinearantriebe mit integriertem Wegmesssystem. F&M Feinwerktechnik, Mikrotechnik, Mikroelektronik 104 (1997) 9, S. 634–636.
[4.45] Welk, C.: Detektion interner sensorischer Eigenschaften von elektrodynamischen Lineardirektantrieben. Diss. Universität Stuttgart IKFF, 2004.
[4.46] Clauß, C.; Schinköthe, W.; Welk, C.: Integrierte Wegmessung in Lineardirektantrieben - Potenziale und Grenzen. Tagung Innovative Klein- und Mikroantriebstechnik, Darmstadt 3./4. 03. 2004, ETG-Fachberichte 96, S.117–122.
[4.47] Würbel, J.: Entwicklung kleiner elektronisch kommutierter Lineardirektantriebe in Flachbauweise. Diss. TU Dresden 1984.
[4.48] Fa. ANORAD Europe B.V.: Firmenschriften, Valkenswaard, Niederlande. (Stammhaus ANORAD Corporation, Hauppauge New York, USA).
[4.49] Fa. ETEL S.A.: Firmenschriften, Motiers, Schweiz.
[4.50] Fa. Sulzer Electronics AG: Firmenschriften, Zürich, Schweiz.
[4.51] Fa. präTEC GmbH: Firmenschriften, Rohr/Thüringen.
[4.52] Fa. SKF Linearsysteme GmbH: Firmenschriften, Schweinfurt.
[4.53] Fa. Philips Industrial Automation Systems: Firmenschriften, Eindhoven, Niederlande.
[4.54] Fa. Vishay Siliconix, www.vishay.com.
[4.55] Gundelsweiler, B.: Dimensionierung und Konstruktion von feinwerktechnischen elektrodynamischen Lineardirektantrieben. Diss. Universität Stuttgart, 2003.

References

[4.56] Clauß, C.; Schinköthe, W.: Integrierte Wegmessung in Lineardirektantrieben - Stand und Ausblick. 50. Internationales Wissenschaftliches Kolloquium der TU Ilmenau, Ilmenau 19.-23.09.2005.

[4.57] Boldea, I.; Nasar, A.: Linear electric Actuators and Generators. Cambridge: University Press 1997.

[4.58] Chen, B.; Lee, T.; Peng K.; Venkataramanan, V.: Hard Disk Drive Servo Systems. Springer London 2006.

[4.59] Boldea, I.; Nasar, A.: Linear Motion Electromagnetic Devices. Taylor and Francis 2001.

[4.60] Basak, A.: Permanent-Magnet DC Linear Motors. New York: Oxford University Press 2002.

[4.61] Holmes, G.; Lipo, T.: Pulse Width Modulation for Power Converters: Principles and Practice. Wiley-IEEE Press 2003.

[4.62] Bose, B.: Power Electronics and Motor Drives: Advances and Trends. Academic Press 2006.

[4.63] Blank, G.; Löwe, B.; Wendorff, E.: Vorrichtung und Verfahren zur ebenen berührungslosen Mehrkoordinatenmessung. DD 215645.

[4.64] CIS Institut für Mikrosensorik e. V. Erfurt: Sachbericht zur Entwicklung und Herstellung eines 2D-Messsystems für hochdynamische Mehrkoordinatenantriebsmodule.

[4.65] Dettmann, J.: Fullerene. Basel, Boston, Berlin: Birkhäuser Verlag 1994.

[4.66] Do Quoc Chinh; Schinköthe, W.: Elektrodynamischer Motor zur Erzeugung von Dreh- und Schubbewegungen. Wirtschaftspatent DD 253 331, 1986.

[4.67] Do Quoc Chinh: Elektromechanische Antriebselemente zur Erzeugung kombinierter Dreh-Schub-Bewegungen für die Gerätetechnik. Diss. TU Dresden, Fakultät Elektronik 1987.

[4.68] Fa. LPKF CAD/CAM Systeme Thüringen GmbH: Schneller, hochgenauer Zwei-Koordinaten-Antrieb. Firmenschrift, Suhl.

[4.69] Fa. Heidenhain: Offenes inkrementales Zwei-Koordinaten-Messgerät. Firmenschrift, Traunreut.

[4.70] Fa. Carl Zeiss Jena GmbH: Firmenschrift zum Encoder Kit L.

[4.71] Freitag, H.-J.: Neue Wege in der Längen- und Winkelmessung. F&M Feinwerktechnik, Mikrotechnik, Mikroelektronik 103 (1996) 4, S. 257–263.

[4.72] Furchert, H.-J.: Zweikoordinatenmotor. DE 30 37 648 vom 04.10.1980.

[4.73] Furchert, H.-J.: Oberflächenmotor. DD 205 330 vom 01.04.1982.

[4.74] Furchert, H.-J.: x-y-Flächenantrieb mit begrenzter ν-Drehung und z-Verschiebung. DD 222 747 A1.

[4.75] Furchert H.-J.: Kugelwanderungsfreie ebene Mehrkoordinatenwälzführung. DD 291 812.

[4.76] Furchert, H.-J.: Dimensionierung und Strukturierung von integrierten Gleichstromflächenantrieben kleiner Leistung für minimale Bauräume. Habilitation, TH Ilmenau 1990.

[4.77] Furchert, H.-J.: Stand und Perspektiven der Mehrkoordinatenantriebe. VDI Berichte Nr. 1269, 1996, S. 175–190.

[4.78] Furchert H.-J.: Gleichstromlinearmotor mit richtungsumkehrender Kraft-Strom-Kennlinie. DD 257 337 A1.

[4.79] Furchert H.-J.: Optimierte Gleichstromlinearmotorbaugruppen für integrierte Mehrkoordinatenantriebe. Quintessenz 1992, Leistungsspektrum des Fachbereiches Maschinenbau- und Feinwerktechnik, Fachhochschule Gießen-Friedberg, S. 8–22.

[4.80]   Halbach, K.: Design of permanent multipole magnets with oriented rare earth cobalt material. Nuclear Instruments and Methods, vol. 169, no. 1, pp. 1–10, 1980.
[4.81]   Holmes, M.; Hocken, R.; Trumper, D.: A Long-Range Scanning Stage Design (The LORS Projekt), ASPE 1996 Annual Conference, Monterrey, Nov. 11–14.
[4.82]   Kallenbach, E.; Furchert, H.-J.; Löwe, B.: Mehrkoordinatenantriebe für die Roboter- und Automatisierungstechnik. 31. IWK der TH Ilmenau 1986. Vortragsreihe B 1, H. 3, S. 131–135.
[4.83]   Kallenbach, E.: Systementwurf-Methoden zum systematischen Entwurf mechatronischer Produkte des Maschinenbaus. Mechatronik Workshop, Braunschweig 1992.
[4.84]   Kalusa, U.: Innovative und konventionelle Positioniersysteme aus der Sicht der Laserbearbeitung und Mikrosystemtechnik, VDI-Berichte 1269, S. 435–444.
[4.85]   Kelby, E. (Jr.); Wallskog, A.J.: Rotary & linear magnetomotive positioning mechanism. US 3745433 vom 02.02.1971.
[4.86]   Kim, W.-J.; Trumper, D.: Linear Motor-Leviated Stage for Photolithography, Annals of the CIRP Vol. 46/1/1997.
[4.87]   Kovalev, S.; Gorbatenko, N.; Nikitenko, J.; Saffert, E.; Kallenbach, E.: Ein magnetisch geführter Präzisionsantrieb mit 6 Freiheitsgraden. 2. Polnisch-Deutscher Workshop der TU Warschau und der TU Ilmenau, Sept. 1998.
[4.88]   Krause, W. (Hrsg.): Konstruktionselemente der Feinmechanik. München, Wien: Carl-Hanser-Verlag 1992.
[4.89]   Löwe, B.: Untersuchungen zum Einsatz von Wegmesseinrichtungen in Zweikoordinatenantrieben ohne Bewegungswandler. Diss. TH Ilmenau 1986.
[4.90]   Molenaar, L.; Zaaijer, E.; van Beek, F.: A Novel Long Stroke Planar Magnetic Bearing Actuator, MOVIC '98, Zurich, Switzerland, Auf. 25–28, Volume 3
[4.91]   Saffert, E.; Schäffel, Ch.; Kallenbach, E.: Regelung eines integrierten Mehrkoordinatenantriebs. 41. IWK der TU Ilmenau, Sept. 1996. Steuerungssystem für integrierte Mehrkoordinatenantriebe. 42. IWK der TU Ilmenau, Sept. 1997. Planar Multi-coordinate Drives. 2. Polnisch-Deutscher Workshop der TU Warschau und der TU Ilmenau: Werkzeuge der Mechatronik, Sept. 1998.
[4.92]   Schäffel, Ch.; Saffert, E.; Kallenbach, E.: Ein neuartiger integrierter Mehrkoordinatenantrieb mit Verdrehsperre durch Feldkräfte. 40. Internationales Wissenschaftliches Kolloquium der TH Ilmenau 1995.
[4.93]   Schäffel, Ch.: Untersuchungen zur Gestaltung integrierter Mehrkoordinatenantriebe. Diss. TH Ilmenau, Fakultät für Maschinenbau 1996.
[4.94]   Schäffel, Ch.; Glet, U.: Absolutes Zweikoordinatenmesssystem mit Drehwinkelerfassung. DE 42 12 990 A1.
[4.95]   Sorber, J.: Der Drehschubmotor - ein Antriebselement für kombinierte Dreh-Hubbewegungen. VDI Berichte Nr. 1269, 1996, S. 191–204.
[4.96]   Sprenger, B.; Binzel, O.; Siegwart, R.: Control of an High Performance 3 DOF e Linear Direct Drive Operating with Submicron Precision, MOVIC '98, Zurich, Switzerland, August 25–28, 1998, Volume 3.
[4.97]   Trumper, D.; Williams, M.; Mguyen, T. H.: Magnet Arrays for Synchronus Machines, IEEE 1993, Ind. Appl. Soc. Annual Mtg., Toronto, Canada, Oct.

[4.98] TU Ilmenau, Institut für Mikrosystemtechnik, Mechatronik und Mechanik und Institut für Prozessmess- und Sensortechnik: Firmenschrift Laserinterferometrisch geregeltes Mehrkoordinaten-Positioniersystem für Nanotechnologien. Institut für Mikrosystemtechnik, Mechatronik und Mechanik: Firmenschrift Feldgeführter High-Speed-Positioniertisch.

[4.99] TU Chemnitz, Institut für Fertigungsmesstechnik und Qualitätssicherung i. G.: Transformations-Messsystem mit Auswertung von Strichkodestrukturen zur absoluten Weg- und Winkelmessung.

[4.100] Wendorff, E.: Integriertes optoelektronisches Mehrkoordinatenmesssystem für integrierte Mehrkoordinatenantriebssysteme der Gerätetechnik. Diss. TH Ilmenau 1986.

[4.101] Williams, M.E.; Trumper, D.L.; Hocken, R.: Magnetic bearing stage for photolithography, Annals of the CIRP, 42 (1), p. 607–610, 1993.

[4.102] Mollenhauer, O.; Spiller, F.: A new XY linear stage for Nanotechnology – The stage can draw 3 diameter true circle. Advanced Photoic Technology Conference and Exhibition, Japan 11/2003.

[4.103] Schäffel, Ch.: Planar Motion Systems and Magnetic Bearings. 4th Polish-German Mechatronic Workshop, Suhl 2003.

[4.104] Schäffel, Ch.: Feldgeführter planarer Präzisionsantrieb. DE 19511973.

[4.105] Peltra, E.: Sawyer Motor Positioning Systems. Proceedings of Conference Applied Motor Control, University of Minnesota, Minneapolis, 1986

[4.106] Kallenbach, E.; Eick, R.; Quendt, P.: Elektromagnete. Stuttgart: Teubner, 1994

[4.107] Wendorf, E.; Kallenbach, E.: Direct drives for positioning tasks based on hybrid steppermotor. Intelligent Motion 1993, Proceedings pp. 38–52

[4.108] Räumschüssel, E.; Lipfert, R.: Nichtlineares Modell eines Linearschrittmotors auf der Basis von Daten aus der Magnetfeldberechnung. 45. IWK TU Ilmenau, 2000, Tagungsband, S. 529–534

[4.109] Dreifke, L.; Kallenbach, E.: Direktantrieb mit internen Sensoren und Magnetflussregelung. 45. IWK TU Ilmenau, 2000, Tagungsband S. 545–550

[4.110] Balkovoi, A.; Kallenbach, E.: Linear Stepping Motor Model for Thrust Analysis and Control. 45. IWK TU Ilmenau, 2000, Tagungsband S. 571–576

[4.111] Kuhn, Ch.: Ein nichtlineares Regelungskonzept für lineare Direktantriebe. 45. IWK TU Ilmenau, 2000, Tagungsband S. 541–544

[4.112] Kallenbach, E.; Schilling, M.: Antriebsmodule als Elemente der Geräte- und Automatisierungstechnik, Feingerätetechnik, Berlin 39 (1990) 4

[4.113] Büngener, W.: Prüfung und Beurteilung der Positions- und Schrittwinkelabweichungen von Hybridschrittmotoren. Dissertation Universität, Kaiserslautern 1995

[4.114] Dittrich, P.: Untersuchungen des Bewegungsverhaltens linearer elektromagnetischer Schrittantriebe der Gerätetechnik. Dissertation TH Ilmenau 1980

[4.115] Eissfeld, H.: Regelung von Hybridschrittmotoren durch Ausnutzung sensorischer Motoreigenschaften. Dissertation TU, München 1991

[4.116] Räumschüssel, E.: Simulation mechatronischer Systeme in Vergangenheit und Gegenwart. Festschrift für Eberhard Kallenbach, TU Ilmenau 2000, S. 65–74

[4.117] Dreifke, L.: Untersuchungen an planaren Hybridschrittmotoren mit Hallsensoren zur Magnetflussregelung und Positionsbestimmung, Dissertation TU Ilmenau 2002

5
Piezoelectric Drives

Hartmut Janocha

5.1 Physical Effect

Certain crystals, such as quartz, feature a physical relationship between mechanical force and electric charge. When the crystal lattice ions are elastically shifted relative to one another due to an external force, an electric polarization can be detected by means of metallic electrodes on the surface. This so-called piezoelectric effect was first scientifically explained by the brothers Jacques and Pierre Curie in 1880. It is described as the direct piezoelectric effect and forms the basis for piezo sensors. The effect is reversible, and is then called the reciprocal or inverse piezoelectric effect. If, for instance, an electric voltage is applied to a disc-shaped piezo crystal, the thickness of the crystal changes due to the reciprocal piezoelectric effect. It is this property that is made use of in piezoelectric actuators and drives [5.4].

In the analytical description of the piezo effect by linear state equations, the electric displacement density D and the electrical field strength E are combined with the mechanical strain S and the mechanical stress T. Which of these parameters are chosen as independent variables (an electrical and a mechanical one), depends on the relevant task. The following possibility is used frequently:

$$D = dT + \varepsilon^T E \tag{5.1a}$$

$$S = s^E T + d_t E . \tag{5.1b}$$

In this system of equations, the piezoelectric charge constant d indicates the intensity of the piezo effect; ε^T is the dielectric constant for constant T and s^E is the elastic compliance coefficient for constant E. The mentioned parameters are tensors of the first to fourth order. A simplification is possible by using the symmetry properties of tensors. Usually, the Cartesian coordinate system in Fig. 5.1a is used, with axis 3 pointing in the direction of the electric polarization P.

All material-dependent parameters can be described by matrices (2nd-order tensors) – d_t in Eq. (5.1b) is then the transpose of matrix d in Eq. (5.1a) – whose elements are marked with double indices. In ε^T, the first index

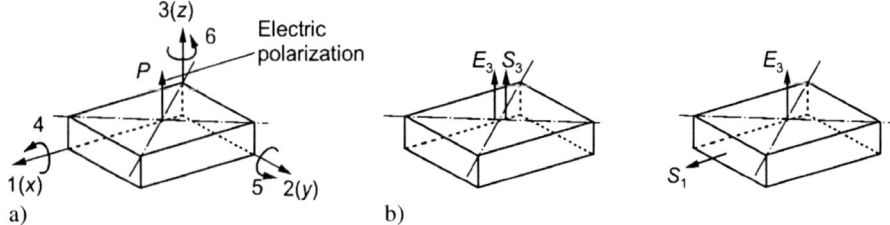

Fig. 5.1. Definition of the axes in piezoelectric materials. **a** The digits 4, 5, and 6 indicate the shear on the axes 1, 2 and 3 or x, y and z; **b** *left*: longitudinal (d_{33}) effect, *right*: transversal (d_{31}) effect

marks the direction of $\mathbf{D}$, the second the direction of $\mathbf{E}$, and correspondingly $\mathbf{S}$ and $\mathbf{T}$ in $\mathbf{s}^E$, and $\mathbf{E}$ und $\mathbf{S}$ in $\mathbf{d}$. The example in Fig. 5.1b is based on the condition that the field strength works in the direction of the polarization 3. The resulting elongation in the left-hand figure points as well in direction 3 (longitudinal effect), in the right-hand figure, however, it works in direction 1 (transversal effect). These two characteristics of the inverse piezo effect are quantified using the charge constants d_{33} and d_{31}.

From the piezoelectric constants it is possible to determine an important parameter of piezo materials, the coupling factor k. For the coupling factor of the longitudinal effect applies for instance

$$k_{33} = \frac{d_{33}}{\sqrt{s_{33}^E \varepsilon_{33}^T}}. \tag{5.2}$$

Since k^2 corresponds to the ratio of stored mechanical energy to supplied electrical energy, achieving actuators with high elongation efficiency requires substances with a large value of k.

In ferroelectric materials, one must add to the linear piezo effect in Eq. (5.1) a strain that depends on the square of the electric field strength. This strain component is negligibly small in traditional materials, but it can be increased systematically in order to reach the strength of the linear piezo effect. This so-called electrostrictive effect is independent of the polarity of the control voltage, and the corresponding diagram $S(E)$ shows a very small hysteresis. The effect is long-term stable (no creep, easily reproducible), while on the other hand, the operational range of temperature is limited to about 30 K, and the effect is not reversible. The electrostrictive effect is presently of less significance for use in transducers.

5.2 Piezoelectric Components

5.2.1 Piezoelectric Materials

Piezoelectric materials can be grouped into the class of natural and synthetic crystals, such as quartz or tourmaline, into one of polymers, such as polyvinylidene fluoride (PVDF) or that of polycrystalline ceramics.

For the production of piezo actuators, sintered ceramics are mainly used, especially lead-zirconate-titanate (PZT) compounds. After sintering, the Weiss' domains of a ceramic body (the regions consisting of crystallites of uniform dipole orientation) will show a statistically distributed orientation, i. e. the macroscopic body is isotropic and has no piezoelectric properties. Only when a strong electrical DC field is applied does the dipole regions become almost completely arranged ("polarized"). After switching off the polarization field, this arrangement remains to a large extent, that is, the ceramic body features a remnant polarization P_r, combined with a permanent elongation S_r of the body (see Fig. 5.2).

PZT ceramics are chemically inactive and can cope with high mechanical loading, but are also brittle and therefore difficult to process. The permissible compressive stress is considerably higher than the tensile stress (see Table 5.1). In the event of distinct tensile load the elements therefore have to be pre-stressed mechanically (Fig. 5.6). Some parameters that are important for actuator construction are significantly different in piezoceramics and quartz. While quartz, for instance, features $k_{11} = 0.09$, ceramics achieve values of up to $k_{33} = 0.7$. However, these values are unfortunately more dependent on temperature in polycrystalline PZT ceramics than in single crystals. Other disadvantages of piezoceramics are creep and the lower long-term stability of the material's properties.

PZT ceramics belong to the group of ferroelectric materials which feature a hysteretic behaviour shown in the diagram $P(E)$ in Fig. 5.2a. The saturation

Table 5.1. Values of the two PZT ceramics PXE 52 and PXE 54 (multilayer ceramics)

		PXE 52	PXE 54	Units
Piezoelectric charge constant	d_{31}	-270		10^{-12} As/N
	d_{33}	580	>450	10^{-12} As/N
Relative permittivity	$\varepsilon_{11}^T/\varepsilon_0$	3000		
	$\varepsilon_{33}^T/\varepsilon_0$	3500	3000	
Compliance coefficient	s_{11}^E	16		10^{-12} m²/N
	s_{33}^E	20		10^{-12} m²/N
Coupling factor	k_{31}	0.39		
	k_{33}	0.74	>0.6	
Compressive strength	T_p	>600	>600	N/mm²
Tensile strength	T_t	≈80	>80	N/mm²
Curie temperature	ϑ_C	165	220	°C
Density	ρ	7.8	7.9	10^3 kg/m³

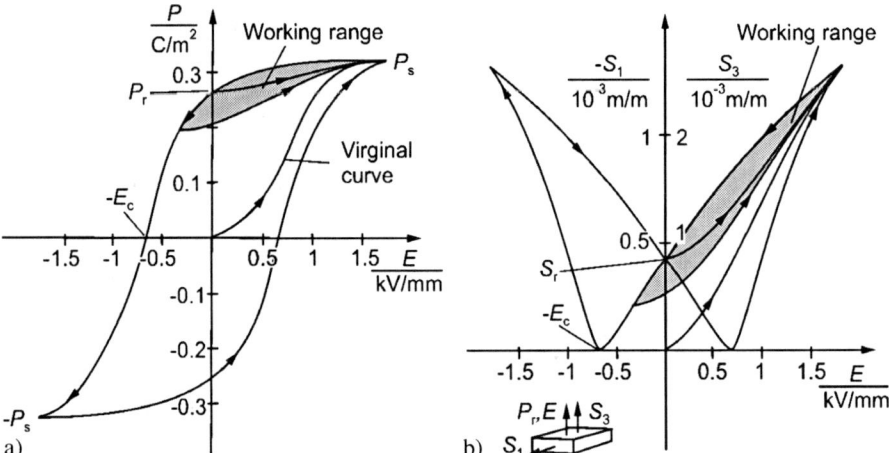

Fig. 5.2. Diagram for a typical piezoceramic for $T = 0$. **a** Characteristic $P(E)$. **b** Actuator characteristic $S(E)$; the actuator's operation cycle starts at point $E = 0$, S_r. The *small shaded areas* describe the unipolar mode and the *large shaded areas* describe the (asymmetric) bipolar mode

polarization P_s, the applied remnant electric polarization P_r and the coercive field strength $-E_c$ are characteristic points indicated on the curve. For actuator operation, the characteristic curve $S(E)$ of the polarized ceramic – the so-called butterfly curve – is significant, see Fig. 5.2b. The maximally achievable strain is limited by the saturation and the repolarization. Precautions must be taken in order to avoid depolarization during actuator operation due to electrical, thermal and mechanical overload. Piezoceramics, for instance, gradually lose their piezoelectric properties even at operating temperatures far below the Curie temperature ϑ_C (depending on the material 120 °C...500 °C, for multilayer ceramics (see below) 80 °C...220 °C). Above ϑ_C it behaves like a normal dielectric or paramagnetic. Thus the operating temperature of piezo transducers should not exceed $\vartheta_C/2$. When in certain applications the operating voltage is applied reversely, it may not exceed 20% of the rated voltage, or depolarization may occur.

Apart from the piezo effect, ferroelectricity and their mechanical and thermal solid-state properties, piezoceramics exhibit even more effects. Due to shifts in temperature, polarization and field strength, charges can occur in ferroelectrics possibly resulting in charges on the surface and an electrical field inside the material. This so-called pyroelectricity can be a disturbance, especially in low-frequency applications.

5.2.2 Piezoelectric Elements

Piezoceramic elements are mainly available now as plates or discs with a quadratic, circular or ring-shaped profile and a thickness between 0.3 and several millimetres, with or without metal electrodes. Most are designed to make use of the longitudinal effect (see Fig. 5.3a), which due to the high d_{33} value, is the strongest effect. When making use of the transversal effect, the actuator stroke depends not only on the d_{31} constant but also on the dimensions of the material, whereby the influence of the quotient s/l on stiffness and stroke is oppositional (see Fig. 5.3b). Besides these two effects, the shear effect that is described by the charge constant d_{15}, is recently being used commercially.

Since the 1980s, multilayer ceramics have grown more important. The so-called green and several tens of micrometers thick ceramic foil is cut into pieces and then coated with an electrode paste, similar to multilayer capacitors. The pieces are then placed on top of each other, pressed and sintered. They form a kind of monolithic object that is used as a finished transducer or as a basis for producing stacks (see Fig. 5.4.). Multilayer ceramics reach the maximum permissible field strength already at a driving voltage of about 100 V or less ("low-voltage actuators"), and thereby achieve the same strain as ordinary piezoceramics do for a driving voltage in the kilovolt range.

In addition, piezoelectric polymers are available as foils with a thickness on the order of several tens of micrometers. Such polymers have been known since 1924; but a major milestone was marked with the discovery of the strong piezo effect in polyvinylidene fluoride (PVDF) in 1969. Piezoelectric PVDF films are produced by mechanically drawing the material and polarizing it in order to form a useful transducer material. The drawing techniques include extrusion and stretching, and while processing the film, the material is subjected to a strong electrical polarization field. Typical for PVDF, charge constants are

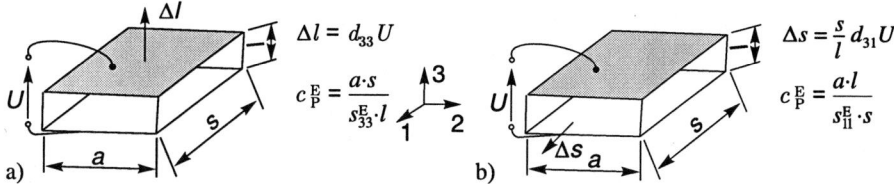

Fig. 5.3. Inverse piezo effect in polarized ceramics. Voltage U is applied in the direction of polarization. **a** Longitudinal effect. **b** Transversal effect (c_P^E: stiffness of the piezoelectric material for constant field E strength)

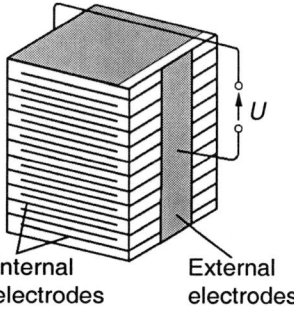

Fig. 5.4. Basic structure of a stack (d_{33} effect) comprised of multilayer piezoceramic

$d_{33} \approx -30\,\mathrm{pC/N}$ and $d_{31} > d_{32} > 0$; the coupling factor k_{33} is about 0.2, and the Curie temperature is near 110 °C.

For applications in the field of electrical microactuators, very thin piezoelectric films are preferably implemented with the help of sputter technologies. Frequently used materials include ZnO, ZnS, and AlN. These are placed on appropriate substrates, for instance, in the form of beams and membranes, whereby it is also possible to produce multilayer designs. A strong anisotropy of the expansion rate leads to a distinct orientation of the polycrystalline layers, so that the piezoelectric values may reach approximately the values of polarized ceramics under optimal precipitation.

5.3 Piezoelectric Actuators

An operating mode of a piezo transducer that lies significantly below its lowest eigenfrequency is called a quasi-static mode. The user can either build a piezo transducer from piezoceramics that are available on the market, or he may benefit from the broad range of available standardized and cased transducers. Figure 5.5 gives an idea of the transducer variety offered by a leading producer.

Fig. 5.5. Examples of piezo transducers (source: Physik Instrumente [5.12])

5.3.1 Stack Translator (Stacked Design)

Structure. The active part of the transducer consists, for instance, of many 0.3 to 1-mm-thin ceramic discs that are coated with metal electrodes, e.g. made of nickel or copper, for applying the operating voltage. The discs are stacked in pairs of opposing polarization and glued together. Highly insulating materials seal the stack against external electrical influences. In other designs, the multilayer ceramics described above are used.

Figure 5.6 features the parallel electrical connection and the series mechanical connection of the stack. Its displacement is the sum of the single element elongations Δl. The applied field and the achieved elongation are in line with the polarization, that is, the piezo constant d_{33} is used (longitudinal effect). The transducer can also handle tractive forces, if pre-stressed with a slotted tube spring as shown in Fig. 5.6 or with an anti-fatigue bolt, as is usually the case.

Static and dynamic behaviour. Equations (5.1) show that an ideal piezoelectric transducer input can be considered as an electric capacitor with the capacitance C and its output as a mechanical spring with the stiffness c_P. This is illustrated in Fig. 5.7a for the d_{33} transducer, but the description holds in principle for all piezo transducers. Since C is in reality always lossy and c_P always has a mass, the amplitude response $|u/F|$ (sensory operation) has an electrically determined lower cut-off frequency f_g and a mechanical eigenfrequency f_o. When operated as an actuator, the electrical input is a voltage, that is, C is constantly recharged, so that f_g has no effect on the amplitude response $|\Delta l/u|$, as shown in Fig. 5.7b.

An analytic expression easy to handle for the static actuator transfer behaviour of the piezo transducer arises from Eq. (5.1b) by substituting the occurring state quantities by their vectorial and scalar equivalent Δl, F and

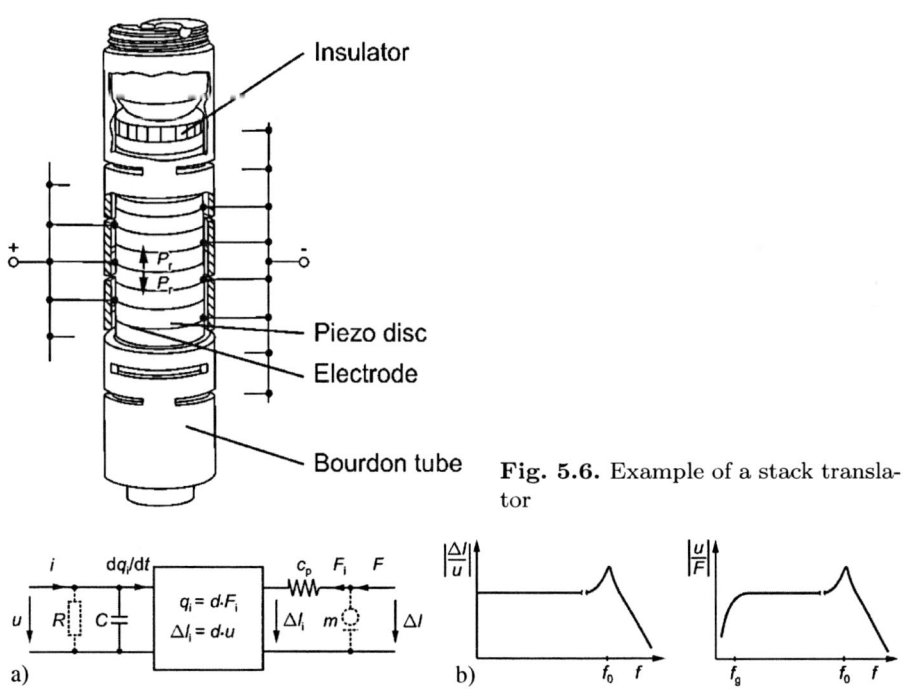

Fig. 5.6. Example of a stack translator

Fig. 5.7. Stack translator. **a** Electromechanical equivalent circuit diagram. **b** Amplitude response of the actuator (*left*) and sensory (*right*) transfer behaviour

u or directly from the equivalent circuit diagram 5.7a without considering the inertial effect of the mass m:

$$F = c_\mathrm{P}(\Delta l - d \cdot u) \, . \tag{5.3}$$

Figure 5.8a describes the graph $F(\Delta l)$ with u as a parameter. There are two distinguished operating points. In the unloaded case ($F = 0$) – depending on the applied voltage – the greatest possible displacement, called no-load stroke Δl_L, is achieved. If the actuator, however, is clamped ideally stiff ($\Delta l = 0$) it generates a so-called clamping or blocking force F_B; this force is the most powerful force that can be achieved in connection with the respective control voltage.

In general, actuators are operated between these two extreme points, whereby two types of load can appear as described in Fig. 5.8b for the case that the actuator is driven unipolarly with the voltage u_max.

- The load is constant, for example a weight F_G. Figure 5.8b shows that the actuator deflection remains constant independently of F_G and only the operating point of the deflection transfers to $-\Delta l_0$.
- The load depends on the displacement, e.g. a spring force $F_\mathrm{F} = -c_\mathrm{F}\Delta l$. Figure 5.8b shows that the operating point of F_F remains unchanged but the actuator deflection is reduced to $\Delta l'_\mathrm{max}$.

Figure 5.9 clarifies this behaviour in the form of the functional dependency $\Delta l(u)$ taking into consideration the characteristic graph in Figure 5.2b. The non-linear behaviour reminds us of the fact that the linear behaviour ("small-signal operation") on which Figs. 5.7 and 5.8 are based is only an approximation of the relevant large-signal mode of actuator operation.

For the quasi-static mode considered here, piezo actuators can be described as slightly damped PT$_2$ elements whose operating frequency is limited by the first eigenfrequency

$$f_0 = \frac{1}{2\pi}\sqrt{\frac{c_\mathrm{P}}{m_\mathrm{eff}}} \, . \tag{5.4}$$

If the transducer is fixed on one side, the effectively moved mass m_eff can be approximated by $m/3$ (m: transducer mass). If loaded by a mass M, m_eff is

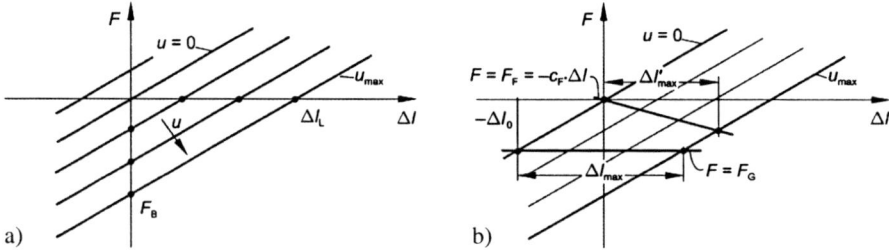

Fig. 5.8. Static characteristic curves $F(\Delta l)$ of a stack translator. **a** General behaviour. **b** Influence of different loads F_F and F_G

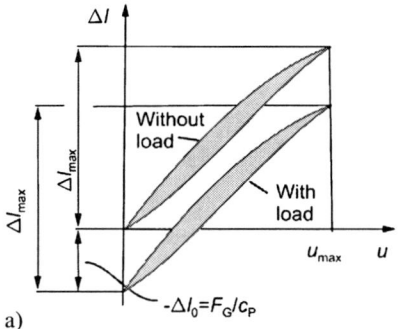 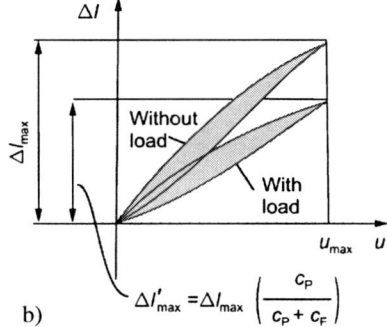

Fig. 5.9. Static characteristic $\Delta l(u)$ of a stack translator. **a** Constant load. **b** Load that depends on the displacement

to be equated with $M + m/3$. For commercial stack translators, f_o lies within the kilohertz range. As a guide, standard transducers can be operated up to 80% of their eigenfrequency.

Due to its capacitive input behaviour (see Fig. 5.7a), piezoelectric transducers exhibit a flow of electrical energy only when the material is experiencing a change in strain. This energy flux corresponds with a shift in charge and subsequently with the current

$$i = C\frac{du}{dt}. \quad (5.5)$$

Note that the implied condition that C is a constant in Eq. (5.5) is only an approximation, since the real transducer capacitance increases with increasing driving voltage.

For a constant supply of current, the length of time required to build up the voltage u in the piezoceramic is therefore

$$t = C\frac{u}{I}. \quad (5.6)$$

This time interval is directly proportional to the transducer capacitance and inversely proportional to the control current; this is crucial in order to design the required control electronics (cf. Sect. 5.5).

For a sinusoidal driving signal, Eq. (5.6) leads to the current amplitude

$$\hat{i} = \omega C \hat{u}, \quad (5.7)$$

which must be driven by an electronic power controller in order to charge and discharge the ceramic capacitance. Since the capacitance of piezo translators of equal dimensions increases proportionally with the square of the number of ceramic layers, the capacitance values of multilayer translators lie in the microfarad range, and the charge and discharge current described in Eq. (5.7) is considerably higher than for those translators whose capacitances lie in the nanofarad range (see Table 5.2).

With Eq. (5.7) it is also possible to estimate the upper cut-off frequency for which an amplifier can provide the maximal output current $i = i_{max}$. In addition, one must consider that some electrical power is transformed into heat due to hysteresis (see Fig. 5.2) during dynamic operation. Under extreme conditions, this may lead to thermal depolarization of the ceramics.

5.3.2 Laminar Translators

In contrast to the stack design, the laminar design is based on the piezo constant d_{31} and the transversal effect. The larger the quotient s/l of the piezoelectric element (see Fig. 5.3b), the greater the effect. This leads to strip-shaped elements with low stiffness. Therefore, several layers of strips are piled up, similar to the stack design to form a so-called laminate improving the mechanical stability. Since the transversal effect is applied, the results are flat transducers that shorten proportionally to the applied voltage, as d_{31} is negative. Table 5.2 lists several characteristic values while an example is presented at the foreground of Fig. 5.5.

Table 5.2. Characteristic values of piezoelectric transducers

	Stack design	Laminar design	Disc design	Stack with displacement amplifier	Units
Nominal displacement	5...90	...45	50...200	...100	µm
Stiffness	18...2000	...15	0.15...0.3	...1.4	N/µm
Eigenfrequency	6...50	...13	1.1...2.5	...2.2	kHz
Compressive strength	...30000	...450	20...50	...50	N
Tensile strength	...3500	...100	...20	...50	N
Nominal driving voltage	150...1500	...1000	...1000	...1000	V
Capacitance	...130	...145	16...70	...70	nF
Coefficient of thermal expansion	0.1...0.8	...0.7	1.0...4.8	...2.0	µm/K

5.3.3 Bending Elements

Bending elements feature the transversal effect as well. They can consist, for instance, of a PZT ceramic (unimorph, rarely defined as monomorphic) mounted onto a piece of spring metal. If the length of the ceramic is altered while the length of the metal remains constant, the element bends in order to compensate the different behaviour, and is therefore phenomenologically quite similar to the thermo-bimetal.

In the disc translator design, the elements consist of circular discs each with a diameter of a few centimetres, which allow a displacement of up to several 100 µm, as shown in the middle of Fig. 5.5. A commercial actuator design uses side-by-side unimorphs in which adjacent ceramics are on opposite sides of a metal substrate. In its simplest form, the metal forms a U shape with two ceramic pieces, see Fig. 5.10a. Versions with 3 or 4 prongs are available to give higher force or greater movement.

The typical bending element is a connection of two thin piezoelectric ceramic strips with or without inactive underlays: bimorph or trimorph. One can distinguish between two designs. The polarization of parallel bimorph or trimorph ceramics is codirectional, see Fig. 5.10b. For $U_2 = U_1/2$ the position of the bender becomes neutral. The series bimorph or trimorph however, consists of two inversely polarized ceramic stripes, so the two piezoceramics – in the case of trimorph – are always driven in the polarization direction.

Compared to stack translators, bending elements feature a greater deflection, a lower stiffness, a smaller blocking force and a lower eigenfrequency.

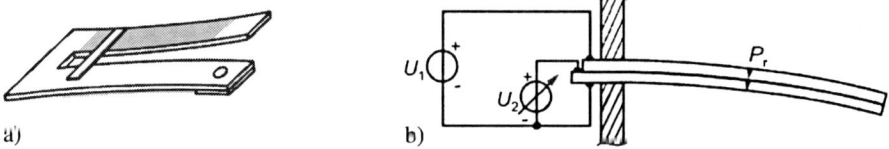

Fig. 5.10. Bending transducer. **a** Unimorph (according to [5.14]). **b** Parallel bimorph. U_1: constant bias voltage, U_2: control voltage ($U_2 < U_1$)

5.3.4 Tube Elements

Tubes are tubular piezoceramics with metallized cylindrical surfaces inside and outside (see Fig. 5.11). Tubes use the transversal effect. If the polarity of the control voltage U matches with the polarization direction, the tube contracts in axial and radial direction, as d_{31} is negative.

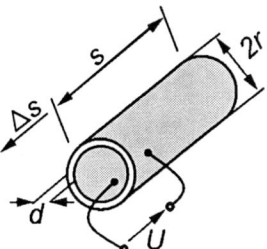

Fig. 5.11. Tubular piezo transducers

Tubes are applied, for example, as clamping devices for precision arbors (see Sect. 5.4.1.1) or as miniaturized pump drives in inkjet printers where there is a volumetric effect caused by the superposition of the radial and axial deformation.

5.3.5 Displacement Amplification

In piezoelectric transducers with displacement amplification, the achieved deflection is increased by constructive means. The stiffness of such a design decreases with the square of the displacement amplification ratio and is therefore much smaller than in the stack design.

This kind of transducer used (for displacements of up to 1 mm with forces of several tens of Newtons) is achieved for instance with elastic joints or hinges. These elastic hinges transform small angular alterations into parallel movements free of backlash. Figure 5.12 illustrates the principle. Figure 5.5 shows in the middle a model.

The highly elastic material region of the displacement amplifier in Fig. 5.22a is locally concentrated, while the designs in Fig. 5.22b and c make use of the global elastic behaviour of metallic materials. The so-called moonie transducer in Fig. 5.22b consists of a piezoelectric disc sandwiched between two metal end caps. With the application of a control voltage in axial direction, the longitudinal as well as the transversal effect come to effect. Due to the cap design, the small radial disc deflection translates into a much larger perpendicular displacement [5.6]. Figure 5.12c shows a design in which a piezo translator and so the d_{33} effect is applied [5.8].

Figure 5.13 shows an entirely different solution. A hydraulic force-displacement transformer functions according to the two-piston hydraulic principle. Leak-free operation is achieved in the presented design through the use of two folding bellows of different effective diameters. This special design keeps the enclosed oil volume small, thereby increasing the stiffness of the whole design and minimizing the amount of error due to thermal fluid expansion.

With the above introduced principle, it is usually possible to implement an amplification factor of up to 10. Greater values are constructively possible but quickly lead to a worsening of the dynamic behaviour of the entire system.

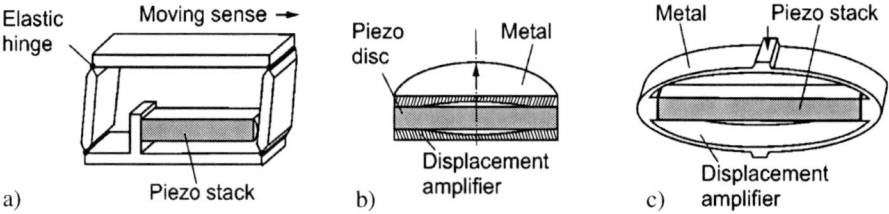

Fig. 5.12. Mechanical displacement amplification. **a** Implementation with elastic hinges (hybrid converter). **b** Moonie transducer. **c** Amplified piezo actuator (APA)

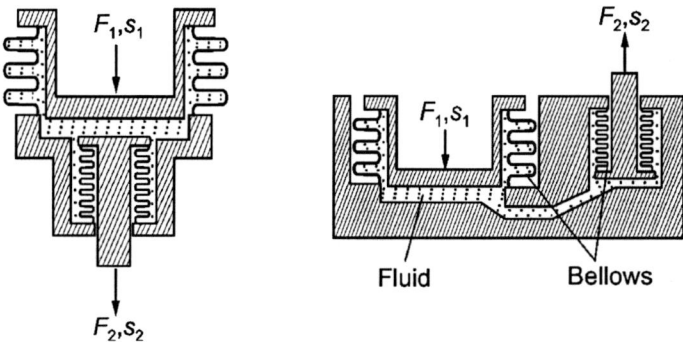

Fig. 5.13. Hydrostatic displacement amplification

Table 5.3. Important features of piezoelectric transducers

Advantages	Disadvantages
– large forces achievable, high stiffness – high electromechanical efficiency – very short response time (range of microseconds) – negligibly low power consumption in static operation – various couplings between the field and the strain axes possible – large selection and availability of different ceramic materials	– characteristic values of the ceramic are dependent on temperature and ageing – piezo effect can be lost by the influence of high temperature, large electrical field strength or mechanical shock – distinct characteristic hysteresis – strong self heating of the ceramic in dynamic operation – high-voltage power supply necessary for the capacitive load (up to several microfarad)

Table 5.2 lists characteristic values of piezo stack transducers and other types.

In Table 5.3 some important advantages and disadvantages of piezo transducers for quasi-static operation are summarised.

5.4 Piezoelectric Motors

When applying the inverse piezo effect, even considerably higher translations or rotations than so far described can be implemented. The appropriate solutions are based upon the fact that displacement or angle increments of a carriage or a rotor generated with the help of piezoelectric driving and clamping elements are summed up quickly until the desired total displacement or angle is reached. There are two different implementation possibilities: The first one moves the carriage or rotor directly by piezo elements that are actuated significantly below the lowest eigenfrequency of the total system (inchworm and walking drives). The other possibility is to convert natural oscillations of piezo elements first into resonant motions of a fixed stator which then activates the carriage or rotor (ultrasonic motors). Both possibilities are hereinafter explained in detail. What they have in common is the fact that the driving force in principle is transferred to the moveable part by frictional engagement.

5.4.1 Inchworm and Walking Motors

5.4.1.1 Inchworm Motor

The most popular representative of the inchworm drive is the inchworm® motor commercially available since 1975. The name originates from the fact that the motion is similar to that of an inchworm. Figure 5.14 describes the principle: a smooth shaft (carriage) that is to be positioned axially is enclosed by three piezo tubes (d_{31} effect). The two external elements are placed on the shaft with very small clearance. When electrically activated, they clamp the rod. The clearance of the tube in the middle is large and when voltage is supplied it extends in the axial direction, i.e. this element is responsible for the forward motion.

The movement is coordinated by an electronic controller as follows. First, tube 1 clamps the shaft (1), cylinder 2 expands (2) and moves the shaft to the left. Afterwards, tube 3 clamps the shaft as well (3), and then tube 1 reopens (4). The shaft is now held by tube 3 and cylinder 2 shortens again (5).

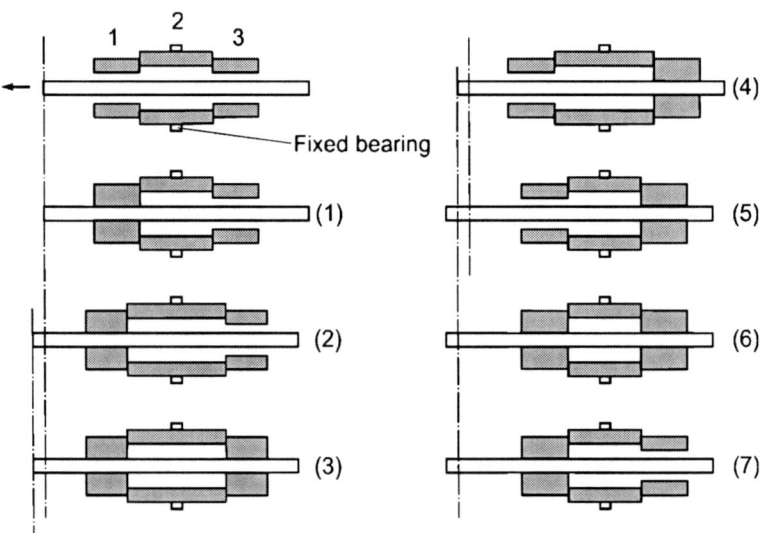

Fig. 5.14. Inchworm motor. Motion and characteristic values (according to [5.11])

Now, tube 1 clamps the shaft (6) and tube 3 reopens (7). Cylinder 2 expands again, and the whole procedure starts anew. The control voltages at the two external elements alternate (phase-shifted) always between two extreme values. The voltage of the feed element, however, proceeds in steps where the height of the steps defines the smallest displacement increment of the carriage and the steepness of the stairs determines its speed.

The fit between carriage and external piezo elements must be kept very precisely; it depends on temperature and gradually wears out. As the link is frictionally engaged a precise positioning can be made only in a closed control loop in combination with a displacement sensor (see Sect. 5.6.1). The displacements of a commercial series of inchworm drives cover a range from 6 to 200 mm where the displacement resolution is 2...4 nm. The maximum speed lies between 0.5...2 mm/s ($v_{max}/v_{min} = 5 \cdot 10^5$), and 5...15 N is mentioned as the thrust force. The application field of these motors is the precise positioning of components in optical devices and equipment [5.11].

5.4.1.2 Walking Motor

The basic elements of the Piezo LEGS^{TM} motor consist of multilayer stacks according to Fig. 5.4. Two electrically insulated stacks which are mechanically linked together very closely form a so-called leg. Figure 5.15a demonstrates how longitudinal as well as bending modes (the darker the shading, the higher the voltage) can be generated by imposing unipolar voltages of equal or unequal magnitude to the two halves of the leg. Therefore, the voltages $u_r(t)$ und $u_l(t)$ in Fig. 5.15b lead to a closed elliptic motion of the stack as indicated in the same figure.

Figure 5.16 explains the emergence of a feed motion with four legs. Two legs form a pair that is driven equally and consequently moves in-sync. A second pair is abridged staggered to the first one and is driven out of phase to this one. This produces phase-shifted elliptical movements of the ends of the stacks by which a carriage that is pressed against the stacks by springs is transported laterally in small steps. This is shown in Fig. 5.16 in single sequences where the level of the effective voltage is much higher than the darker the elements are. One has to consider that the maximum (electrical) operating frequency

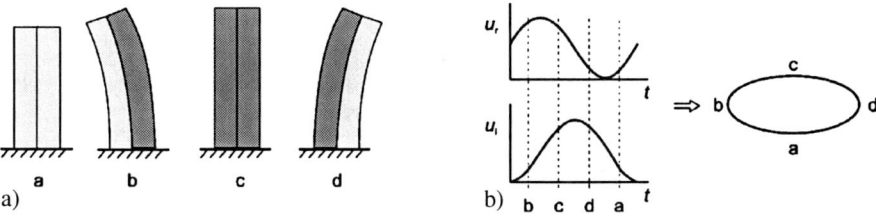

Fig. 5.15. Piezo LEGS^{TM} motor. a Deflections with different driving voltages. b Generation of a closed series of movements (according to [5.13])

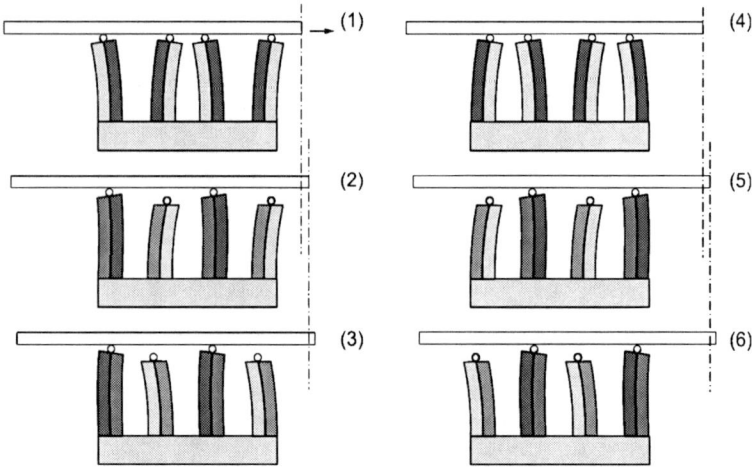

Fig. 5.16. Piezo LEGS™ motor. Series of movements of the walking motor (according to [5.13])

has to remain clearly below the (mechanical) eigenfrequency of the carriage in order to guarantee contact between the friction partners.

A motor implemented for demonstration purposes needs unipolar driving voltages from $0\ldots 42\,\text{V}$ (peak-to-peak value), the increment without load is $3\,\mu\text{m}$. The speed (max. 12.5 mm/s) is adjusted *via* the frequency of the control voltage. The holding force mentioned is 7.3 N, the traverse path of this design is 35 mm [5.13].

5.4.2 Ultrasonic Motors

The operating frequency range of the piezoelectric drives described so far lies significantly below their lowest eigenfrequency that – as well as the higher eigenfrequencies – is determined by the mechanical characteristics of the PZT ceramics used. In contrast, ultrasonic motors use especially the resonant vibrations of an oscillator, often a metallic one, which is activated by piezoelectric components.

Such a resonator arrangement forms the static part of a motor and is called the stator. A suitable electronic circuit makes sure that the stator is induced to vibrate mechanically in the ultrasonic range. The generated frequencies usually lie between 20 and 150 kHz where the amplitudes do not exceed a value of $10\,\mu\text{m}$. Consequently, no human being is able to hear or see the vibrations of the resonator.

A mechanical coupling, achieved with the help of springs between the stator and a movable part of the ultrasonic motor, the rotor or the carriage, make sure that the stator vibrations are transformed – depending on the motor construction – into rotatory or translatory movement of the rotor or the carriage.

This motion is comprised of displacement increments in the nanometer and micrometer range and occurs quite steadily due to the inertia of the rotor despite the impulsive stimulation.

Ultrasonic motors also have the problem of wear at the point where resonator and rotor contact and so the durability of the drive is influenced very strongly. Remedy can be the use of special surfaces that reduce the wear or soft layers that dampen the hard impacts of the resonator on the rotor as well as large contact surfaces to diminish punctual stress. As the transmission of power is effected by friction, ultrasonic motors are also called friction motors.

Corresponding to the type of the stator ultrasonic motors are divided in rod-shaped and ring-shaped resonators. Both types and the relative principles to generate rotor and carriage motions are explained in the following section.

5.4.2.1 Rod-Shaped Resonators

The longitudinal dimension of this type of resonator is significantly greater than the lateral dimension. Appropriate ultrasonic motors that have been developed in the former Soviet Union are also called vibration or micro impact drives. There is a distinction between resonators that are solely stimulated by longitudinal vibrations (monomodal resonators) and those where the overlapping of at least two vibration modes with orthogonal deflections leads to an elliptic motion of the resonator ends (bimodal resonator).

Monomodal resonator. The node of a longitudinal vibration lies in the middle of the rod; the resonator is affixed there with a low level of vibration. The maxima of the deflections are at the ends of the rods ("$\lambda/2$-resonator"). One end is pushed eccentrically against the rotor. Due to the expansion (motion A, see Fig. 5.17a), the end of the resonator meets the comparatively hard rotor. Because of the eccentric target and, in addition, considerably amplified by the chamfering of the resonator end the tip moves aside (motion B). The resulting motion is tangential to the rotor, affecting its drive *via* friction contact. Due to the subsequent contraction, the contact between rotor and stator opens and the elastic deformation of the tip diminishes. This operating principle enables only one possible direction of rotation.

Bimodal resonator. In this case, a longitudinal and a flexural mode of the rod overlap. The longitudinal mode affects motion A indicated in Fig. 5.17b, the flexural mode results in motion B. If the two frequencies are proportional *via* an integer constant, the overlapping results in a closed motion of the resonator end (Lissajous figures) "softly" drive the centrically set rotor. *Via* the phase of the two vibrations the direction of rotation of the rotor as well as its rotation speed can be adjusted. It must be possible to activate the two vibrations independently of each other. Due to the deviation of a real resonator from the ideal form, the vibration modes, rather, are coupled and thus they normally will perform a beat motion. This point is a key problem of resonators that are activated in several modes at the same time.

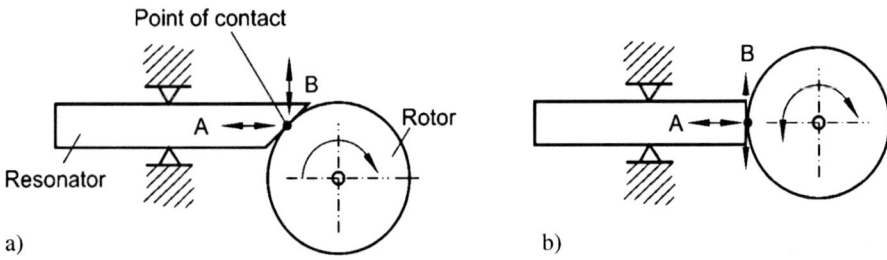

Fig. 5.17. Operating principles of ultrasonic motors based on rod-shaped resonators. **a** Monomodal motor with longitudinally vibrating resonator. **b** Bimodal motor with a longitudinal (A) and a bending vibration (B) of the resonator

The bimodal drive described hereafter is an ultrasonic motor with reversible direction of rotation that can be activated in a monophase [5.1]. The unwanted mode coupling mentioned above has been avoided by calibrating the eigenfrequency of the longitudinal and bending mode activated by a piezo transducer to the proportion 1:2 by an intended influencing of the rod shape. Figure 5.18a shows the two vibration modes of the resonator as a result of FEM simulations.

The control electronics of the motor is described in a simplified manner in Fig. 5.18b. The two sinusoidal phase-coupled control voltages are electronically added and brought to the necessary amplitude with a power amplifier (effective value 200...300 V). The output voltage of the amplifier is applied equally to the piezoceramics to generate longitudinal and bending vibrations. Due to the high frequency difference, each piezo transducer always reacts only to the appropriate spectral component.

By modifying the phase between the two driving signals, a rotation speed was adjusted continuously from 0 to approx. 300 min^{-1} in a test motor. The time lag between applying the electrical control signal and arriving at the final rotation speed amounts to 20 ms. As – unlike the monomodal resonator – the

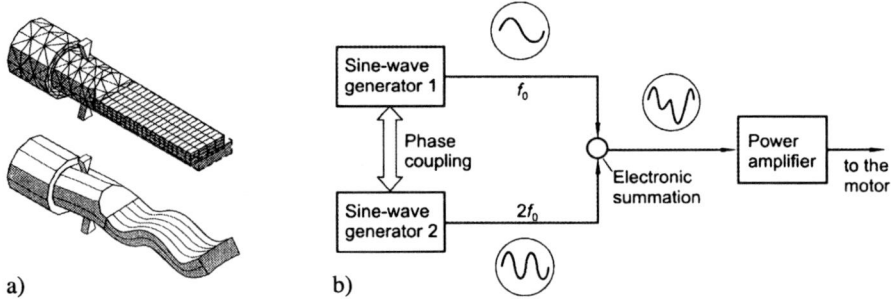

Fig. 5.18. Bimodal ultrasonic motor. **a** FEM simulation of a longitudinal vibration (24 kHz) and a bending vibration (48 kHz). **b** Electric control scheme (according to [5.1])

hard and therefore energy-dissipative collision of the resonator to the rotor does not occur here, high efficiencies and a long durability of this bimodal drive are possible.

The Elliptec motor in Fig. 5.19a is a commercially available design where a simple electronic driver unit excites a 5-mm-high multilayer stack (cf. Fig. 5.4) to ultrasonic vibrations. A main mode with longitudinal and bending components arises as well as a slightly distinct auxiliary mode. They initiate the tip of a particularly formed oscillator to perform an elliptic motion [5.10]. Pressing the tip against a moveable carriage or a rotor for example with a spring causes translatory or rotatory movements (see Fig. 5.19b).

Switching the operation frequency for example from 79 to 97 kHz leads to a change of the main mode and thus to a directional inversion of the motor movement. The voltage amplitude of the piezoceramic is $6\ldots 8$ V, the current rating up to 400 mA (depending on the speed). In the speed range from $0\ldots 300$ mm/s, shearing forces reach a maximum of 0.5 N. The increment is indicated to be 10 µm, the response time <0.1 ms and the length of the motor 25 mm.

A derivative of the rod-shaped oscillator can be seen in Fig. 5.20. Here a rectangular piezoelectric plate polarized in the y-direction serves as resonator (stator). On its face there are two separate electrodes and on the reverse side a common counter electrode. If for example the left half of the plate is activated with a sinusoidal voltage of appropriate frequency (the other half is passive, cf. Fig. 5.20a), a resonant eigenmode is activated in the ceramic body and the result is a two-dimensional standing wave in the x–z plane. The resulting motion path of the so-called friction tip runs linearly and is rotated 45° in the positive x-direction; thus the micro impacts push the carriage in Fig. 5.20b upwards and to the right. An inversion of the direction follows by activating the right half of the plate (the left half is then passive) which causes a modification of the angle between the friction tip trajectory and the x-direction to 135° [5.7].

The presented concept has been transferred into a commercially available product series [5.10]. The smallest increment of types in this series is 50 nm,

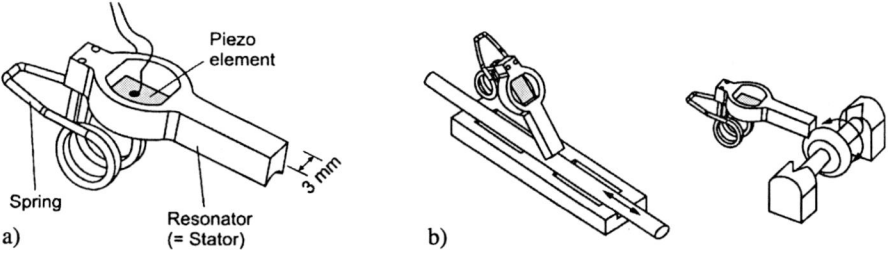

Fig. 5.19. Elliptec motor. **a** Basic assembly. **b** Application as translatory (*left*) and rotatory (*right*) drive [5.10]

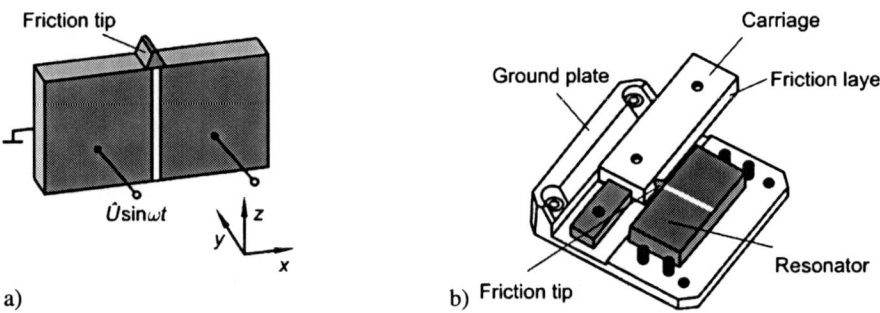

Fig. 5.20. PILine® motor. **a** Construction of the resonator and driving signal. **b** Overall assembly of the linear drive (according to [5.7])

the largest translatory motion is 50 mm. The indicated maximum values are 800 mm/s for the speed of the carriage and 4 N for the pushing or pulling force.

Possible applications of these drives are seen among others in the fields of automotive, peripheral computer devices, toys and optical equipment.

5.4.2.2 Ring-Shaped Resonators

Another operating principle of ultrasonic motors developed in Japan is based on the bending vibrations of an annulus and is explained by means of Fig. 5.21a: Single PZT plates that are divided into two groups (the two groups in Fig. 5.21a are shaded differently for a better understanding) are applied on a stationary elastic stator ring. Each group of plates is polarized in the opposite direction of the neighbouring groups. Thus expansions and contractions occur in the tangential direction (d_{31} effect) upon application of an AC voltage to the plates which firstly generates a standing wave (bending wave) along the ring circumference.

If the second group is activated with an AC voltage of the same frequency but shifted in time by a quarter of a period relative to the first one, the standing wave of this group is spatially displaced by $\lambda/2$ with respect to the first one (cf. similarity with the principle of asynchronous motors). If the

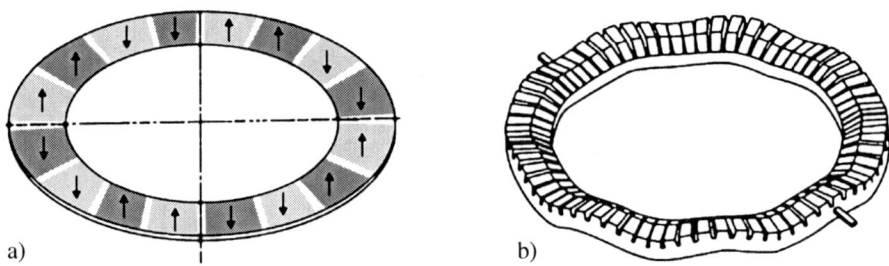

Fig. 5.21. Creation of a travelling wave by means of an annulus-shaped resonator. **a** Principle [5.6]. **b** Example of construction [5.1]

activating frequency is chosen such that for the ring resonator in Fig. 5.21a four peaks and troughs are created along its circumference, a travelling wave is created. Then the surface elements of the PZT plates move in elliptical paths. This results in a transport effect so that a rotor that is pressed against the stator is carried along in the opposite direction of propagation of the travelling wave. As the circulation sense of the travelling wave is reversible, this motor has two directions of rotation.

Figure 5.21b shows a representative resonator design that can be found frequently in commercial ultrasonic motors. The rotor (not shown) lies on the upper side of the ring, the piezo elements (also not shown) are affixed to its underside.

The stator of the ultrasonic motor in Fig. 5.22 consists of two ring-shaped piezoceramic discs that are each composed of eight segments alter-

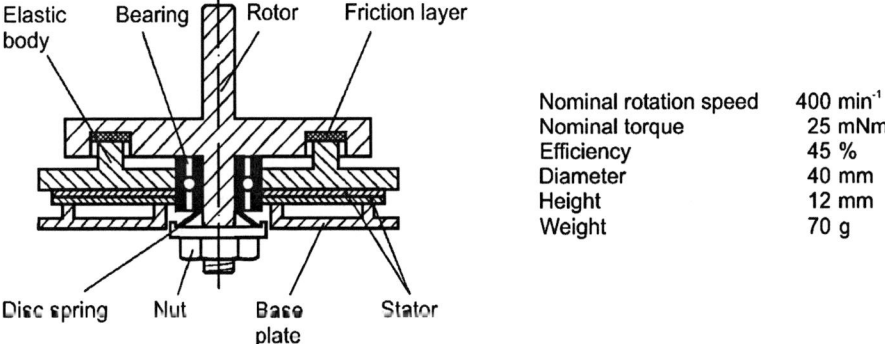

Nominal rotation speed	400 min^{-1}
Nominal torque	25 mNm
Efficiency	45 %
Diameter	40 mm
Height	12 mm
Weight	70 g

Fig. 5.22. Ultrasonic motor. Example of construction and characteristic values (prototype from Matsushita)

Table 5.4. Advantages and disadvantages of ultrasonic travelling wave motors with respect to electromagnetic motors

Advantages	Disadvantages
– response time in the lower millisecond range due to small moment of inertia of the rotor – torques greater than for electromagnetic electric motors of the same size – low rotational speeds possible without gearing – high holding torques when switched off (i.e. without energy supply) – quiet operation without shock – no magnetic leakage fields – identical basic concept for rotatory and translatory drives	– operational behaviour dependent on frictional engagement between resonator and rotor (influence of overload, abrasion, temperature, ...) – less suitable for applications with continuous operation

nately polarized along the circumference and that are firmly attached to each other. Both discs are displaced against each other by a 1/4 segment (i.e. $\lambda/4$) and are activated each with two sinusoidal AC voltages phase-shifted by 1/4 period. An ultrasonic travelling wave occurs that is transferred to an elastic body firmly attached to the discs. The rotor rests on the peaks and rotates against the direction of wave propagation. The preload force between the stator and the rotor is adjusted by means of a disc spring and a nut.

The characteristic values in Fig. 5.22 are typical for commercial ultrasonic travelling wave motors of medium size; large constructions reach torques into the Newton-meter range with rotation speeds around $100\,\mathrm{min}^{-1}$ (see for example [5.15]). Applications such as ring-shaped drives for focussing mirror-reflex cameras and camcorders as well as uses in robots and positioning equipment have become known.

The essential characteristics of ultrasonic travelling wave motors compared to electromagnetic motors are listed in Table 5.4.

5.5 Driving Electronics

For the operation of piezo actuators, electronic power amplifiers are indispensable. The concepts to be considered are among other reasons defined by the fact that piezo actuators essentially constitute capacitive loads. If the non-linearity of the static actuator characteristic should be compensated, it is furthermore advisable to control the actuator output variable (displacement or force) or to use inverse control.

5.5.1 Power Amplifier

In the case of piezoelectric walking motors, the power electronics has to be adapted individually to the respective piezo load. Especially the transient signal behaviour and dependences of the control voltage and/or the eigenfrequencies of the piezoceramics have to be considered exactly. This is why manufacturers usually offer these motors together with an optimized control circuit. Piezo transducers for the quasi-static mode, however, require power amplifiers that are characterized by a constant amplitude response and linear phase response over a sufficiently broad frequency range. Such amplifiers are commercially available in a large variety (see for example [5.9]).

If the short reaction times of quasi-static driven piezo transducers are to be fully effective, the electronic power amplifiers must be capable of delivering high currents at high voltages for short periods of time, see Eqs. (5.6) and (5.7). In practice, two approaches have proven effective: voltage control and charge control. Both control approaches have their specific advantages and disadvantages. Thus, the choice of the "best" amplifier presumes exact knowledge of the application including the electrical and mechanical behaviour of the piezo load. Experience shows that voltage amplifiers can in general be applied more easily and universally ([5.3]).

Concerning power amplifiers, one must decide whether a switching amplifier or an analogue amplifier should be implemented. The features of these amplifier types that are important for piezo transducers in the case of voltage control are listed comparatively in Table 5.5. Important distinctions include the quality ("residual ripple") of the voltage-time signal at the amplifier output and the amplifier efficiency. Analogue amplifiers fulfil the first mentioned criteria much better; the second one is better fulfilled by switching amplifiers. These also offer the possibility to improve the efficiency of the overall system (transducer and amplifier) by recovering the field energy stored in the transducer, see the following example.

Figure 5.23 shows the block diagram of a switching amplifier for the voltage control of piezo transducers. The control signal with the desired voltage-time form in the amplitude range $0 \ldots +10$ V is conditioned with a voltage offset and low-pass filtering at the input stage of the power amplifier. A sensor measures

Table 5.5. Comparison of different kinds of amplifiers for quasi-statically driven piezoelectric drives

	Analogue amplifier Class A	Analogue amplifier Class C	Switching amplifier
Losses in the power transistors	very high even in passive state	high when driving	very low
Recovery of stored field energy	impossible	impossible	possible
Residual ripple of output signal	extremely low	very low	high
Ratio of pulse-to-continuous current[1]	typically 3.14(π)	up to 100	1
Dynamics in small-signal operation[2]	extremely high	very high	low
Extraneous loading of actuator[3]	very low	very low	high
Electromagnetic compatibility	very good	very good	poor, active disturbance
Load range[4] (C_A / C_{ANenn})	100	100	about 5

[1] important for the maximum slew rate of single rectangular pulses for a given amplifier size

[2] without activation of current limiting

[3] load due to portions of the actuator current that are not caused by the input signal such as current ripple and discontinuous charging current

[4] range of load capacitance C_A with respect to the nominal value without the necessity of modifying the control parameters of the amplifier

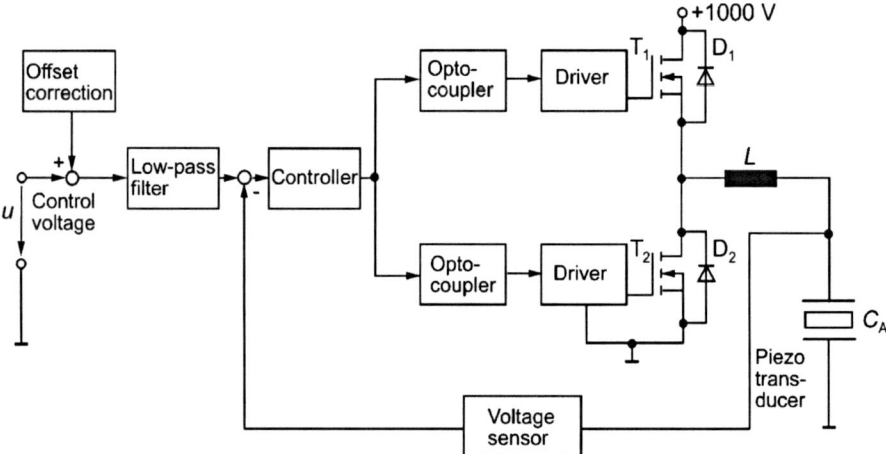

Fig. 5.23. Block diagram of a switching amplifier for the voltage control of piezo transducers

the voltage at the piezo transducer. This control variable is compared to the filtered input signal. An analogue three-level controller generates the value based on the control error. The actuating signal arrives at the driver stage *via* two optocouplers that enable galvanically decoupled control.

The two MOSFETs T_1 and T_2 in the half bridge work in switched mode, i.e. either T_1 is activated for charging the transducer capacitance or T_2 discharges it. This enables low-loss operation compared to analogue amplifiers. Free-wheeling diodes D_1 and D_2 protect the transistors from high induction voltages when switching off the current. The energy stored in the electric field of the transducer capacitance is partially recovered by aid of the inductance L and a free-wheeling diode. To achieve the desired transducer dynamics, an appropriate design of the coil is needed as a poor dimensioning slows down the entire system.

5.5.2 Inverse Control

Many multifunctional materials, i.e. materials that have sensory and actuator properties at the same time such as piezoelectric ceramics, have output–input characteristics with significant hysteresis (see Fig. 5.2). If such transducers are implemented as actuators the large-signal operation is inevitable so that the disadvantageous hysteretic features (non-linearity, ambiguity) become more obvious with growing signal amplitude. A method suitable in many cases for extensively eliminating the hysteretic influence is closed-loop control of the actuator output variable, see the example in Sect. 5.6.1. Another possibility is to compensate the hysteresis in feed-forward control, i.e. sensorless and inherently stable, *via* an inverse control block ("inverse filter").

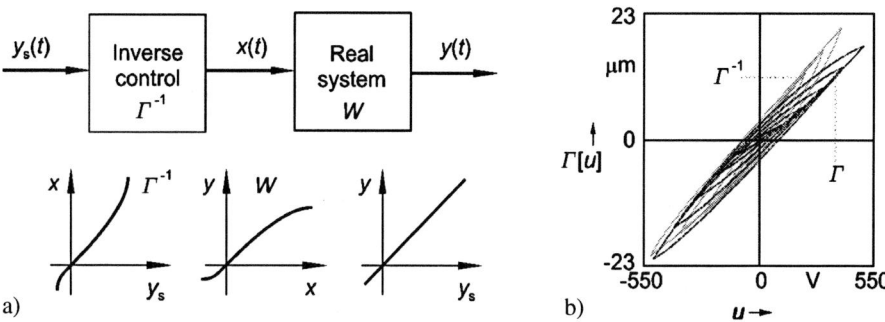

Fig. 5.24. Inverse control. a Basic construction. b Model Γ and compensator Γ^{-1} for a piezo actuator

The core of any inverse controller is a phenomenological model of the hysteretic non-linearities, which should be as exact as possible and which furthermore should be well suited for generating real-time capable control strategies. An analytical description of the hysteresis is possible with so-called hysteresis operators for which the respective transfer elements can be described as a transformation between function spaces. While first and foremost the Preisach hysteresis operator has been used for modelling and inverse control of solid state actuators, the Prandtl-Ishlinskii hysteresis operator has been in use since the middle of the 1990s [5.5]. In contrast to the Preisach operator, it has the advantage that the compensator can be determined with little calculation effort directly from the hysteresis model.

Figure 5.24a describes the principle of inverse control. The main item is the compensator Γ^{-1} of a mathematical model Γ of the non-linearity W. As an example for a piezo actuator Fig. 5.24b shows the operators Γ and Γ^{-1} that have been synthesized with the so-called modified Prandtl-Ishlinski method. Implementing such a filter, the linearity error in the signal transfer due to hysteresis can be reduced by a decimal order of magnitude or more. A digital compensator implemented in hardware in the form of a FPGA (field programmable gate array) makes a practically hysteresis-free actuator operation possible for signals up to 1 kHz; before being processed, the control signal $y_s(t)$ is sampled at 2.5 MHz with 14-bit resolution [5.16]. The necessary calculation capacity is not available in microcontrollers or digital signal processors.

5.6 Implementation Examples

This section should enable the reader to evaluate the possibilities and limits of today's available piezoelectric drives. For this purpose, two commercial products from the field of "linear precision drives" are presented with different intentions: First, the example of a multiaxial positioning system will be used to make clear the high standard already achieved. Afterwards, an inchworm motor example serves to present several problems connected with piezo technology and their solutions. In both cases, the applied design methods are emphasized.

5.6.1 Positioning System with Piezo Drives

During voltage control of piezo transducers, the absolute value of the displacement is not definitely known due to the hysteretic relationship. This behaviour does not affect the accuracy of positioning drives if the position or the displacement is measured exactly. Other solutions rely on inverse control (cf. Sect. 5.5.2) or position control. The latter task implies control in a closed control loop, i. e. it requires a displacement sensor to measure the actual values and a regulator that controls the voltage at the transducer corresponding to the difference between the measured and the command values, see Fig. 5.25.

The actual value is derived from displacement or strain measurement data of the quasi-statically driven piezo transducer. For displacement resolution into the range of some tens of nanometres, usually strain gauges are applied for higher resolutions inductive or capacitive sensors. Advantages of the closed control loop are hysteresis-free positioning, high absolute positioning accuracy, no drift motions, stable positioning despite changing forces and extremely high stiffness.

Figure 5.26 shows the principle of 2-axis positioning systems in serial-kinematic and parallel-kinematic design. Unlike serial kinematics where exactly one actuator and one sensor is associated with each degree of freedom, in parallel kinematics all actuators have a direct effect on a central, moving platform. In the latter case identical dynamic behaviour can be achieved for the x and y axes. Furthermore, parallel kinematics enable the integration of "parallel metrology". As such, all controlled degrees of freedom can be observed simultaneously, and thus guidance errors can be compensated in real-time. The advantages are a much higher trajectory precision, repeating accuracy and flatness of motion.

The piezo actuators in a commercially available parallel-kinematic, 6-axis positioning system operate *via* integrated mechanical displacement amplifiers

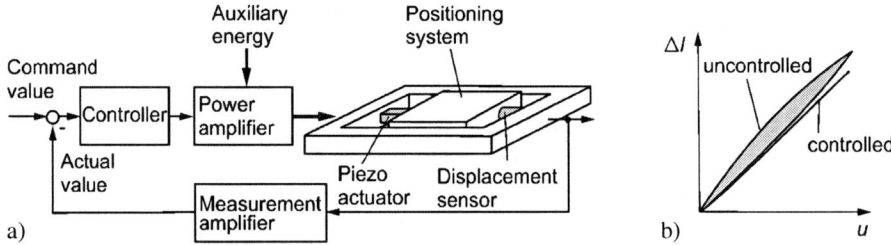

Fig. 5.25. Positioning system with piezo transducer in a closed control loop. **a** System diagram. **b** Static transducer characteristic

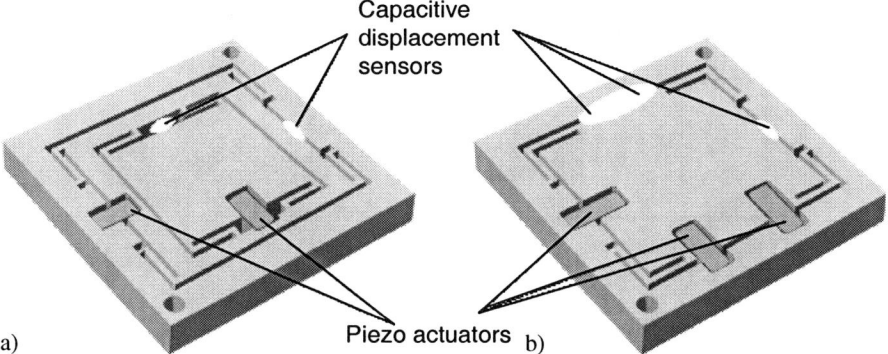

Fig. 5.26. Two-axis positioning systems. **a** Serial kinematics. **b** Parallel kinematics (according to [5.12])

on a multi-hinge parallelogram (cf. Sect. 5.3.4). The position of the moved platform is registered directly by means of four capacitive displacement sensors. The displacements are 100 µm each in the x- and y-directions and 10 µm in the z-direction with a displacement resolution of less than 0.3 nm. In position-controlled operation, the linearity deviation is typically 0.03% and a repeatability of ±2 nm is specified. A 2-kg load is allowed. This precision positioning platform is one of the highest technical achievements today [5.12].

Application examples of precise positioning platforms include scanning electron microscopes, the high-precision adjustment of masks and wafers in the semiconductor industry as well as surface structure analysis and micro manipulation.

5.6.2 Clamping Elements for Inchworm Motors

Commercially available inchworm motors enable displacements of more than 100 mm with resolutions in the nanometer range and thrusting forces up to about 15 N, cf. Sect. 5.4.1. Focussing on the inchworm motor as an example, some aspects that should be considered when aiming for high positioning accuracy and strong thrusting forces are illustrated in this section.

Piezo tubes and thus the d_{31} effect were used in the original inchworm design. When a voltage is applied, both the length and the diameter of the ceramic will change, see Fig. 5.11. This leads in each clamping process of the two external piezo elements also to unwanted axial and lateral motions of the carriage, which are termed as glitches and are several tenths of a micrometer [5.2]. An improvement was reached through the application of stacked piezo rings, i.e. the use of the d_{33} effect in connection with a specially formed collet. This solution reduced the glitches to less than 50 nm. At the same time, their reproducibility improved, and therefore an on-line compensation became possible.

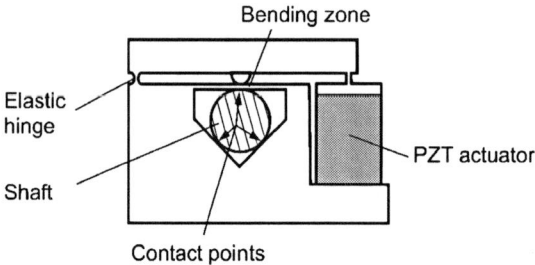

Fig. 5.27. Inchworm motor. Improved clamping mechanism (according to [5.2])

The thrust of the inchworm motor is limited by the friction force that is transferable by the external clamping elements. The friction force cannot simply be increased since its originating clamping force that is switched on and off during the carriage's motion leads to high dynamic mechanical tensions in the PZT ceramic, which can cause fatigue cracks and thus shorten the durability of the motor. By optimizing the manufacturing and test process, the manufacturer can guarantee that the number of cycles to failure of the clamping elements reaches 10^{10} cycles. Furthermore, through constructive measures the critical mechanical tensions can be converted into compressive stresses whose allowable maximum values can be more than one order of magnitude greater than those of the tensions (see Table 5.1).

The principle of a clamping solution encompassing the mentioned findings is described in Fig. 5.27. It is characterized by the following features.

- The PZT elements are linked with the clamping mechanism and the carriage shaft by a bending zone that allows only one direction of motion.
- The carriage has only one degree of freedom, i.e. the clamping process does not cause any disturbing lateral motion (glitches).
- The clamping mechanism is firmly connected with the housing and the piezo elements for clamping and thrusting work independently of each other.

The objectives of further developments of the inchworm motor are an increase in the thrust and driving speed as well as a decrease of the size. Microsystem technology can make important contributions toward these objectives. For example, a miniaturized linear motor has become known in which two micro-technically produced pinions are mechanically contacted and separated cyclically. The thrust generated in this manner is 20 times greater than the thrust transferred by friction in an inchworm motor of the same size and is limited only by the maximum allowed shear stress in the material. The main problem is to manage the contact and separation between the two toothed elements quickly enough, e.g. at frequencies in the kilohertz range.

References

[5.1] Fleischer, M.: Piezoelektrische Antriebe und Motoren. In: Technischer Einsatz Neuer Aktoren (Ed. D.J. Jendritza). Renningen-Malmsheim: Expert-Verlag 1995, pp. 254–266
[5.2] Henderson, D., Fasick, J.: The Inchworm® Piezoelectric Stepping Motor – Advances in Design, Performance and Applications. Proc. 7th Int. Conf. New Actuators, Bremen 2000, pp. 451–455
[5.3] Janocha, H. (Ed.): Actuators – Basics and Applications. Berlin Heidelberg New York: Springer 2004
[5.4] Koch, J.: Piezoxide (PXE) – Eigenschaften und Anwendungen, Heidelberg: Dr. Alfred Hüthig-Verlag, 1988
[5.5] Kuhnen, K.; Janocha, H.: Inverse feedforward controller for complex hysteretic nonlinearities in smart material systems. Control and Intelligent System, Vol. 29, No. 3, 2001, pp. 74–83
[5.6] Uchino, K.: Ferroelectric Devices. Marcel Dekker, New York Basel, 2000
[5.7] Wischnewskiy, W., Kovaler, S., Vyshnevskyy, O.: New Ultrasonic Piezoelectric Actuator for Nanopositioning. Proc. 9th Int. Conf. New Actuators, Bremen 2004, pp. 118–122
[5.8] Cedrat Technologies SA, Meylan/Frankreich www.cedrat.com
[5.9] D*ASSmbH, Saarbrücken/Deutschland www.dass.de
[5.10] Elliptec AG, Dortmund/Deutschland www.elliptec.de
[5.11] EXFO Burleigh Products Group Inc., Victor, NY/USA www.exfo.com
[5.12] Physik Instrumente GmbH & Co. KG, Karlsruhe-Palmbach/Deutschland www.pi.ws
[5.13] PiezoMotor Uppsala AB, Uppsala/Sweden www.piezomotor.com
[5.14] Servocell Ltd. Harlow, Essex/UK www.servocell.com
[5.15] Shinsei Corp., Tokyo/Japan www.tky.3web.ne.jp/~usrmotor/Englisch/html
[5.16] Janocha, H.; Pesotski, D. and Kuhnen, K.: FPGA-Based Compensator of Hysteretic Actuator Nonlinearities for Highly Dynamic Applications. Proc. 10th Int. Conf. on New Actuators, Bremen 2006, pp. 1013–1016

6
Open-Loop and Closed-Loop Control of Electric Fractional-Horsepower Motors

Friedrich Wilhelm Garbrecht

6.1 Introduction

Electric drives are systems designed to create controlled motions. For this reason, electrical energy is converted into mechanical energy by electrical motors. The developments in information technology and microelectronics open new possibilities to effectively control the energy flow in electric motors in such a way that the mechanical power is transformed into a controlled motion following given reference values. Usual modes of operation are open-loop control and closed-loop control. When using open-loop control, the actual value and reference value are not compared and it is therefore unnecessary to measure the actual value. In closed-loop systems, the actual value is measured or observed and compared with the reference value. The controller calculates the control signal from the difference between these two values so that the system deviation disappears. In all cases, closed-loop control is more expensive than open-loop control, and therefore feedback systems should only be applied when high precision is demanded.

Linear voltage regulators are only used in connection with low-power applications to adjust the speed of DC motors. For DC motors with higher power, it is necessary to reduce the losses of power electronics. Therefore, the power semiconductors are only used as switches. In this case, the voltage needed to adjust the required motor speed is generated by pulse-width modulation (PWM). The power semiconductors are controlled by a logic unit or an intelligent control unit. Today, almost only microcontrollers are used as control units. These units are equipped with a timer structure to support the pulse-width modulation. In addition, the microcontroller has a bidirectional counter to analyze the output signals of incremental rotational sensors. Such microcontrollers equipped with suitable software are able to take over all functions of motor control like observing limit values, open-loop control, closed-loop control, dead time generation, etc. By using microcontrollers as master control units and power modules in the power section, the number of components can be reduced to a minimum, leading to a reduction in costs and to higher operational security. In particular, for motors supplied by high voltages, the control signals must be coupled indirectly or by a charge pump to adapt the voltage level of the signals between the control unit and the power semiconductors.

6.2 Circuit Components and Pulse-Width Modulation

6.2.1 Circuits for Indirect Coupling and Voltage Level Adapting

Modern power electronics for the operation of DC motors or rotating field motors consist of two or three controlled half-bridges. They are necessary in order to realize the four-quadrant operation for each motor type. The presented bridge (Fig. 6.1) is designed with field-effect transistors (FETs). They are switched by applying a gate source voltage. There are no problems to control the lower FET of a half-bridge because the source is always connected to the potential V_{Z-} of the DC intermediate circuit. A completely different situation is given at the upper FET of the half-bridge.

Depending on the switching condition (whether the lower or upper FET is conducting), the potential at the source connection of the upper FET can vary between the potentials V_{Z+} and V_{Z-}. Independent of this switching condition there must be a certain voltage between gate and source for the control of the switching states of the upper FET.

For small supply voltages up to 48 V, this voltage can be realized by a voltage division inside the transistor emitter circuit (Sects. 6.2.2.2 and 6.2.2.3). If the intermediate-circuit voltage is much higher, an indirect coupling and/or a voltage adaptation is necessary (bootstrap circuit) (Sects. 6.2.1.1 and 6.2.1.2).

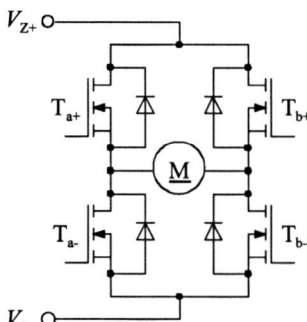

Fig. 6.1. Potential variation at the emitter of the upper transistors of half-bridges

6.2.1.1 Control Signal Transmission by Inductive Pulse Transformers

Figure 6.2 shows the circuit of the inductive pulse transformation for one half-bridge. IGBTs (Insulated Gate Bipolar Transistors) are used as power

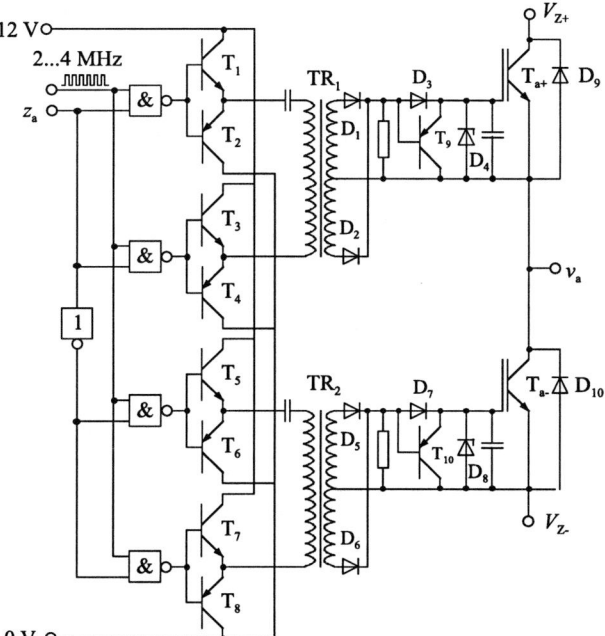

Fig. 6.2. Control signal transmission by inductive pulse transformers

switches. Both IGBTs are controlled by one common control signal. With a slight modification it is possible to control both transistors T_{a+} and T_{a-} by separated control signals z_{a+} and z_{a-}. In this case, it should be taken into account that one transistor is switched off before the other one can be switched on in order to avoid a short circuit in a half-bridge.

6.2.1.2 Bootstrap Circuit

Using bootstrap circuits it is possible to increase the voltage between gate and source to switch the upper FET on by connecting a charged capacitor in series to other voltage sources *via* an electronic switch. By this, the reference voltage for controlling the upper power switch of a half-bridge can be adjusted to the switching state of the lower power switch (Fig. 6.3).

In this circuit, the power transistors T_{a+} and T_{a-} can only be switched by the common control signal z_a. In this case, it must be taken care that T_{a+} and T_{a-} always have different switching states.

When the switching signal z_a is low, the transistor T_1 is conductive. T_1 switches the field-effect transistor (FET) T_3 and the IGBT T_{a-}. As a result, the output of the half-bridge is connected to the potential V_{Z-}. The base potential of the transistors T_5 and T_6 is low due to the low resistance of the transistor T_3. Therefore, T_5 is non-conductive and T_6 is conductive. Since T_6 is conductive, the voltage between gate and emitter of the IGBT is low and

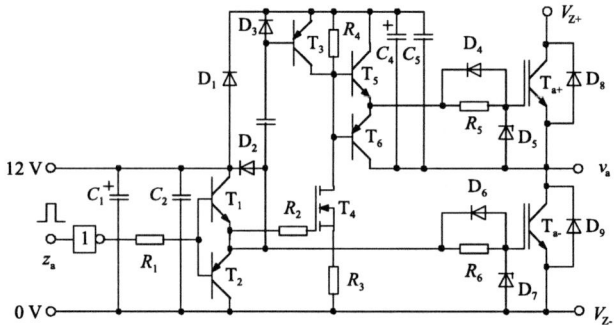

Fig. 6.3. Schematic circuitry of a bootstrap circuit

as a result, T_{a+} is switched off safely. The transistor T_4 is switched off, too. During this process, current flows over the diode D_1 and loads the capacitors C_4 and C_5 with a voltage of nearly 12 V. The energy, stored in the capacitors, is needed to supply the driver circuit of the IGBT T_{a+}. When the control input z_a changes over from L-signal to H-signal, transistor T_1 will be switched off and transistor T_2 becomes conductive. By this, the gate potential of the IGBT T_{a-} is shifted to the emitter potential and T_{a-} gets non-conductive. Now the voltage v_a increases and thus the collector of T_5 reaches a potential about 12 V higher than the potential at the emitter of T_{a+}. Simultaneously, the field-effect transistor T_3 is switched to a non-conductive state. Therefore, the base potential of the transistors T_5 and T_6 is switched to high potential *via* the resistor R_4. Consequently, transistor T_5 becomes conductive and transistor T_6 becomes non-conductive. By this, the potential at the gate of T_{a+} is increased up to the level of the value charged on the capacitors C_4 and C_5. This potential is 12 V higher than the potential at the emitter of T_{a+}. As an effect of the potential difference between gate and emitter T_{a+} is switched conductive. When z_a changes over from L-signal to H-signal, there is a short current pulse through the capacitor C_3 and for the fast switching conductive transistor T_5 a short current pulse flows over the transistor T_4. This causes a fast opening of the IGBT T_{a+}, too.

Only a limited amount of energy can be stored in the capacitors C_4 and C_5. Therefore, the turn-on time of the IGBT T_{a+} must be kept so short that the capacitors are not discharged to a voltage where T_{a+} cannot be safely kept conductive. Therefore, this bootstrap circuit can only be used in those cases where the frequency for switching the half-bridge is sufficiently high for achieving the condition described above.

The drive power of the IGBTs $(T_{a+} \ldots T_{a-})$ is not completely independent of the collector-emitter current, which needs to be switched. Therefore, the bootstrap circuit can only be used to drive the power switches for motors with small power. This fact and the required high switching frequency reduce the usability of the bootstrap circuit. If the user decides to apply these solutions, though, such circuits are offered by many manufacturers on the market.

6.2.1.3 Opto-Couplers

Opto-couplers offer an additional possibility to transfer control impulses from the control circuit to the power section with floating emitter potential (Fig. 6.4).

For their usage it is necessary to transfer the energy required by the driver circuit of the power switch decoupled from the voltages. The simplest solution in this case is to use DC/DC-converters. The operation of transmitting the control signals shall be explained by describing the IGBT T_{a+}. When z_{a+} has L-signal, current flows through the LED of opto-coupler OC_1. As a result, the phototransistor of this opto-coupler is switched to a conductive state and the gate potential of T_{a+} is switched to the emitter potential and T_{a+} becomes non-conductive. With H-signal on z_{a+}, no current flows through the LED of OC_1, and the photo transistor is switched to a high-resistive state. This leads to a voltage between the gate and emitter, that is determined by the z-diode in the driver circuit. By this voltage, the IGBT T_{a+} is switched to a conductive state.

No indirect coupling is required for the IGBT T_{a-} because the emitter always has V_{z-} as reference potential. When there is still an indirect coupling, it is used to avoid a transfer of wire-coupled disturbances from the power section to the control unit. All opto-couplers used in the same circuit should have switching times in a narrow tolerance range to avoid faulty switches in H-bridges and three-phase bridges.

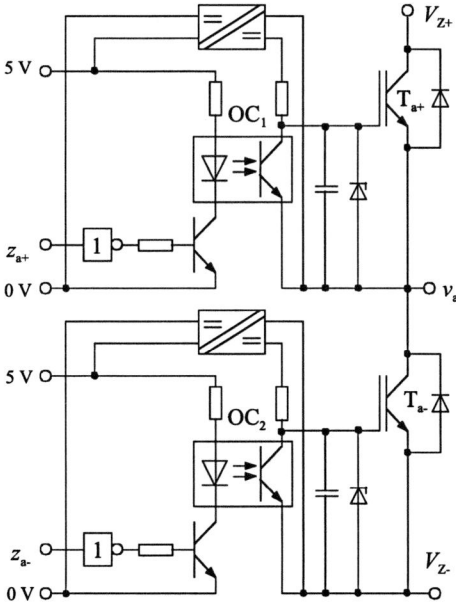

Fig. 6.4. Control signal transmission by opto-couplers

6.2.2 Control Elements for DC and Rotating-Field Motors

Power switches in power electronics can be thyristors, GTO-thyristors, triacs, bipolar transistors, field-effect transistors (FETs) and insulated gate bipolar transistors (IGBTs). In recent applications, FETs and IGBTs are preferred due to their very small gate input power. Drives for higher power are generally run with alternating currents or three-phase currents. IGBTs are used in almost all of these cases due to their very small gate power and a small transition resistance between collector and emitter. By using FETs and IGBTs, switching frequencies beyond the range of audibility can be reached. The switching frequency achieved by thyristors, GTO-thyristors or triacs is only around 500 Hz. An area that is still dominated by thyristors and triacs is the speed control of mains-operated universal motors. Here the voltage control is realizable in a simple way and for a reasonable price.

6.2.2.1 Simple Transistor Circuits for DC Motors

For DC motors with very small power, transistor power amplifiers are often used in linear operation mode. These circuits cause high losses due to the high voltage drop V_{CE} when the motor current and the torque, respectively, are high. Therefore, these circuits with linear operating power switches should only be used when the motor current or the turn-on time are very low. Figure 6.5 shows such a circuit for the one-quadrant operation and in Fig. 6.6 the corresponding circuit for the two-quadrant operation is presented. The circuit for the two-quadrant operation needs a bipolar power supply and, for the complementary stage, a NPN- as well as a PNP-transistor are necessary.

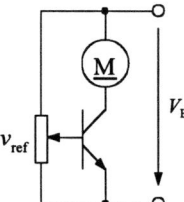

Fig. 6.5. Schematic circuit for a DC motor with low power for one rotational direction (linear operation mode)

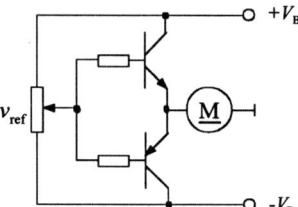

Fig. 6.6. Schematic circuit for a DC motor with low power for both rotational directions (linear operation mode)

6.2.2.2 Full Bridge Circuits for Different Powers

Full bridges are often used as power circuits for DC motors. They allow four-quadrant operation. The fundamental circuit design of a full bridge is shown in Fig. 6.7. When the transistors T_{a+} and T_{b-} are switched on, the current flows through the motor from the left to the right, and when the transistors T_{a-} and T_{b+} are conductive, the current flows in the opposite direction. By changing the pulse width of the control signal it is possible to change the average supply voltage of the DC motor. Because the rotational speed is proportional to the supply voltage for this type of motor, it is possible to adjust the rotational speed *via* the pulse-width modulation.

If the turn-on times are equal for the transistor pairs T_{a+} and T_{b-} as well as for T_{a-} and T_{b+}, the average motor voltage is zero. In connection with a closed-loop control, a holding torque can also be generated.

The diodes connected antiparallel to the power switches serve as free-wheeling diodes. They take over the current flowing through the motor when the power switches are turned off. Since the motor current can flow on through the free-wheeling diodes, voltage peaks can be avoided after switching off the power semiconductors. For a reliable operation, the free-wheeling diodes employed must have a very short reverse recovery time.

The measuring resistor of Fig. 6.7 serves to measure and control the motor current.

In the following, three full bridge applications are presented with different designs of the driver circuit controlling the power switches. All shown circuits are used in practice.

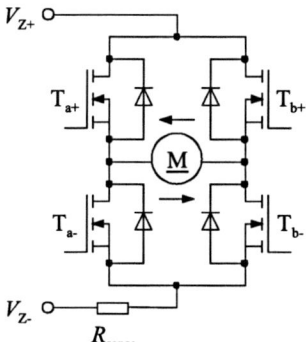

Fig. 6.7. Schematic design of a full bridge circuit

Full Bridge for Low Voltages

For low-voltage motors, it is very simple to design a full bridge for four-quadrant operation. The reason is that the floating reference potential at the upper transistors of each half-bridge, which nearly changes between V_{Z-} and V_{Z+}, can be divided in the voltage drop at the resistor and in the voltage drop

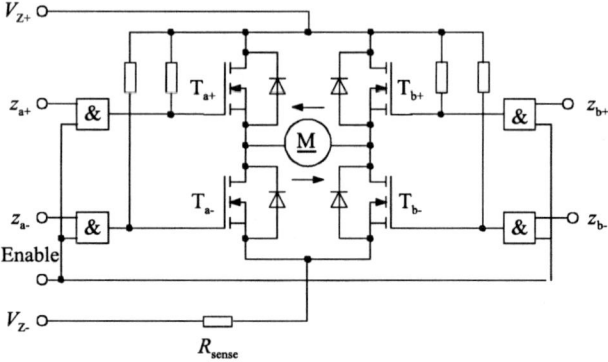

Fig. 6.8. Schematic design of a full bridge circuit with separate control signals for each FET

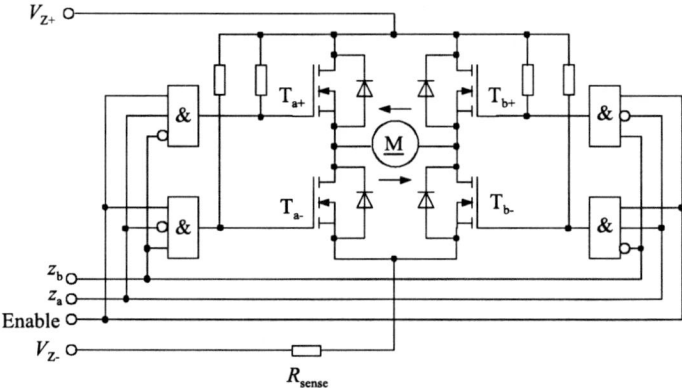

Fig. 6.9. Schematic design of a full bridge circuit with a logical linkage of the gate control signals for each half-bridge

at the switch. When the switch is conductive, the voltage drop is at the resistor, and when the switch is non-conductive, the voltage drop is at the switch. In the simplest setup, it is possible to use logic circuits with open collector output.

Figure 6.8 shows a circuit with individual switching of all power transistors. In this case, interlocking of the control signals is required to avoid a short circuit in the half-bridges. Figure 6.9 shows an example of a circuit with a logical linkage of switching signals. Here, the motor voltage is switched with signal z_a through the transistors T_{a+} and T_{b-} and with signal z_b via the transistors T_{b+} and T_{a-}. In both circuits, the full bridge can be enabled or disabled with the enable signal.

Full Bridge for IGBTs and FETs Connected to AC Mains

Quite often, DC motors with power of more than 100 W are connected to AC mains. Driver circuits like those used for full bridges with low voltage sup-

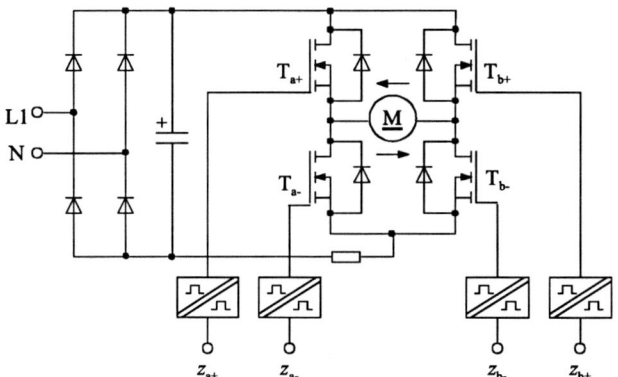

Fig. 6.10. Full bridge circuit for the connection to the AC mains

ply are not usable here because small-signal transistors are not available for high reverse voltages at a reasonable price. In addition, such high collector–emitter voltages would cause very high losses. Consequently, special methods must be applied to transmit the switching signals from the control circuit to the power circuit. These methods include indirect-coupled signal transmission or adapting the signals to the floating reference potential of the power switches. The indirect-coupled transmission can be realized by opto-couplers or inductive transmitters. These possibilities are described in Sect. 6.2.1 for indirect-coupled transmission of the switching signals.

The rectifier and the smoothing capacitor for generating the DC intermediate circuit voltage are also shown in Fig. 6.10.

Full Bridge for Higher Voltages with Thyristors

A full bridge with thyristors is shown in Fig. 6.11 [6.1]. Assuming that the thyristors Th_1 and Th_4 are conductive, the current is flowing from top to bottom through the motor. The commutation capacitors are charged as it is shown at the left side of the capacitors in the circuit diagram. While igniting the thyristors Th_2 and Th_3, the commutation capacitors are connected parallel to the thyristors Th_1 and Th_4. As a result, the cathode potential in these thyristors is higher than the anode potential and the respective thyristors are blocked.

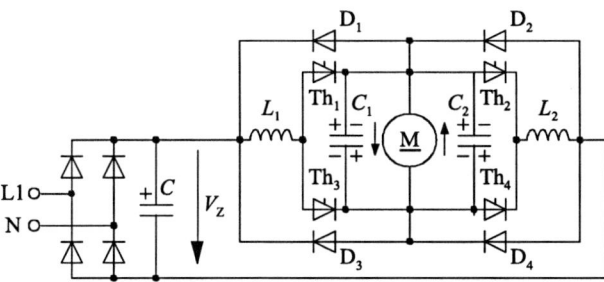

Fig. 6.11. Schematic circuit of a full bridge with thyristors

The commutation of the current from the thyristors Th_1 and Th_4 to the thyristors Th_2 and Th_3 leads to a change of the motor current direction (from bottom to top). The polarity of the capacitor C_1 is reversed through the diode D_1, the thyristor Th_3, and the inductance L_1. Now C_1 has the potential shown on the right side of the capacitor. The polarity of the capacitor C_2 is reversed, too, but by the current through the diode D_4, the thyristor Th_2 and the inductance L_2.

When the thyristors Th_1 and Th_4 are ignited again, the commutation capacitors are switched parallel to the thyristors Th_2 and Th_3. As a result, the cathode potential is higher than the anode potential at these thyristors and consequently they change to the off-state. The polarity of the commutation capacitors changes as shown on the left side of the capacitors. The change of the polarity at C_1 is caused by the current through the diode D_3, the inductance L_1, and the thyristor Th_1. The change of the polarity of C_2 is effected by the current through the diode D_2, the inductance L_2 and the thyristor Th_4. With this commutation the initial state is reached and one operating sequence is ended.

As shown by the operational description, it is possible to change the current direction through the motor by switching the two thyristor pairs Th_1 and Th_4 or Th_2 and Th_3, respectively. By variation of the ratio of the turn-on time of the two thyristor pairs, the rotational speed of the motor can be controlled similar to a transistor full bridge. Using thyristor converters, switching frequencies are lower than in full bridges with transistors. Furthermore, it must be considered that the turn-on time of each thyristor pair must be long enough to change the charge of the commutation capacitors.

6.2.2.3 Three-Phase Bridges with FETs and IGBTs

Three-phase bridges are used to generate three-phase currents from a DC link by using a suitable open-loop control. The six power switches of the three-phase bridge are switched individually. The circuit of such a bridge using FETs is shown in Fig. 6.12. Motors with small power normally have a star connection

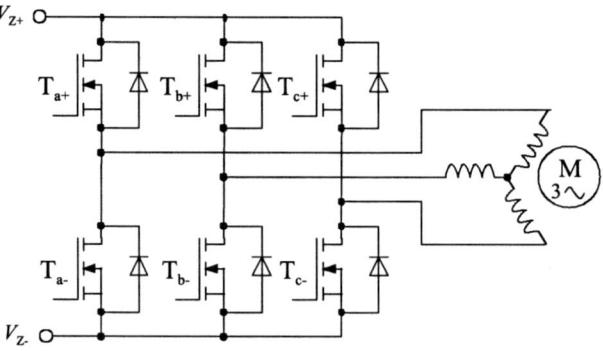

Fig. 6.12. Three-phase bridge

with non-earthed star point. The motor is represented by its windings to show the usual connection to the frequency converter.

Three-Phase Bridges for Low Voltages

Low-power DC motors are replaced more and more by three-phase motors. This is especially the case in automotive applications when cars have an on-board network of 48 V. The essential reason for this is the simple construction and high robustness of the motor as well as the reasonable price of the control and power electronics with highly integrated components. Figure 6.13 shows the simplified schematic of a three-phase bridge for low powers.

As discussed for the full bridge, the reference potential of the upper transistor can vary between the potentials V_{Z+} and V_{Z-}. A collector emitter resistance (or the open collector circuit of an IC) can take over this voltage variation by changing the voltage drop from the collector emitter resistance to the collector resistance and in reverse order, depending on the present reference voltage of the upper transistor.

In the given circuit, the control signals of each half-bridge are logically linked. Hence, in a half-bridge only one transistor can be switched on. Therefore, only the control signals z_a, z_b, and z_c are needed. Within the circuit it must be ensured that the conductive transistor of a half-bridge is turned off before the non-conductive transistor is turned on. It is of course possible to have a separated control signal for each power transistor of the three-phase bridge. With a common signal all power transistors can be enabled.

To design circuits for frequency converters, it is not necessary to use discrete components in many cases because there are integrated power modules available on the market that contain the complete circuit of a frequency converter.

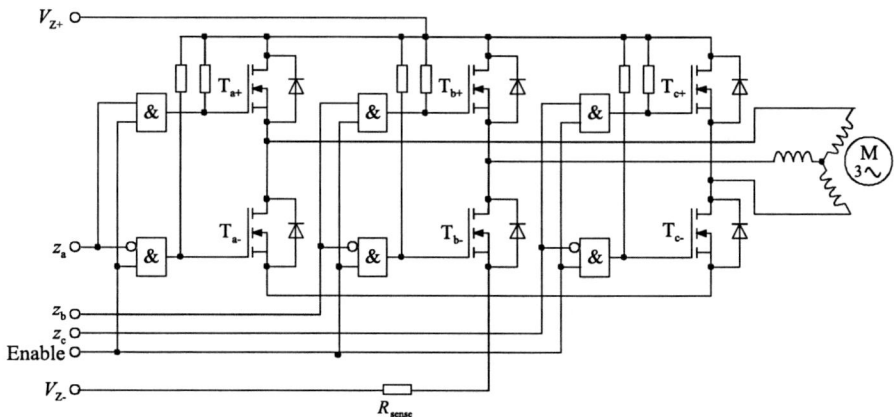

Fig. 6.13. Three-phase bridge for low voltages

Three-Phase Bridges for Connection to AC Supply with FETs and IGBTs

Three-phase motors in the mid-power range (higher than 100 W) are often connected to AC mains. In this case, rectifier voltages of the amplitude of the connected AC main occur and the reference potential for the upper power switch of each bridge branch can vary between 0 V and the potential of the amplitude of the connected AC main.

For the same reasons that have been discussed for the H-bridges in connection to AC mains, the control circuit must be indirectly coupled to the upper power switch or other circuits are necessary to adapt the control signal to the floating reference potential of the upper transistors (three-phase driver circuit). Such circuits are described in Sect. 6.2.1.

Three-phase bridges with six power switches (IGBTs or FETs) integrated in one case are offered by many manufactures. Today, more and more IPMs (intelligent power modules) are offered on the market. In addition to the six power switches and the rectifier, these components contain many protective functions (short-circuit control in each half-bridge, voltage monitoring, temperature checking etc., see Sect. 6.6).

6.2.3 Control Elements for Stepper Motors

The operation of motors for discontinuous motion like stepper motors is described in depth in Sect. 3.3. By setting a determined number of steps and a determined step frequency, these motors can be set with a pre-selected rotational speed at a pre-selected rotor position. Basically, two methods of operation are possible for these motors: unipolar and bipolar operation. In bipolar operation, the current of each phase winding can be reversed. This is not possible in unipolar operation. For each of the two operation methods, stepper motor drivers must be designed differently. The torque reached in bipolar operation is approximately 30% higher than in unipolar operation. To avoid voltage spikes in stepper motor driver circuits, free-wheeling diodes are used.

6.2.3.1 Unipolar Operation

For unipolar operation, stepper motors must have a mid-point tap in their windings. In contrast to bipolar operation, only half the number of power switches is necessary here. Figure 6.14 shows the stepper motor driver for unipolar operation with connected phase windings. The Zener diode ensures a fast reduction of the current in the windings.

6.2.3.2 Bipolar Operation

A mid-point tap on the phase windings is not needed when bipolar operation is applied. However, in this case, each phase needs a full bridge. Figure 6.15 shows the circuit of a stepper motor driver for a two-phase stepper motor with connected phase windings.

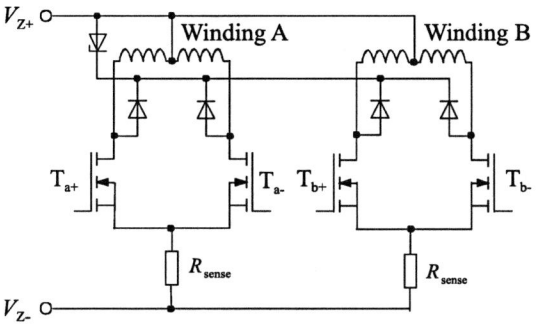

Fig. 6.14. Schematic diagram for the unipolar operation of a two-phase stepper motor

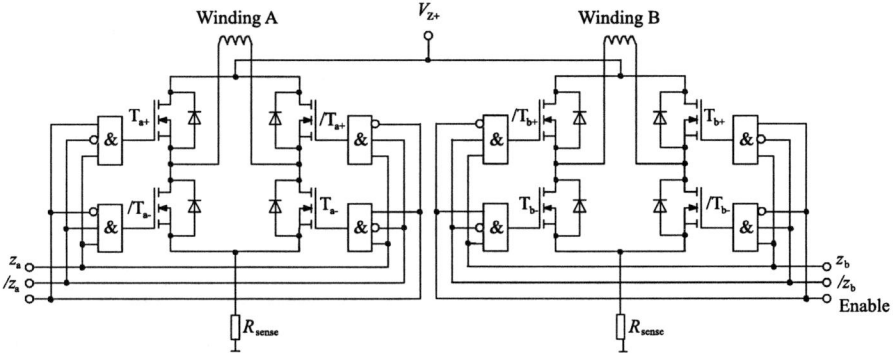

Fig. 6.15. Control element for the bipolar operation of a two-phase stepper motor

6.2.4 Modulation Methods for Three-Phase Motors

Modulation methods serve to generate a mean time value of a certain voltage from a constant DC voltage by variation of the duty ratio when the cycle time is constant.

6.2.4.1 Block Commutation

Using block commutation, the motor windings are connected to block-shaped phase voltages instead of sinusoidal voltages. Figure 6.16 shows the variation in voltage depending on the time for the block commutation. In this diagram, the relation between block voltage and sinusoidal voltage shape is given.

Such block voltage shapes (without pulse-width modulation) were used when the inverter technique was first introduced. With these forms it was possible to limit the switching frequency of the individual thyristors to 50 Hz. The disadvantage of the block voltage shape is the fact that the voltage signals have a high ripple content of unequal multiples of the fundamental term. This can cause disturbances in connection with motors designed for sinusoidal voltage supply. Lately, there are applications of block commutation for rotating-field machines connected to low voltages. For this purpose, the market provides appropriate control circuits to reduce the circuit complexity to

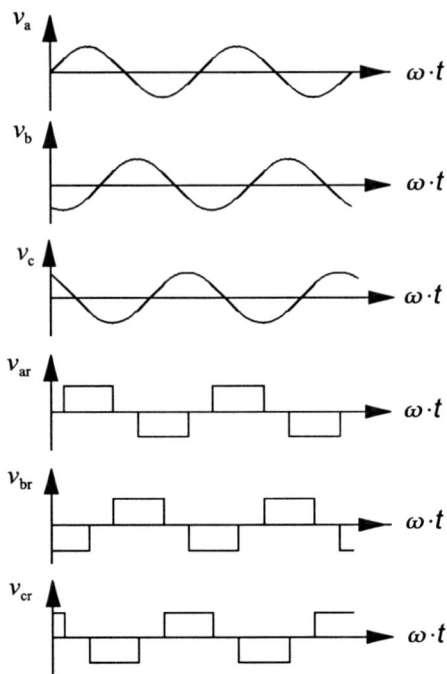

Fig. 6.16. Phase voltages for block commutation

a minimum. In most cases, the power switches consist of FETs. In such applications, the commutation is controlled by sensors which detect the position of the rotor. The open-loop or closed-loop control of rotational speed is achieved by changing the level of the DC link voltage. Circuits for such applications are shown in Sects. 6.3.3 and 6.3.4.

6.2.4.2 Carrier-Based Sinusoidal Modulation (CBSM)

Using the carrier-based sinusoidal modulation method, a stream of pulses is generated by turn-on times of different duration during a constant sinusoidal period where the time integral is equal to the time integral of the sinusoidal function.

This method is derived from the analogue method for creating a pulse-width modulation (PWM) with sinus evaluation. The method compares the sinusoidal reference voltage of a phase of the frequency converter with a triangle voltage. The output signal of the comparator is a rectangular voltage with pulses of different length (Fig. 6.17). The frequency of the triangle voltage should be higher than the frequency of the audible limit for humans.

It must be guaranteed by hardware or software interlocking that in the individual half-bridges only one power switch is conductive. The starting pulse for the non-conductive power switch must not be given until the conductive power switch is definitely turned off after the turn-off pulse has been applied.

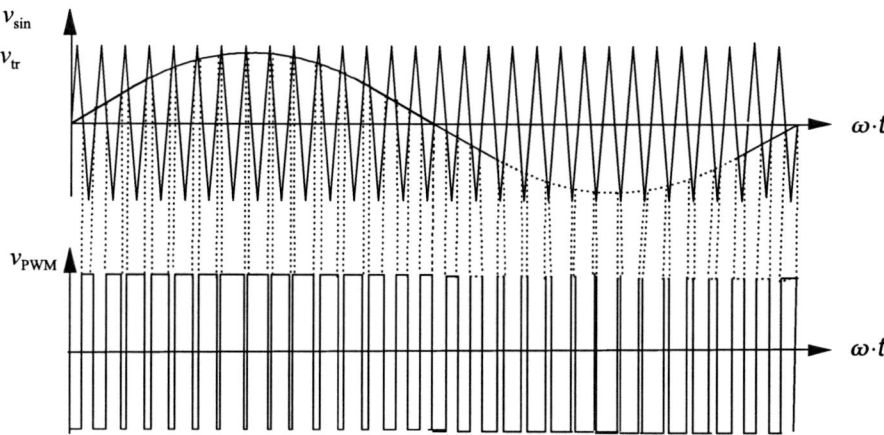

Fig. 6.17. Voltage profile of the carrier-based sinusoidal modulation method

6.2.4.3 Space Vector Modulation

Space vector modulation is another method for generating a pulse-width modulation with sinus evaluation. It assumes that six inverter states are realizable in a three-phase bridge if, at the same switching interval, three of the power switches are turned-off and the remaining three power switches are turned on. Figure 6.18 shows these six inverter states. Two further inverter states can be realized if either all upper power switches in each half-bridge are conductive and the lower power switches are non-conductive or the upper are turned off and all the lower are turned on.

From the three-phase description, these inverter states can be transformed into the two-axis α, β-reference frame by using the relation

$$\begin{vmatrix} v_\alpha \\ v_\beta \end{vmatrix} = \begin{vmatrix} 1 & 0 \\ \frac{1}{\sqrt{3}} & \frac{2}{\sqrt{3}} \end{vmatrix} \cdot \begin{vmatrix} v_a \\ v_b \end{vmatrix}. \tag{6.1}$$

The α, β-reference frame is fixed and coupled to the stator.

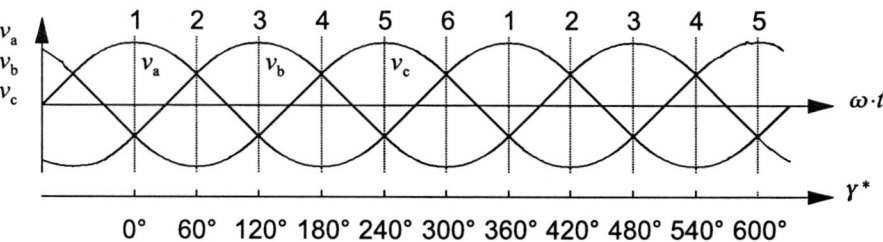

Fig. 6.18. Switching states for the space vector modulation

Additionally, one obtains the nodal equation for the voltage with

$$v_a + v_b + v_c = 0 \ . \tag{6.2}$$

The presentation of the inverter states shown in Fig. 6.18 is given in Fig. 6.19 for the α, β-reference frame. For voltages v_α and v_β given for a specific working point, the voltage space vector $\underline{v}^*$ and the corresponding angle γ^* can be calculated by using the equations

$$|\underline{v}^*| = \sqrt{v_\alpha^2 + v_\beta^2} \tag{6.3}$$

and

$$\gamma^* = \arctan \frac{v_\beta}{v_\alpha} \ . \tag{6.4}$$

In contrast, for a given $\underline{v}^*$ and γ^* it is possible to calculate the voltage components v_α and v_β. Unfortunately, not only the inverter states mentioned above are necessary for the operation of a three-phase motor but also all possible interim values. These are realized by using a time-controlled switching of the adjoining inverter states (Fig. 6.19). In this figure, the same six inverter states are shown, which have already been illustrated in Fig. 6.18. In addition, the inverter states seven and eight are also shown. The inverter state seven means that the upper power switches are turned-on and the inverter state eight implies that the lower power switches are turned-on.

Furthermore, the space-fixed axis of the α, β-reference frame and the directions of the three-phase a,b,c-system are shown in this picture. The turn-on times of the inverter states must be calculated for each sector depending on the length of the voltage vector $\underline{v}^*$ and on the angle γ^*.

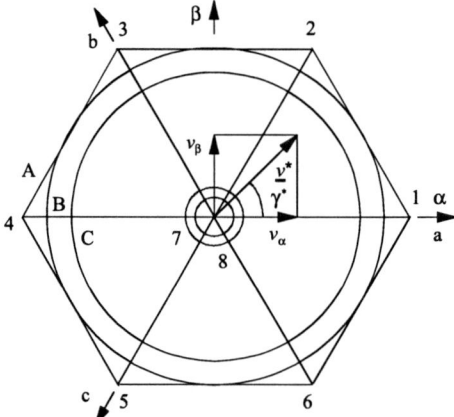

Fig. 6.19. Switching states for the space vector modulation in the stator reference frame with the coordinates α, β

The amplitude of the output voltage is limited by the hexagon. At states one to six it has the value $2/3 \cdot V_Z$. The circle B limits the length of the voltage vector to the value

$$|\underline{v}^*| \leq \frac{V_Z}{\sqrt{3}} \tag{6.5}$$

and the circle C limits the length to the value

$$|\underline{v}^*| \leq \frac{V_Z}{2} . \tag{6.6}$$

The inverter states one to eight listed in Table 6.1 are realizable with the given states of the power switches, where + stands for the turn on state of the upper and the − for the turn on state of the lower power switches.

If a reference voltage space vector stands between two fixed inverter states, then this space vector must be realized by successively turning on the adjoining inverter states for a calculated time. The position of the voltage space vector shown in Fig. 6.19 in segment one between the inverter states one and two is calculated as follows: if the cycle time is T_T and the inverter output voltage shall be limited by the circle B, then the turning-on times of the adjoining inverter states can be determined by the equations

$$t_1 = T_T \cdot \frac{|\underline{v}^*|}{V_Z} \cdot \sqrt{3} \cdot \sin(60° - \gamma^*) \tag{6.7}$$

and

$$t_2 = T_T \cdot \frac{|\underline{v}^*|}{V_Z} \cdot \sqrt{3} \cdot \sin(\gamma^*) . \tag{6.8}$$

If the voltage vector is located inside segment two between the inverter states two and three, then the turning-on times of the adjoining inverter states are calculated using the relations

$$t_2 = T_T \cdot \frac{|\underline{v}^\bullet|}{V_Z} \cdot \sqrt{3} \cdot \sin(120° - \gamma^*) \tag{6.9}$$

Table 6.1. Switching states using space vector modulation

Nr.	a	b	c
1	+	−	−
2	+	+	−
3	−	+	−
4	−	+	+
5	−	−	+
6	+	−	+
7	+	+	+
8	−	−	−

and

$$t_3 = T_T \cdot \frac{|v^*|}{V_Z} \cdot \sqrt{3} \cdot \sin(\gamma^* - 60°) \,. \tag{6.10}$$

For the other segments, similar equations to calculate the conductive times of the adjoining inverter states can be derived. It should be noted that the sum of these must be lower than or equal to the cycle time T_T. If the sum is lower than T_T, one of the inverter states seven or eight must be switched on for the remaining time.

6.2.5 Control Elements

As shown in the previous sections, in applications of electrical drives, the control variables frequently are generated as analogue voltage signals (operational amplifiers) or as numerical values (micro-controllers). Usually, these values are converted to pulse width-modulation (PWM) signals that serve as inputs for the bridges of the power electronics.

6.2.5.1 Pulse-Width Modulation for Full Bridges

Analogue Circuit for Generating PWM Signals

The schematic diagram (Fig. 6.20) for the generation of PWM signals consists of a triangle wave generator and a comparator. In the comparator, the reference voltage is compared with the triangular voltage. Depending on the level of the reference voltage, pulses with different widths are produced. Here, the cases $v_{tr} > v_{ref}$ and $v_{tr} < v_{ref}$ must be distinguished. These pulse widths appear in the output signal of the comparator as positive or negative pulses. A logic circuit generates the control signal z_a from the positive pulses and the signal z_b from the negative pulses. To avoid short circuits, a turn-on delay is necessary between the turn-off signal and the turn-on signal of the power switches of the same half-bridge. Using the OR-function from the z_a and z_b signals, the enable signal is generated. Figure 6.21 shows the signal curves from the circuit in Fig. 6.20. The signal curves show that the frequency of the full bridge is generated on the basis of the frequency of the triangular signal alone. The cycle duration of the triangular signal can be divided into parts of the control signals z_a and z_b, depending on the level of the reference signal.

Today, PWM-signals are mostly generated by microcontrollers [6.2].

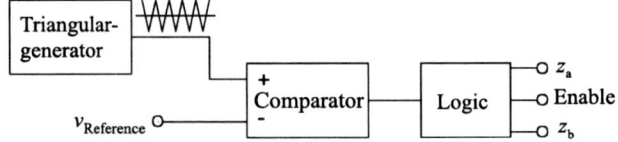

Fig. 6.20. Block diagram for analogue generation of PWM signals

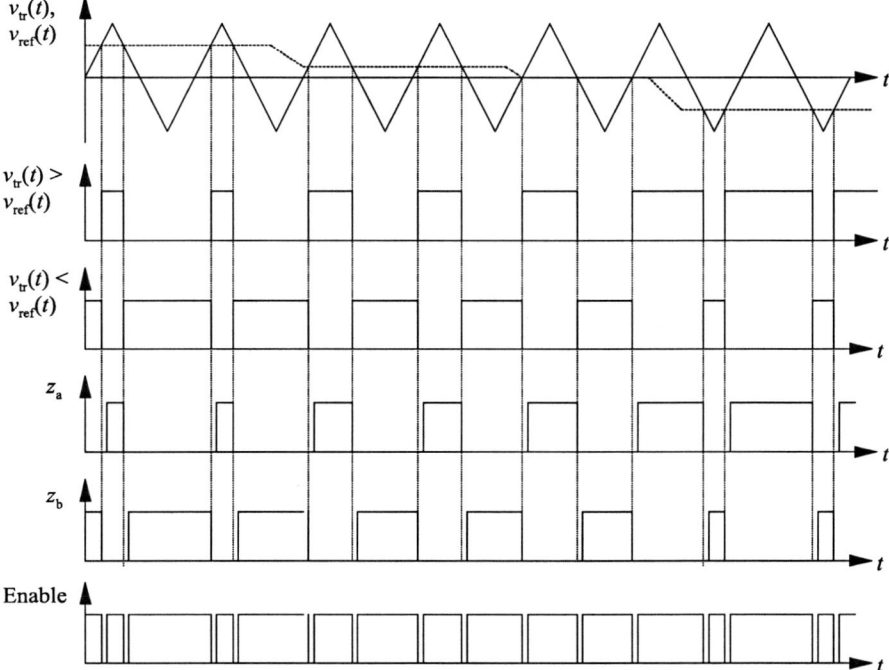

Fig. 6.21. Time behaviour of analogue-generated PWM signals

6.2.5.2 Pulse-Width Modulation for Three-Phase Bridges

Carrier-Based Sinusoidal Modulation (CBSM)

The carrier-based sinusoidal modulation method is the technique mostly used in the past in order to generate pulse-width modulation with sinus evaluation. It is realizable with analogue techniques and by using various digital circuits (e.g. microcontrollers). Not all of these various methods can be discussed here. Therefore, only the analogue technique and one procedure using a microcontroller shall be described.

Analogue Carrier-Based Sinusoidal Modulation

The realization of the analogue carrier-based sinusoidal modulation method for three-phase bridges is described only for one phase. For the other phases, the same design can be used considering the phase shifts of $-120°$ and $-240°$, respectively.

For the realization, a sinusoidal waveform generator with adjustable frequency (e.g. 3 to 100 Hz) is required. In addition, a triangular waveform generator, a comparator and a logic circuit are necessary. In the comparator, the sinusoidal voltage $v_{\sin}$ is compared to the triangular voltage v_{tr}. The elementary structure of the circuit is shown in Fig. 6.22.

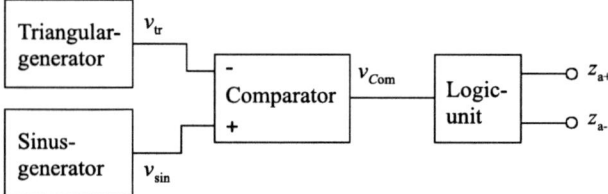

Fig. 6.22. Block diagram of a circuit for analogue realization of the carrier-based sinusoidal modulation method

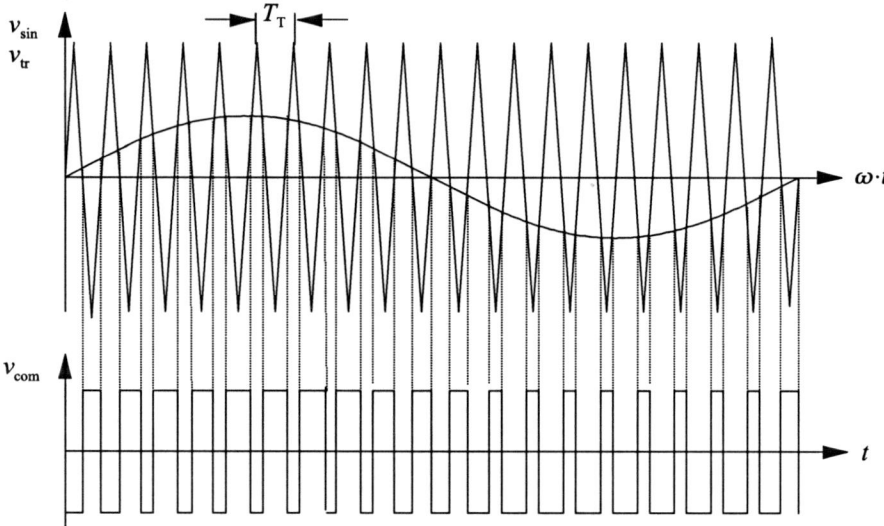

Fig. 6.23. PWM-signal generation by means of the carrier-based sinusoidal modulation method

The time curves of the sinusoidal voltage, triangular voltage and comparator output voltage are shown in Fig. 6.23. The comparator output voltage depends on the sinusoidal reference voltage and on the triangular voltage.

By using the carrier-based sinusoidal modulation method for variable speed adjustment of three-phase motors, only the frequency and the amplitude of the sinusoidal reference voltage must be changed.

Discrete Pulse-Width Modulation

The discrete pulse-width modulation based on the carrier-based sinusoidal modulation method is realized using timers in microcontrollers (microcontroller supported PWM for three-phase bridges). For the computation of switch-on times for the individual switches of a three-phase bridge, the cycle time T_T, the maximum rotating-field frequency f_{max}, and the voltage amplitude must be given as constant values. As values depending on the operating condition, the momentary frequency f and the actual angle γ_1 of the rotating field are given. As usual for three-phase motors, the frequency f is nearly proportional to the effective value of the voltage (see Sect. 6.3.3.2). Based on these values, the switch-on times for the switches T_{a+} and T_{a-} for phase a can be computed by using the following relations

$$T_{Eza+} = \frac{T_T}{2}\left(1 + \frac{f}{f_{max}}\sin(\gamma_1)\right) \tag{6.11}$$

$$T_{Eza-} = \frac{T_T}{2}\left(1 - \frac{f}{f_{max}}\sin(\gamma_1)\right). \tag{6.12}$$

For the two other phases, switch-on times T_{Ezb+}, T_{Ezb-}, T_{Ezc+} and T_{Ezc-} can likewise be computed with Eqs. (6.11) and (6.12) when the arguments of the sine function $(\gamma_x - 120°)$ or $(\gamma_1 - 240°)$, respectively, are used.

6.3 Open-Loop Control of Rotational Speed

6.3.1 Control Elements and Power Sections for Universal Motors

Universal motors are applied predominantly in household appliances, do-it-yourself tools, etc. All devices in this market segment have to be offered to the customer at extremely low prices. This is the reason why inverters and control equipment in these devices must be economical and therefore simple (see also Sect. 2.1.3.5).

6.3.1.1 Half-Wave Phase-Angle Control

Half-wave phase-angle control is the simplest circuit for open-loop rotational speed control of universal motors. Figure 6.24 shows the circuit design in principle.

In this circuit, C_1 and L_1 are components for radio interference suppression, which should always be considered in the circuit design. The loading of the phase-shifter capacitor C_2 is made by R_1 and the adjustable resistor R_2. Triggering of the thyristor Th_1 is achieved with the help of the Diac D_1, the

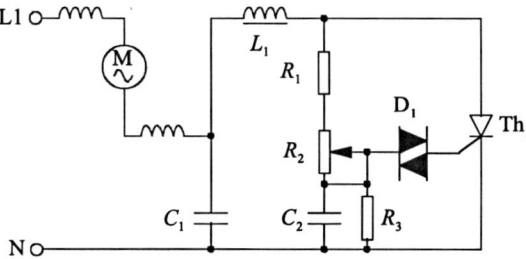

Fig. 6.24. Half-wave phase-angle control

adjustment of the gate trigger time impulse is affected with the adjustable resistor R_2. When the voltage at the capacitor C_2 reaches the ignition voltage of the Diac D_1, then this ignites and supplies the gate current for triggering the thyristor Th_1. As soon as triggering has taken place, the universal motor is connected to the supply voltage during the remaining half-wave. The smaller R_2 becomes, the earlier the thyristor triggers, which corresponds to an increase of motor power.

The circuit of the half-wave phase-angle control can be developed with little effort and is particularly suitable for the motor speed control of small fractional-horsepower drives, as they are demanded, e.g. in hand-drill machines, portable circular saws and in household appliances.

In the same way, full-wave phase-angle controls can be achieved when the thyristor is replaced by a triac.

6.3.1.2 Circuit for Generating Wavy Direct Current from Alternating Current

Using full-wave phase-angle control and connecting a rectifier with filter capacitor at its output, an adjustable DC voltage current without gaps can be generated. Such circuits can be used to operate universal motors or DC motors.

6.3.1.3 Circuit for Rotational Speed Control with Phase-Angle Control IC

The control circuit is implemented in an IC. The operational principle is represented in the block diagram of Fig. 6.25.

The AC mains voltage is connected to the circuit, which serves as synchronizing voltage. Additionally, the supply voltage for the IC is generated from it. By help of the synchronizing voltage, any zero-crossings are detected. Due to these zero-crossings, a ramp generator produces a saw-tooth voltage synchronous to the mains voltage. In a comparator, this is compared with the control voltage. As a function of the height of the control voltage, the ignition impulses are separately produced from each positive and negative half wave for various ignition lag angles. With these impulses, two anti-parallel switched thyristors can be ignited.

Figure 6.26 shows a circuit for rotational speed control with the phase-angle control IC indicated in Fig. 6.25. The rotational speed can be adjusted by the potentiometer R_1.

6.3.2 Open-Loop Rotational Speed Control of DC Motors

The rotational speed of DC motors can be adjusted by many diverse methods. The circuits necessary for it can differ substantially in their characteristics and in their technical effort.

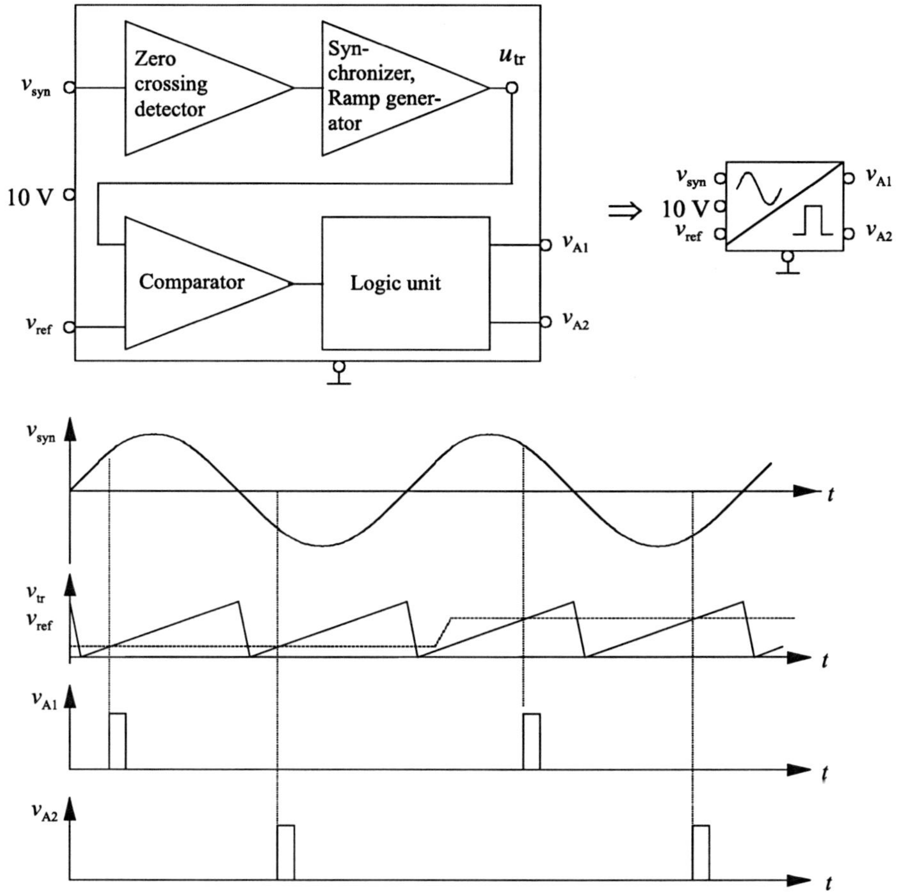

Fig. 6.25. Circuit with phase-angle control IC for the control of the thyristors

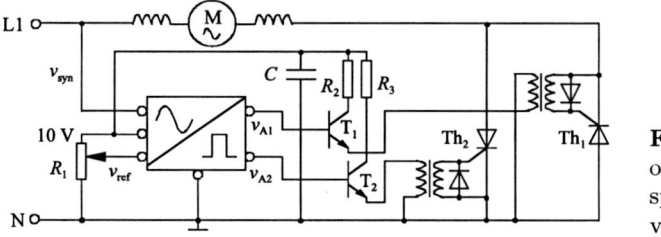

Fig. 6.26. Circuit for open-loop rotational speed control of a universal motor

6.3.2.1 Open-Loop Rotational Speed Control with Simple Transistor Circuits

In these circuits, a transistor connected in series with a DC motor is used, by which the voltage at the motor is adjusted. The desired rotational speed can be adjusted with a voltage divider (Figs. 6.5 and 6.6).

6.3.2.2 Open-Loop Rotational Speed Control in Chopping Operation

In the circuit shown in Fig. 6.27, the transistor is operated only as switch. The control voltage v_{ref}, adjustable at the potentiometer, is converted into pulses with constant frequency and variable pulse duty factor by using the following circuit represented as a functional block in the diagram. The relation between pulse time and period duration is proportional to the control voltage v_{ref}. The free-wheeling diode conducts the current during the turn-off time of the transistor to avoid switching-off peaks in pulse operation. In order to avoid disturbing fluctuations of DC current the period duration of the clock rate should be substantially smaller than the time constant of the motor.

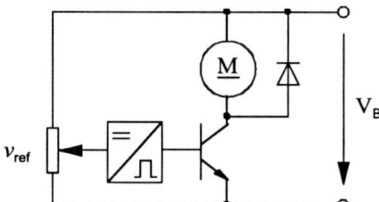

Fig. 6.27. Open-loop rotational speed control in chopping operation

6.3.2.3 Open-Loop Rotational Speed Control with Full Bridges

Figure 6.28 shows the circuit design for the control of rotational speed of DC motors connected to a full bridge. With full bridges, the motor voltage can be switched in both directions. The relation between the switch-on times for the pulses of the two directions of voltage determines the rotational direction and speed of the motor. This relation is given by the control voltage. Besides the DC motor only the power section, the adjustment of the control signal level or the galvanic separation, if necessary, and control equipment is needed. These components are described in detail and individually in Sects. 6.2.1 and 6.2.2. Usually they are combined in one unit (inverter, DC motor driver), as suggested by the dashed line in Fig. 6.28. This unit needs a control signal at the input and as such may serve a control voltage in the range between -10 V and $+10$ V, or for units with digital signal processing in the control section a numerical value between -1024 and $+1023$, for example. At its output, the unit provides a voltage proportional to the input signal.

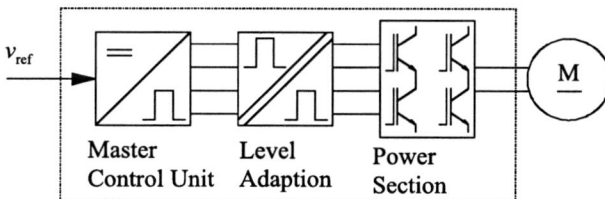

Fig. 6.28. Open-loop rotational speed control of a DC motor with full bridge circuit

6.3.3 Open-Loop Rotational Speed Control of Asynchronous Motors

There are several possibilities to adjust the rotational speed of asynchronous motors. These possibilities differ substantially in their expenditure and in their mode of operation. Only three methods shall be described in the following section. In the cheapest procedure, the supplied power is reduced by phase-angle control and thus the slip of the motor increases. The frequency of the supply voltage is not varied. Frequency change is used by the two other methods (slip compensation and $I \times R$ compensation) by means of frequency converters. These permit a higher control range of rotational speed.

6.3.3.1 Rotational Speed Variation by Changing the Slip

At a constant rotating-field frequency, e.g. at the stiff AC net, the effective value of the input voltage is reduced by phase-angle control. Thus the slip increases with constant load, and the rotor speed is lowered. However, reduction of the input voltage is only possible until the breakdown torque of the motor is reached. This method of speed variation is therefore suitable only when the rotational speed must be varied within small limits. To increase the range of rotational speed variation, the motor must be constructed with high-impedance winding. The disadvantage is that the torque of the motor decreases quadratically with the supply voltage, i.e. the motor torque decreases disproportionately by reducing the supply voltage. A further disadvantage is that the harmonic content of the voltage increases and thus harmonic losses increase (cf. Sect. 2.4.1.6). The design of a circuit for one phase under consideration of these restrictions is shown in Fig. 6.24.

6.3.3.2 Simple Control Circuit with Frequency Converter

The general design of a circuit for open-loop rotational speed control with frequency converters for asynchronous motors is shown in Fig. 6.29. All functional blocks in the dashed border are components of the frequency converter. In the presented case, a control voltage is given which shall be proportional to the rotational speed of the motor. In addition, there is the possibility to specify presets by using the input "Presetting of parameters and options". For

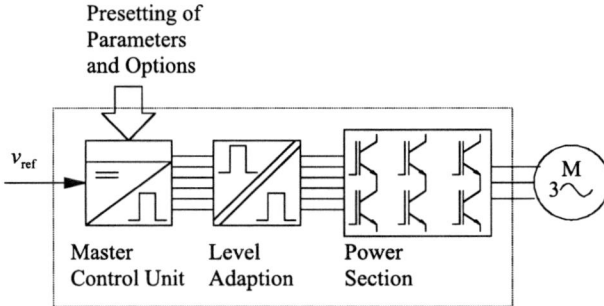

Fig. 6.29. Circuit for the open-loop rotational speed control for asynchronous motors with frequency converters

example, this is the limiting value for the current, frequency ramp of the acceleration and deceleration, and, depending on the chosen control method, the slip frequency or the voltage lifting at low frequencies.

If rated current (for rated torque) in the whole frequency range up to the rated frequency is assumed, then the absolute value of the ohmic voltage drop at the stator winding resistance R_1 is constant. But the relative voltage drop at R_1 increases with respect to the supply voltage v_1 of the motor as a function of decreasing frequency. In order to compensate the voltage drops at the stator winding, the supply voltage is increased in the lower frequency range ($I \times R$ compensation). The increase of voltage is given for the maximum torque by the equation

$$v_0 = R_1 \cdot I_{\text{rated}} \, . \tag{6.13}$$

In the whole range of rotational speed variation the stator voltage is calculated using the relation

$$v_{\text{ref}} = v_0 + \frac{f}{f_{\text{max}}} (v_{\text{max}} - v_0) \, . \tag{6.14}$$

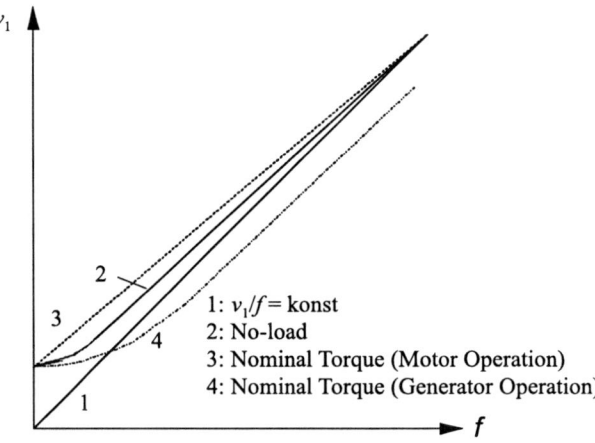

Fig. 6.30. Relationship between frequency and supply voltage for different loads using $I \times R$ compensation

The $I \times R$ compensation results from the static equivalent circuit diagram. As a result, the dynamic behaviour of the motor is only conditionally improved. Figure 6.30 shows the necessary progression of the supply voltage in some load cases as a function of frequency (cf. Sect. 2.4.1.6).

6.3.3.3 Slip-Frequency Compensation

The compensation of the slip frequency is particularly effective in those cases where the motor is always supplied with a constant torque (Fig. 6.31). The compensation method uses the data on the type plate of the motor. From the rated rotational speed, the rated torque and the rated slip can be determined, as well as the associated slip frequency for a certain partial load. The required rotating-field frequency can then be determined from the desired motor rotational speed in consideration of the number of pole pairs and the calculated slip frequency.

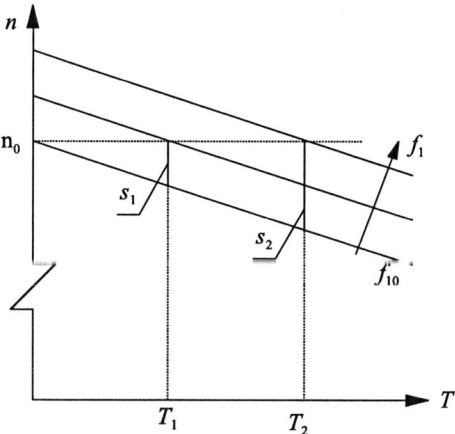

Fig. 6.31. Slip-frequency compensation

6.3.4 Open-Loop Control of Motors with Synchronous Speed

6.3.4.1 Synchronous Motors

Synchronous motors accurately follow the rotary field frequency in their rotational speed. Externally controlled synchronous motors can be operated at an AC net or at a frequency converter.

When a frequency converter is used for open-loop control, then the engine does not need an auxiliary short-circuit winding as it is necessary at the AC net, since the frequency converter can change the frequency along a ramp from zero to the rated frequency. Care must be taken however that the external torques or an additional torque of inertia of the load do not become so large that the motor can no longer follow the frequency ramp.

6.3.4.2 Electronically Commutated Motor

Brushless DC motors (EC-motors) are electronically commutated motors. They are characterized by the fact that the rotating field in the stator may only be switched from the position n to the position $(n \pm 1)$, depending on direction of rotation, when the rotor has reached position n. In order to ensure this forward switching of the rotating field, sensors are necessary that indicate the rotor position. There are three possibilities for detecting the rotor position: firstly, the rotor position can be indicated by a rotor position sensor connected to the motor shaft. Secondly, the motor field sensors can be used, which detect the rotor position in an electric angular dimension. Besides these two methods, "sensorless" techniques can be applied to find the operating values. Shaft-mounted rotor position sensors are mainly used in servo technique, where accurate positioning is always vital. If the demand for positioning accuracy is only low, i.e. for off-the-shelf rotating speed control, motor field sensors might be sufficient. Occasionally, the rotor position is also determined by voltage measurements at the currentless windings.

In the following, the measurement of rotational speed shall be described by the example of a driver with block commutation. In Fig. 6.32 the structure of an open-loop rotational speed control of electronic motors with power section and logic block for the evaluation of the Hall-sensors signals, which are positioned in the stator, is represented. As a function of the signals of these field detectors, the control signals for the power switches in the three-phase bridge are generated by a logic unit. The signal courses of the field sensors, of the control signals for the power switches and the output voltage are indicated in Fig. 6.33. Table 6.2 shows how the control signals for the power switches depend on the output signals of the field sensors.

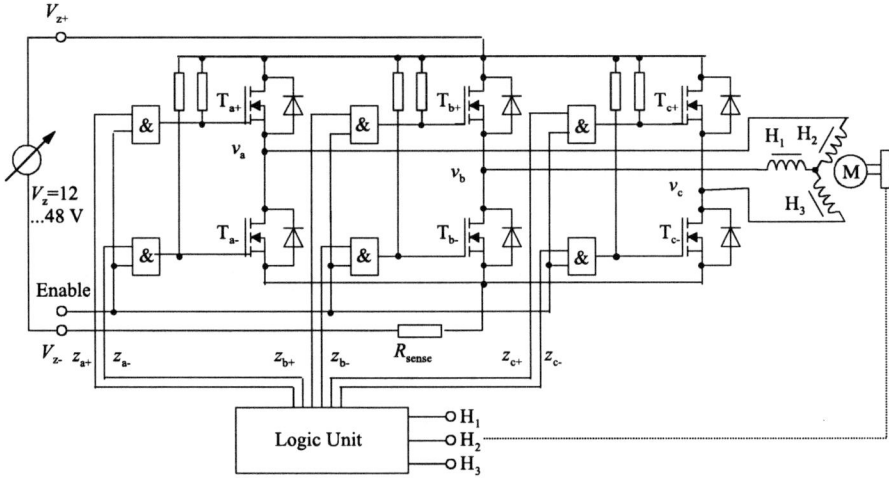

Fig. 6.32. Circuit of an open-loop control for an electronic motor

6 Control of Electric Drives 375

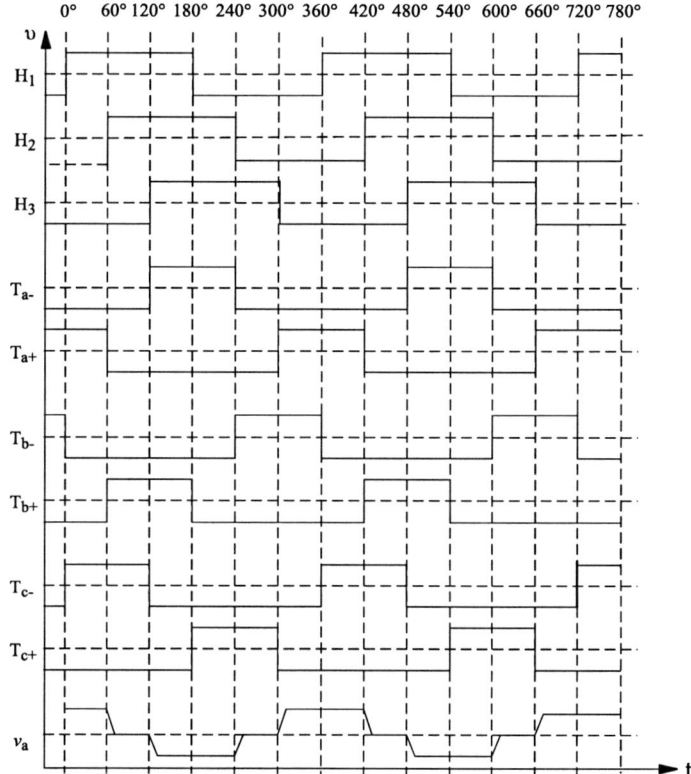

Fig. 6.33. Time-dependent courses of field sensor signals, of the control signal for the power switches and an exemplary output voltage of a three-phase bridge

Table 6.2. Arrangement of the signal correlation between field-sensor signals and control signals of the power switches

Field-sensor signals 1: H-signal 0: L-signal			Switching state of the upper power switches +: turned on −: turned off			Switching state of the lower power switches +: turned on −: turned off		
H_1	H_2	H_3	T_{a+}	T_{b+}	T_{c+}	T_{a-}	T_{b-}	T_{c-}
1	0	0	+	−	−	−	−	+
1	1	0	−	+	−	−	−	+
1	1	1	−	+	−	+	−	−
0	1	1	−	−	+	+	−	−
0	0	1	−	−	+	−	+	−
0	0	0	+	−	−	−	+	−

Since the inverter state directly depends on the rotor position, the rotational speed can be affected only by changing the intermediate circuit voltage V_z.

6.3.5 Positioning and Rotational Speed Open-Loop Control of Stepper Motors

The open-loop rotational speed control of stepping motors can be realized by applying a pulse frequency to the phases of the stepping motor. With each pulse, the rotor of the stepping motor turns one step forward with a constant angle of rotation. The size of the step angle depends on the design of the motor. The positioning control of stepping motors is achieved by the fact that the rotor is stopped after a given number of steps. The number of steps is calculated as the quotient of the angle of rotation, which is necessary in order to achieve the desired position, and the step angle of the motor.

6.4 Positioning and Rotational Speed Closed-Loop Control

A fundamental condition of drive control is that the closed-loop control system behaves stable. The simplest definition of stability means that the output of the system reaches a stationary value when there is a time-limited disturbance variable or a change in the reference input variable.

For simple rotational speed controls with universal motors, the adjustment of the controller parameters can be achieved by the known methods of Ziegler/Nichols or Chien/Hrones/Reswick [6.4–6.6]. For a positioning cascade control with subcontrollers for rotational speed, motor torque and motor current, it is necessary to determine the transfer functions of the controlled system which can comprise a DC motor, an asynchronous motor or a synchronous motor. The closed-loop controllers are adapted to these motors under attention of defined optimizing criteria. Furthermore, a simulation of the control loops is recommendable, because it allows judging the dynamic behaviour of all control circuit signals [6.6]. In the following section, for all three types of motors, the same strategy is applied for determining the controller parameters. This is possible because both the rotational speed system as well as the motor current system have a transfer function of the same structure for all three types of motors. This is given if they have a transfer function dominated by RL-elements.

6.4.1 Closed-Loop Control of Universal Motors

A possible circuit for the closed-loop speed control of a universal motor, which is supplied with alternating current, is shown in Fig. 6.34. A PID-controller is used as an automatic controller. Its output is connected to the control input of a phase shift IC. A tachogenerator serves to measure the rotational speed. The given reference for the rotational speed is named n^*.

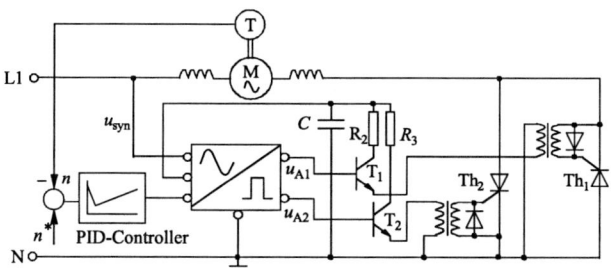

Fig. 6.34. Circuit for closed-loop control of universal motors

6.4.2 Closed-Loop Control of DC Motors

6.4.2.1 Rotational Speed Closed-Loop Control of DC Motors with Full Bridge

Figure 6.35 shows the schematic structure of a rotational speed closed-loop control with a full bridge as power section. This circuit should be used whenever it is important to minimize energy losses of speed-controlled DC motors.

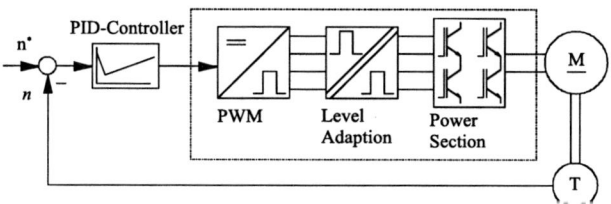

Fig. 6.35. Rotational speed closed-loop control with full bridge as power section

6.4.2.2 Positioning Closed-Loop Control with Subcontrollers for Rotational Speed and Motor Current

Figure 6.36 shows the block diagram of a DC motor with the representation of rotor voltage, back-e.m.f. voltage, stator current and rotor flux.

For the represented DC motor, the following relationships are important from the view of a closed-loop positioning control

$$\begin{aligned} v_q(s) &= i_q(s) \cdot (R + s \cdot L) + e(s), & e(s) &= c \cdot \psi_d \omega(s), \\ T_i &= c \cdot \psi_d \cdot i_q(s), & T_i &= T_l + J \cdot s \cdot \omega(s). \end{aligned} \qquad (6.15)$$

This means: T_i is the internal torque of the motor, T_l is the load torque including friction torque, J is the mass moment of inertia.

By the fixed arrangement of the carbon brushes at the collector, a fixed geometrical allocation between flux and torque generating current is always given in the DC motor. This allocation is selected in such a way that with given flux and current the internal torque is always a maximum. Due to the mechanical commutation of the current, the open-loop control and the closed-loop control of DC motors need the least technical effort.

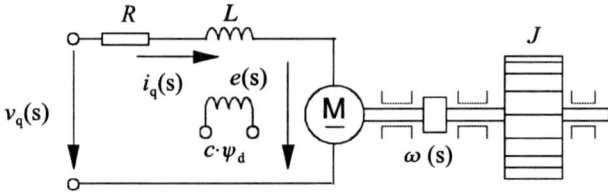

Fig. 6.36. Representation of the DC motor in the Laplace domain

Transfer Function

From Fig. 6.36, the equivalent circuit for the controlled behaviour of the DC motor, indicated in Fig. 6.37, can be obtained.

$$G(s) = \frac{\omega(s)}{v_q(s)} = \frac{K_S}{1 + s \cdot T_2 + s^2 \cdot T_1 \cdot T_2} \tag{6.16}$$

with

$$K_S = \frac{1}{c \cdot \psi_d}, \quad T_1 = \frac{L}{R}, \quad T_2 = \frac{J \cdot R}{(c \cdot \psi_d)^2}. \tag{6.17}$$

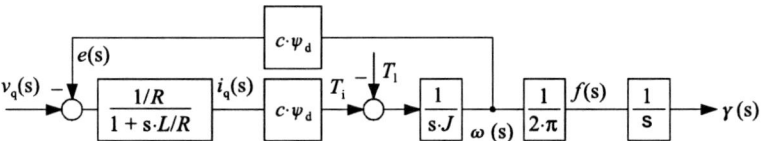

Fig. 6.37. Dynamic model of the DC motor

Block Diagram of Closed-Loop Controlled DC Motor

In order to achieve high control quality, closed-loop controls for drives frequently are designed in cascade structure. These have the characteristic that they already compensate arising disturbances of the drive before these can affect the actual control quality, in this case the position. As an example for this, the closed-loop position control of a DC motor with subcontrollers for rotational speed and motor current is schematically represented in Fig. 6.38. The position controller is a proportional controller (P-controller), while the controllers for rotational speed and motor current are proportional-integral controllers (PI controllers). All values marked with * are reference values. In the indicated example, angle position is controlled.

Using an angle transmitter, the actual value of the angle of rotation is measured and compared with a reference value. This difference between reference value and actual value (error signal) determines the output variable of the position controller, which serves as reference value for the speed controller. At its input, this reference value is compared with the actual rotational

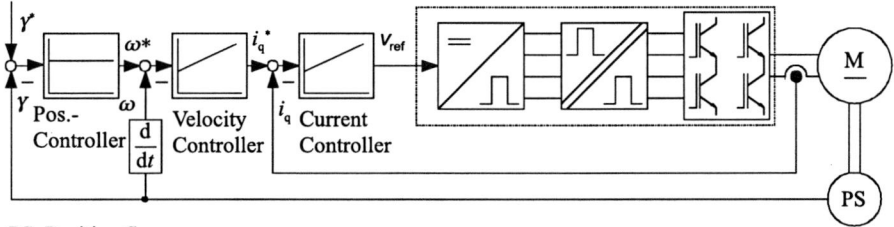

PS: Position Sensor

Fig. 6.38. Positioning cascade closed-loop control with subcontrollers for rotational speed and motor current

speed value, which is gained by differentiation of the position sensor signal. Consequently, the speed controller generates a control signal, which serves the current controller as a new reference value. This determines the error signal from the given reference value and from the current measured at the motor. By use of the error signal, the current controller computes a control signal, which serves as input of the DC motor driver. This makes sure that the motor is not overloaded and that all controller output values are limited to the permissible maximum values for the motor. The controller is only able to process standardized signals. Such standardized signals are, e. g. in analogue technique voltages from -10 V to $+10$ V or in digital signal processing numerical values from -1024 to 1023. Therefore the relevant motor data (rotational speed, current) must be converted into the standardized signals.

In stationary operation, the current controller has a protective function against drive overload. If the torque rises so strongly that the current value exceeds the acceptable maximum value, then the current controller reduces its output value. Thus, the rotational speed is reduced (the reference value no longer is held), but the drive is protected against overload.

For the function of the closed-loop control, it is important that the controllers are correctly adapted to the amplification factor and the time constants of the controlled system. In order to achieve this, the transfer functions of the system must be known. In order to make this evident, a block diagram is indicated together with an equivalent control circuit of the motor, the cascade control and the DC motor driver in Fig. 6.39. From this it can be seen that the transfer function concerning the control loop for the rotational speed is described by Eq. (6.16). The current controller is arranged as a PT_1-element. The transfer function for this control path is

$$G(s) = \frac{i_q(s)}{v_q(s)} = \frac{\frac{1}{R}}{1 + s\frac{L}{R}} . \qquad (6.18)$$

The signal delay, which results from the motor driver, can still be accounted for by an appropriate dead time in the transfer function. For the indicated control path, the controllers are to be adjusted according to the usual criteria used in control engineering [6.4–6.6].

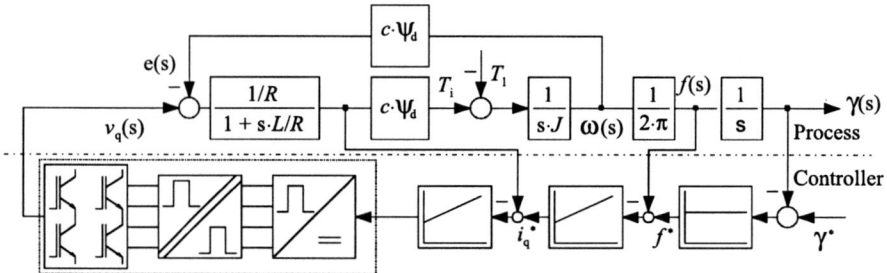

Fig. 6.39. Block diagram of the DC motor control with equivalent circuits of the motor, the cascade control and the DC motor driver

Modern motor closed-loop controllers are realized as sampling controllers. The parameters of these sampling controllers can be computed from the known parameters of the analogue controllers such as amplification factor, reset time and derivative time [6.7]. The sampling interval for the control depends on the time constants of the transfer elements which are to be controlled. For motor control, it is in practice smaller or equal to 1 ms and is therefore substantially larger than the cycle time T_T of the used inverters.

6.4.3 Closed-Loop Control of Asynchronous Motors

6.4.3.1 Simple Rotational Speed Control with Frequency Converters

Figure 6.40 shows the structure of a simple rotational speed control of an asynchronous motor. It consists of a frequency converter with an input for a reference voltage, a proportional plus integral controller (PI controller) and the motor with a tachogenerator.

6.4.3.2 Field-Oriented Control of Asynchronous Motors

In order to be able to use the field-oriented control of the asynchronous motor, a suitable description of transfer behaviour is necessary. In addition to the

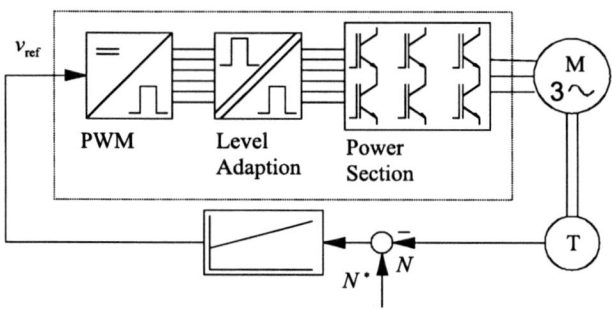

Fig. 6.40. Simple rotational speed control of an asynchronous motor

asynchronous motor, a single-phase equivalent circuit is created. This is represented in a stator-fixed α, β-reference frame and in a rotary d, q-reference frame, respectively, which is oriented towards an electrical quantity (usually the flux). In the d, q-reference frame, the description of the dynamic behaviour and the closed-loop control can take place in a simple manner [6.17, 6.18].

In addition, it is necessary to transfer currents, voltages and flux linkages from the three-phase a,b,c-system into the stator-fixed α, β-reference frame. This is calculated by the equations

$$\begin{pmatrix} i_{1\alpha} \\ i_{1\beta} \end{pmatrix} = \frac{3}{2} \cdot \begin{pmatrix} 1 & 0 \\ \frac{1}{\sqrt{3}} & \frac{2}{\sqrt{3}} \end{pmatrix} \cdot \begin{pmatrix} i_{1a} \\ i_{1b} \end{pmatrix} \tag{6.19}$$

$$\begin{pmatrix} v_{1\alpha} \\ v_{1\beta} \end{pmatrix} = \begin{pmatrix} 1 & 0 \\ \frac{1}{\sqrt{3}} & \frac{2}{\sqrt{3}} \end{pmatrix} \cdot \begin{pmatrix} v_{1a} \\ v_{1b} \end{pmatrix} \tag{6.20}$$

$$\begin{pmatrix} \psi_{1\alpha} \\ \psi_{1\beta} \end{pmatrix} = \begin{pmatrix} 1 & 0 \\ \frac{1}{\sqrt{3}} & \frac{2}{\sqrt{3}} \end{pmatrix} \cdot \begin{pmatrix} \psi_{1a} \\ \psi_{1b} \end{pmatrix} . \tag{6.21}$$

The relation between the angle γ_1 of the rotating field, the rotor angle γ_m and the slip angle γ_2 is shown in Fig. 6.41. The shown u, v-reference frame is firmly connected with the rotor. Furthermore, the currents and flux linkages ψ_{2d} and ψ_{2q} are shown in the coordinate system.

Under consideration of the number of pole pairs, the relation between the different angles of rotation results from Fig. 6.41

$$\gamma_1 = \gamma_2 + p \cdot \gamma_m . \tag{6.22}$$

The relationship between the associated angular speeds is

$$\omega_1 = \omega_2 + p \cdot \omega_m . \tag{6.23}$$

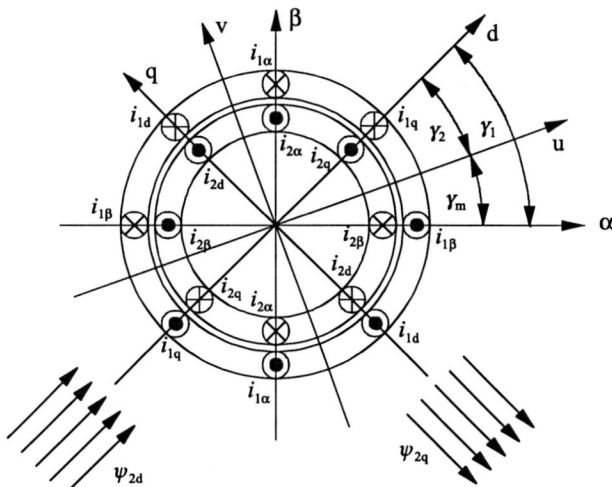

Fig. 6.41. Interrelation between the reference frames of an asynchronous motor and the representation of currents and flux linkages

The conversion from α, β-reference frame into the d, q-reference frame takes place with the rotating transformation for the current with the formula

$$\begin{pmatrix} i_{1d} \\ i_{1q} \end{pmatrix} = \begin{pmatrix} \cos\gamma_1 & \sin\gamma_1 \\ -\sin\gamma_1 & \cos\gamma_1 \end{pmatrix} \cdot \begin{pmatrix} i_{1\alpha} \\ i_{1\beta} \end{pmatrix} = D(-\gamma_1) \cdot \begin{pmatrix} i_{1\alpha} \\ i_{1\beta} \end{pmatrix}. \qquad (6.24)$$

With the same transformation matrix, voltages and the flux linkages can be converted.

Like the currents, voltages and flux linkages, the motor parameters such as resistances and inductances must also be converted by the turns ratio from the three-phase system into the two-axis system.

It is assumed that inductances which couple the flux linkages and the associated motor currents are constant and for the inoperative rotor windings independent from the coordinate system. When defining all phase values, decoupled phases are assumed for simplicity. As can be seen, this is not the case if one determines the flux of the unloaded running motor (without rotor current), e.g. in the direction of the α-axis. It results in

$$\psi_{1a} = (L_H + L_{S1Ph}) \cdot i_{1a} + L_H \cdot i_{1b} \cdot \cos 120° + L_H \cdot i_{1c} \cdot \cos 240°. \qquad (6.25)$$

$L_H + L_{S1Ph}$ is the inductance, which connects one of the three-phase currents with the associated flux. One would measure this inductance, for instance, if current would only flow through one coil. A conversion of Eq. (6.25) and the application of the first Kirchhoff law results in

$$\begin{aligned}\psi_{1a} &= \left(\frac{3}{2} \cdot L_H + L_{S1Ph}\right) \cdot i_{1a} - \frac{1}{2} \cdot L_H \cdot (i_{1a} + i_{1b} + i_{1c}) \\ &= \left(\frac{3}{2} \cdot L_H + L_{S1Ph}\right) \cdot i_{1a}.\end{aligned} \qquad (6.26)$$

Likewise, it follows for the phase b

$$\psi_{1b} = \left(\frac{3}{2} \cdot L_H + L_{S1Ph}\right) \cdot i_{1b}. \qquad (6.27)$$

Equations (6.26) and (6.27) can be represented in matrix syntax as

$$\begin{pmatrix} \psi_{1a} \\ \psi_{1b} \end{pmatrix} = \begin{pmatrix} \frac{3}{2} \cdot L_H + L_{S1Ph} & 0 \\ 0 & \frac{3}{2} \cdot L_H + L_{S1Ph} \end{pmatrix} \cdot \begin{pmatrix} i_{1a} \\ i_{1b} \end{pmatrix}. \qquad (6.28)$$

By using Eqs. (6.19) and (6.21) from Eq. (6.28) follows

$$\begin{pmatrix} \psi_{1\alpha} \\ \psi_{1\beta} \end{pmatrix} = \begin{pmatrix} L_H + \frac{2}{3} \cdot L_{S1Ph} & 0 \\ 0 & L_H + \frac{2}{3} \cdot L_{S1Ph} \end{pmatrix} \cdot \begin{pmatrix} i_{1\alpha} \\ i_{1\beta} \end{pmatrix}. \qquad (6.29)$$

With the simplification

$$L_1 = \left(L_H + \frac{2}{3} \cdot L_{S1Ph}\right) \qquad (6.30)$$

it follows from Eq. (6.29)

$$\begin{pmatrix} \psi_{1\alpha} \\ \psi_{1\beta} \end{pmatrix} = \begin{pmatrix} L_1 & 0 \\ 0 & L_1 \end{pmatrix} \cdot \begin{pmatrix} i_{1\alpha} \\ i_{1\beta} \end{pmatrix}. \tag{6.31}$$

For the correlation between phase resistances, phase currents and phase voltages in the phases a and b of the three-phase system results

$$\begin{pmatrix} v_{1a} \\ v_{1b} \end{pmatrix} = \begin{pmatrix} R_{1Ph} & 0 \\ 0 & R_{1Ph} \end{pmatrix} \cdot \begin{pmatrix} i_{1a} \\ i_{1b} \end{pmatrix}. \tag{6.32}$$

With Eqs. (6.19) and (6.20), Eq. (6.32) provides

$$\begin{pmatrix} v_{1\alpha} \\ v_{1\beta} \end{pmatrix} = \begin{pmatrix} \tfrac{2}{3} \cdot R_{1Ph} & 0 \\ 0 & \tfrac{2}{3} \cdot R_{1Ph} \end{pmatrix} \cdot \begin{pmatrix} i_{1\alpha} \\ i_{1\beta} \end{pmatrix}. \tag{6.33}$$

With the conversion

$$R_1 = \frac{2}{3} \cdot R_{1Ph}, \tag{6.34}$$

it follows for the two-axis system

$$\begin{pmatrix} v_{1\alpha} \\ v_{1\beta} \end{pmatrix} = \begin{pmatrix} R_1 & 0 \\ 0 & R_1 \end{pmatrix} \cdot \begin{pmatrix} i_{1\alpha} \\ i_{1\beta} \end{pmatrix}. \tag{6.35}$$

If the leakage inductances and resistances of the rotor windings are converted to the stator side by the turns ratio, they can be applied in the same way to the single-phase equivalent system by using the relations

$$L_2 = \left(L_H + \frac{2}{3} \cdot L_{S2Ph}^{(1)} \right) \tag{6.36}$$

and

$$R_2 = \frac{2}{3} \cdot R_{2Ph}^{(1)}. \tag{6.37}$$

For the loaded machine (the rotor current is not equal to zero) in the stator-fixed reference frame, the stator and rotor currents can be calculated with the equations

$$\begin{pmatrix} \psi_{1\alpha} \\ \psi_{2\alpha} \end{pmatrix} = \begin{pmatrix} L_1 & L_H \\ L_H & L_2 \end{pmatrix} \cdot \begin{pmatrix} i_{1\alpha} \\ i_{2\alpha} \end{pmatrix} \tag{6.38}$$

and

$$\begin{pmatrix} \psi_{1\beta} \\ \psi_{2\beta} \end{pmatrix} = \begin{pmatrix} L_1 & L_H \\ L_H & L_2 \end{pmatrix} \cdot \begin{pmatrix} i_{1\beta} \\ i_{2\beta} \end{pmatrix}. \tag{6.39}$$

With the inverse of the inductance matrix (Eqs. (6.38) and (6.39)), the currents as a function of the flux linkages can be calculated in the following manner

$$\begin{pmatrix} i_{1\alpha} \\ i_{2\alpha} \end{pmatrix} = \frac{1}{L_1 \cdot L_2 - L_H^2} \cdot \begin{pmatrix} L_2 & -L_H \\ -L_H & L_1 \end{pmatrix} \cdot \begin{pmatrix} \psi_{1\alpha} \\ \psi_{2\alpha} \end{pmatrix} = \begin{pmatrix} K_{11} & K_{12} \\ K_{21} & K_{22} \end{pmatrix} \cdot \begin{pmatrix} \psi_{1\alpha} \\ \psi_{2\alpha} \end{pmatrix} \quad (6.40)$$

and

$$\begin{pmatrix} i_{1\beta} \\ i_{2\beta} \end{pmatrix} = \begin{pmatrix} K_{11} & K_{12} \\ K_{21} & K_{22} \end{pmatrix} \cdot \begin{pmatrix} \psi_{1\beta} \\ \psi_{2\beta} \end{pmatrix} \quad (6.41)$$

with the shortcuts

$$K_{11} = \frac{L_2}{L_1 \cdot L_2 - L_H^2}, \quad K_{12} = K_{21} = -\frac{L_H^2}{L_1 \cdot L_2 - L_H^2}, \quad (6.42)$$
$$K_{22} = \frac{L_1}{L_1 \cdot L_2 - L_H^2}.$$

The voltages of the single-phase equivalent circuit of the asynchronous machine with squirrel-cage rotor in the α, β-reference frame are calculated by the equations

$$v_{1\alpha} = R_1 \cdot i_{1\alpha} + \dot{\psi}_{1\alpha} \quad (6.43)$$
$$v_{1\beta} = R_1 \cdot i_{1\beta} + \dot{\psi}_{1\beta} \quad (6.44)$$
$$0 = R_2 \cdot i_{2\alpha} + \dot{\psi}_{2\alpha} + \psi_{2\beta} \cdot \omega_m \cdot p \quad (6.45)$$
$$0 = R_2 \cdot i_{2\beta} + \dot{\psi}_{2\beta} - \psi_{2\alpha} \cdot \omega_m \cdot p \, . \quad (6.46)$$

The equivalent circuit in the α, β-reference frame using Eqs. (6.43) to (6.46) is represented in Fig. 6.42.

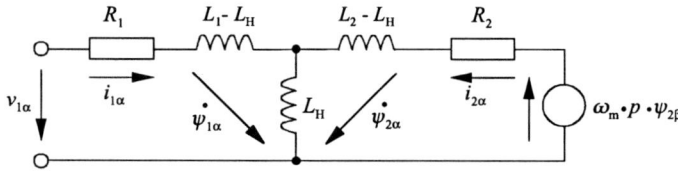

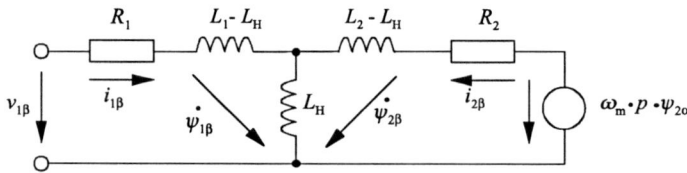

Fig. 6.42. Representation of the asynchronous machine in the stator-fixed α, β-reference frame

For field orientation, the resulting flux space vector of the rotor must always point in the direction of the d-axis of the d, q-reference frame which is rotating with the rotating field of the stator. Therefore, the components $\psi_{2\alpha}$ and $\psi_{2\beta}$ of the rotor flux space vector must be computed. From Eqs. (6.40) and (6.41) we get the following results

$$i_{1\alpha} = K_{11} \cdot \psi_{1\alpha} + K_{12} \cdot \psi_{2\alpha} \Rightarrow \psi_{1\alpha} = \frac{1}{K_{11}} \cdot i_{1\alpha} - \frac{K_{12}}{K_{11}} \cdot \psi_{2\alpha} \quad (6.47)$$

and

$$i_{1\beta} = K_{11} \cdot \psi_{1\beta} + K_{12} \cdot \psi_{2\beta} \Rightarrow \psi_{1\beta} = \frac{1}{K_{11}} \cdot i_{1\beta} - \frac{K_{12}}{K_{11}} \cdot \psi_{2\beta} \, . \quad (6.48)$$

Equations (6.47) and (6.48) are differentiated with respect to time and the expressions for $d\psi_{1\alpha}/dt$ and $d\psi_{1\beta}/dt$ are applied in Eqs. (6.43) and (6.44). It results in

$$v_{1\alpha} = R_1 \cdot i_{1\alpha} + \frac{1}{K_{11}} \cdot \dot{i}_{1\alpha} - \frac{K_{12}}{K_{11}} \cdot \dot{\psi}_{2\alpha} \quad (6.49)$$

and

$$v_{1\beta} = R_1 \cdot i_{1\beta} + \frac{1}{K_{11}} \cdot \dot{i}_{1\beta} - \frac{K_{12}}{K_{11}} \cdot \dot{\psi}_{2\beta} \, . \quad (6.50)$$

The quotient $-K_{12}/K_{11}$ is equal to the expression L_H/L_2. The components of the rotor flux linkage in the α, β-reference frame are obtained by converting Eqs. (6.49) and (6.50). From that follows

$$\dot{\psi}_{2\alpha} = \frac{L_2}{L_H} \cdot \left(v_{1\alpha} - R_1 \cdot i_{1\alpha} - \frac{1}{K_{11}} \cdot \dot{i}_{1\alpha} \right) \quad (6.51)$$

and

$$\dot{\psi}_{2\beta} = \frac{L_2}{L_H} \cdot \left(v_{1\beta} - R_1 \cdot i_{1\beta} - \frac{1}{K_{11}} \cdot \dot{i}_{1\beta} \right) \, . \quad (6.52)$$

The components of the rotor flux linkage in the α, β-reference frame are

$$\psi_{2\alpha} = \left| \underline{\psi}_2 \right| \cdot \cos \gamma_1 \quad (6.53)$$

and

$$\psi_{2\beta} = \left| \underline{\psi}_2 \right| \cdot \sin \gamma_1 \, . \quad (6.54)$$

If Eqs. (6.53) and (6.54) are differentiated with respect to time, this results in

$$\dot{\psi}_{2\alpha} = -\omega_1 \cdot \left| \underline{\psi}_2 \right| \cdot \sin \gamma_1 = -\omega_1 \cdot \psi_{2\beta} \Rightarrow \psi_{2\beta} = -\frac{\dot{\psi}_{2\alpha}}{\omega_1} \quad (6.55)$$

and
$$\dot{\psi}_{2\beta} = \omega_1 \cdot \left|\underline{\psi}_2\right| \cdot \cos\gamma_1 = \omega_1 \cdot \psi_{2\alpha} \Rightarrow \psi_{2\alpha} = \frac{\dot{\psi}_{2\beta}}{\omega_1} . \tag{6.56}$$

For the field orientation of the asynchronous machine, from applied phase voltages and measured phase currents, the rotor flux linkages $\psi_{2\alpha}$ and $\psi_{2\beta}$ are calculated with Eqs. (6.51) to (6.56). From these, the absolute value of the rotor flux linkage and the angle γ_1 between the α-axis of the α, β-reference frame and the d-axis of the d, q-reference frame can be calculated with the formulae

$$\left|\underline{\psi}_2\right| = \sqrt{\psi_{2\alpha}^2 + \psi_{2\beta}^2} \tag{6.57}$$

and

$$\gamma_1 = \arctan\frac{\psi_{2\beta}}{\psi_{2\alpha}} . \tag{6.58}$$

The internal torque can be determined for the single-phase equivalent circuit in the α, β-reference frame with the equation

$$T_i = \frac{L_H}{L_2} \cdot p \cdot (\psi_{2\alpha} \cdot i_{1\beta} - \psi_{2\beta} \cdot i_{1\alpha}) . \tag{6.59}$$

Using the rotating transformation, the single-phase equivalent circuit can be transferred from the α, β-reference frame into the d, q-reference frame. In this system, the relations between the electrical quantities of the asynchronous machine can be calculated by the matrix equation

$$\begin{pmatrix} \dot{i}_{1d} \\ \dot{i}_{1q} \\ \dot{\psi}_{2d} \\ \dot{\psi}_{2q} \end{pmatrix} = \begin{pmatrix} -\omega_0 & \omega_1 & -K_{12} \cdot \omega_g & -K_{12} \cdot \omega_m \cdot p \\ -\omega_1 & -\omega_0 & K_{12} \cdot \omega_m \cdot p & -K_{12} \cdot \omega_g \\ L_H \cdot \omega_g & 0 & -\omega_g & \omega_2 \\ 0 & L_H \cdot \omega_g & -\omega_2 & -\omega_g \end{pmatrix} \cdot \begin{pmatrix} i_{1d} \\ i_{1q} \\ \psi_{2d} \\ \psi_{2q} \end{pmatrix}$$
$$+ \begin{pmatrix} K_{11} & 0 \\ 0 & K_{11} \\ 0 & 0 \\ 0 & 0 \end{pmatrix} \cdot \begin{pmatrix} v_{1d} \\ v_{1q} \end{pmatrix} . \tag{6.60}$$

The shortcuts ω_0 and ω_g used in Eq. (6.60) can be determined with the relations

$$\omega_g = \frac{R_2}{L_2} \tag{6.61}$$

and

$$\omega_0 = K_{11} \cdot \left(R_1 + R_2 \cdot \frac{L_H^2}{L_2^2}\right) . \tag{6.62}$$

The internal torque of the motor must be in equilibrium with torques acting externally. From this condition the following motion equation results

$$J \cdot \frac{d\omega_m}{dt} = p \cdot \frac{L_H}{L_2} \cdot (\psi_{2d} \cdot i_{1q} - \psi_{2q} \cdot i_{1d}) - T_l . \tag{6.63}$$

The derivation of these relations is indicated in [6.2].

By using the differential operator $s = d/dt$, it follows from Eqs. (6.60) and (6.63) that

$$i_{1d} \cdot (s + \omega_0) = \omega_1 \cdot i_{1q} - K_{12} \cdot \omega_g \cdot \psi_{2d} - K_{12} \cdot \omega_m \cdot p \cdot \psi_{2q} + K_{11} \cdot v_{1d} \tag{6.64}$$

$$i_{1q} \cdot (s + \omega_0) = -\omega_1 \cdot i_{1d} + K_{12} \cdot \omega_m \cdot p \cdot \psi_{2d} - K_{12} \cdot \omega_g \cdot \psi_{2q} + K_{11} \cdot v_{1q} \tag{6.65}$$

$$\psi_{2d} \cdot (s + \omega_g) = L_H \cdot \omega_g \cdot i_{1d} + \omega_2 \cdot \psi_{2q} \tag{6.66}$$

$$\psi_{2q} \cdot (s + \omega_g) = L_H \cdot \omega_g \cdot i_{1q} - \omega_2 \cdot \psi_{2d} \tag{6.67}$$

$$J \cdot s \cdot \omega_m = p \cdot \frac{L_H}{L_2} \cdot (\psi_{2d} \cdot i_{1q} - \psi_{2q} \cdot i_{1d}) - T_l . \tag{6.68}$$

Equations (6.64) to (6.68) represent the mathematical model of the asynchronous machine. With these equations, the equivalent circuit indicated in Fig. 6.43 can be drawn.

Equations (6.63) and (6.68) show that the internal torque of the machine reaches its maximum value when the expression $\psi_{2q} \cdot i_{1d}$ is zero. This is the case if the d axis of the d,q-reference frame shows in the direction of the resulting flux. Thus, $\psi_{2q} = 0$. By automatic control, this can be achieved when the stator currents are applied in such a way that the condition $\psi_{2q} = 0$ is fulfilled. In this case, it is spoken of "field-oriented control" or of "vector control".

With $\psi_{2q} = 0$ and, resulting from it, $d\psi_{2q}/dt = 0$, the equivalent circuit (Fig. 6.44) for the field-oriented control of the asynchronous motor results from Eqs. (6.64) to (6.68) with the relations

$$i_{1d} \cdot (s + \omega_0) = \omega_1 \cdot i_{1q} - K_{12} \cdot \omega_g \cdot \psi_{2d} + K_{11} \cdot v_{1d} \tag{6.69}$$

$$i_{1q} \cdot (s + \omega_0) = -\omega_1 \cdot i_{1d} + K_{12} \cdot \omega_m \cdot p \cdot \psi_{2d} + K_{11} \cdot v_{1q} \tag{6.70}$$

$$\psi_{2d} \cdot (s + \omega_g) = L_H \cdot \omega_g \cdot i_{1d} \tag{6.71}$$

$$0 = L_H \cdot \omega_g \cdot i_{1q} - \omega_2 \cdot \psi_{2d} \tag{6.72}$$

$$J \cdot s \cdot \omega_m = p \cdot \frac{L_H}{L_2} \cdot \psi_{2d} \cdot i_{1q} - T_l . \tag{6.73}$$

The fluxes and currents for field-oriented controlled asynchronous motors are represented schematically in Fig. 6.45.

Following Fig. 6.44, the relation for the angular speed of the rotor is

$$\omega_m = \left[(K_{11} \cdot v_{1q} - \omega_1 \cdot i_{1d} + K_{12} \cdot \omega_m \cdot p) \cdot \frac{1}{(s + \omega_0)} \cdot \psi_{2d} \cdot p \cdot \frac{L_H}{L_2} - T_l \right] \cdot \frac{1}{s \cdot J} \tag{6.74}$$

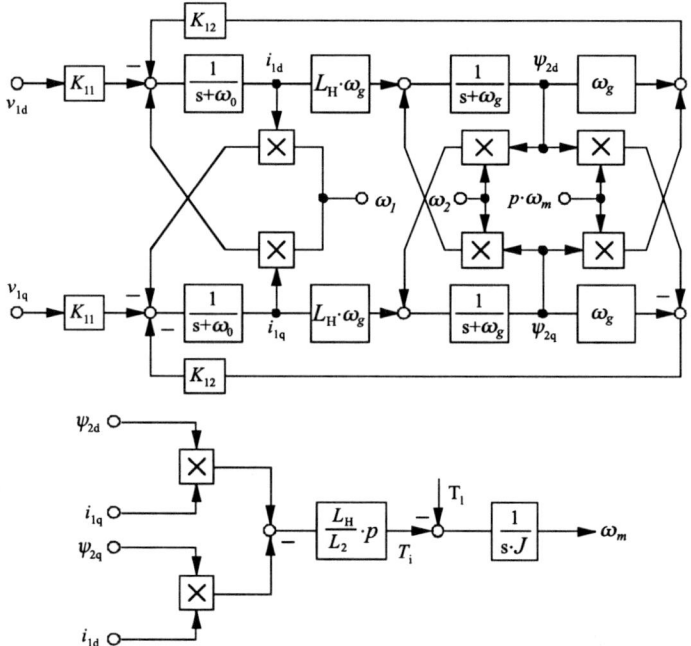

Fig. 6.43. Equivalent circuit of the asynchronous machine

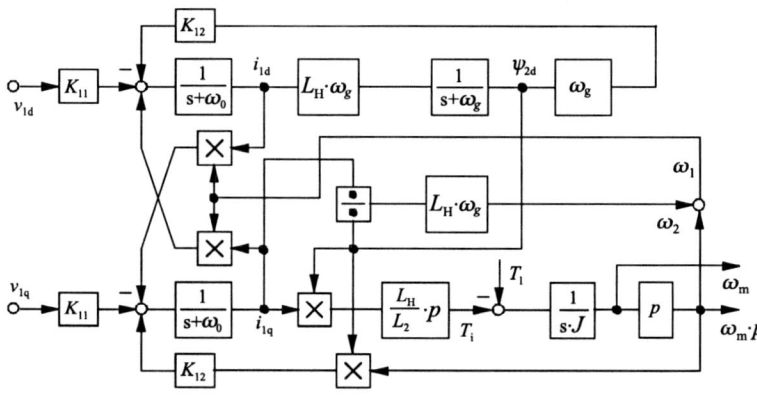

Fig. 6.44. Equivalent circuit for field-oriented control of the asynchronous motor

Because i_{1d} is only the field-generating current and so it does not contribute to the torque generation of the field-oriented controlled asynchronous machine, the expression $\omega_1 \cdot i_{1d}$ is neglected. Furthermore, the load torque is set $T_1 = 0$. In addition, the following simplifications are introduced

$$K_S = -\frac{K_{11}}{K_{12}} \cdot \frac{1}{\psi_{2d} \cdot p}, \quad T_1 = \frac{1}{\omega_0}, \quad T_2 = -\frac{L_2}{L_H} \cdot \frac{J \cdot \omega_0}{(\psi_{2d} \cdot p)^2 \cdot K_{12}}. \tag{6.75}$$

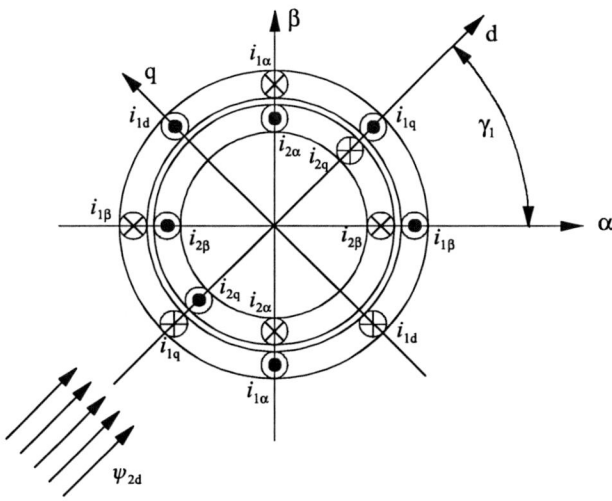

Fig. 6.45. Schematic representation of the fluxes and currents for field-oriented controlled asynchronous motors

With the mentioned simplifications and shortcuts, the following transfer function results for the field-oriented controlled asynchronous machine

$$G(s) = \frac{\omega_m(s)}{u_{1q}(s)} = \frac{K_S}{1 + s \cdot T_2 + s^2 \cdot T_1 \cdot T_2}. \tag{6.76}$$

Figure 6.46 shows the block diagram for the closed-loop position control of an asynchronous motor with field orientation. It presents a cascade position control with subcontrollers for rotational speed and motor current. The rotor fluxes are calculated with Eqs. (6.51), (6.52), (6.55) and (6.56) from applied

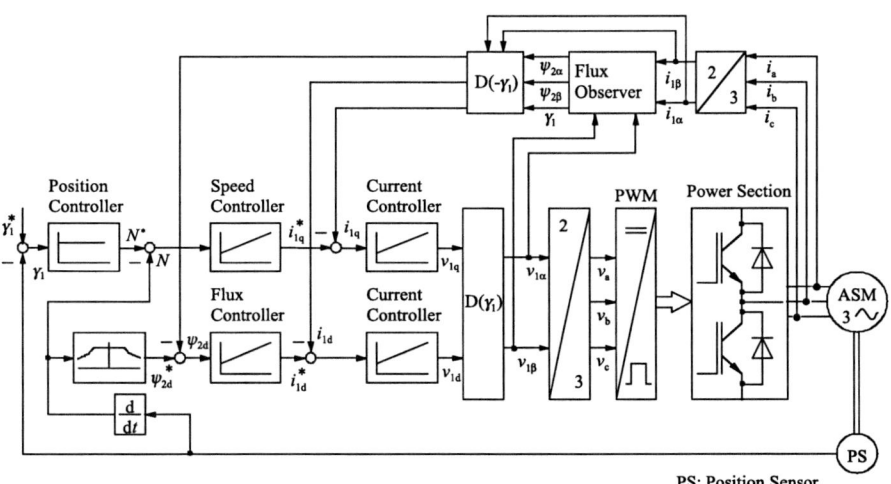

Fig. 6.46. Schematic representation of the field-oriented position cascade control of the asynchronous motor with subcontrollers for rotational speed and motor current

phase voltages and measured currents. The angle γ_1 between the α-axis and the d-axis results from Eq. (6.58). Internal torque can be computed with Eq. (6.59). The conversion of the motor quantities from the three-phase system into the stator-fixed α, β-reference frame and from this into the d, q-reference frame orientated at the rotor flux is calculated with Eqs. (6.19), (6.20), (6.21) and (6.24). These equations can also be used to convert the motor quantities from the d, q-reference frame *via* the α, β-reference frame into the three-phase a, b, c-system.

The controller structure indicated in Fig. 6.46 is one of the most efficient for vector control of the asynchronous motor. Of course, this controller structure can also be used without the position controller, so that only the speed controller and the subcontroller for motor current remain.

6.4.4 Closed-Loop Control of Electronically Commutated Motors

6.4.4.1 Rotational Speed Closed-Loop Control with Block Commutation and Sinusoidal Evaluated Commutation

Electronically commutated motors contain sensors to detect the rotor position. Dependent on the rotor position, the phase currents are commutated for the semiconductor switches. According to the applied modulation process the phase voltages or currents can be generated as rectangular or sinusoidal curves.

Block Commutation

The function of an electronic motor has already been described in Sect. 6.3.4.2 (Fig. 6.32). The rotational speed must be detected for the closed-loop rotational speed control. This can be achieved with a separate sensor or it can be generated from the rotor position sensor. From the control deviation n^*-n, the controller determines an output variable, with which the voltage in the DC voltage link is adjusted. Figure 6.47 shows the structure of such a circuit.

Another method to vary the rotational speed by using block commutation is to change the mean value of the supply voltage for the individual windings by means of PWM and to keep the voltage of the intermediate circuit constant. With this method, a higher dynamic in rotational speed variation can be achieved than by changing the voltage of the intermediate circuit.

Sinusoidal Evaluated PWM Commutation

The sinusoidal evaluated PWM open-loop control (Sect. 6.2.5.2) of electronically commutated motors is only possible with the use of microprocessors or microcontrollers. With the selected method, the rotational speed has no influence on the pulse width. Speed control is to be achieved, as already described in Sect. 6.3.4.2, by changing the intermediate circuit voltage.

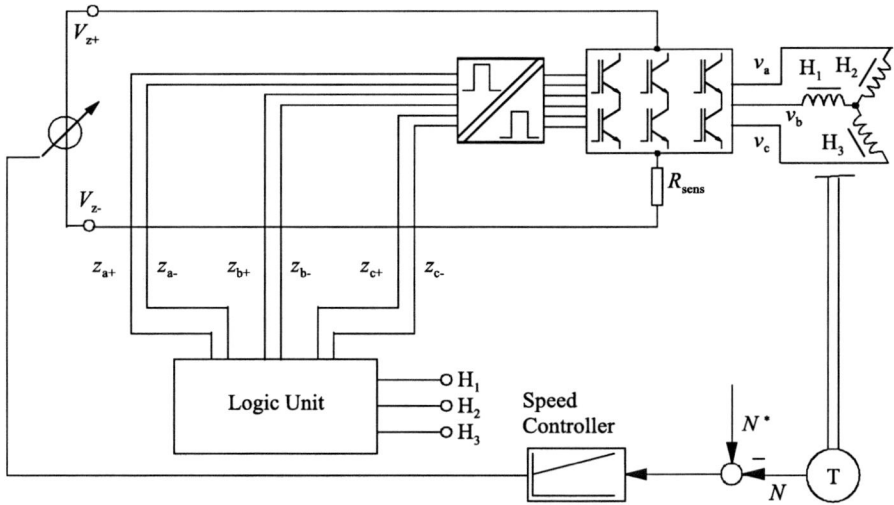

Fig. 6.47. Design of the closed-loop rotational speed control of an electronically commutated motor

Otherwise for rotational speed variation it is possible to change the amplitude of the sinusoidal evaluated voltage for the individual windings by using PWM and to keep the intermediate circuit voltage constant. Like with block commutation with this method it is possible to increase the dynamic of the motor.

The following considerations refer to a motor with one pair of poles. The time T for a full rotation of this motor is determined by the evaluation of the field sensor signal. The following relation leads to the number of pulses with the cycle duration T_T, which is realizable in the time T.

$$k_{\max} = T/T_T \ . \tag{6.77}$$

Switch-on times for the transistors T_{a+} and T_{a-} (Fig. 6.9) during a full rotor rotation can be computed by the relations

$$T_{Eza+} = \frac{T}{2} \cdot \left(1 + \sin\left(\frac{2 \cdot \pi}{T} \cdot k \cdot T_T\right)\right) \tag{6.78}$$

$$T_{Eza-} = \frac{T}{2} \cdot \left(1 - \sin\left(\frac{2 \cdot \pi}{T} \cdot k \cdot T_T\right)\right) \ . \tag{6.79}$$

The computation of switch-on times in accordance with Eqs. (6.78) and (6.79) must take place for $k = 0\ldots k_{\max}$.

After one rotation, the rotation time T is measured again. With the actual value, switch-on times for the next revolution for T_{a+} and T_{a-} are computed again.

For the power switches of the other half-bridges, the determination of switch-on times can take place using same relations, as long as the phase shifts are considered.

6.4.4.2 Positioning and Rotational Speed Closed-Loop Control of Electronically Commutated Motors

Firstly, the control of an electronically commutated motor with permanent magnet excitation and position sensor shall be treated.

From Fig. 6.48, one can determine the stator voltage equation

$$\underline{v}_1 = R_1 \cdot \underline{i}_1 + L_1 \frac{d\underline{i}_1}{dt} + \frac{d\underline{\psi}_1}{dt} . \tag{6.80}$$

For the synchronous motor, the correlation between rotating field and rotor angles is given by the relation

$$\gamma_1 = p \cdot \gamma_m . \tag{6.81}$$

In accordance with Fig. 6.49, the current space vector and all other quantities can also be divided into their components along α- and β-axes. It results

$$v_{1\alpha} = R_1 \cdot i_{1\alpha} + L_1 \cdot \frac{di_{1\alpha}}{dt} + \frac{d\psi_{2\alpha}}{dt} \tag{6.82}$$

$$v_{1\beta} = R_1 \cdot i_{1\beta} + L_1 \cdot \frac{di_{1\beta}}{dt} + \frac{d\psi_{2\beta}}{dt} . \tag{6.83}$$

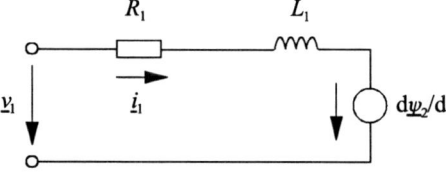

Fig. 6.48. Single-phase equivalent stator circuit of an electronically commutated motor analogue to a synchronous motor with permanent magnetic excitation

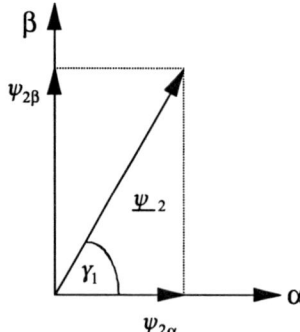

Fig. 6.49. Representation of the space vector of the flux in the stator-fixed reference frame

On the basis of this approach, and using the rotating transformation as given in Eq. (6.24) as well as using the derivatives with respect to time, and neglecting the load angle, it follows within the d,q-reference frame [6.2]

$$\begin{vmatrix} v_{1d} \\ v_{1q} \end{vmatrix} = R_1 \cdot \begin{vmatrix} i_{1d} \\ i_{1q} \end{vmatrix} + \omega_1 L_1 \cdot \begin{vmatrix} 0 & -1 \\ 1 & 0 \end{vmatrix} \cdot \begin{vmatrix} i_{1d} \\ i_{1q} \end{vmatrix} + L_1 \cdot \begin{vmatrix} \frac{di_{1d}}{dt} \\ \frac{di_{1q}}{dt} \end{vmatrix} + \omega_1 \cdot \begin{vmatrix} 0 & -1 \\ 1 & 0 \end{vmatrix} \cdot \begin{vmatrix} \psi_{2d} \\ \psi_{2q} \end{vmatrix}. \tag{6.84}$$

With $\psi_{2q} = 0$ follows from Eq. (6.84)

$$v_{1d} = R_1 \cdot i_{1d} + L_1 \cdot \frac{di_{1d}}{dt} - \omega_1 \cdot L_1 \cdot i_{1q} \tag{6.85}$$

$$v_{1q} = R_1 \cdot i_{1q} + L_1 \cdot \frac{di_{1q}}{dt} + \omega_1 \cdot L_1 \cdot i_{1d} + \omega_1 \cdot \psi_{2d}. \tag{6.86}$$

The equation of motion for this motor with the internal torque T_i is

$$J \cdot \frac{d\omega_m}{dt} = T_i - T_l = \frac{3}{2} \cdot p \cdot \psi_{2d} \cdot i_{1q} - T_l. \tag{6.87}$$

With the transformation of Eqs. (6.85) to (6.87) into the Laplace domain follows

$$i_{1d} = \frac{\frac{1}{R_1}}{1 + s \cdot \frac{L_1}{R_1}} \cdot (v_{1d} + \omega_1 \cdot L_1 \cdot i_{1q}) \tag{6.88}$$

$$i_{1q} = \frac{\frac{1}{R_1}}{1 + s \cdot \frac{L_1}{R_1}} \cdot (v_{1q} - \omega_1 \cdot L_1 \cdot i_{1d} - \omega_1 \cdot \psi_{2d}) \tag{6.89}$$

$$J \cdot s \cdot \omega_m = \frac{3}{2} p \cdot \psi_{2d} \cdot i_{1q} - T_l. \tag{6.90}$$

On the basis of Eqs. (6.88) to (6.90), the dynamic equivalent circuit can be derived (Fig. 6.50).
From Fig. 6.50, the following relation results

$$\omega_m = \left[(v_{1q} - \omega_m \cdot p \cdot i_{1d} \cdot L_1 - \omega_m \cdot p \cdot \psi_{2d}) \cdot \frac{\frac{1}{R_1}}{1 + s \cdot \frac{L_1}{R_1}} \cdot \frac{3}{2} \cdot p \cdot \psi_{2d} - T_l \right]$$

$$\cdot \frac{1}{s \cdot J}. \tag{6.91}$$

To determine the transfer function of the synchronous motor, Eq. (6.91) must be changed by using the following shortcuts for amplification factor and time constants

$$K_S = \frac{1}{p \cdot (i_{1d} \cdot L_1 + \psi_{2d})}, \quad T_1 = \frac{L_1}{R_1}, \quad T_2 = \frac{2}{3} \cdot \frac{R_1 J}{p^2 \cdot (i_{1d} \cdot L_1 + \Psi_{2d}) \cdot \Psi_{2d}}. \tag{6.92}$$

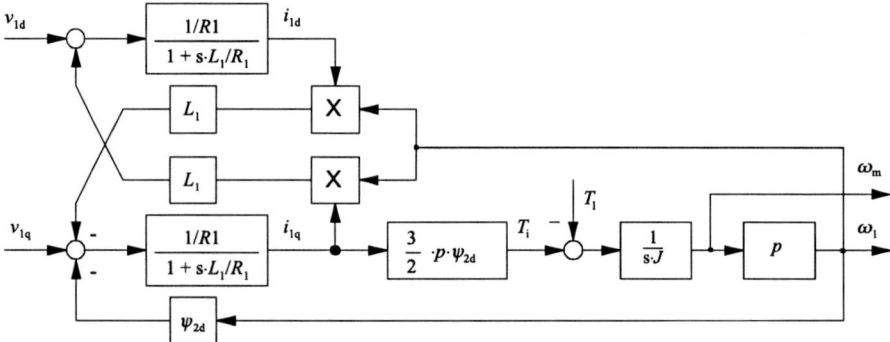

Fig. 6.50. Dynamic equivalent circuit diagram of the electronically commutated motor

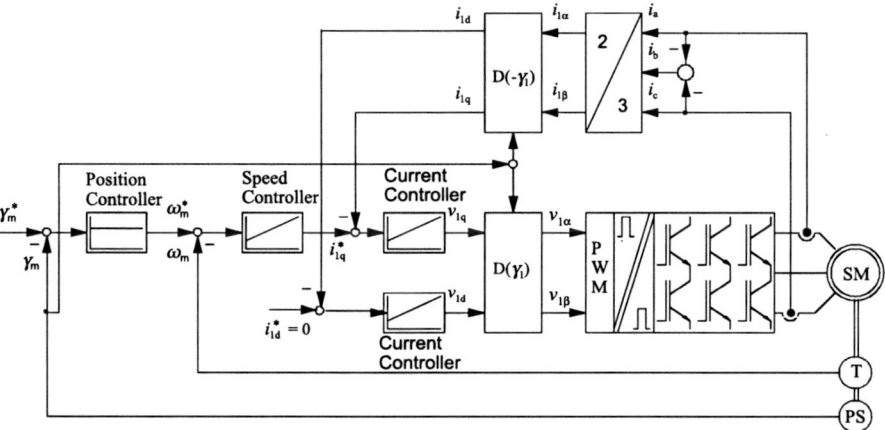

Fig. 6.51. Block diagram for the closed-loop cascade control of position of an electronically commutated motor with subcontrollers for rotational speed and motor current

Thus, the transfer function of the electronically commutated motor is

$$G(s) = \frac{\omega_m(s)}{v_{1q}(s)} = \frac{K_S}{1 + s \cdot T_2 + s^2 \cdot T_1 \cdot T_2}. \qquad (6.93)$$

Figure 6.51 shows the block diagram for the closed-loop cascade control of position of an electronic motor with subcontrollers for rotational speed and motor current.

6.5 Information on the Practice-Oriented Adjustment of Controllers and Simulation of Rotating Speed Closed-Loop Control Circuits

The previous sections refer to procedures for the determination of the controller parameters that are already applied in conventional control technique. Here, a procedure for the simulation of closed-loop rotational speed control is demonstrated that is directly applicable without high complexity.

In connection with the setting-up of closed-loop control operation of electric drives, frequently, the rotational speed control loop causes difficulties. Therefore, one possibility will be described to find the controller parameters. With these, stable behaviour of the control loop can be achieved in most cases.

As shown in the previous sections, the transient characteristic between the voltage v_{1q} and rotor angular speed is always given by the transfer function

$$G(s) = \frac{\omega_m(s)}{v_{1q}(s)} = \frac{K_S}{1 + s \cdot T_2 + s^2 \cdot T_1 \cdot T_2} = \frac{\frac{K_S}{T_1 \cdot T_2}}{\frac{1}{T_1 \cdot T_2} + s \cdot \frac{1}{T_1} + s^2} \,. \quad (6.94)$$

For all individual types of motors, different relations only appear for the computation of the amplification factor K_S and for the time constants T_1 and T_2. With the usual time constants, the motors all show an oscillating behaviour with decaying oscillation, as described by the transfer function (Eq. 6.94). In order to control this process, a proportional plus integral controller (PI-controller) is used for this application with the transfer function

$$G_R(s) = \frac{v_{1q}(s)}{(\omega_m^*(s) - \omega_m(s))} = K_R \cdot \left(1 + \frac{1}{s \cdot T_N}\right) \,. \quad (6.95)$$

Then, operation of the controller can be started with the parameters

$$K_R \approx \frac{1}{K_S} \quad \text{and} \quad T_N \approx (T_1 + T_2) \,. \quad (6.96)$$

For this purpose only an estimation of these process parameters is necessary.

Most servo controllers available on the market are frequency converters with an integrated controller structure. These have the additional capability of data acquisition. It so becomes possible to record the dynamic behaviour of the process data like the reference variable, the control signal and the output of the system. From these recordings, evaluation information can be obtained to change the controller parameters in order to improve the behaviour of the control loop.

In addition, the possibility exists to first simulate the control loop for the optimization of the controller parameters. To achieve this, the transfer functions from motor and controller must be discretized and converted into difference equations. For the rotational speed closed-loop control system, this is done by the z-transformation. In addition, the frequency of the undamped oscillation must be known for the transmission system

$$\beta = \frac{1}{\sqrt{T_1 \cdot T_2}}, \tag{6.97}$$

the damping factor

$$\alpha = \frac{1}{2 \cdot T_1} \tag{6.98}$$

and the frequency of the damped oscillation

$$\omega_e = \sqrt{\beta^2 - \alpha^2}. \tag{6.99}$$

Thus, the transfer function results from Eq. (6.94)

$$G(s) = \frac{\omega_m(s)}{v_{1q}(s)} = \frac{K_S \cdot \beta^2}{\beta^2 + 2 \cdot \alpha \cdot s + s^2}. \tag{6.100}$$

From this, the z-transfer function follows by application of the z-transformation

$$G(z) = \frac{\omega_m(z)}{v_{1q}(z)} = \frac{b_0 + b_1 \cdot z^{-1} + b_2 \cdot z^{-2}}{1 + a_1 \cdot z^{-1} + a_2 \cdot z^{-2}} \tag{6.101}$$

with the parameters

$$b_0 = 0 \tag{6.102}$$

$$b_1 = K_S \cdot \left[1 - e^{-\alpha \cdot T_0} \cdot \left(\frac{\alpha}{\omega_e} \cdot \sin(\omega_e \cdot T_0) + \cos(\omega_e \cdot T_0)\right)\right] \tag{6.103}$$

$$b_2 = K_S \cdot e^{-\alpha \cdot T_0} \cdot \left(e^{-\alpha \cdot T_0} + \frac{\alpha}{\omega_e} \cdot \sin(\omega_e \cdot T_0) - \cos(\omega_e \cdot T_0)\right) \tag{6.104}$$

$$a_1 = -2 \cdot e^{-\alpha \cdot T_0} \cdot \cos(\omega_e \cdot T_0) \tag{6.105}$$

$$a_2 = e^{-2 \cdot \alpha \cdot T_0}. \tag{6.106}$$

From the z-transfer function, the difference equation results with the right shift set of the z-transformation

$$\begin{aligned}\omega_m(k) = &\, b_0 \cdot v_{1q}(k) + b_1 \cdot v_{1q}(k-1) + b_2 \cdot v_{1q}(k-2) \\ &- a_1 \cdot \omega_m(k-1) - a_2 \cdot \omega_m(k-2).\end{aligned} \tag{6.107}$$

The discretisation of the transfer function of the PI-controller is achieved by forming differences, which can be done with a very low sample period in relation to the time constants of the process which shall be controlled. The z-transfer function results

$$G_R(z) = \frac{v_{1q}(z)}{(\omega_m^*(z) - \omega_m(z))} = \frac{q_0 + q_1 \cdot z^{-1}}{1 - z^{-1}} \tag{6.108}$$

with the parameters

$$q_0 = K_R \tag{6.109}$$

$$q_1 = K_R \cdot \left(\frac{T_0}{T_N} - 1\right). \tag{6.110}$$

From Eq. (6.108), the difference equation follows with the right shift set of the z-transformation

$$v_{1q}(k) = v_{1q}(k-1) + q_0 \cdot (\omega_m^*(k) - \omega_m(k)) + q_1 \cdot (\omega_m^*(k-1) - \omega_m(k-1)). \tag{6.111}$$

With Eqs. (6.107) and (6.111), a simulation routine can be developed for the rotational speed control system derived from the block diagram in Fig. 6.52 using a general programming language. In this diagram an additional disturbance variable $n(z)$ is introduced. For simulation of the control loop, software available on the market can also be used.

The simulation method is shown for a rotational speed control system with the amplification factor $K_S = 1.6\,1/\text{V} \cdot \text{s}$ and the time constants $T_1 = 0.05\,\text{s}$ and $T_2 = 0.09\,\text{s}$. Thereof, the adjusting parameters $K_R = 0.625\,\text{V} \cdot \text{s}$ and $T_N = 0.14\,\text{s}$ result. The sample time amounts to $T_0 = 0.001\,\text{s}$. The results of the simulation are represented in Fig. 6.53. It shows a movement procedure with a starting ramp, a phase with constant rotational speed and a braking ramp. Additionally, a disturbance variable $n(z)$ is added in the range of constant rotational speed. The entire control procedure lasts 1.5 s

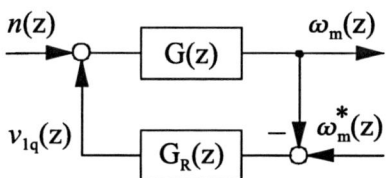

Fig. 6.52. Block diagram for the simulation of a rotational speed-control loop

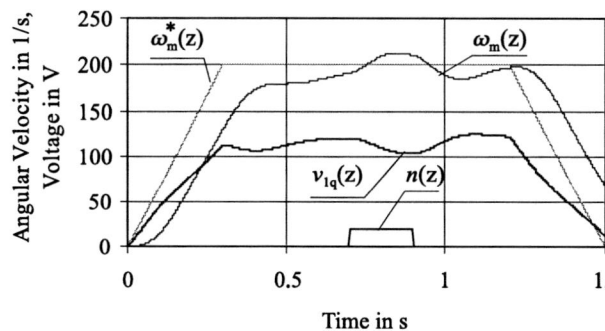

Fig. 6.53. Simulation results of the rotational speed-control loop with simplified determination of the controller parameters

and makes very high demands upon the control functions. The dynamic control quality can be improved for the selected example, if after this first adjusting step the amplification factor is increased and/or the integral time is reduced [6.19].

Further information about the behaviour of the control loop can be obtained when its reference transfer function is set up in the z-domain with Eqs. (6.101) and (6.108). With the parameters for the rotational speed control system, the poles can be computed. Their position provides information on the behaviour of the control loop.

6.6 Structure of Drive Electronics for Three-Phase Alternating Current Motors Using Circuits with Large-Scale Integration

Three-phase alternating current motors are used increasingly for small drive powers. The reasons are their excellent characteristics concerning the closed-loop rotational speed and position control and because they have low maintenance costs and a long life. This trend is favoured by the fact that, increasingly, components with large-scale integration of power and control electronics are available on the market. Thus, the structure of hardware is highly simplified, which contributes substantially to cost reduction and to operational reliability. In Fig. 6.54 the structure of such drive electronics for three-phase motors is indicated schematically.

As Fig. 6.54 shows, such drive electronics only consist of two components. These are a µController as master control unit and an intelligent power module briefly designated as IPM. These are to be connected in the indicated way to the motor [6.20, 6.21].

The microcontroller measures the necessary actual values of the drive and determines the necessary control variables for the connected motor from these based on predefined open-loop and/or closed-loop control algorithms.

For the power section, only an IPM is necessary in such a structure. This contains the three-phase bridge, here developed with IGBTs, and is here, for low-power applications, connected to AC mains. For the smoothing of the intermediate circuit voltage produced by an internal rectifier, an external filter capacitor is to be attached. In the IPM, a module is integrated for galvanic separation of the control unit and the power section as well as for the level adjustment of the gate signals to the changing emitter potentials of the IGBTs T_{a+}, T_{b+} and T_{c+}. In addition, a special control voltage, usually 15 Vdc, is necessary. This is produced by an internal switch mode power supply (SMPS).

For safe operation and for protection against fault switching, fault manipulations and overload, the IPM is additionally equipped with a number of protective functions. Different operating conditions are monitored by comparators with Schmitt trigger behaviour concerning the exceeding and/or the undershooting of permissible operating conditions. If these are arranged within

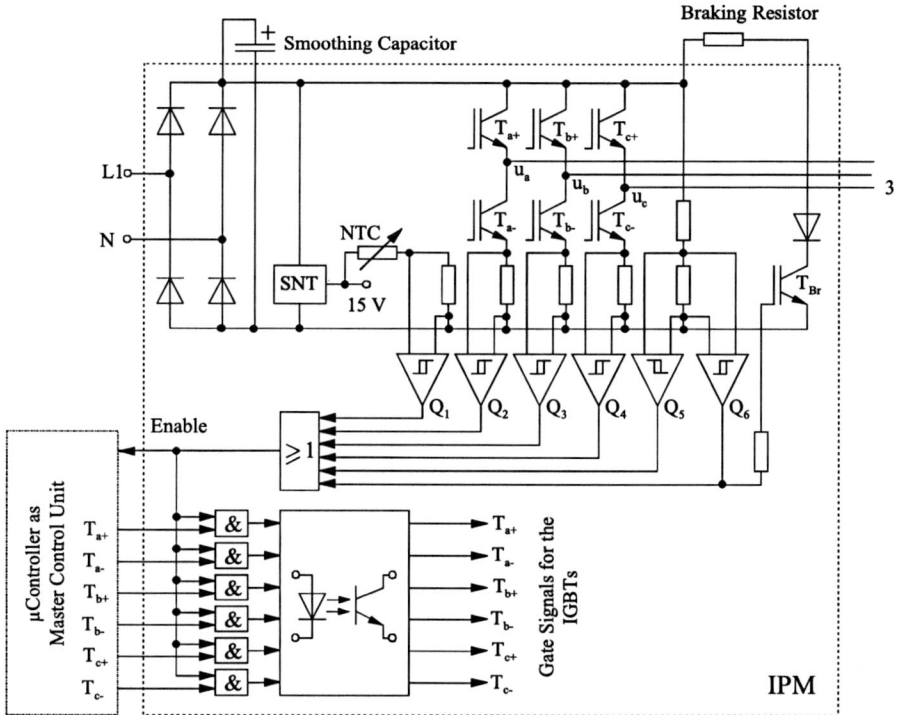

Fig. 6.54. Schematic structure of drive electronics for three-phase alternating current motors using large-scale integrated components for control and power electronics

the permissible range, all comparators provide an H signal. From these, an ENABLE signal (valid signal) is generated *via* an OR linkage. As soon as one of the monitored operating conditions exceeds and/or undershoots the permissible range, the ENABLE signal changes from H signal to L signal and sets the control signals for all IGBTs to L signal in the IPM. Thus, all IGBTs change into the non-conductive condition. At the same time, an interrupt service routine for the introduction of suitable procedures can be activated in the microcontroller by this signal change.

In the representation in Fig. 6.54, temperature is controlled by an NTC (Q1). With the help of shunts, the individual half-bridges are controlled for short circuits (Q2 to Q4). By means of a voltage divider, the control of the intermediate circuit voltage for undershooting (Q5) and/or exceeding (Q6) of fixed limit values take place.

The example selected here shows the structure of an IPM only schematically. These devices are offered on the market by many manufacturers. For this reason the characteristics can vary strongly. In some IPMs, a measuring device for the phase currents i_a, i_b and i_c is already realized. The user has the possibility to select which IPM is best suited for his application.

Different types of motors (ASM, PMSM) can be attached to a circuit, as schematically represented in Fig. 6.54, without changing the hardware. By the software installed on the microcontroller, different modes of operation (open-loop and closed-loop control of rotational speed, closed-loop control of position) can be realized with different kinds of motors.

Generating Control Signals in the Microcontroller for the Power Switches of the Three-Phase Bridge

The generation of control signals for the power semiconductors of the three-phase bridge is described by use of the example of space vector modulation. Space vector modulation is described in Sect. 6.2.4.3. It is to be demonstrated here in example how these signals for the segment $0° \leq 60°$ are produced. The switching times t_1 and t_2 of the switching status 1 and 2 are determined with Eqs. (6.7) and (6.8). These depend on the voltage space vector $\underline{v}^*$ and on the angle γ^*. Depending on the size of these two values, $t_1 + t_2$ is mostly smaller than the cycle time T_T. Thus the time t_P results in

$$t_P = T_T - t_1 - t_2 . \tag{6.112}$$

During this period t_P a zero vector must be inserted into the switching cycle (only either the upper or the lower power switches of the three-phase bridge are conductive).

Here, for the time t_P the zero-vector is always realized by conductive power switches T_{a-}, T_{b-} and T_{c-}. Which switching status in the range $0° \leq 60°$ is to be realized can be taken from Table 6.1.

Current microcontrollers have timers with symmetric output signals for the realization of pulse-width modulation. The time-dependent output signals of the timer and of all control signals for the power switches in the three-phase bridge are presented in Fig. 6.55

The cycle time is computed with the relation

$$T_T = 2 \cdot (\text{TICO} - \text{TC}) \cdot \frac{1}{f_{\text{TIMER}}} . \tag{6.113}$$

TICO and TC are the final value and initial value of the timer process, and f_{Timer} is the clock frequency of the timer. The values for turning on and off the power switches T_{a+}, T_{b+} and T_{c+} for the range $0° \leq 60°$ can be calculated by the equations

$$X_{\text{TaPWM}} = \text{TICO} - \frac{t_1 + t_2}{T_T} \cdot (\text{TICO} - \text{TC}) \tag{6.114}$$

$$X_{\text{TbPWM}} = \text{TICO} - \frac{t_1}{T_T} \cdot (\text{TICO} - \text{TC}) \tag{6.115}$$

$$X_{\text{TcPWM}} = \text{TICO} . \tag{6.116}$$

The cycle time T_T and the switching times of the individual switching states are shown in Fig. 6.55. The same considerations undertaken here for the segment $0° \leq 60°$ must be made for the other five segments.

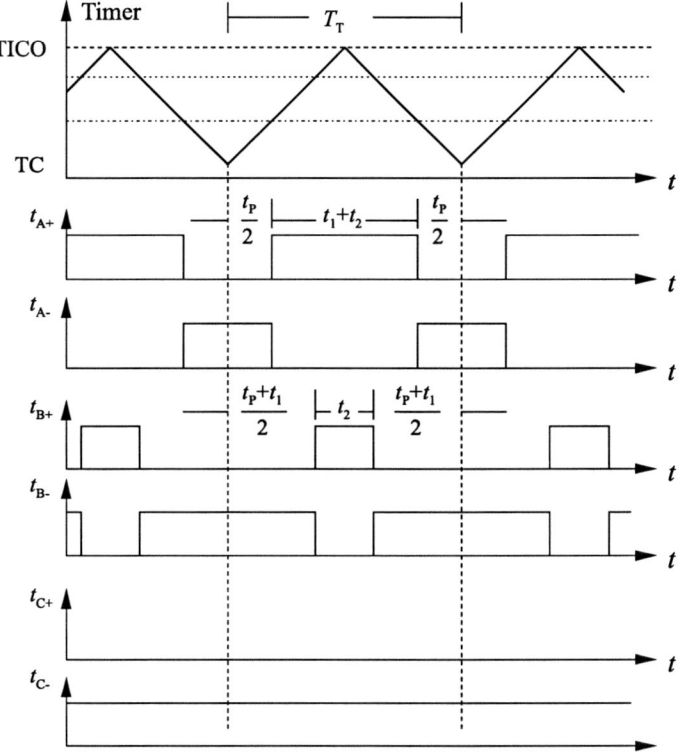

Fig. 6.55. Time-dependent output signals of the timer and of all control signals for the power switches in the three-phase bridge by applying space vector modulation

Most of the μControllers planned for this application still have a dead band generator, with which it is possible to adjust a dead time between turning off and turning on the power switches of the same half-bridge.

6.7 Sensorless Determination and Closed-Loop Control of Rotational Speed, Rotor Position and Internal Torque

In order to realize economical and robust drives with closed-loop control for rotational speed and position with moderate precision demands, one more and more abandons on the application of special sensors and determines, depending on type of motor, the internal torque, the rotational speed and/or the rotor position from the applied phase voltages and the measured currents by using a dynamic mathematical model of the motor. This technique is known as sensorless closed-loop control. A condition for this is that all relevant motor parameters for the computation are available.

In the following sections, solutions will be shown how the rotor position and internal torque for DC motors, permanent magnet synchronous and asynchronous motors can be computed from applied phase voltages and measured phase currents. The precision of the attainable results depends essentially on how exactly the motor parameters, the phase voltages supplied to the motor and the motor currents can be determined. In addition, it is important to consider the operational changes of these parameters in the computation.

DC Motor (PMGM)

The relations between the rotational speed of the DC motor, the given rotor voltage, the measured rotor current, the internal torque and the motor parameters can be taken from the equivalent control circuit of the DC motor in Fig. 6.37. From this follows the equation

$$i_q(s) = [v_q(s) - e(s)] \cdot \frac{\frac{1}{R}}{1 + s \cdot \frac{L}{R}} = [v_q(s) - c \cdot \psi_d \cdot \omega(s)] \cdot \frac{\frac{1}{R}}{1 + s \cdot \frac{L}{R}}. \tag{6.117}$$

Using the syntax of the differential operator $s = \mathrm{d}/\mathrm{d}t$, the differential equation follows from Eq. (6.117)

$$i_q(t) + \frac{\mathrm{d}i_q(t)}{\mathrm{d}t} \cdot \frac{L}{R} = [v_q(t) - c \cdot \psi_d \cdot \omega(t)] \cdot \frac{1}{R}. \tag{6.118}$$

By discretisation of Eq. (6.118) with the sampling time T_0 and the relation $t = k \cdot T_0$ follows

$$i_q(t) = i_q(k \cdot T_0), \quad \frac{\mathrm{d}i_q(t)}{\mathrm{d}t} = \frac{i_q(k \cdot T_0) - i_q((k-1) \cdot T_0)}{T_0},$$
$$v_q(t) = v_q(k \cdot T_0), \quad \omega(t) = \omega(k \cdot T_0). \tag{6.119}$$

With the simplified syntax $k \cdot T_0 \to k$ and $(k-1) \cdot T_0 \to (k-1)$, the following difference equation results

$$i_q(k) + \frac{i_q(k) - i_q(k-1)}{T_0} \cdot \frac{L}{R} = [v_q(k) - c \cdot \psi_d \cdot \omega(k)] \cdot \frac{1}{R}. \tag{6.120}$$

A conversion of Eq. (6.120) gives

$$\omega(k) = \frac{v_q(k) - R \cdot i_q(k) - L \cdot \frac{i_q(k) - i_q(k-1)}{T_0}}{c \cdot \psi_d}. \tag{6.121}$$

If the angular speed of the motor is known, the rotor angle can be computed from it by the equation

$$\gamma(k) = \gamma(k-1) + \omega(k) \cdot T_0. \tag{6.122}$$

The internal torque can be computed in accordance with Fig. 6.37 with application of the discretisation

$$T_i(k) = c \cdot \psi_d \cdot i_q(k). \tag{6.123}$$

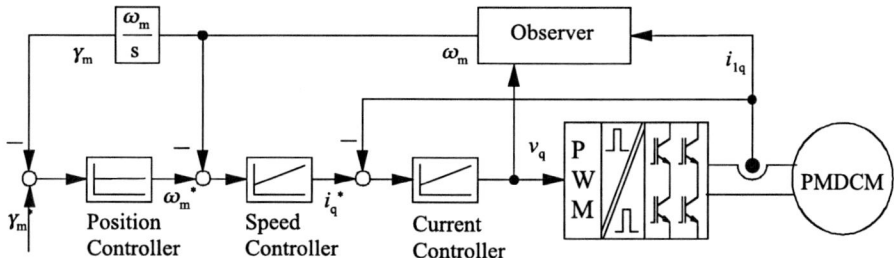

Fig. 6.56. Block diagram for the sensorless closed-loop position control with subcontrollers for rotational speed and motor current of a DC motor with permanent magnet excitation

Figure 6.56 shows the block diagram for the closed-loop position control with subcontrollers for rotational speed and motor current of a DC motor with permanent magnet excitation. The angular speed of the rotor can be computed with motor parameters using Eq. (6.121). By integrating the angular speed, the rotor position can be calculated by Eq. (6.122). Furthermore, it is also possible to determine internal torque using Eq. (6.123).

With consideration of the motor efficiency, the torque at the motor shaft can also be computed from the internal torque.

*Permanent Magnet Excited Synchronous Motor
(PMSM, Brushless DC Motor, EC Motor)*

The following considerations refer to a two-pole, permanent magnet excited synchronous motor. If rotational speed and rotor angle for this motor have to be determined by the supplied phase voltages and the measured phase currents without special sensors, it is a precondition that the rotor position is detected at the beginning of the procedure. In order to determine this there are several possibilities. The simplest uses stator inductance dependent on the rotor position. At the rotor position and the stator voltage indicated in

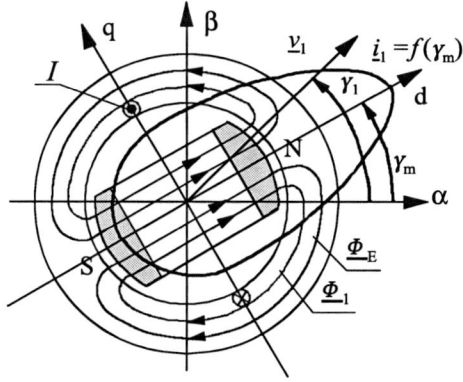

Fig. 6.57. Schematic representation of the magnetic field lines of flux and the current locus with a rotating voltage space vector of constant length

Fig. 6.57, the flux of the permanent magnet and the flux caused by the armature current of the stator winding add up. Thus, the magnetic circuit of the motor in the d-direction generates a smaller inductance than in the q-direction orthogonal to the d-direction. A condition for this technique is that the rotors show distinct unsymmetries in their ferromagnetic structure along the rotor circumference and that this is also detectable by current measurement.

According to Ampers law, the magnetic circuit as indicated in Fig. 6.57 leads to

$$w \cdot I + H_M \cdot l_M = H_{Fe} \cdot l_{Fe} + H_\delta \cdot l_\delta . \quad (6.124)$$

In Eq. (6.124) are

l_M entire length of the magnet material in direction of magnetization,
l_{Fe} entire length of the iron path,
l_δ entire length of the lines of flux in the air gap ($l_\delta = 2 \cdot \delta$ doubled air gap length)
I current toward the fixed d-axis

With $H_{Fe} \cdot l_{Fe} \ll H_d \cdot l_d$ follows from Eq. (6.124)

$$H_\delta \cdot l_\delta = H_M \cdot l_M + w \cdot I . \quad (6.125)$$

Is Eq. (6.125) divided by $A \cdot B$ follows

$$\frac{H_\delta \cdot l_\delta}{B \cdot A} = \frac{H_M \cdot l_M + w \cdot I}{B \cdot A} = R_\delta . \quad (6.126)$$

In order to simplify, it is assumed for Eq. (6.126) that the same flux density is present in the permanent magnet material, in the iron and in the air gap and also it is simplifying assumed that all have approximately the same effective cross section. In practice, this condition is not always achieved.

The relation between the magnetic resistance R_M and the inductance of the magnetic circuit results in this case in

$$L_1 \sim \frac{1}{R_\delta} = \frac{B \cdot A}{H_M \cdot l_M + w \cdot I} . \quad (6.127)$$

If the voltage vector v_1 is rotated by 180°, it follows that

$$H_M \cdot l_M - w \cdot I = H_{Fe} \cdot l_{Fe} + H_\delta \cdot l_\delta . \quad (6.128)$$

With the same considerations as made in (Eqs. (6.124)–(6.127)), the inductance of the magnetic circuit in the case of the new rotor position results in

$$L_2 \sim \frac{1}{R_\delta} = \frac{B \cdot A}{H_M \cdot l_M - w \cdot I} . \quad (6.129)$$

For the measurement, the rotor must be blocked. At the position where the stator voltage space vector produces the smallest inductance due to the stator, the largest current flows with the smallest phase shift between the stator voltage and stator current. Thus, for a blocked rotor, a rotating stator voltage space vector produces the current locus represented in Fig. 6.57 and the rotor angle can be clearly determined on the base of the current locus.

For further considerations, the single-phase equivalent circuit diagram of the synchronous motor is used. This is represented in Fig. 6.58. Here, the relation $X_d = \omega \cdot L_1$ can be used to compute the voltage drop of $jX_d \cdot i_1$ at the inductance.

The allocation of stator voltage vector $\underline{v}_1$, stator current vector $\underline{i}_1$ and synchronous generated internal voltage vector $\underline{v}_P$ is indicated in Fig. 6.59 as both real and imaginary part.

The allocation of stator voltage, stator current and synchronous generated internal voltage, indicated in Fig. 6.59, is for any rotary field angle γ_1 to be transferred into a stator-fixed α, β-reference frame. For this system, the voltages $v_{1\alpha}$ and $v_{1\beta}$ and the currents $i_{1\alpha}$ and $i_{1\beta}$ can be calculated from the applied phase voltages v_a, v_b, and v_c as well as from the measured phase currents i_a, i_b, and i_c by Eqs. (6.19) and (6.20). From the currents $i_{1\alpha}$ and

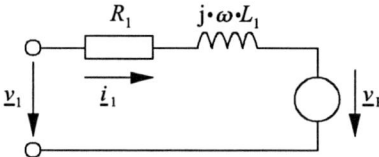

Fig. 6.58. Single-phase equivalent circuit of the synchronous motor with permanent magnet excitation

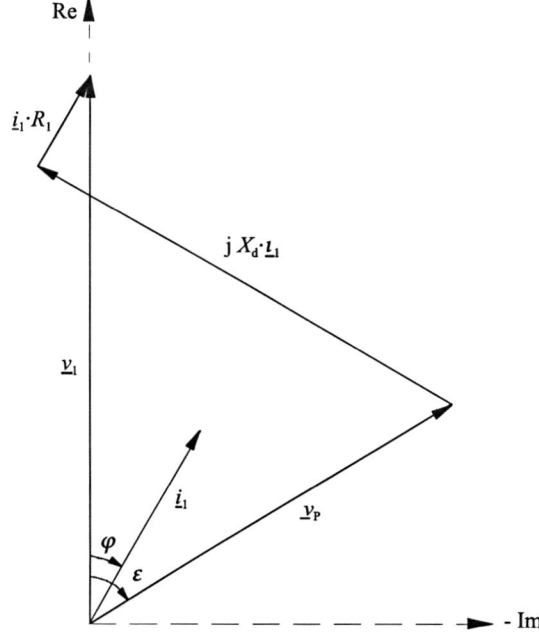

Fig. 6.59. Allocation of the space vectors of the stator voltage, the stator current and the synchronous generated internal voltage in the complex coordinate system

$i_{1\beta}$, the resulting current space vector can be calculated with the equation

$$|\underline{i}_1| = \sqrt{i_{1\alpha}^2 + i_{1\beta}^2} \,. \tag{6.130}$$

Figure 6.60 shows the space vector diagram for a loaded synchronous motor considering the stator voltage vector $\underline{v}_1$, stator current vector $\underline{i}_1$ and a synchronous generated internal voltage vector $\underline{v}_P$ as well as the rotational angle of the rotor.

From Fig. 6.60, the power of the synchronous motor results in

$$P = \underline{v}_1 \cdot \frac{v_P}{X_d} \cdot \sin(\gamma_1 - \gamma_m) \,. \tag{6.131}$$

From Eq. (6.131) for the internal torque follows

$$T_i = \frac{P}{\omega_m} = \underline{v}_1 \cdot \frac{v_P}{X_d \cdot \omega_m} \cdot \sin(\gamma_1 - \gamma_m) \,. \tag{6.132}$$

Thus, the largest theoretical internal torque results in the case of a load angle of $\gamma_1 - \gamma_m = 90°$. A synchronous motor is usually operated with a load angle of approximately 70°. If the load angle $\gamma_1 - \gamma_m$ is kept constant and the synchronous generated internal voltage $\underline{v}_P$ adjusts itself for the specific

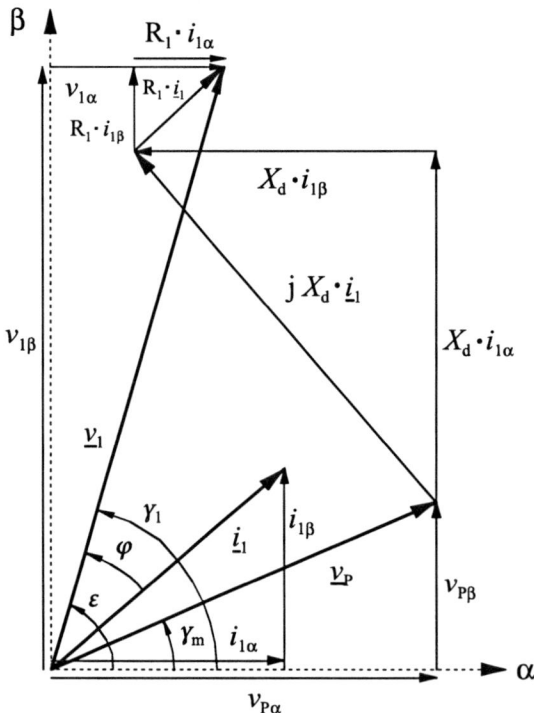

Fig. 6.60. Space vector diagram (voltage and current space vectors) of the synchronous motor

operation, then the internal torque can be affected by the length of the stator voltage space vector $\underline{v}_1$.

From the space vector diagram in Fig. 6.60, the following relation can be read

$$\underline{v}_1 = \underline{i}_1 \cdot R_1 + j X_d \cdot \underline{i}_1 + \underline{v}_P .\tag{6.133}$$

By separating the voltage and the current space vector into the components of the stator-fixed α, β-reference frame, the following equations result from (6.133)

$$v_{1\alpha} = R_1 \cdot i_{1\alpha} + X_d \cdot i_{1\beta} + v_{P\alpha} \tag{6.134}$$

and

$$v_{1\beta} = R_1 \cdot i_{1\beta} + X_d \cdot i_{1\alpha} + v_{P\beta} .\tag{6.135}$$

The synchronous reactance X_d can be calculated from the inductance L_1 and from the angular frequency of the rotating field by the relation

$$X_d = j \, \omega_1 \cdot L_1 .\tag{6.136}$$

If $v_{1\alpha}$ and $v_{1\beta}$ are applied to the motor and $i_{1\alpha}$ and $i_{1\beta}$ are calculated by the coordinate transformation from the measured phase currents, the voltages $v_{P\alpha}$ and $v_{P\beta}$ can be computed with the relations

$$v_{P\alpha} = v_{1\alpha} - R_1 \cdot i_{1\alpha} - j X_d \cdot i_{1\beta} \tag{6.137}$$

and

$$v_{P\beta} = v_{1\beta} - R_1 \cdot i_{1\beta} - j X_d \cdot i_{1\alpha} .\tag{6.138}$$

In accordance with Fig. 6.60, the angles γ_1 and γ_m are computed from the stator voltage components and from the synchronous generated internal voltage components by the equations

$$\gamma_1 = \arctan \frac{v_{1\beta}}{v_{1\alpha}} \tag{6.139}$$

and

$$\gamma_m = \arctan \frac{v_{P\beta}}{v_{P\alpha}} .\tag{6.140}$$

For the practical calculation, the following relations result from Eqs. (6.137) to (6.140), if they are discretized in the same manner as Eqs. (6.119) and (6.120),

$$v_{P\alpha}(k) = v_{1\alpha}(k) - R_1 \cdot i_{1\alpha}(k) - j X_d(k) \cdot i_{1\beta}(k) \tag{6.141}$$

$$v_{P\beta}(k) = v_{1\beta}(k) - R_1 \cdot i_{1\beta}(k) - j X_d(k) \cdot i_{1\alpha}(k) \tag{6.142}$$

$$\gamma_1(k) = \arctan \frac{v_{1\beta}(k)}{v_{1\alpha}(k)} \tag{6.143}$$

$$\gamma_m(k) = \arctan \frac{v_{P\beta}(k)}{v_{P\alpha}(k)} .\tag{6.144}$$

The angular velocity of the rotor follows from the equation

$$\omega_m(k) = \frac{\gamma_m(k) - \gamma_m(k-1)}{T_0}. \tag{6.145}$$

For the reasons specified above, a simple closed-loop position control is best suited to the structure represented in Fig. 6.61. From the applied phase voltages and the measured phase currents the angles γ_1 and γ_m can be calculated with Eqs. (6.143) and (6.144) specified above. In the circuit in Fig. 6.61, the desired load angle ε^* and the desired rotor angle γ_m^* can be specified.

For the realization of the given function, $\omega_m = f(\gamma_m)$, the circuit configuration in Fig. 6.61 is unsuitable because here no control of the current takes place and so the danger of motor overload exists. In order to avoid this, a consideration of the synchronous motor in the d, q-reference frame is recommendable. In this, the torque can be affected directly via the presetting of the torque-generating current component i_{1q}. Because it concerns a permanent magnet excited motor, the flux-generating current i_{1d} is set to zero.

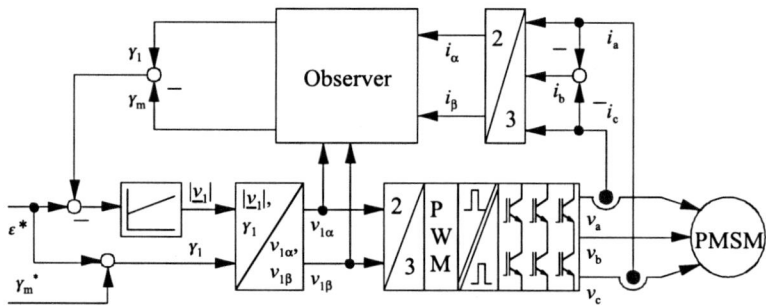

Fig. 6.61. Block diagram for a load angle and a closed-loop position control of a permanent magnet excited synchronous motor (PMSM)

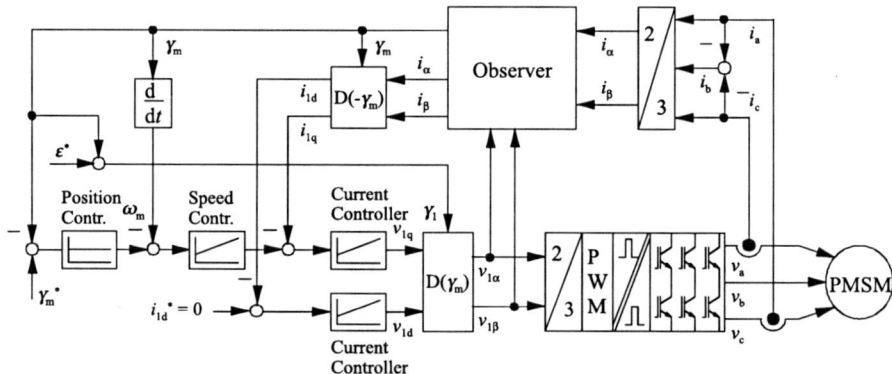

Fig. 6.62. Block diagram of a closed-loop position control with subcontrollers for rotational speed and motor current

The block diagram of a closed-loop position control with subcontrollers for rotational speed and motor current is shown in Fig. 6.62.

As for the circuit in Fig. 6.61, the angle γ_m is to be computed by the observer. By an inverse rotating transformation as in Eq. (6.24), the actual values of the voltages v_{1d} and v_{1q}, as well as of the currents i_{1d} and i_{1q} can be calculated. Then, in consideration of the rotating-field angle γ_1 the values v_{1d} and v_{1q} provided by the controllers must be converted into the voltage values $v_{1\alpha}$ and $v_{1\beta}$ necessary for the frequency converter. This is done with Eqs. (6.139)–(6.143).

$$\begin{pmatrix} v_{1\alpha} \\ v_{1\beta} \end{pmatrix} = \begin{pmatrix} \cos\gamma_1 & -\sin\gamma_1 \\ \sin\gamma_1 & \cos\gamma_1 \end{pmatrix} \cdot \begin{pmatrix} v_{1d} \\ v_{1q} \end{pmatrix} = D(\gamma_1) \cdot \begin{pmatrix} v_{1d} \\ v_{1q} \end{pmatrix}. \tag{6.146}$$

Asynchronous Motor

Also for the asynchronous motor, the possibility exists for computing the rotational speed from the supplied voltages and the measured currents by the parameters of the motor with the help of system theory. For this purpose, the single-phase equivalent circuit in the representation of the stator-fixed α, β-reference frame and the flux oriented d, q-reference frame must be used. These representations are illustrated in Figs. 6.42 and 6.63.

Starting point for the computation of the rotational speed from the terminal quantities is the internal torque computation in the d, q-reference frame. This can be determined from Fig. 6.41 and, with consideration of the number of pole pairs, results in

$$T_i = p \cdot (\psi_{2q} \cdot i_{2d} - \psi_2 \cdot i_{2q}). \tag{6.147}$$

The currents i_{2d} and i_{2q} are not measurable and must be substituted. For this, the relations

$$i_{2d} \cdot R_2 = \omega_2 \cdot \psi_{2q} - \dot{\psi}_{2d} \tag{6.148}$$

and

$$i_{2q} \cdot R_2 = -\omega_2 \cdot \psi_{2d} - \dot{\psi}_{2q} \tag{6.149}$$

are taken from the single-phase equivalent circuit of the asynchronous motor represented in Fig. 6.63.

A conversion of Eqs. (6.148) and (6.149) results in

$$i_{2d} = \frac{\psi_{2q}}{R_2} \cdot \omega_2 - \frac{\dot{\psi}_{2d}}{R_2} \tag{6.150}$$

and

$$i_{2q} = -\frac{\psi_{2d}}{R_2} \cdot \omega_2 - \frac{\dot{\psi}_{2q}}{R_2}. \tag{6.151}$$

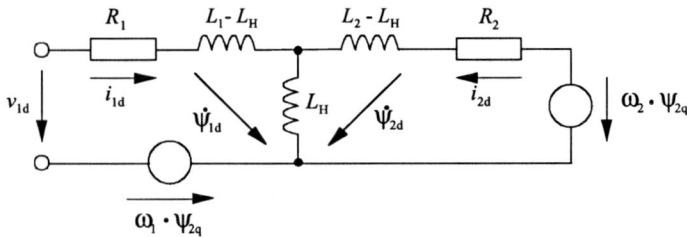

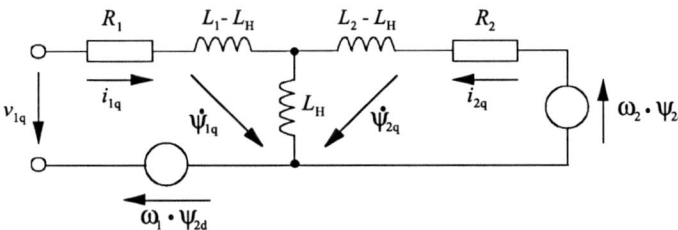

Fig. 6.63. Representation of the single-phase equivalent circuit of the asynchronous motor in the d, q-reference frame

Equations (6.150) and (6.151) are inserted in Eq. (6.147). This results in

$$T_\text{i} = \frac{p}{R_2} \cdot \left(\omega_2 \cdot \psi_{2q}^2 - \psi_{2q} \cdot \dot{\psi}_{2d} + \omega_2 \cdot \psi_{2d}^2 + \psi_{2d} \cdot \dot{\psi}_{2q} \right) . \qquad (6.152)$$

The derivatives of the fluxes with respect to time are assumed to be small compared to the fluxes themselves and therefore are disregarded. Thus follows

$$T_\text{i} = \frac{p}{R_2} \cdot \omega_2 \cdot (\psi_{2q}^2 + \psi_{2d}^2) = \frac{p}{R_2} \cdot \omega_2 \cdot |\psi_2|^2 . \qquad (6.153)$$

A conversion of Eq. (6.153) gives

$$\omega_2 = \frac{T_i}{|\psi_2|^2} \cdot \frac{R_2}{p} . \qquad (6.154)$$

The absolute value of the flux in Eq. (6.154) can also be computed from the flux components $\psi_{2\alpha}$ and $\psi_{2\beta}$, so that it results in

$$\omega_2 = \frac{T_i}{\left(\psi_{2\alpha}^2 + \psi_{2\beta}^2\right)} \cdot \frac{R_2}{p} . \qquad (6.155)$$

The fluxes in Eq. (6.155) must be computed from the terminal quantities. This can be achieved with Eqs. (6.51) to (6.56), using the motor parameters and Eq. (6.42).

The internal torque in the α, β-reference frame can be computed with the relation Eq. (6.59). Thus follows from Eq. (6.155)

$$\omega_2 = R_2 \cdot \frac{L_H}{L_2} \cdot \frac{\psi_{2\alpha} \cdot i_{1\beta} - \psi_{2\beta} \cdot i_{1\alpha}}{\psi_{2\alpha}^2 + \psi_{2\beta}^2} . \qquad (6.156)$$

With the relations Eqs. (6.55) and (6.56) the equation for the computation for the angular frequency of the slip follows from Eq. (6.156)

$$\omega_2 = R_2 \cdot \frac{L_H}{L_2} \cdot \frac{\dot{\psi}_{2\alpha} \cdot i_{1\alpha} + \dot{\psi}_{2\beta} \cdot i_{1\beta}}{\dot{\psi}_{2\alpha}^2 + \dot{\psi}_{2\beta}^2} \cdot \omega_1 \ . \tag{6.157}$$

If in practice such an angular speed evaluation is carried out in the discrete time domain with a constant sample period T_0, then, with the simplified syntax from Eqs. (6.119) and (6.120) and from Eqs. (6.51), (6.52) and (6.157) following relations can be derived

$$\dot{\psi}_{2\alpha}(k) = \frac{L_2}{L_H} \cdot \left(v_{1\alpha}(k) - R_1 \cdot i_{1\alpha}(K) - \frac{1}{K_{11}} \cdot (i_{1\alpha}(k) - i_{1\alpha}(k-1))/T_0 \right) \tag{6.158}$$

$$\dot{\psi}_{2\beta}(k) = \frac{L_2}{L_H} \cdot \left(v_{1\beta}(k) - R_1 \cdot i_{1\beta}(K) - \frac{1}{K_{11}} \cdot (i_{1\beta}(k) - i_{1\beta}(k-1))/T_0 \right) \tag{6.159}$$

$$\omega_2(k) = R_2 \cdot \frac{L_H}{L_2} \cdot \frac{\dot{\psi}_{2\alpha}(k) \cdot i_{1\alpha} + \dot{\psi}_{2\beta}(k) \cdot i_{1\beta}}{\left(\dot{\psi}_{2\alpha}(k)\right)^2 + \left(\dot{\psi}_{2\beta}(k)\right)^2} \cdot \omega_1(k) \ . \tag{6.160}$$

From the angular frequency of the slip $\omega_2(k)$ and the angular speed of the rotating field $\omega_1(k)$, the angular speed $\omega_m(k)$ of the rotor can be computed with

$$\omega_m(k) = \frac{1}{p} \cdot (\omega_1(k) - \omega_2(k)) \ . \tag{6.161}$$

From the angular speed of the rotor and the sample period, the rotor angle results as well

$$\gamma_m(k) = \gamma_m(k-1) + \omega_m(k) \cdot T_0 \ . \tag{6.162}$$

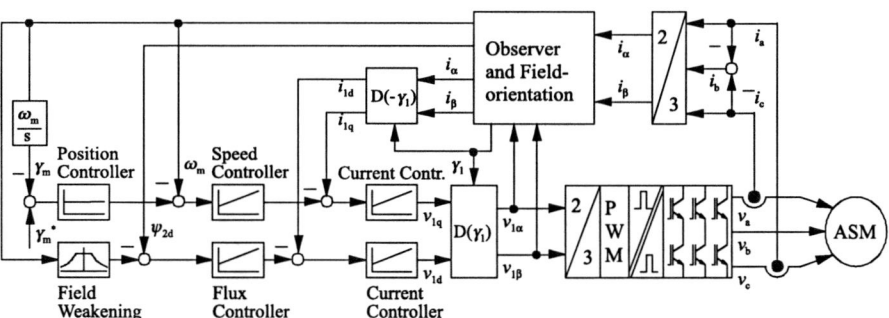

Fig. 6.64. Block diagram of the field-oriented closed-loop control of an asynchronous motor with position controller and subordinated controllers for rotational speed and motor current control without use of special sensors

Figure 6.64 shows the block diagram for the field-oriented closed-loop control of an asynchronous motor. A closed-loop position control with subcontrollers for rotational speed and motor current without use of special sensors is represented. Here, angular velocity and rotor angles are computed by an observer from applied phase voltages and measured phase currents. In the interior control circuit, the control of the position *via* the torque generating current i_{1q} takes place and in the outer circuit the control of the rotor flux *via* the flux generating current i_{1d} is realized. As is to be inferred from the representation, operating in the field weakening range is also possible. Operating within the field weakening range is necessary in order not to exceed the rated power at rotational speed above the rated rotational speed. In addition to the rotational speed and rotor angle, the internal torque of the motor can also be computed with the equations mentioned above by the measured and/or computed motor quantities [6.28, 6.29, 6.30].

Consideration of Operating Conditions

The motor can heat up critically during operation. Because copper shows very high temperature coefficients, it changes resistances of stator and of rotor windings according to Eq. (6.163).

The temperature-dependent resistance of the windings can be computed by following function

$$R_\vartheta = R_{20} \cdot (1 + \alpha_{Cu} \cdot (\vartheta - 20\,°C)) \qquad (6.163)$$

with

α_{cu} Temperature coefficient in 10^{-6}/grd of copper ($\alpha_{cu} = 3030$),
δ Temperature in °C,
R_δ Resistance at the temperature δ,
R_{20} Resistance at the temperature 20 °C.

List of the Most Important Symbols

1 Parameter Designations

i	[A]	Current
i_d	[A]	Flux generating current in the d, q-reference frame
i_q	[A]	Torque generating current in the d, q-reference frame
J	[Nms^2]	Moment of inertia
j	[–]	Runtime variable in the intermittent time interval ($t = j \cdot T_0$)
j	[–]	Marking of reactive parts
k	[–]	Runtime variable in the intermittent time interval ($t = k \cdot T_0$)
K_s		Amplification factor of the controlled system (process)
K_R		Amplification factor of the automatic controller
L	[Vs/A]	Inductance
L_H	[Vs/A]	Main inductance
L_S	[Vs/A]	Leakage inductance
L_{S1}	[Vs/A]	Stator inductance of a phase measured in Y-connection
L_{S2}	[Vs/A]	Rotor inductance of a phase measured in Y-connection and converted to stator side
T_i	[Nm]	Internal torque
T_l	[Nm]	Load torque
N	[min^{-1}]	Revolutions per minute
p	[–]	Number of pole pairs
P	[W]	Power
R	[V/A]	Resistance
R_{1Ph}	[V/A]	Stator resistance of a phase, measured in Y-connection
R_{2Ph}	[V/A]	Rotor resistance of a phase, measured in Y-connection and converted to stator side
s	[s^{-1}]	Differential operator
t	[s]	Time
T_0	[s]	Sample period
T_T	[s]	Cycle time
T_1	[s]	Time constant
T_2	[s]	Time constant
v	[V]	Voltage
V_Z	[V]	Output voltage of the DC intermediate circuit
γ_1	[grd]	Angle between the d, q-reference frame and the α, β-reference frame
γ_m	[grd]	Mechanical angle of the rotor related to the α, β-reference frame
γ_2	[grd]	$\gamma_1 - p \cdot \gamma_m$ angle of the d, q-reference frame referring to a reference frame rotating with $p \cdot \omega_m$
ω	[s^{-1}]	Angular velocity
ω_1	[s^{-1}]	Angular velocity of the d, q-reference frame in relation to the α, β-reference frame

ω_2 [s^{-1}] Angular velocity of the slip $\omega_1 - p \cdot \omega_m$
ω_m [s^{-1}] Angular velocity of the rotor in relation to the α, β-reference frame
ψ [Vs] Magnetic flux linkage
$\underline{i}$ [A] Space vector of the current
$\underline{v}$ [V] Space vector of the voltage
$\underline{\Psi}$ [Vs] Space vector of the magnetic flux linkage

2. Index designations

a, b, c Marking for phase values
α, β Marking for components of an orthogonal, stator-fixed reference frame
d, q Marking for components of an orthogonal reference frame, which is firmly oriented towards an electrical quantity (flux, current or voltage) and/or its reference value
u, v Components of an orthogonal, rotor-fixed reference frame
"1" Marking for stator quantities
"2" Marking for rotor quantities
"*" Marking for reference values

References

[6.1] Bystron, K.: Leistungselektonik, Carl Hanser Verlag, 1979
[6.2] Garbrecht, Schaad, Lehmann: Workshop der professionellen Antriebstechnik, Franzis Verlag, 1996
[6.3] Johannis, R.: Handbuch des 80C166, Siemens AG, Berlin und München
[6.4] Reuter, M.: Regelungstechnik für Ingenieure, Vieweg Verlag, 1989
[6.5] Orlowski, P. F.: Praktische Regelungstechnik, Springer Verlag 1999
[6.6] Mann, H., Schiffelgen, H., Froriep, R.: Einführung in die Regelungstechnik, Hanser Verlag, 1997
[6.7] Garbrecht, F. W.: Digitale Regelungstechnik, vde-verlag 1991
[6.8] Garbrecht, F.W.: Process Simulation by Using Discrete State Models, OPTIM '98 Conference, Proceedings, Vol. 2, S. 563-565, Transylvania University of Brasov
[6.9] Brosch, P. F.: Moderne Stromricherantriebe, 3. Auflage, Vogel Buchverlag, Würzburg 1998
[6.10] DIN 41750: Begriffe für Stromrichter
[6.11] Heumann, K.: Grundlagen der Leistungselektronik, B. G. Teubner, Stuttgart 1995
[6.12] Lander, C. W.: Power Electronics, Mc Graw Hill-Book Company, Third Edition, London 1993
[6.13] Kiel, E., Schumacher, W.: Der Servocontroller in einem Chip, Elektronik 1994, H. 8, S. 48 - 60
[6.14] Schönfeld, R.: Elektrische Antriebe, Springer-Verlag, Berlin-Heidelberg 1995

[6.15] Schröder, D.: Elektrische Antriebe Bd. 1 Grundlagen, Springer-Verlag, Berlin-Heidelberg 1994
[6.16] Schröder, D.: Elektrische Antriebe Bd. 2 Regelung von Antrieben, Springer-Verlag, Berlin-Heidelberg 1994
[6.17] Texas Instruments: Field-oriented Control Three-Phase AC-Motors. Application Note Literature Number BPRA073, 1998
[6.18] Mäder. Regelung einer Asynchronmaschine unter alleiniger Verwendung an den Klemmen messbarer Größen; Dissertation, TH Darmstadt, 1981
[6.19] Horn, Dourdoumas. Regelungstechnik; Pearson Studium, 2004; ISBN 3-8273-7059-0
[6.20] International Rectifier: www.irf.com, Design Support/Application/Motion Control/Technology Solutions/Integrated Power Moduls
[6.21] International Rectifier: www.irf.com, Design Support/Application/Motion Control/Technology Solutions/Gate Driver ICs
[6.22] Michrochip: www.Michrochip.com, Application Design Centers/Motor Control Solutions/Application Solutions/Variable Speed Brushless DC Motor/Application Notes
[6.23] Michrochip: www.Michrochip.com, Application Design Centers/Motor Control Solutions/Application Solutions/Brushed DC Motor/Application Notes
[6.24] Michrochip: www.Michrochip.com, Application Design Centers/Motor Control Solutions/Application Solutions/Stepper Motor/Application Notes
[6.25] Infineon: 16 BIT CMOS Microcontroller Product, Field Oriented Control of a Single DC Link Shunt, Application Note V 1.0, 2004
[6.26] Texas Instruments: Digital Signal Processing Solution for Permanent Magnet Synchronous Motor, Application Note Literature Number BPRA044, 1997
[6.27] Beineke, Kiel, Nürnberger: Initialisierung des Kommutierungwinkels für lineare und rotative Synchron-Direktantriebe mit Inkrementalgeber, Tagungsband IPC/SPS/DRIVES 2003, Nürnberg, S. 811–822
[6.28] Texas Instruments: Digital Signal Processing Solution for AC Induction Motor, Application Note Literature Number BPRA043, 1996
[6.29] International Rectifier: www.irf.com, Design Support/Application/Motion Control/Application Solutions/Industrial AC Servo Motors
[6.30] International Rectifier: www.irf.com, Design Support/Application/Motion Control/Application Solutions/Industrial Sensorless Motors

7
Magnetic Bearing Technology

Wolfgang Amrhein, Siegfried Silber

7.1 Introduction

Magnetic bearing drives open new application domains that are either impossible or severely limited using drives with conventional bearings. In recent years it has been possible to reduce substantially the manufacturing costs of such systems in significant areas and thus to achieve a higher degree of acceptance in the market. Based on the current state-of-the-art technology, and considering the constantly declining price/performance ratio of electronic components, it can be assumed that in the future the application of magnetic bearing technology will increase in special industrial applications; however, magnetic bearing technology will not achieve the wide usage and the lower price levels of slide bearing and ball bearing technologies in standard applications.

Characteristics

The application domains of magnetic bearing drives are directly related to their characteristics. The following are some of the most important features[1]:

+ Contactless rotation
+ No lubricants required
+ No mechanical wear and tear
+ Largely maintenance-free
+ High reliability
+ Large range of speeds
+ High efficiency (especially with permanent magnetic designs)
+ Low vibration
+ Low noise level
+ High static stiffness
+ Loss-free compensation of external static forces

---

[1] Partially dependent on the type of design.

+ Electronic imbalance compensation during operation
+ Process data acquisition (e. g. for early fault detection)
+ Electronical control of stiffness and damping (and thus of the dynamic system behavior)

The following are the disadvantages compared with standard slide and ball bearing technologies:

− Low level of force densities
− Low dynamic stiffness
− Large space requirement
− High price

Application Domains

Magnetic bearing technology is particularly preferred in application domains where conventional bearings are taxed beyond their technical limits. This is the case, for example, in applications using pumps or compressors where contamination of the pumped medium with bearing grease or abraded particles is to be avoided by means of hermetically capsulated, bearingless pumps without sealing rings. This includes medical blood pumps and pumps for the semiconductor, chemical and pharmaceutical industries. The market segment of magnetic bearing turbo-molecular pumps is also of particular significance; these pumps are used for applications in high-vacuum technology where pressure levels of about 10^{-10} mbar are encountered. They are also used in etching and CVD processes, for ion implantation in areas of semiconductor technology, in electron beam microscopy, for gas analysis and in the manufacture of flat-screen monitors and optical layers.

Other examples with a high requirement on speed are ultra-centrifuges for mechanical separation processes in biochemistry, molecular biology, virology or diagnostics, milling or grinding spindles for low heat dissipation, high-

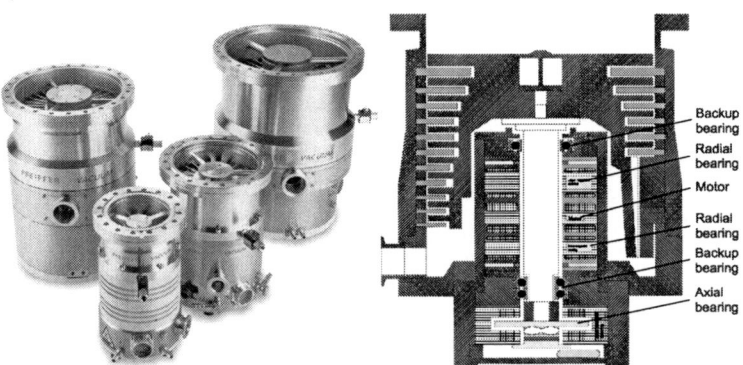

Fig. 7.1. Designs of magnetic bearing turbo-molecular pumps (Sources: © Pfeiffer Vacuum GmbH (*left*); © Leybold Vacuum GmbH (*right*))

precision workpiece machining, turbo-compressors and expanders, and high-speed drives in spinning machines.

Types of Design

Magnetic bearing drives represent complex electromechanical systems with high research and development requirements. The complexity arises from the interplay between the different scientific disciplines involved in the development process of these systems. In particular this includes knowledge of motor and magnetic bearing technology, control technology, electronics, sensorics and rotor dynamics.

Depending on their design, magnetic bearings have an axial, radial or conical air gap. In some respects, the principles of operation of magnetic bearings are quite different from one another. They can be divided into the following classes:

1. **Passive magnetic bearings**
   - **Permanent magnet bearings** with
     – Purely permanent magnet components,
     – Permanent magnet and ferromagnetic components or
     – Permanent magnet and superconducting components.
   - **Electrodynamic eddy current bearings** with
     – Metallic conductors or
     – Superconductors.
2. **Active (controlled) magnetic bearing systems**
   - **Electromagnetic bearings** with
     – Electromagnetic or
     – Permanent magnet bias magnetization of the air gap designed as
       - Unipolar bearings (unipolar bias flux) or
       - Hetero-polar bearings (hetero-polar bias flux).
   - **Bearingless motors** (motors with integrated magnetic bearing windings) designed as
     – Synchronous motors (e.g. permanent magnet motors, reluctance motors) or
     – Asynchronous motors.

The following section gives a brief overview of some of the most significant types of design of both passive and active magnetic bearings and describes their function and features with the help of examples. For the history and further study of the principles of magnetic bearing technology, please refer to [7.3, 7.6, 7.7, 7.10, 7.11, 7.13, 7.14, 7.18, 7.24].

7.2 Passive Magnetic Bearings

The term 'passive magnetic bearings' covers magnetic bearings that do not require any external current feed for achieving the function of the bearing in one or more degrees of freedom. Depending on the pairing and arrangement of the permanent magnet, ferromagnetic or superconducting bearing components, the effect of the forces can be achieved either in an attractive or repulsive manner. Pairing combinations using ferromagnetic pole arrangements always lead to attractive behavior.

Passive magnetic bearings basically do not permit complete bearing function in all degrees of freedom and are thus usually combined with active (i.e. electrically controlled) bearing systems.

7.2.1 Permanent Magnet Bearings

Due to their low values of stiffness and damping, permanent magnet bearings are suitable primarily for small electric drives. Here we distinguish two basic concepts: bearings only made of permanent magnet material and those that consist of both permanent magnet and ferromagnetic components. Figure 7.2 illustrates some construction variants for both radial and axial bearing designs.

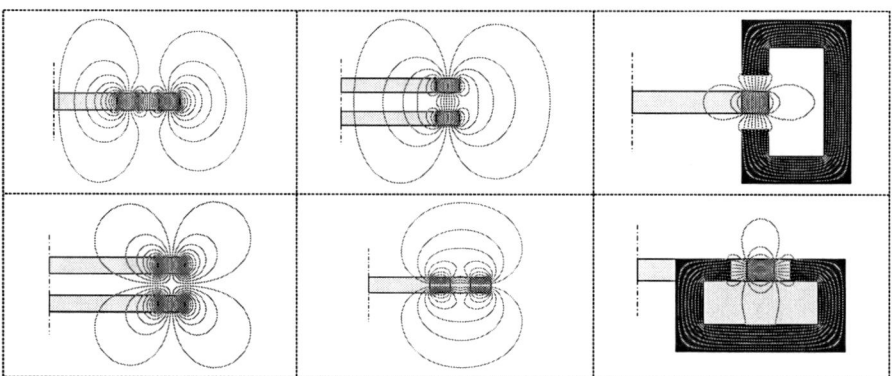

Fig. 7.2. Construction principles of permanent magnet bearings and their force characteristics. *First row*: Radial bearing designs: Repulsive/attractive/attractive (with ferromagnetic poles) force characteristics. *Second row*: Axial bearing designs: Repulsive/attractive/attractive (with ferromagnetic poles) force characteristics

Physical Model of Permanent Magnet Rings

An anisotropic permanent magnet blank is manufactured from crystalline powder, pressed into shape in a strong magnetic field, and finally sintered under inert gas at a temperature above the Curie point. During the sintering process the crystals largely retain the orientation assumed at the time of pressing and align themselves in an anti-parallel manner. The magnet thus has no field effect to the outside.

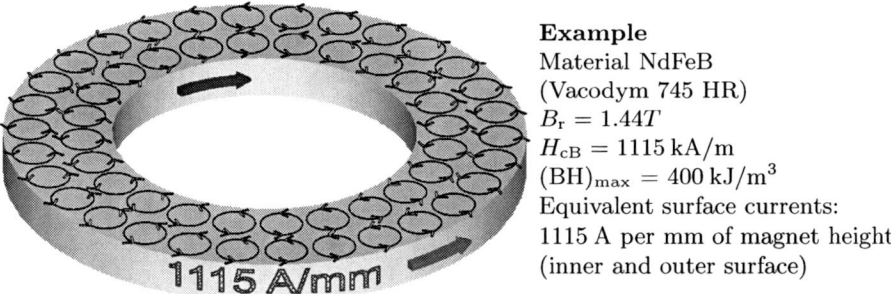

Example
Material NdFeB
(Vacodym 745 HR)
$B_r = 1.44\,T$
$H_{cB} = 1115\,kA/m$
$(BH)_{max} = 400\,kJ/m^3$
Equivalent surface currents:
1115 A per mm of magnet height
(inner and outer surface)

Fig. 7.3. Equivalent surface currents arising from the interaction of the aligned orbits of the elementary charges (schematic boundary surface model)

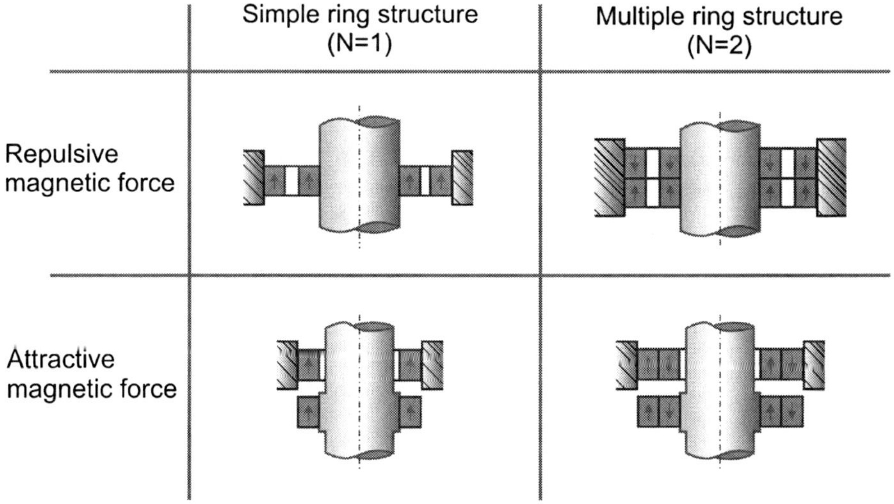

Fig. 7.4. Various configurations of radial permanent magnet bearings

The self-rotation (spin) of the electrons provides the ferromagnetism responsible for the development of the magnetic field. The elementary magnetic spin moments resulting from the electron movements are aligned in a parallel manner with the help of an adequately large magnetizing impulse. Figure 7.3 illustrates the result of the magnetization in a schematically simplified manner with the help of representation of the aligned direction of motion of the elementary charges [7.3].

With the new types of NdFeB materials having coercive field strengths H_{cB} of over $1100\,kA/m$, adequate surface currents at the inner and outer surface of the ring amount to more than 1.1 kA per mm of magnet height. For the sake of comparison, DC surface currents of this order can be achieved electromagnetically with zero loss only in superconducting materials. Retrofitting with an adequate ultra thin copper strip would be unfeasible due to the thermal load.

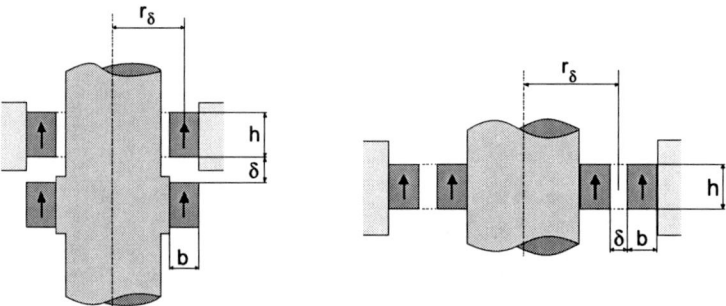

Fig. 7.5. Geometric parameters of the ring bearing

Examples of Radial Bearings

Figure 7.4 illustrates a few possible arrangements of radially stable (and axially unstable) bearings with axially magnetized permanent magnet rings, which are discussed in greater detail in [7.1–7.5,7.8]. Here we assume that the rings are mounted over non-ferromagnetic materials (e. g. highly alloyed chromium steels, non-ferrous metals or plastics).

The geometry of the bearings is described with the help of five parameters: magnet height h, ring width b, average air gap radius r_δ, air gap δ, and the number of ring sets N (compare Fig. 7.5).

Radial Stiffness

The radial stiffness s_r is a function of these geometric parameters and the remanent flux density B_r:

$$s_r = dF_r/dr = s_r(B_r, h, b, r_\delta, \delta, N) \ . \tag{7.1}$$

The following equation provides a dimensionless representation of the radial stiffness [7.5]:

$$s_r^* = \frac{s_r}{V} \frac{\delta^2}{\sigma_{Br}} \ . \tag{7.2}$$

V is the volume of the magnetic rings including the enclosed air gap. The reference pressure σ_{Br} is defined as the pressure with an air gap flux density $B = B_r$, whereby B_r corresponds to the remanent flux density of the magnetic material. The following equation applies to σ_{Br}:

$$\sigma_{Br} = \frac{1}{2\mu_0} B_r^2 \ . \tag{7.3}$$

With the introduction of dimensionless geometrical parameters, the radial stiffness s_r^* can now be expressed as a function $s_r^* = s_r^*(h/\delta, b/\delta, r_\delta/\delta, N)$. For many practical designs, the condition $r_\delta \gg \delta$ applies, so that in these cases the radial stiffness can also be described by the proximity function:

$$s_r^* \cong s_r^*(h/\delta, b/\delta, N) . \tag{7.4}$$

Table 7.1 together with Figs. 7.6 and 7.7 show the stiffness values obtained analytically for repulsive and attractive single and double ring bearings [7.5]. A comparison of these diagrams shows that the right array of curves is a result of the reflection of the left array about the 45° axis (assuming identical scales of the axes). The repulsive and attractive arrangements thus show the same values for the radial stiffness when we exchange the ring height and the ring width while maintaining the average air gap radius and air gap. This applies both to single and multiple ring arrangements.

The standardized maximum radial stiffness is 31% higher for double ring bearings than for single ring designs. Table 7.1 also illustrates that with a stack number greater than two the specific radial stiffness increases only marginally.

Using the following equation:

$$s_r = s_r^* \frac{V \sigma_{Br}}{\delta^2} . \tag{7.5}$$

Table 7.1. Optimal geometrical shapes for repulsive bearings with a stack number 1 to 5

Stack number N	Standardized magnet height $\frac{h}{\delta}$	Standardized magnet width $\frac{b}{\delta}$	Maximum s_r^*
1	1.19	1.19	0.0191
2	1.53	1.15	0.0249
3	1.72	1.135	0.0257
4	1.78	1.13	0.0263
5	1.81	1.125	0.0268

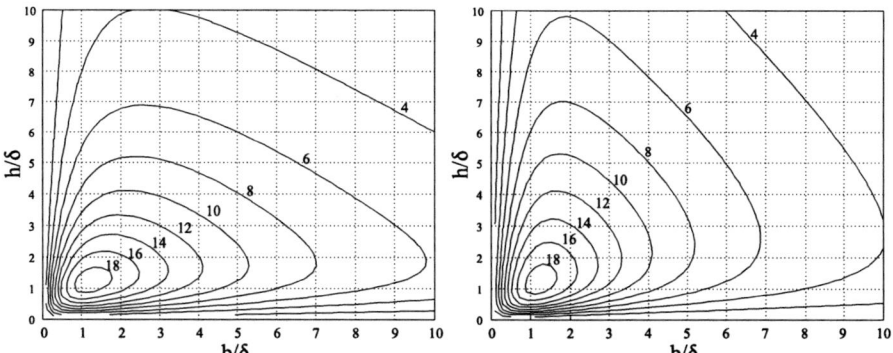

Fig. 7.6. Values of dimensionless radial stiffness $s_r^*(h/\delta, b/\delta, N = 1) \times 10^3$ for repulsive (*left*) and attractive (*right*, [7.2]) single ring bearings

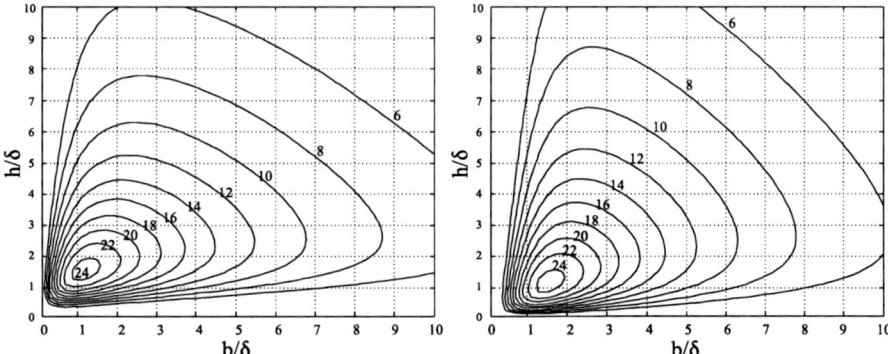

Fig. 7.7. Values of dimensionless radial stiffness $s_r^*(h/\delta, b/\delta, N = 2) \times 10^3$ for repulsive (*left*) and attractive (*right*, [7.2]) double ring bearings

the radial stiffness can be quickly estimated from the diagrams in Figs. 7.6 and 7.7 for a special application.

Axial Stiffness

Permanent magnet bearings have the characteristic that the stability achieved in one direction ties in with an instability in another direction [7.3, 7.6, 7.11]. This unstable behavior is the result of negative stiffness. The following relation applies to pure permanent magnet radial bearing designs (without the use of ferromagnetic materials):

$$s_z = \mathrm{d}F_z/\mathrm{d}z = -2s_r \ . \tag{7.6}$$

Here s_z is the stiffness in the axial direction.

Due to polarization, the following equation applies to radial permanent magnet bearings using ferromagnetic materials in parts:

$$s_z < -2s_r \ . \tag{7.7}$$

The destabilizing forces working axially on the magnet rings come up as soon as one ring moves away from the center plane of the second ring. Figure 7.8 illustrates typical bearing behavior [7.9] for a sample construction. This shows the declining radial stiffness with increasing distance from the center zone, the initially increasing destabilizing axial force, and the nearly linear behavior of the axial stiffness within a small axial offset. In order to avoid axial forces higher than $|F_0|$ at the time of startup, the axial offset on both sides is mechanically restricted to z_0 in practical designs.

Repulsive vs. Attractive Bearings

Radial bearing arrangements with repulsive characteristics, as a rule, offer the advantage of simple assembly and disassembly of the rotor (see Fig. 7.4). Compared to arrangements with attractive characteristics, the disadvantage

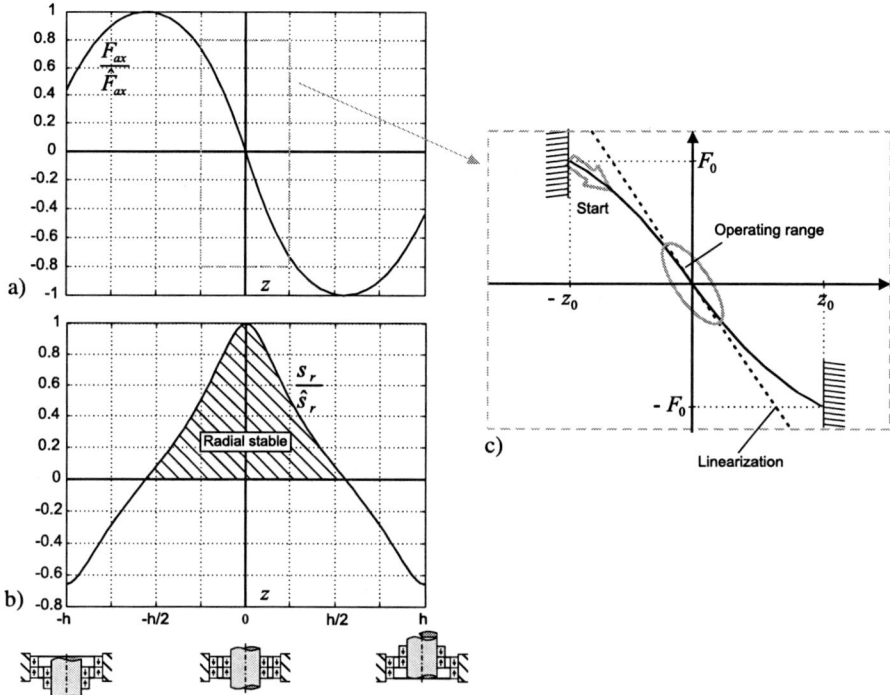

Fig. 7.8. Typical bearing behavior with axial offset of the magnet rings from the center position. a Axial force. b Radial stiffness. c Mechanical offset limit

of repulsive bearings can be the mutual reduction of the field strength within the magnet rings. Thus it must be ensured that the demagnetization of the ring material is ruled out by an appropriate design. Special care must be taken with the application of NdFeB compositions in higher temperature ranges.

Radial bearings with attractive characteristics (see Fig. 7.4) cause somewhat high axial tensile forces. However, the forces are compensated with the use of a second radial bearing identical in construction, so that this feature does not result in any disadvantage.

Damping

Permanent magnet bearings typically demonstrate very low damping values. The vibration characteristics, however, can be improved by means of mechanical and electrical damping measures [7.3]. Mechanical damping can be achieved by fixing the stator rings to the seat of the bearing, not rigidly but with the help of vibration-damping elements. These are fasteners with high internal friction. However, the utilisable frequency range of such mechanical vibration absorbers is severely limited to some extent. Eddy current damping systems can be used as an alternative. These are electrically conducting mate-

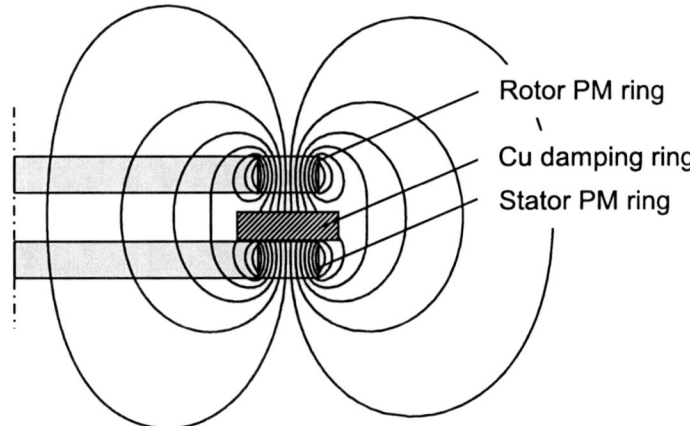

Fig. 7.9. Attractive radial ring bearing with passive electrical damping

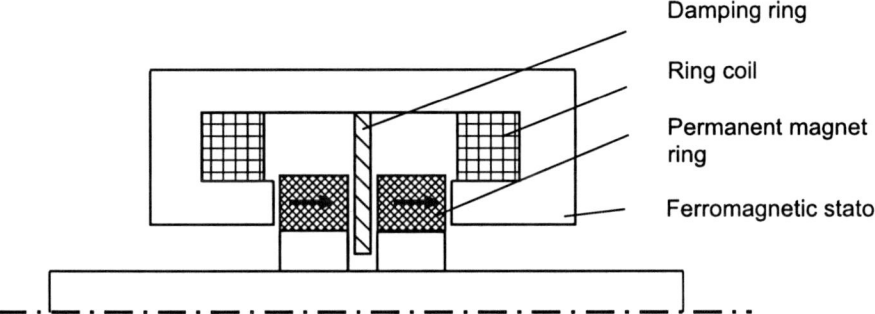

Fig. 7.10. Passive radial permanent magnet bearing integrated in an electromagnetic axial bearing

rials (e. g. Cu or Al rings) that are introduced into the air gap or leakage field of the bearing rings or of additional permanent magnet auxiliary rings. Figure 7.9 illustrates a schematic arrangement with a copper ring as the damping element.

An active system damping can also be achieved by means of electrically controlled bearing components. In such cases, the level of damping is adjusted in connection with the position control of the unstable degree of freedom of the passive bearing. Figure 7.10 illustrates an example of an integrated bearing system consisting of a passive radial permanent magnet bearing with associated eddy current damping and an electromagnetic axial bearing [7.3].

7.3 Active Magnetic Bearing Systems

In contrast to passive magnetic bearings, in active magnetic bearing systems the levitation of the rotor is controlled electronically. The advantages include improved vibration control and adjustable levels of stiffness and damping. Thus electronics can be used to control the dynamic system response. By adjusting the stiffness during run-up to speed, rotor resonances can be suppressed, or by offsetting the axis of rotation in the main inertia axis, rotor imbalances can be compensated electronically.

Dual magnet bearing systems are of particular significance in the area of applications for small drives: electromagnetic bearings and bearingless motors. The following section presents an overview of different types of designs and their features.

7.3.1 Electromagnetic Bearings

7.3.1.1 Magnetic Bearings with DC Bias Magnetization

Electromagnetic bearings are available in different designs, e.g. radial, axial and combined bearings, which can be designed as heteropolar bearings or homopolar bearings, respectively. Homopolar bearings show a uniform sign of the bias flux along the circumferential air gap.

Figure 7.11 illustrates the principle of construction of a classic electromagnetic bearing – in this case, a heteropolar bearing – for stabilizing two radial degrees of freedom. Maxwell forces come into play orthogonally on the pole surface. These comply approximately with the following square relation in the unsaturated operation mode:

$$F = \frac{A}{2\mu_0}B^2 \tag{7.8}$$

with A: area of a pole surface element, B: normal component of the flux density, F: resulting force orthogonal to the area of the pole surface element.

Linearization of Characteristics

Unfavorable control behavior results from the nonlinear force-current function that has a saddle point in the operating range around the point of origin. A substantially improved force-current characteristic is achieved by the insertion of a DC bias flux in the air gap and an opposite arrangement of the stator-sided bearing components. Therefore in the model of Fig. 7.11 a bias current i_B is superimposed on the control current i_C. As an alternative, the same magnetomotive force can also be achieved by means of separate, magnetically coupled control and bias magnetization coils. As Fig. 7.11 illustrates, both methods result in a linear characteristic with a larger gradient. The gradient k_i is almost constant for small modulation amplitudes around the point of origin and can be adjusted with the help of the bias current i_B.

Due to the square characteristic of the force-current relation, usually only one half of the parabolic curve is used in connection with the electrical supply of the pole winding. Depending on the design of the magnetic circuit, in case of very large modulation or bias amplitudes, ferromagnetic saturation can

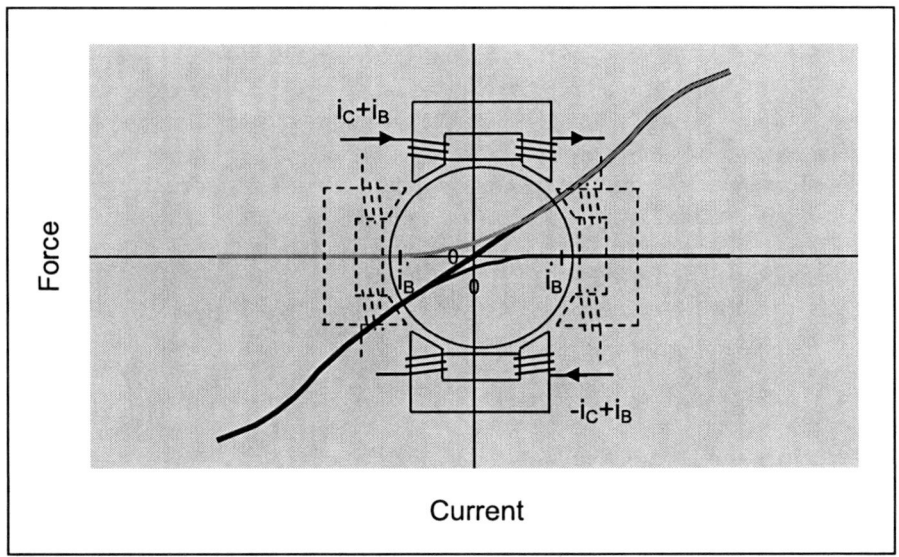

Fig. 7.11. Linearization of the force-current characteristic with the introduction of a bias current i_B

occur. This leads to a flattening of the curve in the border regions of the force-current characteristic.

Similarly, on the assumption of $i_M = i_B =$ const. the force/distance function $F(\delta_C)$ for small offsets δ_C from the center position can be linearized. Assuming constant magnetomotive force, it is permissible to define a force/distance constant k_δ in a manner analogous to that of the force-current constant k_i.

The bearing force can thus be represented as the following linear function for the limited operational range:

$$F(i_C, \delta_C) = k_i i_C + k_\delta \delta_C \tag{7.9}$$

with

$$i_C = i_M - i_B \tag{7.10}$$

and

$$\delta_C = \delta_0 - \delta_M . \tag{7.11}$$

In the above equations, i_M is the magnet current, i_C the control current, i_B the bias current, δ_M the magnetic air gap, δ_C the control distance and δ_0 the air gap for center position of the rotor. It is practicable to select the signs of the movement equation opposite to that of the previous current function so that the magnetically forced offset δ_C of the rotor occurs in the direction of the force [7.10].

Dynamic Behavior of the Control Path

The dynamic behavior of the control path in the principal axes is described by means of the following relation, taking into account Eq. (7.9):

$$m\ddot{\delta}_C = F(i_C, \delta_C) \, . \tag{7.12}$$

The destabilizing influence of the positive feedback of the air gap dependent force component $k_\delta \delta_C$ can be recognized from the block diagrams in Figs. 7.12 and 7.13.

In voltage-controlled systems the differential equation of motion is supplemented by the following voltage differential equation:

$$u_C = Ri_C + L\frac{di_C}{dt} + k_u \frac{d\delta_C}{dt} \, . \tag{7.13}$$

We assume here that the control and bias currents flow in magnetically coupled but electrically separated windings.

The following equation applies to a state representation:

$$\dot{\boldsymbol{x}} = \boldsymbol{A}\boldsymbol{x} + \boldsymbol{b}u \tag{7.14}$$

consisting of

$$\boldsymbol{x} = \begin{pmatrix} \delta_C \\ \dot{\delta}_C \\ i_C \end{pmatrix}; \quad \boldsymbol{A} = \begin{pmatrix} 0 & 1 & 0 \\ \frac{k_\delta}{m} & 0 & \frac{k_i}{m} \\ 0 & -\frac{k_u}{L} & -\frac{R}{L} \end{pmatrix} \quad \text{and} \quad \boldsymbol{b} = \begin{pmatrix} 0 \\ 0 \\ \frac{1}{L} \end{pmatrix} \, . \tag{7.15}$$

The vector $\boldsymbol{u}$ contains the adjustable parameters of the system and includes the voltage u_C in the case of voltage control.

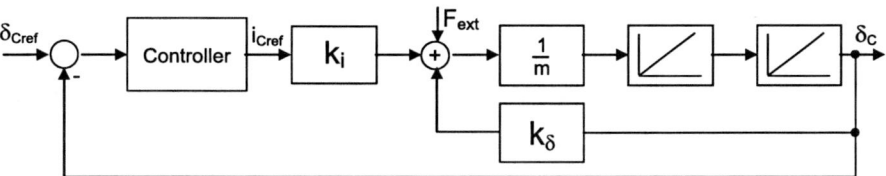

Fig. 7.12. Current-controlled position control

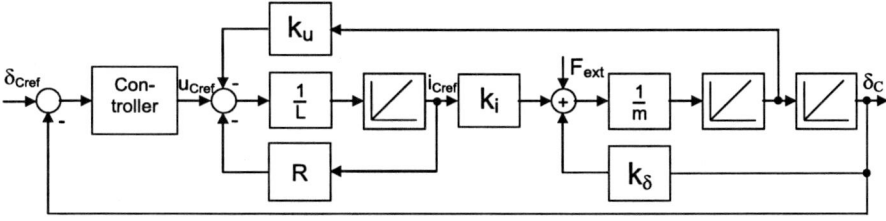

Fig. 7.13. Voltage-controlled position control

Considerations for Position Control

In order to ensure a stable operation, the destabilizing influence of the coupling of $k_\delta \delta_C$ must be compensated by means of a position-dependent adjustment current i_C. The minimum control gain for this purpose is k_δ/k_i. As shown in the signal flow diagrams in Figs. 7.12 and 7.13, the mechanical path is characterized by a double integral behavior, which means that basically a controller with PD characteristics is adequate to achieve system stability. In practical designs, however, the PD controller is largely unsuitable as a result of the limited adjusting range of the power electronics and also due to the partly distinct sensor signal noise in connection with switched power converters. Thus in industrial applications PID structures are often used, and in cases of special rotor-dynamic requirements, even significantly more demanding controller algorithms are implemented.

As shown in Figs 7.12 and 7.13, position control can be achieved with either current control or voltage control. The method of current control is based on a quasi-ideal current amplifier and thus requires high dynamics of the power converter. In contrast to current control, where the electromagnetic force is directly adjusted by taking the characteristic $F(i_C, \delta_C)$ into account, in voltage control the system response is additionally influenced by means of current-dependent and also partly by temperature-dependent bearing parameters. It is thus recommended, particularly for the range of higher degrees of modulation, to take these dependencies into account in the mathematical design of the bearing model.

Voltage control is often preferred over current control since, on the one hand, no high-resolution current sensor is required for the current adjustment and, on the other, the control concept already takes into account the characteristic of the electrical path. Figures 7.12 and 7.13 illustrate the basic structure of the control circuits for both methods [7.10, 7.11].

From the control point of view, not one (as considered so far) but five degrees of freedom are relevant to the design of magnetic bearing systems. While it is adequate to use simple models with punctual concentrated mass for the consideration of the rotor movement along the longitudinal axis of the shaft, the movements in the longitudinal directions of the other axes or around the other axes cannot be treated separately due to the coupling by gyroscopic effects or elastic structures. In certain designs, however, it is possible to use approximately decoupled subsystems for the analytical investigations and thus to simplify substantially the control structures. For a detailed study of the rotor-dynamic aspects, see the references in [7.10, 7.14, 7.27].

Radial and Axial Bearing Designs

Figure 7.14 illustrates a drive system positioned electromagnetically in five degrees of freedom, consisting of two radial bearings, one axial bearing and two backup bearings. The latter provide an emergency function in the event of a possible system crash.

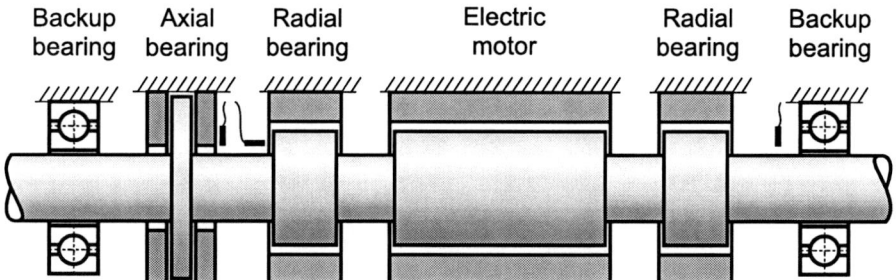

Fig. 7.14. Electric drive with a complete electromagnetic bearing system

Figure 7.15 presents two basic designs of electromagnetic radial bearings. These variants differ not only in the arrangement of the flux planes but also with respect to the bias flux direction in the air gap. While in the example on the left the sign of the bias flux changes along the stator circumference (heteropolar bearing), the middle example is designed to have a homopolar flux with respect to the flux direction. The advantage of the heteropolar bearing is that its design is easier to manufacture, whereas that of the homopolar bearing is its lower iron losses in the rotor. The latter is very useful, for example, in applications with high speed requirements. However, due to the wide pole gaps along the circumference, even the design at the center is not free from harmonics in the air gap.

A typical design for an electromagnetic axial bearing is illustrated on the right in Fig. 7.15, a homopolar design with ring coils. While the static bias flux of this bearing design distributed homogenously along the circumference leads in principle to very low rotor losses, there is an eddy current problem with the selection of the lamination planes in the stator and the rotor due to the dynamic changes of the control currents. Thus the rotor disk and also the stator are frequently manufactured of solid materials, e.g. soft magnetic composites (SMC), due to cost considerations.

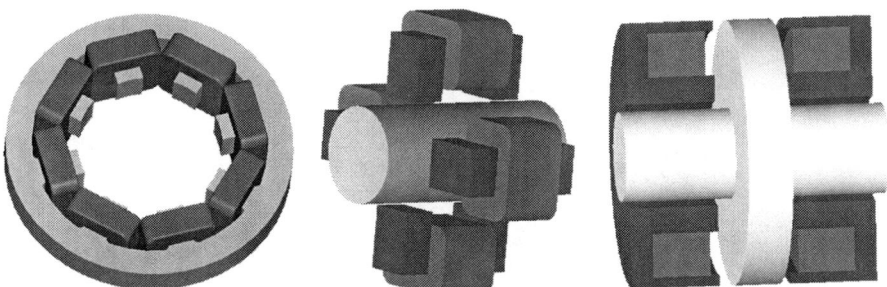

Fig. 7.15. Design forms of radial (*left, center*) and axial (*right*) bearings (*left*: heteropolar bearing; *center and right*: homopolar bearings)

7.3.1.2 Magnetic Bearings with Permanent Magnet Bias Magnetization

The DC bias magnetization discussed in the previous section can cause large copper losses. This is particularly true when, as in the case of hermetically encapsulated pumps (canned pumps), large air gaps need to be magnetized. In such applications the bias current losses constitute a large part of the total losses and thus contribute significantly to heating the drive system. More favorable alternatives from the energy point of view are systems where the air gap is biased by using permanent magnet materials. Very high energy densities can be achieved with the use of NdFeB permanent magnet materials with relatively low material costs (compare Sect. 7.2.1).

Homopolar Radial Bearings

From the viewpoint of minimizing the electric power requirement, it is beneficial to separate the flow of the electromagnetic flux and the permanent magnet flux in parallel circuits with an overlap in the air gap as shown in Fig. 7.16. For comparison: with completely overlapping magnetic circuits (which correspond to a serial arrangement of permanent magnets and electromagnetic coils) the electrical magnetomotive force has to cover not only the magnetic potential difference in the air gap but also the potential difference across the permanent magnets which in use of rare earth materials have very low permeability values. An advantage of such a serial assembly would be its higher frequency range.

Figure 7.16 illustrates the three-dimensional flux circuits developed in a homopolar arrangement. While the principal flux directions of the permanent magnet circuit lie in the $r-z$ planes, the electromagnetic control fluxes flow in the $r-\varphi$ planes of the front and rear laminated sheet packages. The permanent magnet ring is magnetically connected to both stator lamination stacks *via* ferromagnetic spacer rings. The rotor, in this case the shaft, serves as magnetic feedback.

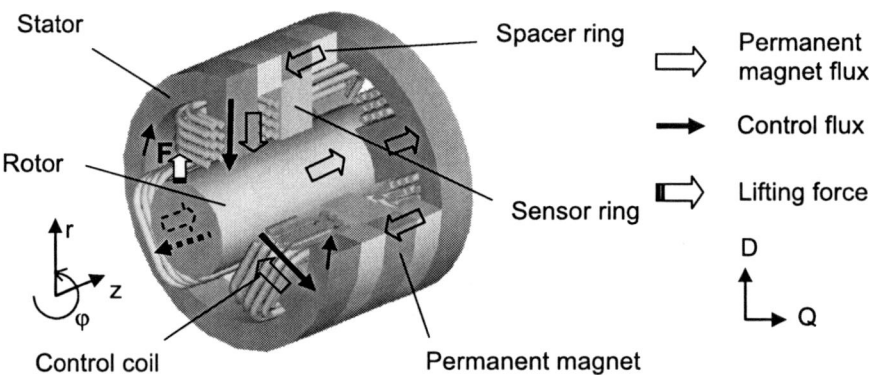

Fig. 7.16. Permanent magnet excited homopolar bearing (schematic representation)

The laminated sheet packages shown in Fig. 7.16 illustrate only three winding poles. The movements along the winding axes are thus coupled. The decoupling of the current parameters required for control can be achieved with the help of a transformation which transfers the winding system into an orthogonal stator-fixed D, Q coordinate system (Clarke's transformation). The conversion between the two systems can be carried out by the following equation:

$$\begin{bmatrix} i_{S1} \\ i_{S2} \\ i_{S3} \end{bmatrix} = \begin{bmatrix} 1 & 0 \\ -\frac{1}{2} & \frac{\sqrt{3}}{2} \\ -\frac{1}{2} & -\frac{\sqrt{3}}{2} \end{bmatrix} \begin{bmatrix} i_D \\ i_Q \end{bmatrix} . \tag{7.16}$$

Heteropolar Radial Bearings

A simpler construction can be attained with a heteropolar structure (see Fig. 7.17) [7.12]. In this example the permanent magnet as well as the electromagnetic flux flows lie in one common plane. The overlap of the flux components in the air gap occurs in the sections of the winding poles in the same manner as in the previous example. The flux density distribution for a rotor force acting downwards is represented in the middle picture in Fig. 7.17. In this example, the upper winding pole is almost field-free due to the flux components flowing in opposite directions, while the two lower winding poles have very high flux densities.

The picture on the right illustrates an industrial design of a permanent magnet excited hetero-polar bearing. The permanent magnets are located in the recesses of the pole legs. They are demarcated laterally with the help of thin saturation bars.

Low-loss Compensation of Static Forces

The compensation of static forces such as the weight of the rotor or the average radial force in centrifugal pumps demands quite high electromagnetic counterforces and can substantially constrain the operating range of the bearing.

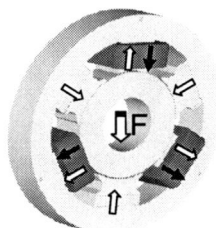
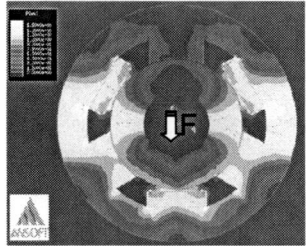

Fig. 7.17. Permanent magnet excited heteropolar bearing. *Left*: flux directions in the poles; *center*: flux density distribution; *right*: industrial design (*center*: calculated with Ansoft Maxwell 3D; *right*: Source: © Levitec GmbH)

Radial bearings excited by permanent magnets offer the possibility of offsetting the rotor from the geometric center to such an extent that the external process forces are countered by the attractive forces of the permanent magnets. In this case, the electrical control power serves only to control minor deviations from the new axis of rotation and can be negligibly small in instances of pure or predominantly static bearing loads.

From Mechanical to Intelligent Bearings

In contrast to mechanical bearings the magnetic bearing can also perform monitoring functions. The continuously monitored process parameters (rotor position, rotor speeds, currents, etc.) can provide information regarding the status of the drive system at any time. This facilitates the timely diagnosis of faults and also enhances process security. Examples are the detection of winding short-circuits, pump wheel deformations or rotor imbalances [7.26]. Diagnostic capabilities are principally available with all active magnetic bearings.

7.3.2 Bearingless Motors

7.3.2.1 Bearingless Permanent Magnet Motors

Bearingless motors represent relatively recent motor technology. Worldwide research has been conducted in this field since the beginning of the 1990s. Meanwhile a few series products have also been developed for several application areas like in the semiconductor and medical industries.

A significant feature of bearingless motors is the bearing winding system integrated into the motor [7.15, 7.16]. Figure 7.18 illustrates the principle of construction of the winding sets using an example of a four-pole permanent

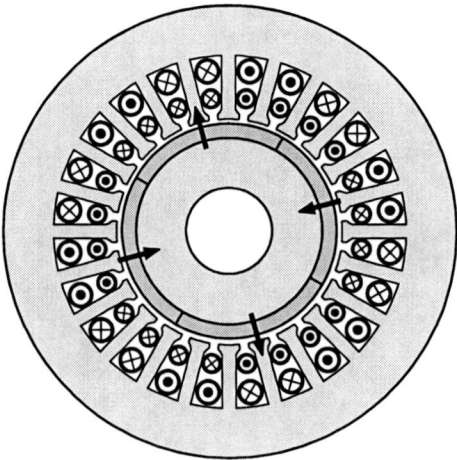

Fig. 7.18. Winding system of a bearingless four-pole rotating-field motor. *Outside*: three-phase four-pole torque winding; *inside*: three-phase two-pole bearing force winding

magnet motor. In order to achieve extensive decoupling between the generation of the radial bearing forces and the torque, the windings are designed with different numbers of poles.

The following relation is used to determine the number of pole pairs:

$$p_{\text{mot}} = p_{\text{mag}} \pm 1 \tag{7.17}$$

with p_{mot} the number of pole pairs of the torque winding and p_{mag} the number of pole pairs of the bearing force winding.

Generation of Radial Bearing Forces and Torque

An understanding of the development of the radial forces and the torque is important to comprehend the principle of operation. The bearingless motor of Fig. 7.19 with sinusoidal distributed magnetomotive forces in the stator is particularly useful in order to explain the physical correlations. Figure 7.19 illustrates a two-pole permanent magnet motor with a two-pole (a, b) and also a four-pole rotating-field winding set (c, d).

Two types of forces are distinguished: Lorentz forces, resulting from the armature currents in the magnetic fields and Maxwell forces arising from the interface of two media having different permeability values.

Figure 7.19a shows torque development resulting from the tangential Lorentz force vectors. Figure 7.19b depicts the distribution of the Maxwell forces acting in a radial manner. It becomes apparent that the force components compensate each other for a symmetric field distribution and a centered rotor as a result of their opposing directions, so that there is no resulting radial force acting on the rotor. The behavior in Fig. 7.19c, d is different. Both the Lorentz and the Maxwell forces create adjustable radial force vectors[2], but produce no torque in contrast to the configuration illustrated in Fig. 7.19a.

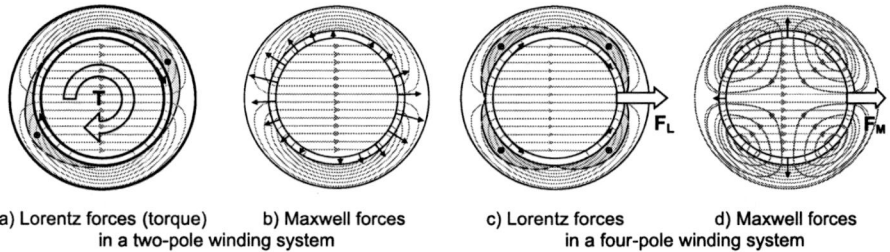

a) Lorentz forces (torque) in a two-pole winding system
b) Maxwell forces
c) Lorentz forces in a four-pole winding system
d) Maxwell forces

Fig. 7.19. *Lorentz* and *Maxwell forces* in a two-pole permanent magnet excited motor

---

[2] Annotation: In general the Lorentz and Maxwell overall force vectors have different directions. Figure 7.3–7.9c, d show a special combination of orientations of the magnetomotive force distribution and the permanent magnet fields which in this case lead to the same directions.

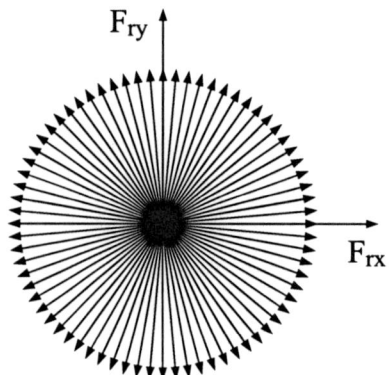

Fig. 7.20. Circular force locus on turning rotor (fundamental wave motor)

These illustrations demonstrate that substantially a separate control of the torque and radial bearing forces can be achieved by winding sets differing in one pole pair. Ideally, a sinusoidal permanent magnet field combined with a sinusoidal magnetomotive force distribution produce a circular locus of the radial force vector when the rotor is turned and the currents in the bearing force winding are kept constant (see Fig. 7.20). In contrast to the control of magnetic bearings, whose ferromagnetic rotors have no distinct magnetic direction of preference, the magnetomotive force of the radial bearing force winding of bearingless motors must be controlled in relation to the radial and the angular rotor position.

Bearingless Motors with Concentrated Windings

Winding systems with a low number of concentrated windings (tooth coil windings) have clear cost advantages over designs with a large number of distributed windings. Figure 7.21 illustrates variants with eight (a) and four (b) coils as alternatives to the fundamental wave motor of Fig. 7.18 shown above. The design on the left in Fig. 7.21 shows a three-phase motor with a four-pole alternating field torque winding and a two-pole rotating-field radial bearing force winding. To ensure startup in the four zero torque angular positions, the ferromagnetic stator poles are asymmetrically designed to achieve an auxiliary reluctance torque.

The winding arrangement of the motor in Fig. 7.21a with mechanically separated torque and bearing force windings can be transferred to the simpler four coil winding system in Fig. 7.21b by the following equation:

$$\begin{bmatrix} i_\mathrm{a} \\ i_\mathrm{b} \\ i_\mathrm{c} \\ i_\mathrm{d} \end{bmatrix} = \begin{bmatrix} 1 & -1 & -1 \\ 1 & 1 & 1 \\ -1 & 1 & -1 \\ -1 & -1 & 1 \end{bmatrix} \begin{bmatrix} i_1 \\ i_2 \\ i_3 \end{bmatrix}. \qquad (7.18)$$

Thus there are four phases, each consisting of one concentrated coil. By means of the current components $i_1 \ldots i_3$ in each coil, the torque (i_3) and the radial

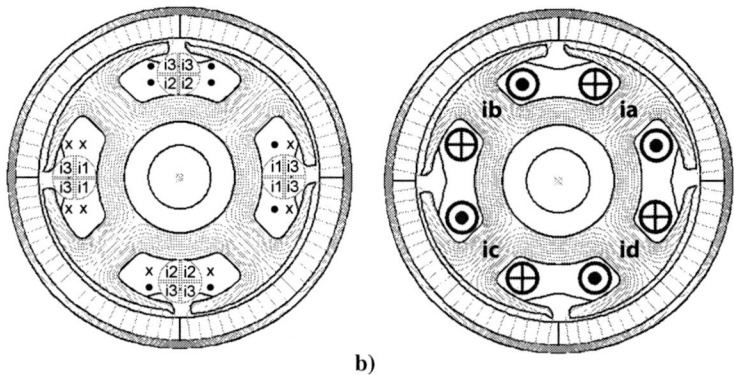

Fig. 7.21. Design of bearingless motors with concentrated coils. **a** With separate windings for torque and radial bearing forces. **b** With only one set of windings

bearing force components (i_1, i_2) are generated simultaneously in accordance with Eq. (7.18).

Three degrees of freedom can be stabilized actively with drives like these in a very cost-effective manner. However, these motor designs are no longer fundamental wave motors, but motors with high harmonic contents of flux linkages [7.19–7.23,7.28]. Typically the force locus of these motors shows an elliptical form in contrast to the circular radial bearing force characteristics of fundamental wave motors shown in Figs. 7.18 and 7.20.

Passive Stabilization in Three Degrees of Freedom

The magnetic bearing part of the motor has to stabilize five degrees of freedom for the contactless positioning of the rotor. A few applications such as pumps and fans permit a flat disk-shaped construction of the stator and rotor with a small axial length compared to the air gap diameter (see Fig. 7.22). If the rotor is moved by external forces axially or in tilting directions, then with appropriate magnetic design, reluctant restoring forces are generated

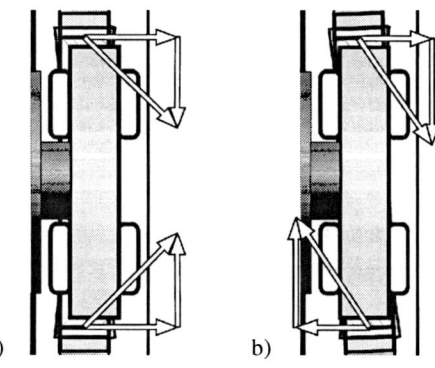

Fig. 7.22. Passive stabilization of the external rotor of a bearingless disk-shaped motor in the axial direction (**a**) and in one of the tilting directions (**b**) by reluctance forces

that stabilize the rotor passively in three degrees of freedom. Thus in the case of such a bearingless disk motor, it suffices to control only two radial degrees of freedom for the aspired stabilization of the rotor position. In this manner, very cost-effective bearingless drive concepts can be implemented with comparatively low mechanical and electrical complexity.

Control

The block diagram in Fig. 7.23 illustrates the control principle of bearingless permanent magnet motors using an example of a three-phase arrangement. The digital signal processor (DSP) performs the control of the radial rotor coordinates, the control of the angular position, the speed or the acceleration, the control of the stator currents, the evaluation of the rotor position, angle, speed and acceleration, the supervision of the operating states of the motor and the power electronics, and to some extent the execution of additional special functions (such as the electronic compensation of the rotor imbalance by means of offsetting the rotation axis in the principal axis of inertia).

The rotor position coordinates are controlled in two orthogonal axes x and y. The functional relation between the radial bearing forces and the currents is required for the modeling of the system and the design of the control algorithms [7.18]. The following function applies to the radial bearing forces in both principal axes:

$$\begin{bmatrix} F_{rx} \\ F_{ry} \end{bmatrix} = \boldsymbol{F}_r(\boldsymbol{i}, \boldsymbol{x}, \varphi_r) \tag{7.19}$$

with the rotor offset

$$\boldsymbol{x} = \begin{bmatrix} x \\ y \end{bmatrix} \tag{7.20}$$

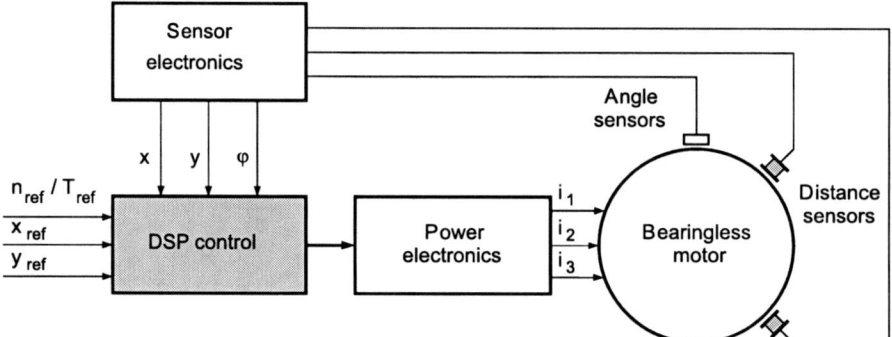

Fig. 7.23. Block diagram of the electronic control of a three-phase bearingless motor

and the current vector

$$i = \begin{bmatrix} i_1 \\ i_2 \\ i_3 \end{bmatrix}. \qquad (7.21)$$

Equation (7.19), as a rule, cannot be represented in a closed mathematical form due to the highly nonlinear correlations. Thus it is appropriate to linearize the force equation around the operating point i_0 and x_0:

$$F_\mathrm{r}(i, x, \varphi_\mathrm{r}) = F_\mathrm{r}|_{i_0, x_0, \varphi_\mathrm{r}} + \left.\frac{\partial F_\mathrm{r}}{\partial x}\right|_{i_0, x_0, \varphi_\mathrm{r}} \Delta x + \left.\frac{\partial F_\mathrm{r}}{\partial i}\right|_{i_0, x_0, \varphi_\mathrm{r}} \Delta i. \qquad (7.22)$$

The radial force consists of three components: a static and a motion-dependent as well as a current-dependent component. The following assumptions have been made to simplify the bearing force function:

1. Selection of the operating point (i_0, x_0) such that the static force component becomes zero;
2. Only very small offsets Δx occur in the closed control loop.

Based on these assumptions, we can specify the following approximate functions for the control of the orthogonal radial bearing forces:

$$\begin{bmatrix} F_{\mathrm{r}x} \\ F_{\mathrm{r}y} \end{bmatrix} = \left.\begin{bmatrix} \frac{\partial F_{\mathrm{r}x}}{\partial i_1} & \frac{\partial F_{\mathrm{r}x}}{\partial i_2} \\ \frac{\partial F_{\mathrm{r}y}}{\partial i_1} & \frac{\partial F_{\mathrm{r}y}}{\partial i_2} \end{bmatrix}\right|_{i_0, x_0, \varphi_\mathrm{r}} \begin{bmatrix} \Delta i_1 \\ \Delta i_2 \end{bmatrix} + \left.\begin{bmatrix} \frac{\partial F_{\mathrm{r}x}}{\partial i_3} \\ \frac{\partial F_{\mathrm{r}y}}{\partial i_3} \end{bmatrix}\right|_{i_0, x_0, \varphi_\mathrm{r}} \Delta i_3. \qquad (7.23)$$

A matching design of the substantial motor leads to independence between the bearing forces and the torque function for operation in the non-saturated region. For this case it is permissible to ignore the last term in Eq. (7.23) and to build the following inverse function for $i_0 = 0$ and $x_0 = 0$ (operating point with centered rotor):

$$\begin{bmatrix} i_1 \\ i_2 \end{bmatrix} = \left.\begin{bmatrix} \frac{\partial F_{\mathrm{r}x}}{\partial i_1} & \frac{\partial F_{\mathrm{r}x}}{\partial i_2} \\ \frac{\partial F_{\mathrm{r}y}}{\partial i_1} & \frac{\partial F_{\mathrm{r}y}}{\partial i_2} \end{bmatrix}\right|^{-1}_{i_0, x_0, \varphi_\mathrm{r}} \begin{bmatrix} F_{\mathrm{r}x} \\ F_{\mathrm{r}y} \end{bmatrix}. \qquad (7.24)$$

The rotor angle-dependent force-current functions of the matrix can be obtained, for example, by means of finite element calculations and are used for the configuration of the $K_\mathrm{m}(\varphi)$ matrix of Fig. 7.24.

Analyses of motors with a higher number of phases can be conducted in a manner similar to that done for the previous designs [7.23, 7.27]. Figure 7.25 illustrates an example of a bearingless five-phase motor with only one coil per phase. In contrast to the four-phase motor design of Fig. 7.21 the five coil arrangement allows a rotating-field operation of the motor. Asymmetrically designed ferromagnetic stator poles are no longer required.

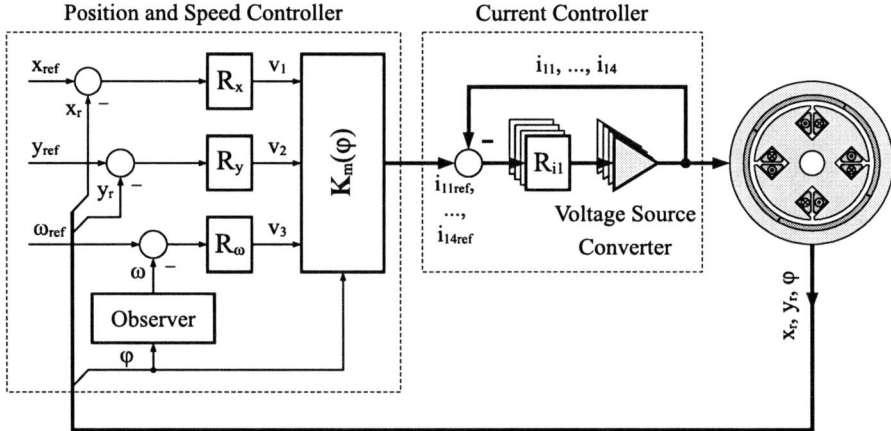

Fig. 7.24. More detailed block diagram of the electronic control of a four-phase bearingless motor

Fig. 7.25. Bearingless motor with five coils. *Left*: cross section; *right*: mechanical construction

Examples of Practical Designs

A typical example of application of a bearingless drive is illustrated in Fig. 7.26. It involves a bearingless pump without slide sealing rings, as required for liquids of high purity or chemically aggressive solutions. The housing of the centrifugal pump is hermetically encapsulated and includes the magnetically suspended pump wheel. The radial rotor position is controlled by means of the stator windings, which are located outside the (non-ferromagnetic) pump housing.

An industrial design of a bearingless pump is illustrated on the left in Fig. 7.27. This pump has separate windings for the development of torque

7 Magnetic Bearing Technology 441

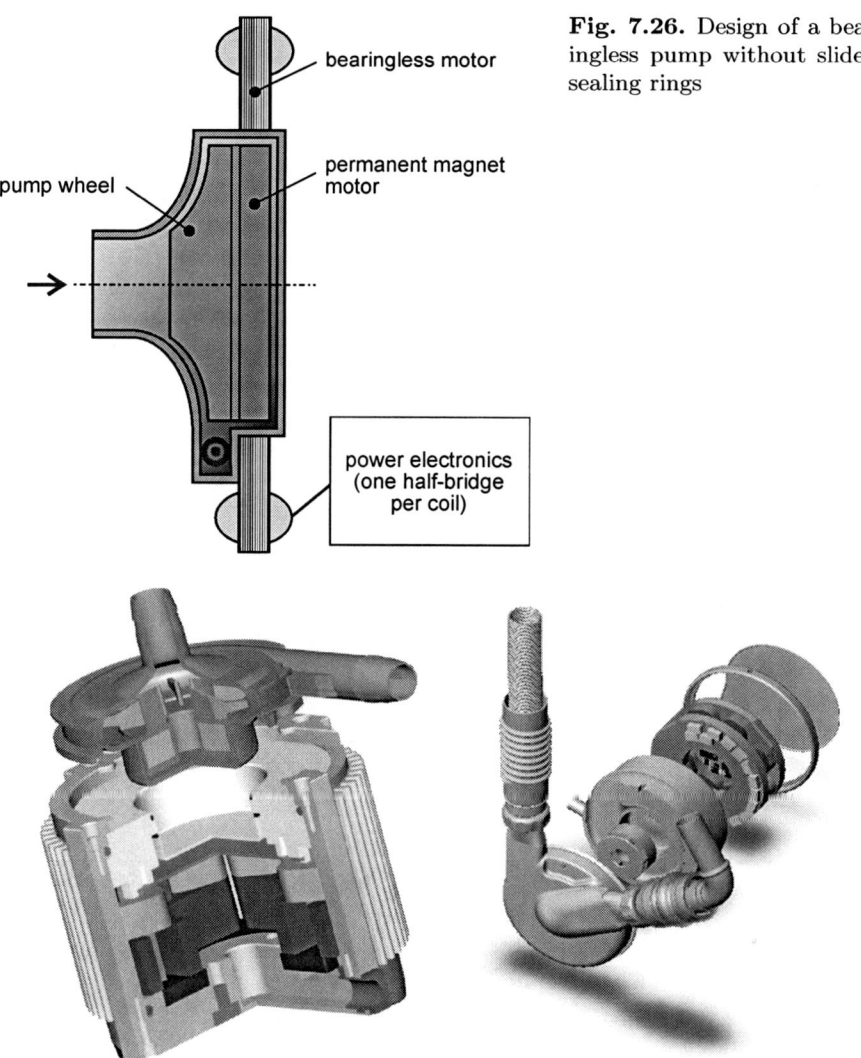

Fig. 7.26. Design of a bearingless pump without slide sealing rings

Fig. 7.27. Industrial designs of bearingless pumps. *Left*: pump with replaceable pump attachment (Source: © Levitronix GmbH); *Right*: implantable ventricular assist device (heart pump) (Source: © 2001 Thoratec Corporation)

and radial bearing forces. The hydraulic part of the pump is hermetically encapsulated. It can be used as a low-cost disposable part, e.g. for medical applications like cardiovascular surgeries (blood pump) demanding a high degree of sterility and hemocompatibility, by simply plugging it in and out.

The picture on the right in Fig. 7.27 illustrates another application; it shows the prototype of an implantable heart pump. This is a completely integrated high reliability system designed to be fault-tolerant; the signal electronics as well as the power electronics and the sensor system are realized as redundant sub-systems so that even in the event of the failure of one component the pump can maintain its operation.

Bearingless segment motors[3] Another type of basic bearingless drive design, the bearingless segment motor, is exemplified in Fig. 7.28. This group of motors is typically characterized by four or more separate stator segments, where each of these segments contributes to radial levitation forces and motor torque generation [7.29, 7.30]. Classical heteropolar electromagnetic radial bearings (cf. Fig. 7.11) show a very similar mechanical stator design but excite no tangential forces on the rotor and therefore no torque. Although the mechanical setup becomes quite simple, the characteristic of force generation is nonlinear and generally speaking there are additional disturbing effects like cogging torque, radial reluctance forces and torque ripple which have to be taken into account during the motor design process.

The passive stabilization of three degrees of freedom by reluctance forces (cf. Fig. 7.22) also holds true for this discoidal motor concept. Furthermore, it is possible to separate the phase currents into components generating torque and radial bearing forces in a similar manner as described in Eq. (7.18).

The special arrangement of the stator segments together with the interspaces between these components typically lead to higher numbers of permanent magnet pole-pairs in comparison to bearingless motors presented before. Furthermore, it is obvious that the stator design leads to marginal flux linkages between the windings of adjacent stator elements, resulting in very low mutual inductances and couplings.

Fig. 7.28. Basic bearingless segment motor design featuring different segment and winding configurations for illustration

---

[3] The research work leading to this motor type was funded by the Austrian Science Fund (FWF) under the contract number P17523-N07.

Figure 7.29 shows a prototype of a twelve-pole segment motor with four stator elements and coils. A quadratically designed supporting frame holds the stator segments and guarantees the structural rigidity of the construction. The rotor is composed of circularly shaped permanent magnet segments providing sinusoidal air gap fields.

Figure 7.30 shows the typical course of the force curve of a turning rotor at constant stator currents. The force locus turns out to be an ellipse.

The most important advantage of the bearingless segment motor is the possibility of adapting the number and size of the stator segments to the

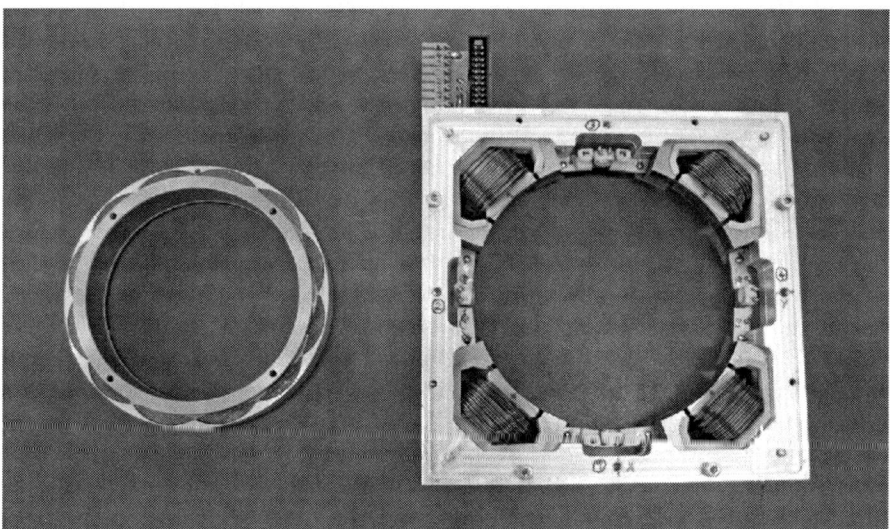

Fig. 7.29. Bearingless segment motor prototype using only four stator segments and four stator coils

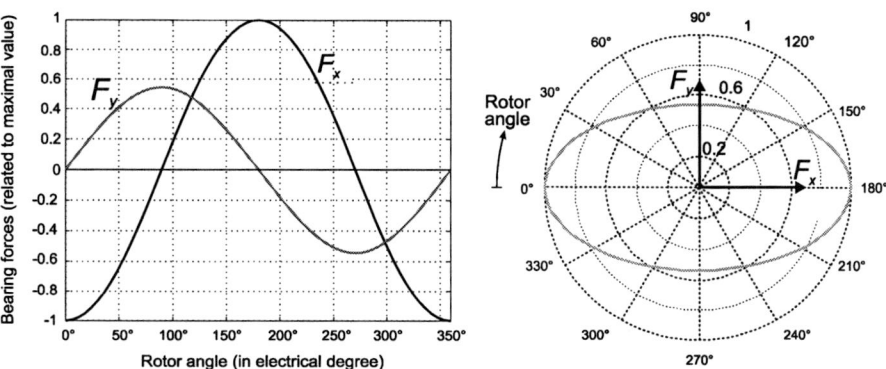

Fig. 7.30. Angle dependent bearing forces of a bearingless segment motor (when only one stator coil is energized at rated current) and its force locus

specific requirements of the application. This can lead to very low system costs and weight. For example, slice rotors with large diameters and low radial load and torque specifications can be driven cost-effectively by only a low number of stator segments separated from each other by large interspaces.

The design examples shown in the figures above represent alternating field motors. With stator segment numbers larger than four rotational field motors can also be realized (cf. Fig. 7.25).

7.3.2.2 Bearingless Asynchronous Motors

Asynchronous motors are suitable for bearingless applications only to a limited extent. If, as in the case of bearingless hermetically encapsulated pumps, the application requires large air gaps compared to the rotor diameter, then the magnetic flux needed for developing the torque and bearing forces can be created efficiently only by means of high-energy permanent magnetic materials. Creating the flux electromagnetically would lead to high reactive power and copper losses. Thus bearingless asynchronous motors are suitable in principle only for applications that permit small motor air gaps.

The design of the rotor construction of bearingless asynchronous motors requires special consideration. Additional voltages are induced in the short-circuit cage with the generation of the magnetic fields for the building up of radial bearing forces; these induced voltages lead to interfering eddy currents and resultant braking torques. In order to avoid these effects, special rotor winding configurations are necessary [7.16, 7.25]. An example of such an arrangement is illustrated in Fig. 7.31. The conducting rods are distributed on several phases such that the rods of one phase are connected on both sides to a short-circuit ring that is insulated from the rings belonging to the other phases. The peripheral distances between the rod groups of one phase are chosen in such a manner that no resultant eddy currents and therefore no braking

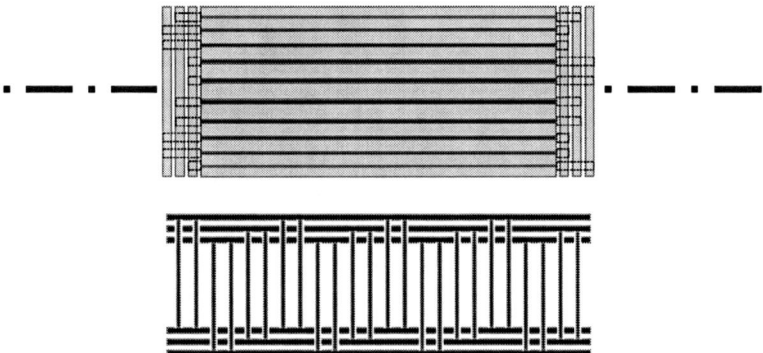

Fig. 7.31. Rotor short-circuit winding system of a bearingless asynchronous machine. *Below*: squirrel cages unwound in the plane

torques are caused by the radial bearing windings in the squirrel cages. The selective suppression becomes possible by reason of the different pole pair numbers of the torque and bearing force windings. Due to the complicated construction of the squirrel cages, such short-circuit windings are sometimes manufactured in the form of wound windings.

As a result of the high manufacturing costs associated with these measures, the bearingless asynchronous motor is typically not suitable for cost-critical series applications. Their application so far is primarily limited to specialized applications.

References

[7.1] J.K. Fremerey: Radial Shear Force Permanent Magnet Bearing System with Zero-Power Axial Control and Passive Radial Damping, Proc. of the First International Symposium on Magnetic Bearings, Springer-Verlag, Zurich, June 6–8,1988, p. 25–32

[7.2] M. Lang: Berechnung und Optimierung von passiven permanentmagnetischen Lagern für rotierende Maschinen, Dissertation, Technische Universität Berlin, Germany, 2003

[7.3] J.K. Fremerey: Course Luftlagerungen, Technische Akademie Esslingen, February 2000

[7.4] M. Lang, J.K. Fremerey: Optimization of permanent-magnet bearings, 6th International Symposium on Magnetic Suspension Technology, Torino, Italy, 2001

[7.5] M. Lang: Optimierung von permanentmagnetischen Lagern, 5. Workshop Magnetlagertechnik, Zittau – Kassel, September 20–21, 2001, p. 87–95

[7.6] S. Earnshaw: On the Nature of the Molecular Forces which Regulate the Constitution of the Luminiferous Ether, Trans. Cambridge Phil. Soc. 7, 1848, p. 97–112

[7.7] H. Kemper: Schwebende Aufhängung durch elektromagnetische Kräfte; eine Möglichkeit für eine grundsätzlich neue Fortbewegungsart, ETZ 59, 1938, p. 391–395

[7.8] J.-P. Yonnet: Permanent Magnet Bearings and Couplings, IEEE Transactions on Magnetics 17, 1981, p. 1169–1173

[7.9] G. Jungmayr, W. Amrhein, W. Angelis, S. Silber, H. Grabner, D. Andessner: Design of a Hybrid Magnetic Bearing, 8th International Symposium on Magnetic Suspension Technology, Dresden, Germany, September 26–28, 2005, p. 209–213

[7.10] G. Schweizer, A. Traxler, H. Bleuler: Magnetlager, 1st Edition, Springer Verlag, Berlin/Heidelberg/New York, 1993

[7.11] R. Schoeb, T. Schneeberger: Theorie und Praxis der Magnetlagertechnik, Skriptum, Johannes Kepler Universität Linz, Austria, 2004

[7.12] M. Reisinger, W. Amrhein, S. Silber, C. Redemann, P. Jenckel: Development of a Low Cost Permanent Magnet Biased Bearing, 9th International Symposium on Magnetic Bearings, Lexington, Kentucky, USA, August 3–6, 2004, p. 113–118

[7.13] P.K. Sinha: Electromagnetic Suspension: Dynamics and Control, IEE Control Engineering Series, V.30, Peter Peregrinus Ltd., London, 1987
[7.14] A. Chiba, T. Fukao, O. Ichikawa, M. Oshima, M. Takemoto, D.G. Dorrell: Magnetic Bearings and Bearingless Drives, 1st Edition, Elsevier Newnes, Oxford, 2005
[7.15] J. Bichsel: Beiträge zu lagerlosen elektrischen Motoren, ETH Dissertation No. 9303, Zurich, Switzerland, 1990
[7.16] R. Schoeb: Beiträge zu lagerlosen Asynchronmaschinen, ETH Dissertation No. 10417, Zurich, Switzerland, 1993
[7.17] A. Chiba, T. Deido, T. Fukao, M.A. Rahman: An Analysis of Bearingless AC Motors, IEEE Transaction on Energy Conversion, Vol. 9, No. 1, March 1994, p. 61–68
[7.18] W. Amrhein, S. Silber: Lagerlose Permanentmagnetmotoren, Elektrische Antriebe - Grundlagen, D. Schröder, Springer Verlag, Berlin/Heidelberg/New York, 2nd Edition, 2000, p. 357–384
[7.19] W. Amrhein, S. Silber: Bearingless Single-Phase Motor with Concentrated Full Pitch Windings in Interior Rotor Design, 6th International Symposium on Magnetic Bearings (ISMB), Paul E. Allaire, David L. Trumper (Eds.), Cambridge, August 1998, p. 486–496
[7.20] S. Silber, W. Amrhein: Bearingless Single-Phase Motor with Concentrated Full Pitch Windings in Exterior Rotor Design, 6th International Symposium on Magnetic Bearings (ISMB), Paul E. Allaire, David L. Trumper (Eds.), Cambridge, August 1998, p. 476–485
[7.21] W. Amrhein, S. Silber, K. Nenninger, G. Trauner, M. Reisinger: Developments on Bearingless Drive Technology, JSME International Journal Series C, Volume 46, No.2 June 2003, p. 343–348
[7.22] Silber Siegfried, Amrhein Wolfgang, Bösch Pascal, Schöb Reto, Barletta Natale: Design Aspects of Bearingless Slice Motors, 9th International Symposium on Magnetic Bearings, Lexington, Kentucky, USA, August 3–6, 2004, paper No. 3
[7.23] Grabner Herbert, Wolfgang Amrhein, Siegfried Silber, Klaus Nenninger: Nonlinear Feedback Control of a Bearingless Brushless DC Motor, 6th IEEE International Conference on Power Electronics and Drive Systems (PEDS), Kuala Lumpur, Malaysia, November 28 to December 1, 2005, p. 366–371
[7.24] J.W. Beams: High Rotation Speeds, Journal of Appl. Phys., 8, 1937, p. 795–806
[7.25] U. Blickle: Die Auslegung lagerloser Induktionsmaschinen, ETH Dissertation No. 13180, Zurich, Switzerland, 1999
[7.26] E. Knopf, M. Aenis, R. Nordmann: Aktive Magnetlager – Ein Werkzeug zur Fehlererkennung, Kolloquium Aktoren in Mechatronischen Systemen, March 11, 1999, organized by SFB 241, TU Darmstadt, Fortschritt Berichte VDI, Reihe 8, Nr. 743, VDI-Verlag, Düsseldorf, Germany, 1999
[7.27] H. Grabner: Dynamik und Ansteuerkonzepte lagerloser Drehfeld-Scheibenläufermotoren in radialer Bauform, Dissertation, Johannes Kepler Universität Linz, Austria, May 2006
[7.28] W. Gruber, W. Amrhein, S. Silber, H. Grabner, M. Reisinger: Analysis of Force and Torque Calculation in Magnetic Bearing Systems with Circular Airgap, 8th International Symposium on Magnetic Suspension Technology (ISMST8), Dresden, Germany, September 26–28, 2005, pp. 167–171

[7.29] W. Gruber, W. Amrhein: Design of a Bearingless Segment Motor, 10th International Symposium on Magnetic Bearings (ISMB10), Martigny, Switzerland, August 21–23, 2006

[7.30] W. Gruber, W. Amrhein, M. Haslmayr: Segment Motor with Five Stator Elements – Design and Optimization, International Conference on Electrical Machines and Systems 2006 (ICEMS2006), Nagasaki, Japan, November 20–23, 2006

8
Mechanical Transfer Units

Werner Krause

It is the function of mechanical transfer units to adapt the energy of motion, provided by the drive, to the demanded parameters and to direct it to the downstream working element, assembly or device. Therefore, gearings, couplings and shafts including the related bearings are required.

The fulfillment of a given transfer function, which represents the interrelationship of motions of input and output link, is the main task of gearings. Input as well as output are mechanical variables. Thus gearings are also known as mechanical transformers, especially for rotational speed and torques. Furthermore, it may be their task to guide a gearing element to predefined positions or points of the element on predefined trajectories. They are then called guide gearings.

Couplings connect shafts to transfer rotations and torques. Shafts are line segments, which carry parts of the device and their weight and operation forces. Couplings are carried by bearings, which also assure the position of moving parts in space.

Consecutively an overview of transfer elements which are most often required in small electric drives will be given, to allow an unerring choice. For their dimensioning and design as well as for further mechanical elements required to assemble a drive, please refer to further reading [8.1, 8.2].

Please consider that all mechanical transfer elements significantly influence the behavior of motion of the entire drive. They have a mass, subject to influences of friction, attenuation and elasticity. Their closure often causes backlash. To conceive these properties is the precondition for dynamic analysis of transfer elements and description of motion performance already during conception. The task of dynamic analysis is to determine the necessary force to cause a prescribed state of acceleration from known applied force and mass allocation. Applied forces are forces at the working element, resistive forces, forces of mass and elasticity. They cause reactions in couplings, links and bearings as a result of the given limitation of movement by constraints of single elements. Accordant impacts are caused by inertia forces, which are opposed by the elements to a change of the state of motion as a kinetic resistance.

Inertia forces emerge from non-linear functions of motion of input or transfer unit, but also from vibration as result of backlash, elastic deformation or friction. Especially at high working speed and with resonances between input and natural frequencies, vibrations are significant.

Modeling, simulation and optimization of the dynamic behavior of drives are tasks of device and machine dynamics.

8.1 Gearings

Gearings are mechanical devices to transfer motions and forces or to lead points of a body on defined paths. They consist of moveable elements, linked to each other, of which the mutual possibility of motion are defined by the nature of the joints. One element of the gearing is always the basic body (base); the minimum number of bodies and gearings is three, respectively.

8.1.1 Kinds of Gearings

Related to the function, a distinction is drawn between transfer gearings and guide gearings. The fulfilment of a given transfer function, which represents the interrelation of motions of input and output link, is the main task of transfer gearings. Inputs as well as outputs are mechanical quantities.

A transfer gearing may be symbolized according to Fig. 8.1. If the transmission function is linear, that is $\psi = k\varphi$, it is called steady transmitting gearing with constant transmission ratio. Besides toothed gearings, screw mechanisms, friction wheel- and belt drives are applicable for steady transmission of rotational speed. All gearings with nonlinear transmission function are summarized as unsteady transfer gearings, e.g. cam mechanisms or step mechanisms.

Table 8.1 gives an overview of typical transfer functions and examples of related gearings.

Form and position of point paths characterize the usage of guide gearings, so that the terms input link, output link and transfer function are not used typically. The rectilinear translation of a point P by cam mechanisms or circle shearing motion of a body K by a parallel crank mechanism are typical examples (Fig. 8.2).

Kinds of gearings may be classified by their design, i.e. by characteristic elements. You may differentiate eight groups of gearings, which are called basic gearings (Table 8.2).

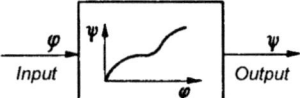

Fig. 8.1. Block diagram of transfer gearings

8 Mechanical Transfer Units

Table 8.1. Typical transfer functions and examples of gearings

transfer function	form of input motion (samples) rotation	form of input motion (samples) translation	transfer function	form of input motion (samples) rotation	form of input motion (samples) translation
s, ψ	gear train	screw mechanism	s, ψ	geneva mechanism	cam step mechanism
s, ψ	double crank	belt drive	s, ψ	cam dwell mechanism	cam mechanism
s, ψ	cam mechanism	slider-crank mechanism	s, ψ	geared linkage	belt drive

Table 8.2. Systematization of guide gears by characteristic elements

Group of gearings (characteristic element)	Examples	Group of gearings (characteristic element)	Examples
Coupling drives (stiff elements, revolute pair, prismatic pair)		Wedge mechanism (stiff elements, coupled by wedge surface)	
Cam mechanism (cam, yoke follower, cam pair)		Screw mechanism (motion screw, screw nut)	
Gear train (gear wheel, e.g. cylindrical gear, external gearing, internal gearing conical gear, worm)	outer gear / inner gear	Flexible drive (belt, band, wire, chain)	
Friction body drive (force-coupled friction bodies, e.g. disc, cones)		Pressuring medium mechanism (gaseous, liquid pressuring medium)	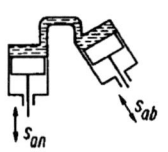

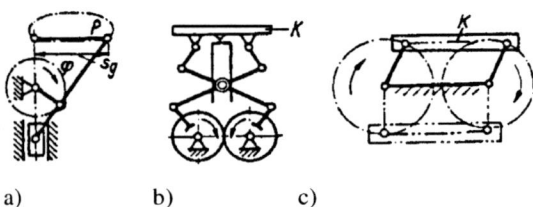

Fig. 8.2. Examples of guide gearings. **a** Rectilinear translation of point P on a path s_g by a centric slider-crank mechanism. **b** Direct rectilinear translation of a body K by a gearing with symmetric double drive. **c** Circle shearing motion of a body K by a parallel crank mechanism

Different basic mechanisms may be coupled serially or in parallel or their effects may be superposed. They are called combined mechanisms and, described according to the involved forms of mechanisms, e. g. geared linkage.

Consecutively the most important gearings for small drives are discussed.

8.1.2 Gear Trains

Gear trains transform rotational speeds and torques between two or several shafts. The gearing of the train effects a matching in form and permits an interlocking and skid-free transmission of motion and force. To characterize the transmission of motion the average transmission (resp. transmission ratio) i or the current transmission i_0 is considered:

$$i = \frac{n_1}{n_2} = \frac{d_2}{d_1}, \quad i_0 = \frac{\omega_1}{\omega_2}; \tag{8.1}$$

index 1 drive end (input of gearing), index 2 output end (output of gearing), n rotational speed, d reference circle diameter of wheels, ω angular speed.

8.1.2.1 Classification

Aspects of systematization are base frame alignment of the wheels, number of gearing steps, position of axles and basic shapes of the wheel body in the first instance. For classification criteria that reference the toothing, i. e. involute or cycloid tooth system, spur toothing, helical toothing etc., may be consulted.

Base Frame Alignment of the Wheels

Single level gear trains have three elements and satisfy the definition of gearings (refer to Sect. 8.1.1). They consist of two wheels (1, 2) and a fixed connection of axes (ligament s) (Fig. 8.3a). If the ligament is unmovable, that is it is connected firmly to the base, it is called a stand gear.

The gearing is called an epicyclic gear if it circulates, that is the ligament is connected revolvable to the base, because at least one wheel rotates with

8 Mechanical Transfer Units   453

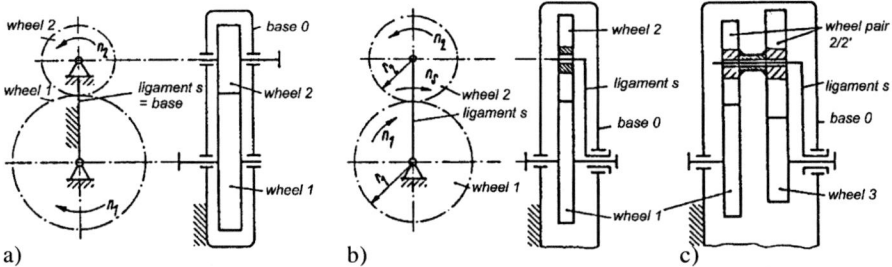

Fig. 8.3. Derivate of epicyclic gears from stand gears (refer to Table 8.4). **a** Single level stand gear. **b** Single level epicyclic gear. **c** Two level epicyclic gear

the ligament (Fig. 8.3b, c). In general the base-connected wheels are called central or sun wheels; those on the rotating ligament are called circulating or planet wheels (from there also planetary gearing).

Number of Gearing Steps

In single level gear trains rotation speed and torque are transformed only once between input and output (Fig. 8.4a), in multi level gear trains several times. Additionally you differentiate, depending on whether the axles of input and output are aligned or not, returning and non-returning gearings (Fig. 8.4b, c). A special case is the wheel chain (Fig. 8.5) where several externally geared wheels are arranged in a successive chain. The even gearing transfers the

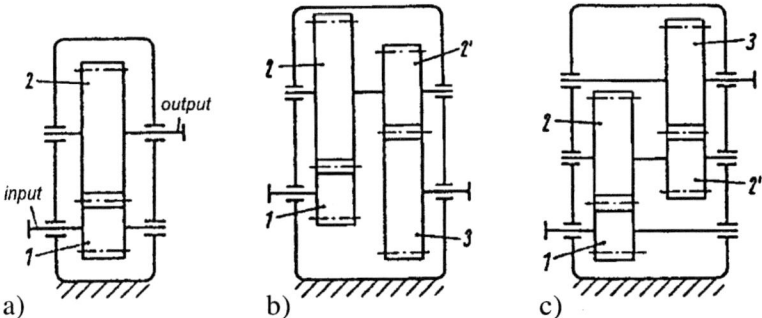

Fig. 8.4. Cylindrical gear box. **a** Single level. **b** Two level, returning. **c** Two level, non-returning

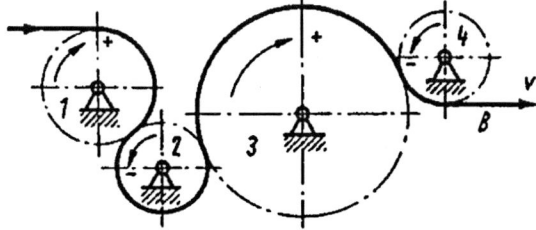

Fig. 8.5. Wheel chain

rotation from the downstream wheel to the upstream wheel without transformation but with reverse direction of rotation. All reference circle diameters rotate with the same speed v, as if a band B is pulled through.

Position of Axles and Basic Shapes of Wheel Bodies

Cylindrical gears have parallel axles and the basic geometric shapes of the paired wheels are cylinders (Fig. 8.6a). The axles of bevel gears intersect (b); basic shapes of the wheel bodies are orbital cones. Intersecting axles exist in crossed helical gears as well as in worm gears. In worm gears the crossing angle is 90° generally. The basic geometric shapes of the paired wheel bodies are cylinder and globoid (Fig. 8.6c), while the basic shape of both wheels in crossed helical gears is a cylinder (Fig. 8.6d). A special cylindrical gear is the rack-and-pinion gear (Fig. 8.6e).

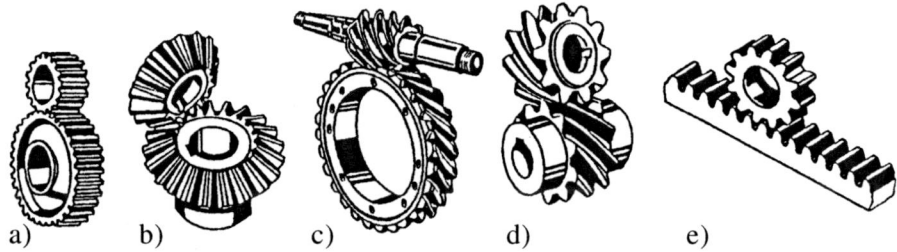

Fig. 8.6. Forms of gear trains. **a** Cylindrical gear. **b** Bevel gear. **c** Worm gear. **d** Crossed helical gear. **e** Rack-and-pinion gear

8.1.2.2 Gears

According to the demand of steady transmission of motion, configuration and arrangement must not occur randomly, but have to fulfil defined kinematic and geometric requirements. Those arise from the basic rules of toothwork. The fulfilment assures constant momentary transmission i_0 according to Eq. (8.1) as well as profile overlap $\varepsilon_\alpha > 1$, that is the latest ending of contact of one pair of shoulders by the time the following pair engages exactly with each other.

Forms of Profile and Toothing

Randomly formed shoulders are not reasonable for practical use, even if they fulfil the basic rules of toothwork. They are hardly producible. Regular-shaped shoulders are practical. The involute profile became technically important (Fig. 8.7). In clockwork, gearings of simple devices and transmissions into higher speed cycloid gearings are applied (cp. Sect. 8.1.2.3).

Besides the profile form, the shoulder line course, which defines the usual forms of toothing, is of interest for gearing (Fig. 8.8).

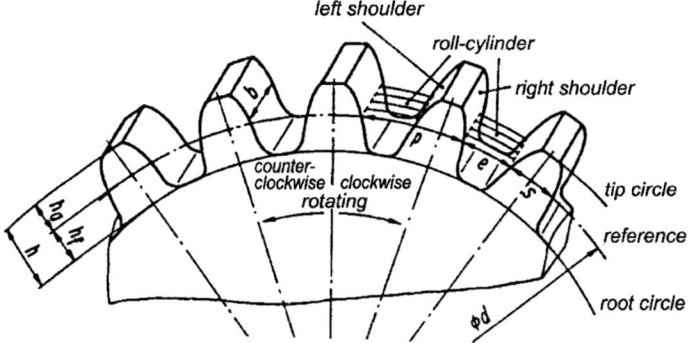

Fig. 8.7. Specifying parameters of gear trains (spur gears with involute profile)

Nomenclature and Determinants

The basic terms and notions on gear wheels are determined in DIN 3960 and DIN 58405 and shown in Fig. 8.7 for spur gears. Using gear tooth forming, the circumference of a gear is divided into z equal parts according to the number of teeth. The distance between two successive, equally directed shoulders is called pitch. If it is measured on the circumference of a reference circle of diameter d between two left or two right shoulders the pitch is named reference circle pitch p. Reference circle diameter d, pitch p and number of teeth z are linked as follows:

$$pz = d\pi \tag{8.2}$$

or rather

$$d = \frac{pz}{\pi} . \tag{8.3}$$

The ratio d/z is called modulus (diametral pitch) m (partition of diameter) and you get

$$d = mz \tag{8.4}$$

or rather

$$p = m\pi . \tag{8.5}$$

Pitch p merges the thickness of one tooth s and the width of the gap e:

$$p = s + e . \tag{8.6}$$

All of the three dimensions are measured in arc length on the reference circle. Further determinants are addendum h_a, measured from reference circle to tip circle, dedendum h_f, measured from reference circle to root circle and height of tooth h as the sum of addendum and dedendum. For standardized gearing these dimensions are denoted depending on module.

For module m only standardized values should be used (Table 8.3).

You have to differentiate the basic dimensions of cylindrical gearing with angular gears between normal pitch p_n and reference circle pitch p_t (Fig. 8.9).

Table 8.3. Modulus m in mm for cylindrical gears (according to DIN 780, values in line 1 for favored use to assure interchangeability)

Line 1	0.05	0.06	0.08	0.1	0.12	0.16	0.2	0.25	0.3	0.4	0.5	0.6	0.7
Line 2	0.055	0.07	0.09	0.11	0.14	0.18	0.22	0.28	0.35	0.45	0.55	0.65	0.75
Line 1	0.8	0.9	1.0	1.25	1.5	2.0	2.5	3	4	5	6	8	10
Line 2	0.85	0.95	1.125	1.375	1.75	2.25	2.75	3.5	4.5	5.5	7	9	

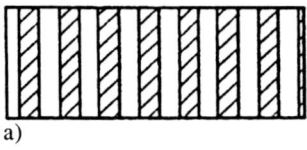

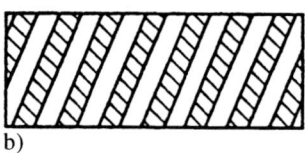

Fig. 8.8. Forms of toothing. a Spur gear. b Angular gear

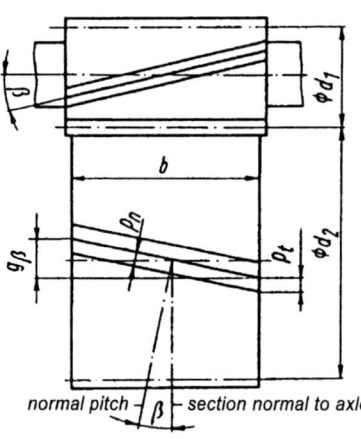

Fig. 8.9. Cylindrical gearing with angular gears (profile overlap g_β projected in drawing plane)

Normal pitch is the distance between two left or right shoulders, measured on the casing of a partial cylinder normal to the direction of shoulder

$$p_n = m_n \pi . \tag{8.7}$$

Reference circle pitch is the distance between two left or two right shoulders, measured in the front section (section normal to axles) of the partial cylinder. The coherence of normal and reference circle pitch is given by the cant angle β:

$$p_t = \frac{p_n}{\cos \beta} . \tag{8.8}$$

Accordingly you have to differentiate the front modulus m_t and the normal modulus m_n. It is said to be

$$m_t = \frac{m_n}{\cos \beta}.\qquad(8.9)$$

The normal modulus has to be a standardized value according to Table 8.3.

Helical gears have different properties from rectilinear gears in relation to geometry of toothing, distribution of forces and operating behavior. Thus helical gears are smooth running because the shoulders of a pair of teeth do not engage abruptly but gradually. An adequate accuracy of the wheels is required. The overlap increases as well, because the cleft overlap ε_β is added to the profile overlap ε_α [8.1].

Reference Profile

Contemplating a gear wheel with infinite radius you achieve a gear rack-and-pinion gearing. The involute merges into a straight line and the shoulders become planes. Due to the simple and exactly producible form the rack profile was defined as the basic profile for involute gearing and is called the reference profile. If the tool is derived from it, all wheels may be geared in a such way, so that gears work together properly regardless of their number of teeth. To satisfy the different requirement, two reference profiles with different tooth heights are standardized (Fig. 8.10).

The reference profile after DIN 867 is preferably applied to modulo $m \geq 1$ mm in mechanical engineering. The reference profile after DIN 58400 in contrast is used for modulo $m < 1$ mm and satisfies the requirements of mass production (especially sufficient overlap and high backlash of tooth point at large gear tolerance).

Limits of Gear Geometry, Extremely Small Numbers of Teeth

Producing a gear by roll milling, a part of the root shoulder between reference and root circle that is necessary for transmission of motion may be cut off

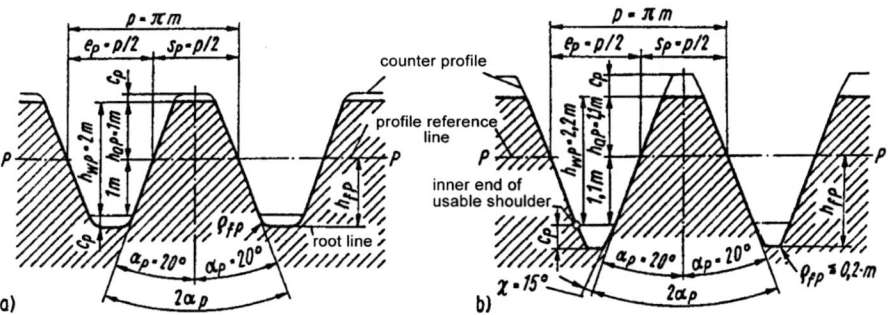

Fig. 8.10. Reference profile with counter profile. **a** According DIN 867. **b** According DIN 58400

(Fig. 8.11). The whole root of the tooth becomes weak and the tooth rating is decreased. This disadvantageous occurrence is termed undercut. If just no undercut appears, the number of teeth is called the marginal number of teeth z_{min}. Practically this limit may be diminished to $z'_{min} = (5/6)z_{min}$ without deterioration of contact ratio. Marginal numbers of teeth of $z_{min} = 17$ and $z'_{min} = 14$, respectively, arise for reference profile with $h_a = 1.0\,\mathrm{m}$ or rather $z_{min} = 19$ and $z'_{min} = 16$ for a profile with $h_a = 1.1\,\mathrm{m}$.

The undercut may be avoided during production of involute gears by roll milling cutters if the tool is displaced sufficiently from the wheels reference circle. This operation is called shift of profile. By using the existing limits for undercut and tapering of teeth the number of teeth displayed in Fig. 8.12 may be realized.

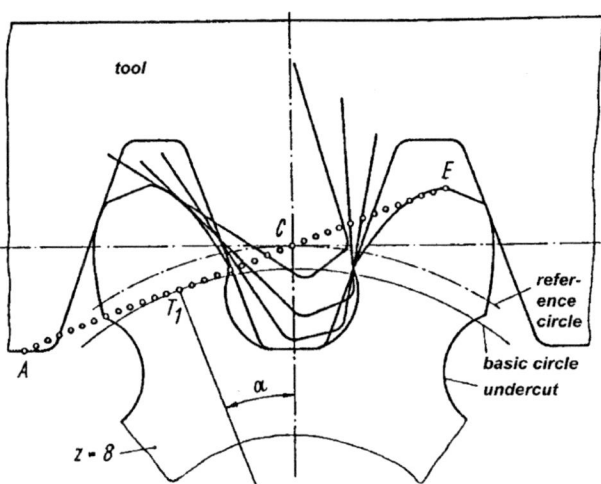

Fig. 8.11. Formation of undercut

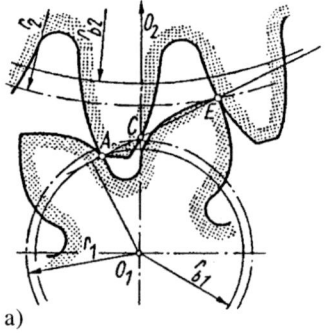

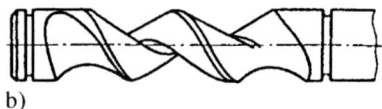

Fig. 8.12. Minimal number of teeth. **a** $z = 5$ for spur gear. **b** $z = 1$ for angular gearing

Besides prevention of undercut and adaptation of centre-to-centre distance to a given value, shift of profile may be utilized to advance the smoothness by increasing overlap and enhancing bearing strength [8.1] (q.v. DIN 3992 and DIN 3994).

8.1.2.3 Cylindrical Gears

Cylindrical gears with involute gearing (spur gears) are the most common gear mechanisms. They are used for transmission of rotation and torques between parallel shafts for very small constructions with modulo from 0.05 mm and outputs of only a few watts as well as for high outputs and rotational speeds up to 100 000 rpm. Peripheral speed of the wheels may reach 200 m/s.

Cylindrical gears are manufactured as single-level gears for transmissions up to $i = 8$ and as two-level gears up to $i = 35$ (cp. Figs. 8.3 and 8.4). The calculative centre-to-centre distance for single-level spur gears is

$$a_\mathrm{d} = m\frac{(z_1 + z_2)}{2} = \frac{(d_1 + d_2)}{2}. \tag{8.10}$$

The transmission arises from

$$i = \frac{n_1}{n_2} = \frac{d_2}{d_1} = \frac{z_2}{z_1}. \tag{8.11}$$

The overall transmission in multi-level gearings is the product of transmissions of every stage.

$$i_\mathrm{total} = i_\mathrm{I} \cdot i_\mathrm{II} \cdot \ldots \cdot i_\mathrm{n}. \tag{8.12}$$

The transmission for the two level gearing according to Fig. 8.4b, c arises from

$$i_\mathrm{total} = \frac{n_1}{n_3} = \left(\frac{n_1}{n_2}\right)\left(\frac{n_{2'}}{n_3}\right) \tag{8.13}$$

$$= \left(\frac{d_2}{d_1}\right)\left(\frac{d_3}{d_{2'}}\right) = \left(\frac{z_2}{z_1}\right)\left(\frac{z_3}{z_{2'}}\right). \tag{8.14}$$

The gears 2 and 2' are fixed on the same shaft, thus $n_2 = n_{2'}$.

The division of total transmission into partial transmission is undertaken in multi-level power gearings in general under the condition of minimization of the total volume of all wheels [8.1, 8.2, 8.11]. To simplify the production of cylindrical gears for propulsion gearings in precision mechanics with generally small outputs an equal transmission in every stage is desired. Thus the transmission of a stage i_i of a n-level gearing is calculated by $i_\mathrm{i} = \sqrt[n]{i_\mathrm{total}}$.

- The efficiency of every stage is about 94–99%.

In epicyclic gears the calculation formulas are more complicated [8.1]. Furthermore, additional rolling power follows from the relative rolling speed in these gears: this works as interior power and cannot be released outwards. They may reach a multiple of input power and lower the efficiency in comparison with stand gears.

For completion of small and micro motors, high transforming cylindrical gears are required. For stand gears (cp. Sect. 8.1.2.1) blank and plugging constructions are implemented (Table 8.4). Due to the frequent repetition of parts, plugging construction features the advantage of automated assembling. The Wolfrom gear is, among epicyclic gears, characterized by good transmission properties and compact construction (Table 8.4/2.2).

Table 8.4/2.3 shows a single-level epicyclic gear with bull gears directly engaged with each other without an interposed epicyclic wheel. This is possible because of the elastic design of the smaller, externally geared wheel 1. Ligament and epicyclic gear are represented by the eliptic body 2, which is called wave generator and which bears the elastic wheel 1 at two opposed points of the circumference against wheel 3, that is attached to base. The rolling elements 1 and 2 diminish friction. This gearing became popular in different configurations as harmonic drive. It becomes more important in small drives. Transmission $i = n_2/n_1 = z_3/(z_3 - z_1)$ reaches values up to $i = 320$, but may be much higher in special cases. In the gearing shown the minimal difference of number of teeth has to be $z_3 - z_1 = 2$ or a multiple of two. Using an eccentric as wave generator it would be 1, using an arch triangle it has to be 3, etc.

A further construction of a single-level epicyclic gear is the cyclo-gear [8.1].

Cylindrical gearings with non-involute gear became important only for limited applications. Besides cycloidal toothing, which is in part still used in film hoist gearing and rack-and-pinion gearings, pinion stick toothing can be found in simple counting trains, toys etc. and circular arc toothing (Fig. 8.13) can be found in products with low input power, e. g. in watches, indicator gearings in manometers and devices for electromechanically timed relays. For these areas of application an efficiency as high as possible is required, which is better obtained with these special gearings. Furthermore, gearings in watches have to transmit torques accurately during engagement, too, which involute gearings do not allow [8.1].

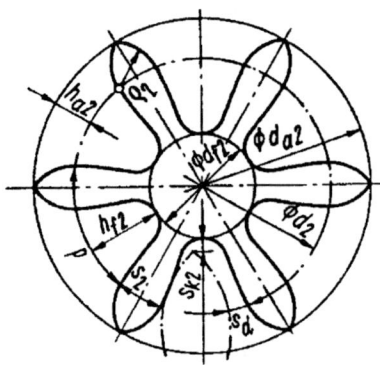

Fig. 8.13. Circular arc toothing (clockwork gearing)

Table 8.4. Type of construction of highly translating cylindrical gears

Type of gearing	Properties
1. Cylindrical stand gear 1.1 Blank construction 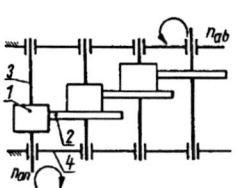	• Gear 1 and wheel 2 fixedly on shafts 3 arranged, pivoted between blanks 4 • Transmission $i_{\text{total}} = i_\mathrm{I} \cdot i_\mathrm{II} \cdot \ldots \cdot i_n$ with $i_\mathrm{I} = z_2/z_1$ etc. Advantages: – Low bearing friction – Flat construction Disadvantages – Kit with different transmissions hardly realizable, because of infrequent repetition of parts
1.2 Plugging construction 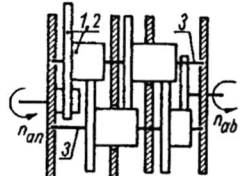	• Gear-wheel combination 1, 2 pivoted on fixed, continuous axles 3 • Transmission: see construction of blank Advantages: – Kit with huge number of frequent repetition of parts possible Disadvantages: – Bending of long axles – Overhung bearing of driving axle – Bigger losses by friction in bearing
2. Cylindrical epicyclic gear (the numbers of teeth have to be inserted into the equations)	
2.1 Epicyclic gear with two-level planet wheel 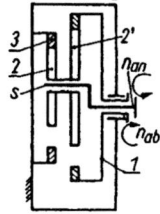	• Input ligament s, planet wheels 2 and 2' fixedly connected and rotatable arranged on ligament s, ring gear 3 fixed frame, output ring gear 1 • Transmission $i = z_1 z_2 / [(z_1 - z_{2'})(z_1 - z_3)]$ Advantages: – High transmission ratio in one level possible Disadvantages: – Internal toothing required (difficult to produce) – The higher i the lower efficiency
2.2 Wolfrom epicyclic gear 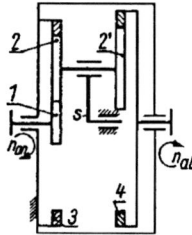	• Input sun wheel 1, planet wheels 2 and 2' rotatable connected to ligament s, ligament s not loaded by torque, ring gear 3 fixed to frame, output ring gear 4 • Transmission $i = z_2 z_4 (z_1 + z_3) / [z_1(z_2 z_4 - z_3 z_{2'})]$ Advantages: – High transmission ratio in one level possible Disadvantages: – Internal toothing required – the higher i the lower efficiency

Table 8.4. (continued)

Type of gearing	Properties

2.3 Harmonic drive	• Input-elliptic wheel 2, output-elliptic wheel 1, ring gear 3 fixed to frame
	• Transmission $i = z_3/(z_3 - z_1)$
	Advantages:
	  – High transmission ratio in one level possible
	  – Compact construction possible
	Disadvantages:
	  – Internal toothing required
	  – Elastic wheel 1 difficult to produce, requires special technology and materials

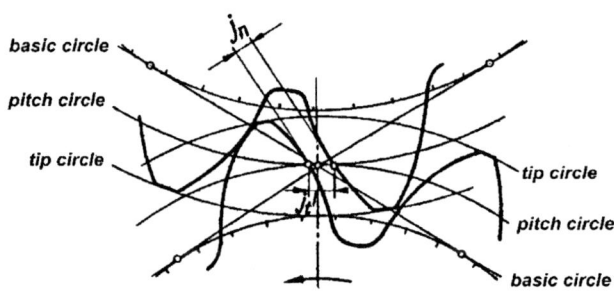

Fig. 8.14. Flank clearance of cylindrical gearings

Tolerances of Toothing, Adoption of Gearings

Depending on the intended use, running smoothness, minimal attrition, isogonal transmission of rotation and replaceability of gear elements are demands on gear trains. Furthermore only assembling of gearing without reworking is economical. This requires tolerated dimensions and a system of fits for gear transmissions. To define a certain fit of gearings a defined flank clearance is required for manufacturing and operating reasons (difference in production and assembling, possibility to apply lubricant etc.) besides tip backlash c (cf. Fig. 8.10), which indicates the distance between tip and root circle of the mating gear. Rotary flank clearance j_t and contact flank clearance j_n are differentiated (Fig. 8.14), whereas

$$j_t = \frac{j_n}{\cos \alpha} \qquad (8.15)$$

is essential. α represents the angle of engagement (cf. Figs. 8.10 and 8.11).

For gearings with modulo m from 0.2 to 3 mm DIN 58405 defines a number of fields for base tangent length extent dependent on quality to generate a certain flank clearance, labeled with small characters h, g, f, e, d etc. Centre-to-centre distance is assigned to field J.

DIN 3961 to 3969 contain analogical specifications for gearings in machine building with $m \geq 1$ mm [8.11].

For devices in mechanical engineering, minimal flank clearance of about 20–25 μm for centre-to-centre distance up to 20 mm and of about 35–50 μm for centre-to-centre distance up to 125 mm must not be under-run if avoidance of uneven running is necessary. Backlash has to be defined considering operating temperature and conditions for power gearings [8.1, 8.11].

As an example for indication according to DIN 58405: matching of two gears of quality 7 with base tangent length extend after field e in tolerated centre-to-centre distance 7J and mandatory approval by double flank gear test (accumulative deviation: designation in table of parameters of drawing): $7J/7eS''$.

Material Selection

The material selection for gears in small drives primarily involves economic considerations. Low production costs, low abrasion and good resistance to moisture etc. are important for the choice of materials, because the share of material costs is low compared to labour costs.

Steel is used if gears must have high stability despite low dimensions, whereas the connection shall be realised with plane shafts by interference fit (tight fit) or also if high accuracy of production and few abrasion are required (e. g. construction steels S275JR to E 360 or heat-treated steels C22E to C60E, 34 Cr4, 42CrMo4) – see advice on p. 503.

Slow turning wheels, which generally have to transmit low forces, are made out of copper alloy. Compared with steel, these materials possess better working properties and corrosion resistance. Furthermore they are non-magnetic (e. g. CuZn40Pb2, CuZn40). Sometimes CuSn6 (tin bronze) is also used, e. g. because of its resistance in gears, which encounter water (water meter, etc.) and further in the application of gears subjected to impacts.

Plastics offer several advantages. They function vibration-reducingly and elastically compensate variations of toothing by their low coefficient of elasticity. These gears operate quietly, but are sensitive to influences of moisture and temperature, which affect dimensional stability. Major examples are polyamide (PA) and polyoxymethylene (POM), and to some extent also moulded thermosetting materials from phenolic resin with fabric inlay (fabric-base laminate). When mounting fabric-base laminate gears, the temperature must not rise above 100 °C and these gears must not be paired because of enhanced abrasion. The operating temperature of PA and POM shall not rise above 80 °C. They may be paired with each other if load is not too high. Except for wheels of PA and POM, wheels of different materials have to be paired to minimize abrasion.

Lubrication

Metallic wheels or those of polyamide with modulo $m < 1$ mm and mostly light load are lubricated only once before assembling with grease or oil. Similarly for power gears with circumferential speed of wheels $v_u < 1$ m/s grease is sufficient for lubrication. For industrial gearings with $v_u \geq 1\,\text{m/s}$ splash lubrication has to be planned; that is, wheels with teeth are dipping into an oil bath in the frame. Because of the great variety of lubricants [8.1] consultation of experts is recommended for extreme operating conditions.

Calculation of Bearing Strength

Gears have to be dimensioned often in such a way that no claims occur (tooth breakage, pitting, abrasion or galling), whereas the durability depends on the choice of material, its hardness and surface quality, surface pressure, lubrication, etc. Operating conditions (e. g. impulsive impacts), deviation of toothing and deformation of gears additionally influence operating conditions.

Already definition of transmission i and with it number of teeth z affect load capacity, because whole-number transmission (e. g. $i = 3.0$) in one gearing level always causes pairing of the same flanks and with it the same flaws. This involves unequal abrasion especially at untempered flanks and excitation of vibration at higher circumferential speed. For avoidance the transmission of every stage should not be chosen in whole numbers. Prime numbers for larger number of teeth accomplish this demand.

- Estimated dimensioning of stability after Bach's relation is sufficient for untempered cylindrical gears of metallic gear materials with modulo range below 1mm, which are bound to procedures of mass production in fabrication and used in small drives (marginal load) (Fig. 8.15):

$$F_t \leq bp C_{\text{border}}, \tag{8.16}$$

with $F_t = \frac{2M_d}{d}$ (d reference circle diameter) and $M_d = 9.55 \cdot 10^6 \frac{P}{n}$; M_d in Nmm, P in kW and n in rpm.

$C_{\text{border}} \approx 0.07 \sigma_{\text{b\_allowed}}$ is applied for metallic materials for profiles with $h_a = 1.0\,\text{m}$ (Fig. 8.10), size of tooth root $s_f \approx 0.52p$ and height of tooth $\bar{a} \approx 0.64p$. $C_{\text{border}} \approx 0.1 \sigma_{\text{b\_allowed}}$ is applied for profiles with $h_a = 1.1\,\text{m}$, $s_f \approx 0,72p$ and $\bar{a} \approx 0,82p$; with $\sigma_{\text{b\_allowed}} = \sigma_{\text{bW}}/S \approx (0,3\ldots 0,5) R_m/S$; values for σ_{bW} and R_m see [8.1, 8.2]; safety factor $S = 2\ldots 4$.

Bending moment is $M_b = F_{ta}\bar{a}$, considering the tangential force $F_{ta} = M_d/r$ to affect the outmost point of tooth tip. If the force F_t, related to the reference circle, is approximately considered to equal tangential force F_{ta}, then $F_t \leq bs_f^2 \sigma_{\text{b\_allowed}}/(6\bar{a})$ with $M_b \leq W_b \sigma_{\text{b\_allowed}}$ and section modulus $W_b = bs_f^2/6$ for cross section of tooth root.

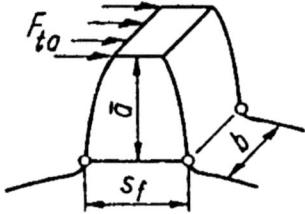

Fig. 8.15. Assumed load of tooth at rough calculation after Bach

By introducing the ratio of tooth width $\lambda = b/m = 5 \ldots 20$, this relation may be used for the approximate calculation of modulo m. With $p = m\pi$

$$m \geq \sqrt{\frac{F_t}{(\lambda \pi C_{\text{border}})}} \geq \sqrt[3]{\frac{2M_d}{(z\lambda \pi C_{\text{border}})}} \,. \qquad (8.17)$$

For helical gears p has to be substituted by front pitch $p_t = m_n \pi / \cos \beta$ (m_n standard modulus).

- For power gears (high load) exact calculations of bearing strength of tooth root and tooth flank must be made after approximate dimensioning of wheels after DIN 3990. They are shown simplified according to conditions of gearing for precision mechanics in [8.1, 8.2].
- To calculate the load capacity of plastic wheels, which have, unlike metal, no endurance strength but only finite fatigue life, some peculiarities have to be considered, which can also be found in [8.1, 8.2].

No trusted calculation basis exists for abrasion strength.

Operating Performance

- *Noise performance.* Fundamentals for appropriate methods to decrease noise are measurements (acoustic pressure and frequency spectrum of noise) at different operating conditions, to account for the major noise cause. The subsequent described results of noise analyses refer exclusively to the most frequently used device gearings with roll-milled, involute geared cylindrical spur wheels (modulo $m \leq 1$ mm), which are produced without additional reoperation and surpassing effort in production and assembly due to requirements of the mass market and linked matters of efficiency and hence characterized by relatively high toothing deviation.

It has to be determined categorically, that considerably noise reduction is a challenging problem, because noise of gearings in large parts depends on manufacturing and assembling tolerances. Furthermore it is known, that great efforts are necessary to lower the relatively low level of noise in small gearings compared with industrial gearings of general machine building.

Factors which lower noise are, besides appropriate lubrication, among other things low surface roughness, compliance of close assembling tolerances and reduction of adversarial effect of tip flange engagement, caused by toothing deviations (low centre-to-centre distance, to ensure $\varepsilon_\alpha \geq 1$, use of tip overlapping method because of evolving favorable tip flange curve [8.1, 8.2]). But they are of minor importance compared to material pairing. Dependence of noise behavior on resulting coefficient of elasticity is shown in Fig. 8.16. Noise behavior clearly improves with decreasing value E_{res}. Additional noise reduction is achieved for $E_{\text{res}} > 2000 \, \text{N/mm}^2$ if the wheel with lower coefficient of elasticity does the driving.

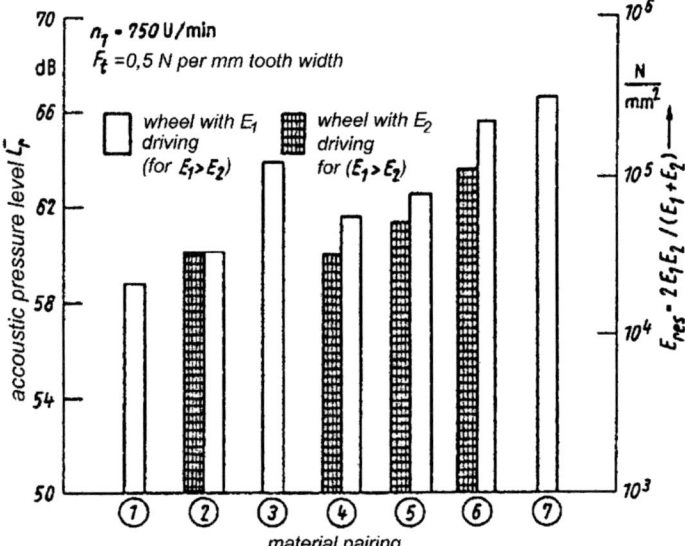

Fig. 8.16. Influence of material pairing on noise behaviour [8.1, 8.10] (testing wheel: $z_1 = 43$; $z_2 = 61$; $m = 0.5$ mm; profile with $h_a = 1.0$ m). 1 PA/PA, 2 PA/Fbl–Fbl/PA, 3 Fbl–Fbl, 4 St/PA–PA/St, 5 St/Fbl–Fbl/St, 6 St/Br–Br/St, 7 St/St, (PA – polyamide, Fbl – fabric base laminate, St – steel, Br – brass), $\bar{L}_r$ median acoustic pressure, measured in area of predominantly direct noise on cladding hemisphere with radius r, $n_1 = 750$ rpm

You have to consider furthermore the dependency of subjective noise impression on frequency and duration of noise. Pulsed or fluctuating noise is sensed louder than steady sounds [8.8].
- Transmission accuracy of rotation angle and behavior of power loss [8.10].

8.1.2.4 Crossed Helical Gears

Tooth traces of paired wheels in crossed helical gears have the same direction of pitch (both with left or both with right pitch) unlike cylindrical gearings. Centre-to-centre distance arises from wheels' helix angles β_1 and β_2:

$$a = 0,5 m_n \left(\frac{z_1}{\cos \beta_1} + \frac{z_2}{\cos \beta_2} \right) . \tag{8.18}$$

Transmission is

$$i = \frac{n_1}{n_2} = \frac{z_2}{z_1} = \left(\frac{d_2}{d_1} \right) \left(\frac{\cos \beta_2}{\cos \beta_1} \right) . \tag{8.19}$$

This means that transmission $i \neq 1$, e.g. $i = 2$ is possible for crossed helical gears of similar diameter in contrast to cylindrical gears.

The ratio of contact of two engaged flanks are considerably more adverse for crossed helical gears than for cylindrical gears. Besides roll sliding, screw sliding appears, and point contact emerges alongside the pitch. This implicates higher surface pressure and hence more abrasion. Crossed helical gears are only suitable for low power in comparison with worm wheels, but their insensitivity to small deviations of axle crossing angle and small extensions of centre-to-centre distance is advantageous (Fig. 8.17).

- The efficiency is highly dependent on the helix angle of wheels. It may represent 50–95%. The most convenient ratios result from $\beta_1 = \beta_2 = \Sigma/2 = 45°$. Crossed helical gears may be completed self-locking.

Material Selection, Calculation of Bearing Strength

Sliding abrasion dominates in crossed helical gears with axle angle $\Sigma > 25°$ due to point contact and heavy sliding. This requires the use of resistant materials and adequate lubrication. Plain flank surface of the harder wheel is fundamental, too. Especially resistant are boron-treated steel wheels.

At least one-time lubrication before starting but better splash lubrication in grease is recommended for rolling speeds v_g up to 0.5 m/s or circumferential velocity v up to 1 m/s. For higher circumferential speeds oil splash lubrication is recommended. Galling may be avoided by oils with slight extreme pressure additives. But thereby rolling abrasion may increase [8.1].

The amount of transmissible power is limited. Hence in general only approximate calculation of bearing capacity as defined in Sect. 8.1.2.3 is taken, which is considered to be sufficient for gearings in small drives.

For toothing tolerances for single crossed helix gears see Sect. 8.1.2.3. For gearing fits presently no standardised specifications are available.

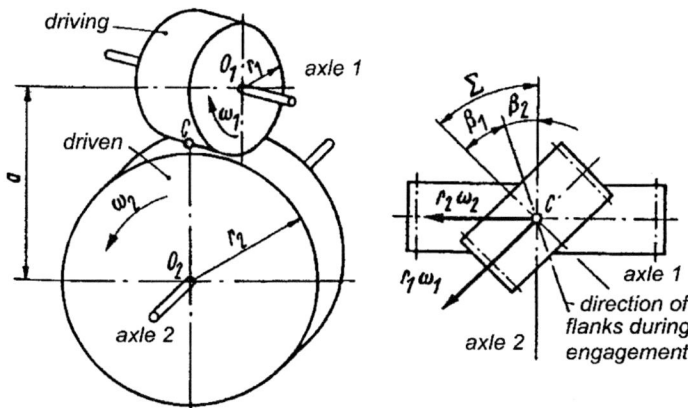

Fig. 8.17. Crossed helical gears [8.1, 8.2]

8.1.2.5 Worm Gears

As already mentioned in Sect. 8.1.2.1, wheel bodies may be cylinders and globoids. The following four combinations are possible: cylindrical worm–globoid wheel (Fig. 8.18), cylindrical worm–cylindrical wheel (Fig. 8.19), globoid worm–cylindrical wheel and globoid worm–globoid wheel [8.11].

The small wheel is mostly constructed as a cylindrical worm. It has a trapezoid thread with number of starts g (generally $g = 1\ldots 5$) according to the number of teeth z_1. For the large wheel (worm wheel) usually a globoid wheel is used. In small drives a simple crossed helix gear is adequate, primarily if it involves low burdened gearings.

Furthermore, a big variety of flank forms and worm wheel shapes come into operation, especially at minor demands on load capacity and accuracy as well as for transmissions into high speeds (Fig. 8.20). The already mentioned combination with a simple crossed helix gear is shown in Fig. 8.20a. Embodiments, with trapezoid or triangular form of worm tooth in axle section are shown in Fig. 8.20b, c. The wheel has a sawtooth profile inclined into direction of motion and with rounded tips in front section. The worm is mostly produced by milling with a screw-cutter with trapezoid profile or by turning. The wheel is fabricated by roll milling, cutting tools are used if sheet metal is the source material. Injection molding is used for plastic materials. Figure 8.20d shows a simplified version, in which the worm is composed of a cylindrical body wrapped by a screw-shaped wire. The worm wheel may be an involute geared spur wheel or a wheel of thin stamping.

For combination cylindrical worm and globoid wheel and crossed helical gear:

Centre-to-centre distance

$$a = \frac{(d_1 + d_2)}{2}. \tag{8.20}$$

Transmission

$$i = \frac{n_1}{n_2} = \frac{z_2}{g} = \frac{z_2}{z_1}. \tag{8.21}$$

Reference circle diameter

$$d_{m1}(= d_1) = \frac{z_1 m_n}{\sin \gamma_m}; \quad d_2 = m_t z_2. \tag{8.22a}$$

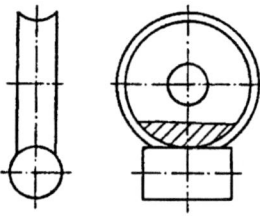

Fig. 8.18. Combination cylindrical worm–globoid wheel

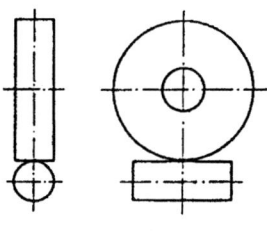

Fig. 8.19. Combination cylindrical worm–cylindrical wheel

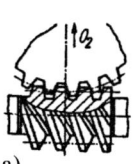
a)
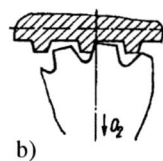
b)

c)
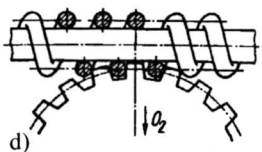
d)

Fig. 8.20. Varieties of cylindrical worm gears. a Pairing cylindrical worm – crossed helical gear. b Worm tooth with trapezoid section. c With triangular section. d Cylindrical worm body, wrapped by a screw-shaped wire – round wire profile

Central inclination angle

$$\tan \gamma_m = \frac{d_2}{(ud_{m1})}; \quad u = \frac{z_2}{z_1}. \tag{8.22b}$$

Self-locking appears, if $\gamma_m \leq \rho$ (ρ = angle of friction). Modulo in normal section and modulo in front section linked by $m_n = m_t \cos \gamma_m$, in which m_n have to comply with the data of Table 8.3.

Worm gears are preferably used for transmissions into slow going (i_{max} = 100 for each level). They may transmit higher power than, e. g. crossed helical gearings (Sect. 8.1.2.4) and move at low noise, but are sensitive to changes of centre-to-centre distance.

- Efficiency is relatively low. It lies in the range of 18–90% maximum.

Material Selection, Calculation of Bearing Strength

Because of high sliding speed, low adhesion factor, high surface quality and minor galling tendency have to be examined for material selection. For general aspects see Sect. 8.1.2.3.

Worms. They are generally produced machinably from construction steels (among others S235JR and S275JR), light metal or copper alloy or molded from plastic. Heat treated steels without surface curing have to be preferred at impulsive impacts (among others C45, 34CrMo4, 42CrMo4), for power gearings cementation steels (among others 16MnCr5 or C15, case-hardened) as well as the aforesaid heat treated steels, but reverbatory or induction cured, are used [8.1, 8.11]; see also advice on p. 503.

Worm wheels. For simple crossed helical gears, which are preferably used in small drives and devices, selection of material, lubrication and design occur as described in Sect. 8.1.2.3. It is the same for machinable produced globoid wheels. However, you have to consider that wheels from light metal alloy or zinc alloy have to be designed wider than wheels from cast iron.

As for crossed helical gearings, it is not necessary to calculate bending stress of worm and wheel teeth, because abrasion limit defines the bearing strength. For small gearings with $v_g \leq 8\,\mathrm{m/s}$ only approximate calculation of bearings strength according to Sect. 8.1.2.3 is required.

In DIN there are so far no norms for toothing tolerances and gearing fits. It is suggested to use the tolerances for toothing in cylindrical gearings after DIN 3961 as well as DIN 58405 (Sect. 8.1.2.3)

8.1.2.6 Bevel Gears and Crown Gears (Face Gears)

Axles of bevel wheels in bevel gearings intersect under an angle Σ in point O (Fig. 8.21a), the pitch cones of jointly acting wheels contact in the common surface line $\overline{OC}$. They roll up on each other without sliding.

From the relations of Fig. 8.21:

Angle of axles
$$\Sigma = \delta_1 + \delta_2 . \tag{8.23}$$

Transmission
$$i = \frac{n_1}{n_2} = \frac{z_2}{z_1} = \frac{r_2}{r_1} = \frac{\sin \delta_2}{\sin \delta_1} . \tag{8.24}$$

The reference cone angles δ_1 and δ_2 arise from

$$\cot \delta_1 = \frac{\left(\dfrac{z_2}{z_1} + \cos \Sigma\right)}{\sin \Sigma} \quad \text{and} \tag{8.25a}$$

$$\cot \delta_2 = \frac{\left(\dfrac{z_1}{z_2} + \cos \Sigma\right)}{\sin \Sigma} . \tag{8.25b}$$

These equations band together in the common case of $\Sigma = 90°$ as

$$\tan \delta_1 = \frac{z_1}{z_2} \quad \text{and} \quad \tan \delta_2 = \frac{z_2}{z_1} . \tag{8.25c}$$

Because of teeth alignment on conic wheel bodies, pitch, tooth gauge, gap width, tooth height, etc., are not constant but vary across tooth width b.

- Efficiency is about 96%.

Kinematically clean running is only achieved if both cone tips are located in the axles' point of intersection. Bevel gearings have to be mounted exactly in an axial direction. Bevel gearings may be applied for transmissions up to $i = 6$. For higher transmissions combinations with cylindrical gearings are used (Fig. 8.21b) up to $i = 40$ for two-level and up to $i = 250$ for three-level constructions. For lower requirements concerning bearing strength and

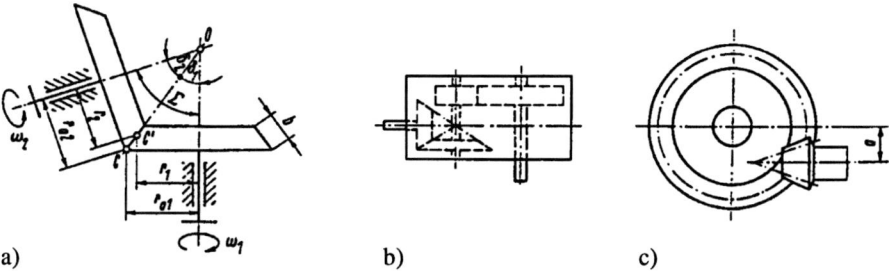

Fig. 8.21. Bevel gearing. **a** Denotations. **b** Cylindrical bevel gearing. **c** Screw bevel gearing (**a** axle offset) [8.11]

running smoothness, rectilinear toothed bevel gears are used (cp. Fig. 8.6b). For high demands hardened spiral bevel gears are applied, and also in screw bevel gearings for interbreeding axles (Fig. 8.21c). Running smoothness and two sided bearing of bevel is advantageous. Because of complicated production and assembly their application should be avoided in small drives as well as the application of cross geared bevel wheels.

The border case of externally toothed bevel wheels, whose half taper angle comes to 90°, is the planar bevel wheel (Fig. 8.22). Its reference face is a plane perpendicular to the wheel axle and it is planar toothed, located on the front surface of the wheel. The pairing of a planar bevel wheel and a bevel wheel is called planar bevel gearing, whose axle angle is generally $\Sigma = 90° + \delta$. If a cylinder wheel (generally the pinion realized as crossed or rectilinear spur gear) is paired with a planar gear, whose tooth system matches the pinion tooth system, a planar spur gearing arises. In precision mechanics the planar wheel is called crown wheel and the respective gearing is called crown gearing (Fig. 8.23), which may be produced profitably.

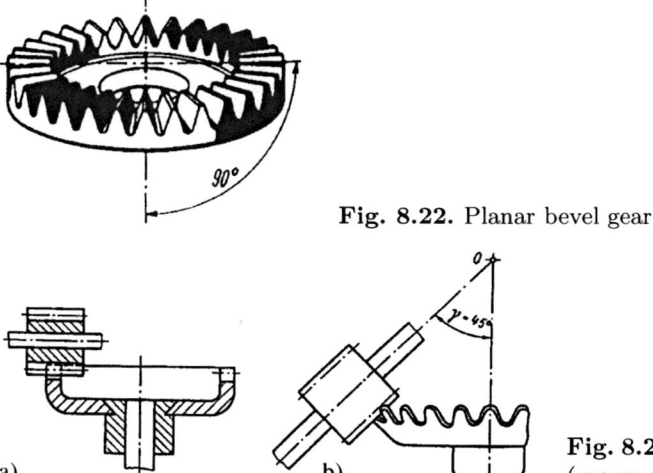

Fig. 8.22. Planar bevel gear

Fig. 8.23. Planar spur gearing (crown gearing) [8.1]

Material Selection, Calculation of Bearing Strength

Selection of material and lubrication arises with respect to the same factors as for spur gears (Sect. 8.1.2.3). Gears must be processed with short driving collars, to attain lower distances from bearing (at one-sided bearing). At very small diameters of shafts, pinion shafts have to be preferred (pinion and shaft from one part). For thermoplastic bevel wheels their low resistance has to be considered [8.1, 8.3].

For calculation of bearing strength, equivalent spur gears with virtual number of teeth z_v are used [8.1] and the existing relations of spur wheels are applied (Sect. 8.1.2.3).

DIN 3965 contains predefinition of toothing tolerances and gearing fits for modulo $m \geq 1$ mm. For modulo below 1 mm no DIN-norms are available.

8.1.3 Flexible Drives

Flexible drives are used if larger distances between input and output have to be bridged or if spatial circumstances do not allow one to use different gearings. They are characterized by simple assembly compared with gear trains and require no or only low maintenance efforts. You differentiate between force-fit (string and band gearings, flat belt and v-belt drives) and form-fit belt drives (toothed belt and chain gearing). Force-fit gearings, whose belts are structureless, work vibration- and impact-reducing and operate on low noise level, while form-fit gearings have structured belts, and thereby may transmit more power and so do not need high pre-stressing, so that shafts

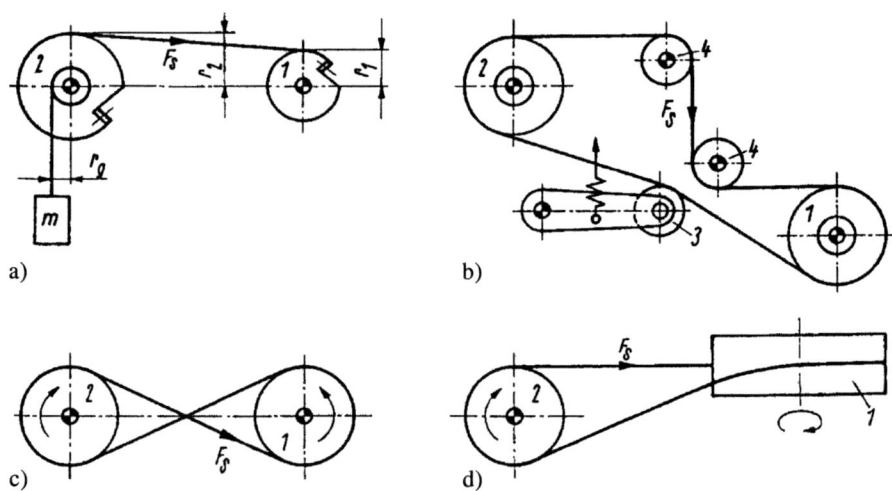

Fig. 8.24. Possible arrangements of belt drives [8.1, 8.2]. **a** Two-dimensional arrangement. **b** With deflexion and tensioning roller. **c** Crossed. **d** Semicrossed – **a** open belt, **b–d** closed belt

and bearings are less stressed. As a result of the polygon effect ([8.1, 8.2])
a more or less uneven rotation comes out. Figure 8.24 shows schematically
different arrangements of belt drives, the belts being designed open (a) or
closed (b–d) and gear elements arranged two- (a, b, c) or three dimensionally
(d).

For power gears predominantly flat belts, v-belts, toothed belts and chains
are used. For low-power gears and guide gearings simple cords, cables, bands
or wires are applied (Table 8.5). For isogonal transmission of motion, belt
drives with force closure are not applicable because of their slip.

If open belts are used as in registers, positioning devices (linear axles
after [8.4]) and adjusting devices (scale drive), special attention must be paid
to the fixing of belt ends at driving and driven wheel, because bearing point
must not disturb transmission of motion (Fig. 8.25).

Torques, which have to be transferred by gearings with closed belts are
depending on angle of enlacement.

Calculation

For geometric dimensions of gearings as in Fig. 8.26 the following factors
apply:

Transmission
$$i = \frac{n_1}{n_2} \approx \frac{d_2}{d_1} \quad \text{(slip occurs)}. \tag{8.26}$$

Angle of inclination
$$\sin \alpha = \frac{(d_2 - d_1)}{2e}, \tag{8.27a}$$

Angle of enlacement
$$\beta_1 = 180° - 2\alpha; \quad \beta_2 = 360° - \beta_1. \tag{8.27b}$$

Belt length
$$L = 2e \cos \alpha + 0.5\pi (d_2 + d_1) + (d_2 - d_1)\frac{\pi\alpha}{180°}. \tag{8.28}$$

Centre-to-centre distance e may not be calculated from the equation for L as
α is unknown. For approximation of e see [8.1].

Generally the correct clamping of the belt has to be observed during construction of gearings. For closed belts this can be achieved by variation of belt length or centre-to-centre distance as well as by application of clamping devices (Fig. 8.24b). Only for highly elastic materials (rubber strings or wire spirals) as well as for new types of elastic flat belts are no special measures necessary.

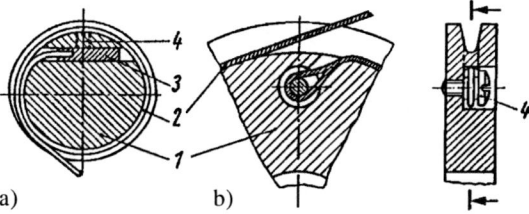

Fig. 8.25. Attachment of belts on wheel. a Steel band. b Cord, 1 wheel, 2 belt, 3 interface, 4 set screw or shoulder screw

Table 8.5. Flexible drives – classification and application

- **Belt drives with force closure**

Cord and band gearings are used in small drives if only low power has to be transferred as a continuous rotation to one or many shafts with a larger distance to each other and if pivoting with angles of rotation $>360°$ has to be implemented. Cords are used for small and medium tensile forces. They are made from hemp $(1\ldots3\,\text{mm})$, catgut $(0.7\ldots2\,\text{mm})$ and from silk and plastic (potentially braided). For higher forces leather (diameter $4\ldots8\,\text{mm}$) or wire spirals and wire ropes or aramid fibers are selected. Bands are produced from woven cotton or silk, from rubber, leather or, at marginally permissible elongation, from steel and at risk of corrosion also from, e.g. phosphor bronze.

Flat belt gearings are implemented at high circumferential speeds and transmission of high power. They are characterized by convenient elastic performance (impact and vibration decreasing).

- Efficiency is about 96–98%.

Transmission is practicable up to $i = 8$ and, if special belts are used, up to $i = 15$. Leather, plastics and textiles serve as belt material; Figure **a** and **b**.

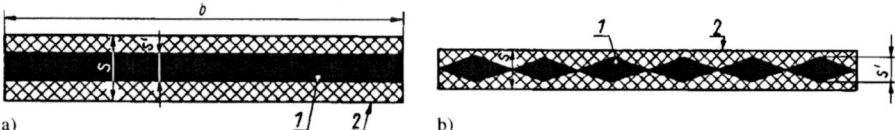

a Leather-polyamide belt. **b** Textile belt, 1 pulling layer, 2 tread

V-belt and o-belt gearings are characterized by pulling of the belt into the trapezoid channel of the belt pulley under load contrary to flat-belt gearings (Figure a). As a result of the wedge effect, high friction forces arise at flanks at relatively low prestressing. Advantages are low stress in bearings and high transmissions at short centre-to-centre distances. Maximum transmission is $i = 15$. Higher bending losses, milling and heating affect adversely compared to flat belts. Allowed belt speeds for v-belt gearings are lower than for flat belt gearings.

V-belts (Figure b–d) consist of fiber strings (artificial silk, polyester fiber and the like), pulling elements (package cord, cable cord) and rubber pads, which encase the fiber strings and profile the v-belt. A medial position between v-belts and flat belts is taken by finned belts (poly-v-belt). They are characterized by flat design and small bending radius and are applicable for small wheel diameters and high transmission ratios, e.g. for washing machine drives [8.11]. O-belts are, like bands, made of rubber, leather or plastic predominantly without pulling strings, or special highly elastic mixtures, e.g. polychloroprene rubber, which is highly non-ageing and resistant to abrasion. At higher load and spatial redirection, belts with pulling strings (f) are also used.

Table 8.5. (continued)

- Efficiency is 92–96%.

Fields of application for v-belts are practically unlimited. They are used for small drives in precision engineering and domestic appliances, e.g. in rotary pumps and ventilators, up to heavy duty drives.

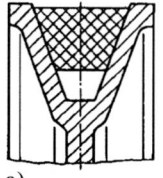
a)
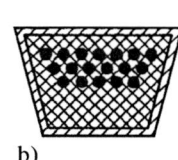
b)
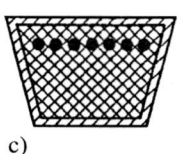
c)
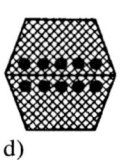
d)
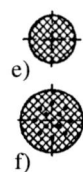
e)
f)

a V-belt slice. **b** Normal v-belt, package cord design. **c** Normal v-belt, cable cord design. **d** Twin-v-belt. **e,f** O-belt without and with pulling strings

- **Flexible drives with form closure**

Synchronous belt gearings [8.1, 8.4] combine the advantages of belts (low mass, high circumferential speed, low noise, maintenance-free) and of chains (non-slip, low prestress). Tangential force is transferred by form closure, while belt teeth engage the disc's gaps. Regarding design and technology, two types of toothed belts are differentiated:
Tooth belts made from rubber (polychloroprene, NBR, HNBR and the like) mostly feature pulling strings from glass fiber or aramide. Toothing is protected against tearing by special tissue layers. These belts, produced by vulcanization procedures, are standardized in ISO 5296 and ISO 13050 and in special standards of automotive engineering. Belts after ISO 5296 are available in six divisions (inch system, **1.**); more divisions are usual in trade.
Tooth belts made from polyurethane are produced with pulling belts from flexible steel wires or aramid by means of cast or extrusion processing. These belts are defined in DIN 7721 and provided in more divisions besides those, shown in (**2.**, metric system).
Besides those standard tooth belts with trapezoid profile, there are belts with high performance profiles **3**. They are characterized by enlarged tooth volume and strengthened, bending compliant pulling strings, that cause a noticeable progression in performance (see also figure down right).
Tooth belts are preferably applied, if non-slip run is required, e.g. in drives of office machines, in robotics as well as in control drives with transmissions up to $i = 10$.

- Efficiency is about 98%.

Table 8.5. (continued)

1. Tooth belts (after ISO 5296)		2. Tooth belts (after DIN 7721)		3. High performance profiles AT trapezoid-, HTD circle-, S parabola form			
Division p_b code	in mm	Division p_b code	in mm	Division p_b code	in mm		
MXL	2.032	T 2.5	2.5	AT 3	3.0	HDT 3M	3.0
XL	5.080	T 5	5	AT 5	5.0	HDT 5M	5.0
L	9.525	T 10	10	AT 10	10.0	HDT 8M	8.0
H	12.700	T 20	20	AT 20	20.0	HDT 14M	14.0
XH	22.225					HDT 20M	20.0
XXH	31.750					S 2M	2.0

p_b – tooth belt pitch

Subsequent examples for arrangements of tooth belt gearings are shown, which demonstrate miscellaneous applications:

S 3M	3.0
S 4.5M	4.5
S 5M	5.0
S 8M	8.0
S 14M	14.0

toothed belt profiles:

1 standard profile
2 trapezoid profile
3 HTD-profile

Chain gears [8.1, 8.11] share the advantage of non-slip work, no pre-stressing is required so that only low axle and bearing loads and more power may be transferred. Uneven rotation as a result of polygon effect is disadvantageous and causes accelerating forces and vibrations. Furthermore they may be used generally only at low circumferential speeds.

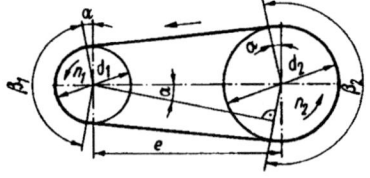

Fig. 8.26. Gears with closed belt: e centre-to-centre distance; d_1, d_2 diameter of disc; α angle of inclination; β angle of enlacement; 1, 2 driving, driven disc

8.1.4 Screw Mechanisms

Screw mechanisms are, depending on the number of screw pairs, different in single screw mechanisms with screw, revolute or prismatic pair, and dual screw mechanisms with two screw pairs and one revolute or prismatic pair, whose transfer function is defined by the lead difference of both screw mechanisms. The derived term of "differential screw mechanism" is not similar to the correct definition of differential gear and therefore has to be avoided. Depending on the type of friction in screw mechanism further classification in slide- and roller-screw drives is possible (Table 8.6).

Calculation

Calculation principle for screw mechanisms may be deduced from basic geometric relations and balance of power in screw threads [8.1].

8.1.5 Coupling Drives

In coupling drives (linkages) fixed coupling elements are connected to either frame mounted or generally planar and spatial movably guided elements; they are thereby connected in inevitable motional and force coupling.

Coupling Drive with Four Elements

Gearings of rotational trees (four elements, four revolute pairs, Table 8.7), slider-crank chains (with single prismatic pair, Table 8.8), scotch-yoke chain (two neighboring joints are prismatic pairs) and slider-loop chain (two opposed prismatic pairs [8.1, 8.7]) are of basic relevance. Depending on what kind of ratio of element length l accordant to the law of *Grashof*

$$l_{\min} + l_{\max} \gtreqless l' + l'' \qquad (8.29)$$

is realized and which elements are used as frame elements, different types of gears are generated.

The main area of application of four-element coupling drives is the realization of transfer functions between two elements which are fixed to the frame in mechanism systems. According to types of gearing in Tables 8.7 and 8.8 two characteristic transfer functions are assigned to every type, which may be influenced by variation of element lengths.

Another area of application opens up by technical use of crosspoint curves, which are described by certain points of the couple like, e. g. the film pull-down mechanisms (Fig. 8.27). According to requirements of point location in the coupling plane, different curves come up, which may be analyzed and drawn by means of computer engineering [8.7].

Table 8.6. Screw mechanisms – classification and application

1. Slide screw drive	Required for high positioning accuracy
1.1 Single screw drives	$i = \varphi_1/s_2 = 360°/P$ with $P = 2\pi r \tan\psi$, P, r – lead, radius of screw thread
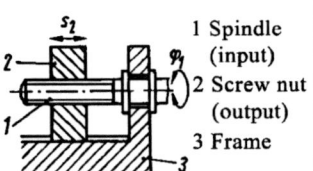 1 Spindle (input) 2 Screw nut (output) 3 Frame	Standardized metric or trapezoid screw thread Efficiency $\eta = 10\ldots30\%$ (improvement by threads with higher angle of inclination ψ and lower profile angle α).
1.2 Dual screw drive	Design engineering: • Free from backlash • Free of constraint coupling a) Bracing by force closed docking
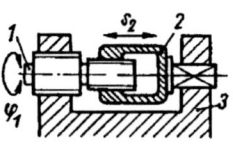 1 Spindle (input) 2 Screw nut (output) 3 Frame	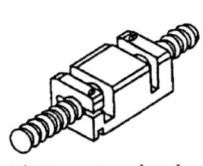 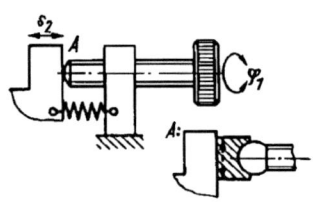 b) Lapping [8.5] • Materials: Spindle (construction and heat-treated steel) Nut (copper alloy, cast iron) Application: construction of devices and apparatus, positioning systems, office machines, measuring devices.
2. Roller screw drive	Pre-finished assemblies Efficiency $\eta = 90\ldots95\%$, Applicable for high precision, free from backlash by the use of double nut
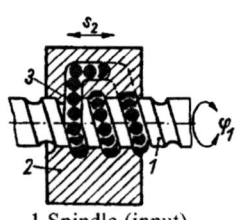 1 Spindle (input) 2 Screw nut (output) 3 Roller body with recirculation tube	Differentiation: • Spherical screw mechanism (for applications in precision engineering, see figure) • Castor screw mechanism (for high load but low rotational speeds) • Planetary castor screw mechanism (up to $n = 5000$ rpm at high load) [8.1, 8.6] Application: positioning systems, measuring devices, scientific toolmaking, machine-tool.

8 Mechanical Transfer Units 479

Table 8.7. Gearings of rotational trees (selection)

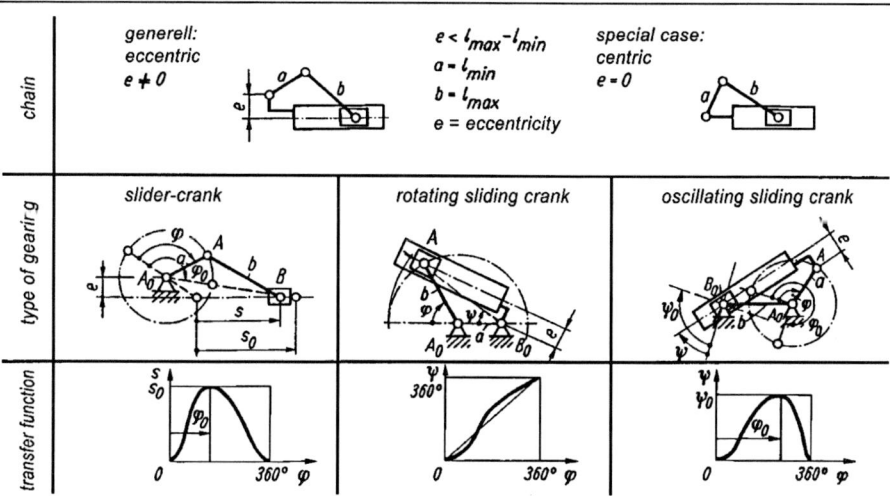

Table 8.8. Gearings of slider-crank chains (selection)

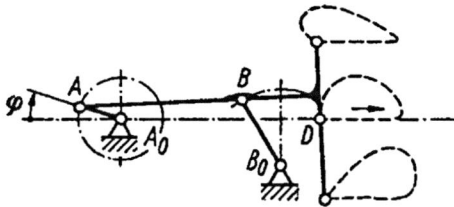

Fig. 8.27. Cross point curve with approximated rectilinear translation for development of pull-down claw [8.1, 8.7]

Multiple Element Coupling Drive

They must be used if four-element coupling drives may not accomplish appointed transfer courses or forms of cross point curves. You have to consider, that increasing of elements always involves higher effort as well as extension of backlash, elasticity in mechanism and maintenance problems (lubrication, abrasion) [8.1, 8.7].

Performance

The characteristic of coupling drives results from the irregular transfer function, that causes accelerated motion of further elements even at constant angular speed of input element. This fact affects burdens of elements and joints, induces vibrations in gearing, passes alternating loads to the frame, generates running noise and influences the even motion of the drive motor *via* input crank. Thus the kinematic transfer function alters in its temporal behavior. Criteria were developed which integrate the variables responsible for running behavior.

Angle of transmission μ may be considered on quasi-static conditions, which, e.g. appears at crank-and-rocker mechanism between coupler and rocker line (Table 8.7) and at minor values ($\mu < 40 \ldots 30°$) indicates an adverse transmission of force and motion. Dynamic running criteria allow better assessment of a mechanism [8.7], which facilitates consideration of speed, acceleration, forces and mass effects besides geometric conditions.

8.1.6 Cam Mechanisms

Cam mechanisms are gearings with at least one cam element, which is connected to the next gear element by a cam joint. The contour of cam element is realized according to the required function of motion.

Aspects of categorization for cam mechanisms are basic geometric shape and form of motion of cam element, form of motion of encroaching element as well as design, behavior of motion and maintenance of contact in cam joint.

The basic form of planar and spatial cam mechanisms consists of ligament, cam element and encroaching element. In Fig. 8.28a ligament 1 is frame, cam element is input element 2 and roll-lever is output element 3, which materializes the function of motion. This form of gearing is mainly used as transfer

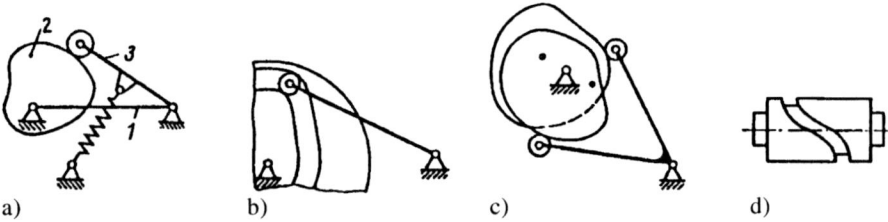

Fig. 8.28. Cam mechanism. **a** With roll-lever. **b** With notch curve. **c** Double cam with double roll-lever. **d** Cylindrical or reel curve

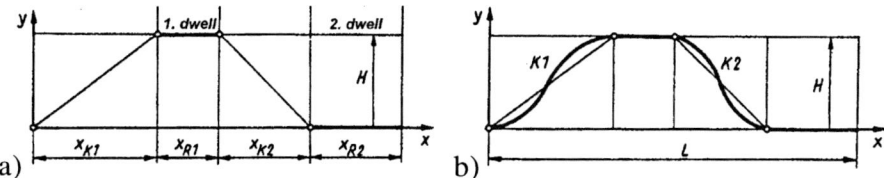

Fig. 8.29. Motion diagrams. a With targeted dwells. b Entire function with course of transition curves K1, K2 ($L = \text{Sum} x$)

gearing; transfer function demands are practically limited by producibility of the cam element and the realization of reliable and wear resistant contact. Cam elements are generally manufactured as flat discs (Fig. 8.28a–c) or as cylinders (Fig. 8.28d); for further variants see [8.1, 8.7].

Pairing of cam and encroaching element must assure compelled motion. Use of springs for force generation in cam joint (Fig. 8.28a) as well as notch and double cams (Fig. 8.28b–d) are appropriate solutions.

Cam mechanisms have to be constructed predominantly as transfer gearings for a given function of motion. This task usually demands defined initial and final positions for a single phase of motion. In Fig. 8.29a a transfer function is given exemplarily, which has the characteristics of a reversed motion with periodically recurring dwell. Progression of motion in motion phase between dwell periods arise from necessary realization of kinematic–dynamic laws if hitchless and jerk-free transmission, small cam stress and reasonable transfer behavior with low inertial forces have to be reached. Transition curves K1 and K2 between dwell have to be defined accordingly in the example (Fig. 8.29b) [8.1, 8.7].

Therefore, a multitude of transfer functions was developed (e.g. power function, sinusoidal function), which fulfil the different requirements. The choice of an appropriate transfer function and its implementation in the form of a cam element is the major phase of cam element construction. Further major dimensions of a cam mechanism are selectable to a certain extent [8.1, 8.7].

Performance

It is mainly influenced by projected cam form, so by shape and sequence of motion phases and their transition, which have to be designed by the right choice of transition curves having regard to transfer task. Slight maxima of acceleration as well as hitchless and jerk-free motion should be targeted. Quality of production, especially attained surface finish, influences the generation of running noise and, because of vibrations as a consequence of deficient surface quality, directly abrasion and persistence.

8.1.7 Stepping Gears

Stepping gears are uneven transforming mechanisms, whose transfer function may be characterized as a unidirectional motion with periodically recurring dwell. Dwell may be momentary or a rest. Geneva mechanism, star wheel gearing and cam step mechanism are important for small drives (Table 8.9).

Geneva Mechanism

It consists, according to Table 8.9, No. 1, of driver 1, to whose axle locking cylinder 2 is attached, and the Geneva wheel 3. This mechanism is an application of sliding crank. Motion of Geneva wheel with radially encroaching driver is hitchless but not jerk-free. Consequential force of inertia must be controlled, especially at high rotational speeds, by accurate fabrication, clean surfaces and adequate lubrication. Jerk may be reduced by appropriate upstream gearings (double crank, gearings with elliptic toothed wheels) [8.1, 8.7].

Geneva mechanism is used in film cameras and projectors, in automats for film base bonding in microelectronics as well as in drives for changer magazines for pens in plotters. In these and other application fields the step-time ratio ν is of interest, which equals the proportion of step-time t_s to total time T (duration of step motion, period). In Geneva mechanisms without upstream gearing therefore ω_{an} = const., step time ratio is also computed from angle, which is passed by driver for rotation of Geneva wheel, and full angle of rotation, that is one complete turn of driver. As the slot number z of Geneva mechanism influences the step angle, ν may be computed for involute toothed Geneva mechanisms from relation $\nu = t_s/T = t_s/(t_R + t_s) = (z-2)/2z$ (t_R dwell time).

Star Wheel Gearing

The variants in Table 8.9, No. 2 consists of driving wheel 1 with pinion rack 2 and catch 3 as well as star wheel 4, whose toothing is interrupted by two latches. The step-time ratio of this arrangement is greater than one (further forms see [8.7]).

Cam Step Mechanism (Table 8.9, No. 3)

Driving element is a cylinder curve 1, whose step section is situated in a small angle of cylinder. Encroachment conditions between wheel 2 and cylinder curve are mainly characterized by sliding friction and adverse contact surface, wherefore abrasion is higher and this gearing is only used for accessory applications.

8.2 Couplings

Couplings connect shafts and transfer rotations and torques. Connection of input axle with output shafts postioned in axle direction may occur permanently or temporarily as well as with constant or variable relative position by form- or force-closed pairing of shaft's end by force transferring adapters.

8 Mechanical Transfer Units   483

Table 8.9. Stepping gears (selection) [8.6]

No.	Description	Schema	Application	Remark
1	Geneva mechanism		Film projector (35-mm and 70-mm-film) bonding automat (film base bonding in microelectronics) changer magazines for pens in plotters	High speed in step middle, driver encroachment not jerk-free; precise production necessary; hardening and beveling of concurrent planes; sufficient lubrication is recommended, step-time ratio dependent on slot number 1 driver; 2 cylinder protection; 3 Geneva wheel
2	Star wheel gearing		Packing-, printing-, coil-winding-machines and everywhere, where periodic dwell and movement with constant speed are required alternating	Toothing may also be designed as involute gearing; step-time ratio are selectable to a certain extend, constant speed in step middle 1 driving disc with 2 rack and pinion gearing 3 cylinder protection 4 star wheel
3	Cam step mechanism		For toys and similar accessory applications	The smaller the step-time ratio, the larger forces, abrasion and clamping danger (self-locking) 1 cylinder curve 2; toothed wheel

Firm couplings are used for simplification and easement of device assembling and require exact alignment as well as constant axial distance of shafts. These preconditions are seldom fulfiled. Therefore, self-aligning couplings are used at shaft displacements (radial, axial, angular or combinations, Fig. 8.30), which often also damp rotary vibrations, torque impacts and noise by the use of elastic elements. Couplings, which are primarily used for this purpose and especially applied in mechanical engineering, are also called rotary-elastic couplings.

If rotation shall be transferred only temporarily, couplings are used which are called switching couplings in case they may be handled from outside. If

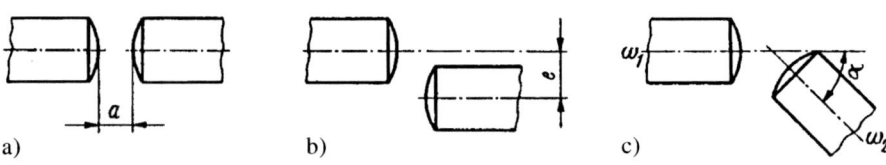

Fig. 8.30. Displacements of axles. **a** Axial displacement. **b** Radial displacement. **c** Angular displacement

temporary coupling and uncoupling occur depending on operating parameters, e.g. rotational speed, torque, direction or angle of rotation, couplings are called clutch or automatic couplings.

8.2.1 Firm Couplings

Firm couplings connect shafts permanently and inflexibly and may not be released during operation. Concerning the structure, sleeve and muff couplings for small torques and flange couplings for transmission of higher torques (Figs. 8.31 and 8.32) are to be distinguished. Dimensioning of casing and shell couplings, which allow small radial size, results from chosen joining elements (screw-, pin- or clamp-connection [8.1, 8.2]).

By flange couplings, which are generally provided with centering (Fig. 8.32), torques are mostly transferred by friction, which is generated by plugging screws.

Firm couplings are seldom used in small drives and devices, because necessary accuracy of position is often not required for functioning and sometimes also not realizable and not intended due to reasons of economic fabrication. They are primarily used in mechanical and electro-mechanical engineering, if bumping and alternating torques should be transferred or if high axial forces and bending momenta appear [8.1, 8.11].

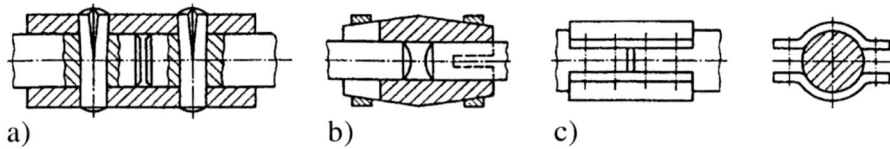

Fig. 8.31. Sleeve and muff couplings. **a** Sleeve coupling with cross pins. **b** With split sleeve. **c** Principle of muff coupling

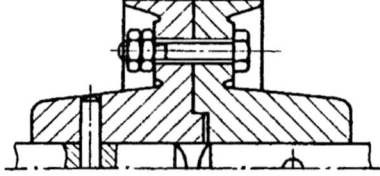

Fig. 8.32. Flange coupling

8.2.2 Self-aligning Couplings

In self-aligning couplings coupling parts are able to compensate displacements caused by production, assembling (Fig. 8.30) or maybe function.

Axial-aligning couplings require alignment of shafts; only axial displacements may be compensated (Fig. 8.30). In Fig. 8.33a, b two versions of sleeve couplings with axial-alignment for small torques are shown, which normally do not require strength calculation or are dimensioned according to joining element if necessary [8.1, 8.2]. Usual designs of flange couplings with axial compensation are shown in Fig. 8.34. Because of production caused deviation in torsion proof bolt gearings with multiple tappets (Fig. 8.34b), transferred force distributes unevenly on tappet bolts and causes high fluctuation of load in shafts and bearings, so that these couplings should only be used in accessory applications.

For the same reason it is assumed in strength calculation concerning surface pressure and bending (shearing strain may be neglected) of bolt, that only 75–80% of the tappets are involved in force transmission.

Bolt coupling in Fig. 8.34c is provided with alternately arranged tappets, which engage with elastic interlayer, whereby shock and vibration absorbance is effected simultaneously.

Couplings with radial alignment (crosswise moveable coupling) are necessary if shafts are displaced against each other for a distance e (Fig. 8.30b) or if danger of displacement appears during operation. For transmission of lowest torques rubber or plastic tubes (Fig. 8.35a) or a helical spring (Fig. 8.35b) are suitable. Tappet couplings in Fig. 8.35c emerge from bolt couplings (Fig. 8.34), if radial sliding of tappet bolt is assured. If backlash-free operation

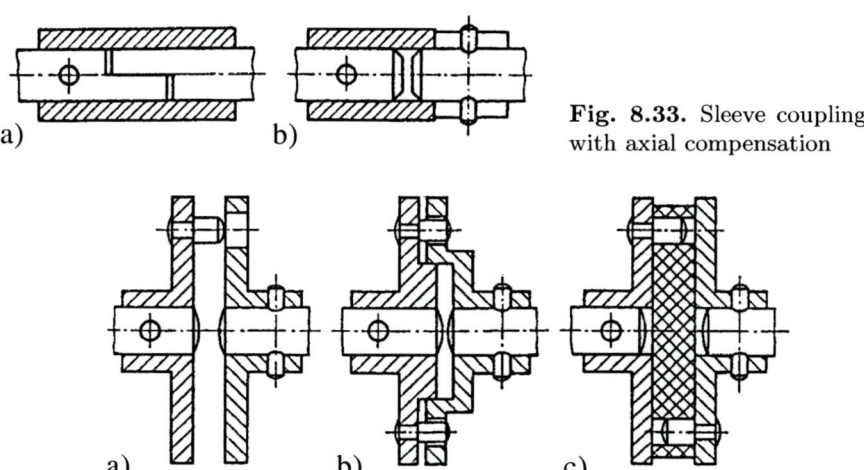

Fig. 8.33. Sleeve coupling with axial compensation

Fig. 8.34. Flange couplings with axial alignment. **a** With single tappet bolt. **b** With multiple tappet bolts and centering boss. **c** With multiple tappet bolts and elastic, stewing interlayer (coupling in **a**) shown disengaged

in both directions of rotation has to be ensured, both tappet surfaces have to be developed flexibly. Disadvantages of these couplings are heavy abrasion by permanent relative motion of coupling parts and variation of output angular speed ω_2.

This variation is avoidable if two tappet couplings are arranged with 90° offset, like cross disk coupling (Fig. 8.35d), which in consequence of its shaping is also applicable for transmission of high torques. From couplings mentioned here with radial alignment, solutions, especially those shown in Fig. 8.35a, b, may be used for compensation of angular displacements and solutions, shown in Fig. 8.35b–d, may be used for compensation of marginal axial displacements.

Couplings with angular alignment are used in combinations of shafts, which are inclined by an angle α (Fig. 8.30c). The case of two interbreeding shafts may be put down to two shafts intersecting a third connecting shaft (Fig. 8.36). Besides already mentioned shafts (Fig. 8.35a, b) for small torques, different flexible couplings are used as couplings with angular alignment. Also for this type of coupling angular speed of drive varies, that may be described by level of inequality U. If shaft 1 (Fig. 8.36a) rotates with constant angular speed $\omega_1 = d\varphi_1/dt$, a changing angular speed $\omega_2 = d\varphi_2/dt$ arises at shaft 2. In Fig. 8.36b ratio of angular speeds for different angles α is shown. By appropriate combination of two flexible couplings (e. g. for parallel input and output shafts, Fig. 8.36c) inequality may be eliminated by symmetric arrangement of joints.

Spherical joint couplings mostly arise from flexible connection of spherical core and cylindrical outside part (Fig. 8.37) by adequate tappets and are appropriate for small torques.

The fold bellow coupling (Fig. 8.38) with soldered on flexible metal tube (torsional stiff) is very popular. A major field of application is connection of rotational measuring systems.

At low angular displacement and very small axial variation the elastic properties of a spring ring may be used (Fig. 8.39).

Torsionally flexible couplings are compliant by built-in elastic elements to torques, so that they may damp impacts and vibrations (cp. Fig. 8.34c).

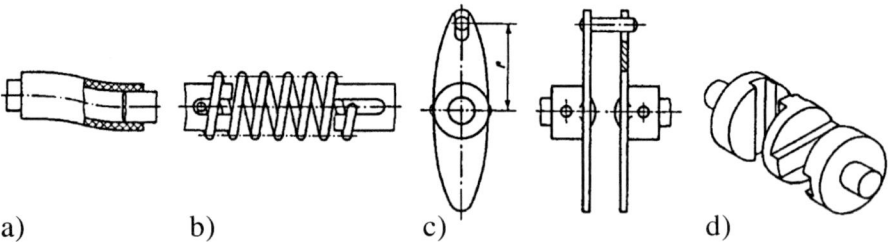

a) b) c) d)

Fig. 8.35. Coupling with radial alignment. **a** Rubber or plastic tube. **b** Helical spring, often only jammed on axle. **c, d** Tappet coupling

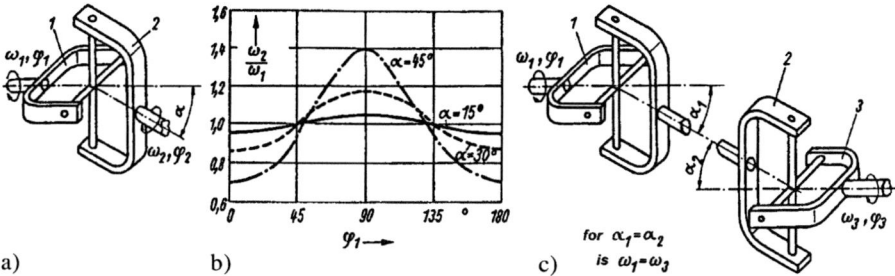

Fig. 8.36. Principle of flexible coupling (cardan joint coupling). **a** Single joint. **b** Characteristics of angular speed ω at single joint. **c** Dual joint

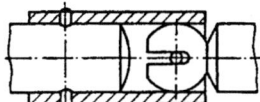

Fig. 8.37. Spherical joint coupling

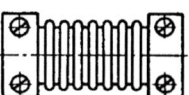

Fig. 8.38. Fold bellow coupling

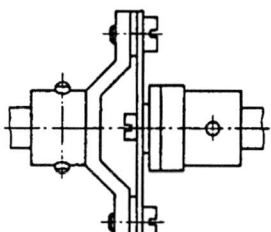

Fig. 8.39. Spring ring coupling

8.2.3 Clutches

Clutches are necessary if for functional reasons the transmission of motion has to be interrupted. Shafts are generally connected by form or force closure, operation of coupling, that is the movement of one part of coupling or generation of pressing force occurs from outside. For small torques mechanical or electromagnetic actuation are often chosen.

- *Clutches with form closure* are shown in Fig. 8.40a with characteristics of torque T_d as well as rotational speed $n_{1,2}$ at input and output during engagement. Clutch should be switched only in stand or constant velocity (synchronous switching), because torque impacts (jerky acceleration) may cause damage of coupling and drive elements (Fig. 8.41).
- *Clutches with force closure* accelerate the downstream element of coupling during switch time $t_s = t_3 - t_1$ to rotational speed n_B. Thereby the jerky encroachment diminishes (Fig. 8.40b, c). Therefore, these couplings may

be switched at different rotational speeds (asynchronous switching). After switching the coupling initially only transmits a slide torque M_R. Due to additional load n_1 decreases; in contrast n_2 increases. After reaching synchronous run at t_2 due to static friction a higher torque M_H (for further acceleration as well as for compensation of load moment) may be transferred.

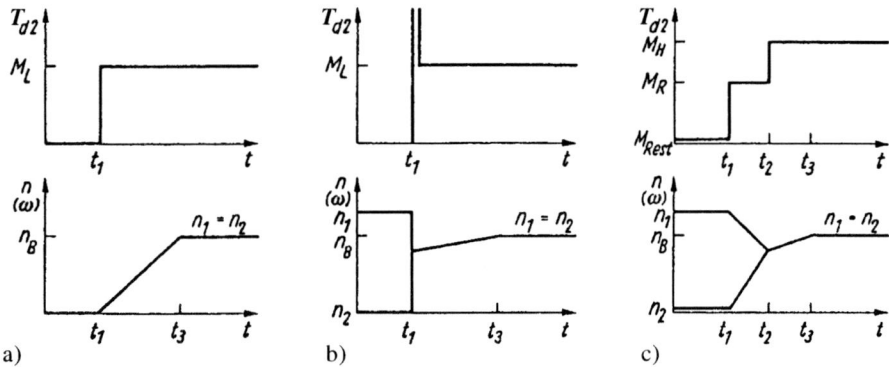

Fig. 8.40. Characteristics of torque and rotational speed for clutches (for non-controlled drives). **a** Form closure and synchronous switching. **b** Form closure and asynchronous switching. **c** Force closure, t_1 time of switching in **a** and **b** or initiation of switching in **c** respectively; t_3 termination of switching; n_1, n_2 rotational speed at input and output; n_B operational rotational speed; T_{d2} output torque; T_H torque of static friction; T_L torque of load; T_R slide momentum; T_{Rest} remaining torque

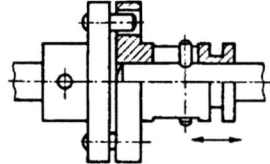

Fig. 8.41. Clutch with form closure

Force closure basically occurs due to solid body friction. Friction surface are annular planes, cone or cylinder shells. Most simple friction clutch is single disk coupling (Fig. 8.42a). It requires only short shifting travel and has very short and exact shift times, especially as dry clutch. Frictional heat is led away well. Disadvantageous in comparison with other couplings are large diameter and higher moment of inertia.

Single disk clutches as mechanically, hydraulically or pneumatically operated couplings are mainly used in machine and vehicle construction. They are manifoldly applied as electromagnetic operated couplings (magnetic clutches, mostly with fixed coil (Fig. 8.42b), but sometimes also with rotating coil) in small drives [8.1, 8.2].

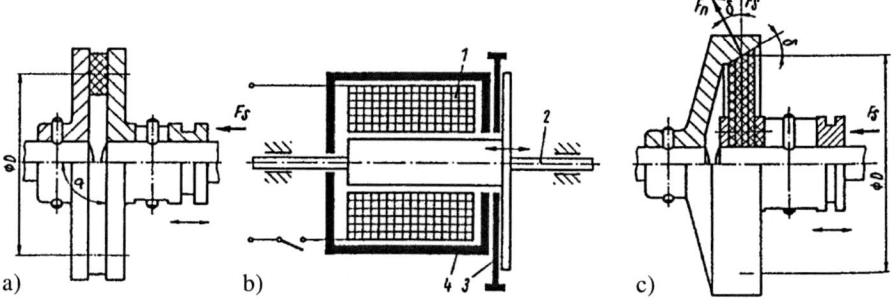

Fig. 8.42. Clutches with force closure. **a** Single disk coupling. **b** Magnetic coupling with fixed coil (principle). **c** Conic coupling – 1 coil; 2 drive; 3 output; 4 coil casing; F_n normal force; F_S switch-, operational force

Couplings with short switching times (<1 ms), like those required in peripheral devices of data processing for very small torques, are represented by electrostatic, rapidly switching couplings and magnetic powder couplings. For transmission of torque they use electrostatic forces or smooth magnetisable powder [8.1]. Also directly integrated or flanged on magnetic clutches in electric motors, so-called armature stop brakes become more important especially for positioning tasks, to ensure exact standstill of drive axle in power-off state [8.1].

With the same pressing force F_S higher torques may be transferred in comparison with disc couplings because of conical designed friction surface of conical clutches (Fig. 8.42c).

Calculation of transmittable torque of clutches results from

$$T_d = \frac{D\mu i F_S}{2 \sin \delta} \qquad (8.30)$$

(D, F_S, δ see Fig. 8.42c; friction coefficient for common pairing of materials see [8.1, 8.2]).

Self-Switching Clutches

The switching process is initiated by alteration of operating conditions at these clutches, namely as a function of torque, rotational speed, direction of rotation or angle of rotation.

- *Torque-dependent clutches* are used, e. g. for load-limiting of input or output of a device or machine. Such a clutch is designed for maximum torque, at whose overstepping connection is disengaged. Clutches are mostly constructed as friction clutches (slip clutches, Fig. 8.43a, b) with given or adjustable torques, whereby the same calculation rules are applied as for switchable clutches.
- *Speed dependent clutches* connect or disconnect shafts at increasing or decreasing rotational speed by using the variation of centrifugal force. They are necessary if an electric motor has a small starting torque and may be

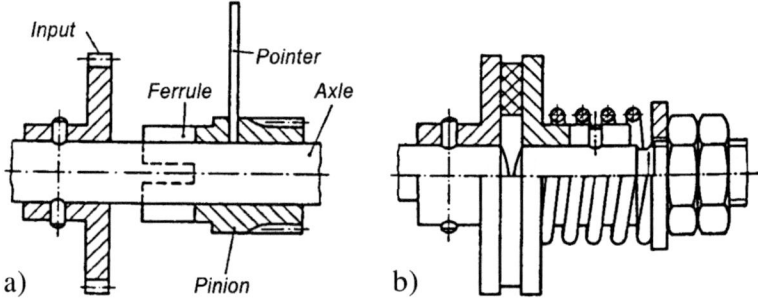

Fig. 8.43. Torque dependent clutch. **a** With fixed rated torque by split sleeve (for adjusting of a pointer). **b** Single disc clutch with adjustable torque by modification of spring prestressing

connected to output only at achieving operational speed. Depending on whether the coupling force should effect above or beneath a certain rotational speed n, engaging clutches with $n_{\text{eff}} > n$ and disengaging clutches with $n_{\text{eff}} < n$ have to be differed. Constructive design mostly results from placing two or several radially movable flyweights on circumference (mass m). Together with restoring springs parts of coupling are connected by friction forces (engaging clutches, Fig. 8.44a) or disconnected (disengaging clutches, Fig. 8.44b) at achieving operational speed. From ratio of forces using the example of engaging clutch (Fig. 8.44c) *friction* torque T_R may be calculated:

$$T_R = \mu F_{\text{eff}} ri = \mu (F_f - F_F) ri \geq T_d \ . \qquad (8.31)$$

By defining constructive conditions (e.g. coupling radii r and r_S, number i of flyweights or coupling jaws, friction coefficient μ, elastic force F_F) dimensioning of clutches is possible considering $F_f - F_F$ with centrifugal force $F_f = m r_S \omega^2$ [8.1, 8.2].

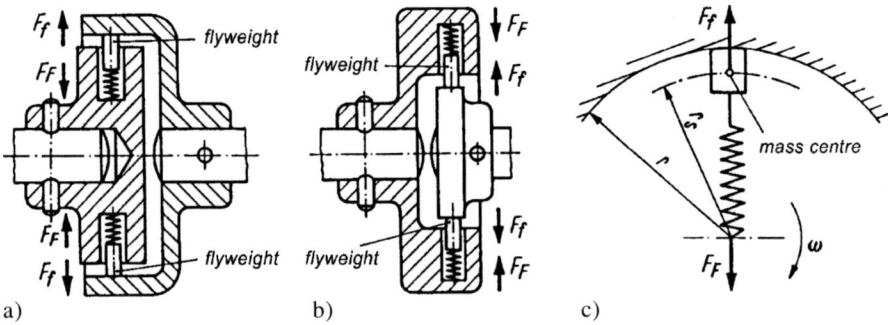

Fig. 8.44. Speed dependent clutch (centrifugal force clutch). **a** Engaging clutch. **b** Disengaging clutch. **c** Ratio of forces – F_f centrifugal force; F_F elastic force

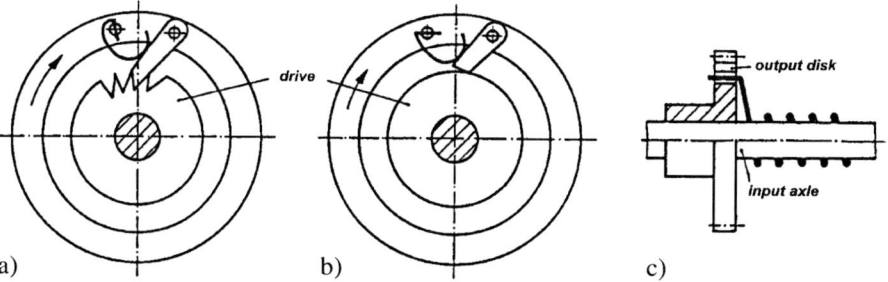

Fig. 8.45. Overrunning clutches. **a** Directed tooth lock. **b** Directed friction lock. **c** Sling spring clutch

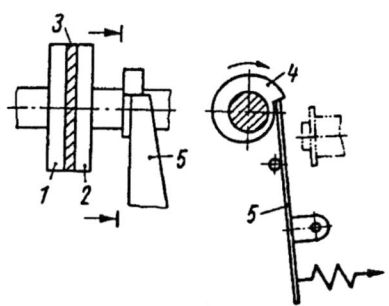

Fig. 8.46. Rotation angle dependent clutch; single revolution clutch with magnetic release as slide clutch with fence, 1 input, 2 output, 3 friction surface, 4 stopper, 5 stopper piece

- *Overrunning clutches* allow transmission of torque in only one direction and permit, e. g. forward motion of driven part in drive direction. Constructive design mostly results from use of directed tooth locks (form closure, Fig. 8.45a) or directed friction locks (force closure, Fig. 8.45b), whereby in the former case a lock sound appears in direction of free run. This sound may be avoided by so-called mute locks with controlled ratchet or at least reduced by special tooth forms. Sling spring clutches (Fig. 8.45c) are constructed simply and therefore used often. Shaft and nave are connected by a nave fixed spring, that is shoved on axle, and has a smaller inner diameter than the axle. In one direction of rotation shaft is taken along, because the spring contracts and seizes on shaft, while it detaches during rotation in reverse direction. Transmittable torque depends on overall dimensioning, but especially on number of spring turns. If free movement in free run is required, difference of inner spring diameter and outer axle diameter is chosen small.
- *Rotation angle dependent clutches* release connection after run through of a specified rotation angle. They are constructed as single revolution clutch for angle of 360° and designed as slide clutches for small torques (Fig. 8.46). Mostly such tasks may be accomplished better by rotatory stepping motors.

8.3 Axles and Shafts

Axles and shafts are used for transmission of weight force of rotating bodies, for absorption of forces, which result from the device function and for support in bearings. Shafts in comparison with axles also transfer rotation and torque. Shafts always revolve, while axles may be realized revolving (axle of vehicle) or stagnant (axle of bicycle). Axles are mostly stressed by bend and as the case may be by tensile or pressing force, shafts in contrast also by torsion. As a result of rotational motion in revolving axles and shafts even at constant shear force, variable bending stress (orbital bend) appears. Torque of shafts is often unsteady, whereas threshold value of stress may be regarded as an extreme condition. Only in few cases alternating torque occurs with shafts' frequency of rotation.

- With respect to function conditioned differences, dimensioning and design of axles and shafts follow the same aspects. Below, for simplicity, only shafts are treated.

8.3.1 Calculation of Conceptual Design

Operative forces have major influence on design of shafts. If all action and reaction forces and resulting maximum forces according to rules of statics are defined, a rough calculation may be executed as initial point (see [8.1, 8.2]).

You calculate on the assumption of bending moment and torque of the most burdened point. The equivalent stress in steel shafts can be calculated according to the hypothesis of energy for change of shape at combined strain as

$$\sigma_v = \sqrt{\sigma_b^2 + 3(\alpha_0 \tau_t)^2} \leq \sigma_{\text{büb}}, \quad (8.32)$$

due to the variable bending stress and supposing a variable torsional strain with strain factor $\alpha_0 = 1$ according to [8.1, 8.2]. With strains $\sigma_b = T_{b\,\text{max}}/W_b$ and $\tau_t = T_d/W_t$ and section modulus of circle cross section towards bend $W_b = (\pi/32)d^3 \approx d^3/10$ and towards torsion $W_t = (\pi/16)d^3 = 2W_b \approx d^3/5$ follows

$$d \approx 2{,}17 \sqrt[3]{(T_{v\,\text{max}}/\sigma_{\text{büb}})}. \quad (8.33)$$

Maximum equivalent strain with respect to Eq. (8.32) and $\alpha_0 = 1$ is

$$T_{v\,\text{max}} = \sqrt{T_{b\,\text{max}}^2 + \frac{3}{4}(\alpha_0 T_d)^2}. \quad (8.34)$$

If distance of bearings and consequently bearing strength and bending moment are not defined at the beginning of the concept, the required diameter of shaft has to be calculated exclusively from torque resulting from power and rotational speed.

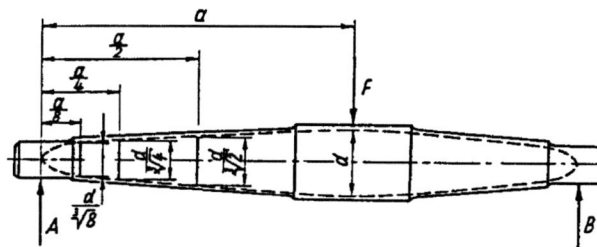

Fig. 8.47. Construction of regraded shaft — cubic parable $(y = kx^3)$

Section modulus toward torsion of circle cross section $W_t \approx d^3/5$ results from $\tau_t = T_d/W_t \leq \tau_{\text{tüb}}$:

$$d \approx \sqrt[3]{\frac{5T_d}{\tau_{\text{tüb}}}}, \qquad (8.35)$$

with $T_d = 9.55 \times 10^6 P/n$ and T_d in Nmm, P in kW, n in rpm.

$\sigma_{\text{büb}} = 30 \ldots 60\,\text{Nmm}^{-2}$ and $\tau_{\text{tüb}} = 12 \ldots 25\,\text{Nmm}^{-2}$ is chosen as approximation for shafts from steel depending on shape (notch effect), material (stability) and distance of bearings (deformation).

With this value d a constructive design is made. The simplest form is a continuous smooth shaft, for which necessary stops for bearings etc. are realized by expanding elements [8.1, 8.2]. With higher powers to transfer, shafts have to be regraded in such a way that shaft encloses a cantilever of similar bending strength (Fig. 8.47).

8.3.2 Recalculation

Shaft may be dimensioned considering the approximately calculated diameter according to requirements of bearing and arrangement of additional input and output elements, whereby conditions of production and assembly suitability has to be respected [8.5]. After that, strain recalculation for all imperiled cross sections and checking of deformation (bending, inclination in bearings) follows. For fast running shafts calculation of vibration is necessary, too.

Recalculation of Existing Strain

Recalculation is necessary at position of maximum bending and equivalent moment and at vibrational stresses as a result of periodically changing forces or circulating bend, and additionally at all notches (gradation of shaft, cross holes, flutes, cut-ins, hub seats, etc.); detailed description [8.1, 8.2].

Recalculation of Deformation

Deformation may derogate function of shaft or attached functional element, e. g. bearing, toothed wheels, rotor of electric motors etc. Tolerable deformation is limited in many cases and may be decisive for dimensioning of shaft. Gauges of deformation are deflection f and inclination β at definite locations

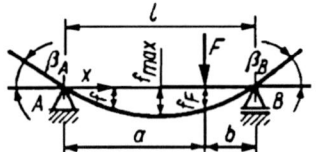

Fig. 8.48. Deformation by shear force F
f bending at any position; f_F bending by force F;
f_{max} maximal bending

of shaft (Fig. 8.48). These gauges may be calculated with differential equation of elastic line, $f'' = -T_b/(EI)$. The solution of this equation for typical cases of bearing and loading is shown in [8.1, 8.2].

Tolerable deformation strongly depends on a particular application, so that generalized values may only be indicated conditionally. Shaft bending in electric motors, for example, must not represent more than 20–30% of the intended air gap between rotor and stator. This again depends on size and type of motor. In small motors with power below 100 W it ranges between 0.05 and 0.03 mm. Exact values may be found in sections about specific types of motors.

Tangential deviation of shaft in bearing may cause overstressing of bearing point (edge loading). Allowed thresholds are conditioned by construction of bearing, but tumbler bearings are best (cp. Sect. 8.4).

Recommended values are:

- Slide bearing with fixed bearing $\beta_{zul} = 3 \times 10^{-4}$,
- Slide bearing with adjustable bearing $\beta_{zul} = 1 \times 10^{-3}$,
- Roller bearing (except tumbler bearing) $\beta_{zul} = 1 \times 10^{-3}$.

Acceptable values for transposition is generally $\varphi_{zul} = 0.25°/m$ for shafts from steel.

Calculation of Vibration

As a result of material elasticity and deadweight of shaft as well as attached masses this assembly is a vibratory mass-spring system. Circulating bending vibrations and additionally torsional vibration may appear. Circulating bending vibrations arise from unbalance of circulating masses. Torsional vibrations evolve from periodically acting torques from input or output; therefore they do not exclusively depend on frequency and amplitude of masses directly attached to shaft but also on up-stream and down-stream mass-spring systems. For evaluation of torsional vibration behavior of a shaft the whole input and output assembly must be considered. This task belongs to machine dynamics [8.12]. Below only circulating bending vibrations are discussed.

If exciter frequency approaches system's resonant frequency, amplitudes attain dangerous levels for operation. Natural frequency ω_0 of a spring-mass system with mass m and spring constant c according to Fig. 8.49 is

$$\omega_0 = \sqrt{c/m}. \qquad (8.36)$$

Spring stiffness of a shaft may be calculated from $c = 48EI/l^3$ (see [8.1, 8.2]).

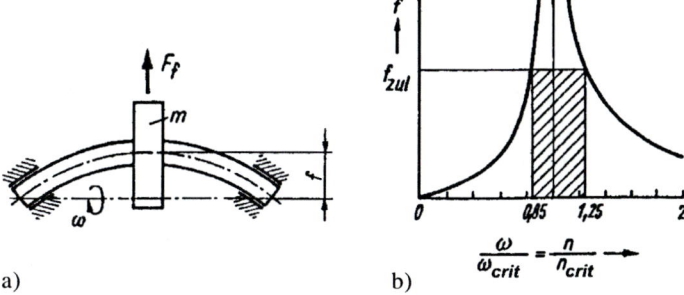

Fig. 8.49. Deflection f of shafts. **a** Displacement of mass centre. **b** Deflection as a function of rotational speed ratio n/n_{krit}, F_f centrifugal force; n_{krit} critical rotational speed

Rotational speed, at which resonance appears, is called critical rotational speed n_{krit}. It may be calculated from ω_0 from Eq. (8.37) or a customized quantity (Eq. 8.38):

$$n_{krit} = \left(\frac{30}{\pi}\right)\omega_0 , \qquad (8.37)$$

$$n_{krit} = 300K\sqrt{\frac{1}{f_G}} , \qquad (8.38)$$

with n_{krit} in rpm; ω_0 in 1/s and f_G in cm.

Thereby f_G is static bending exclusively caused by weight force of the shaft freely bedded on both sides. Bearing factor K considers realization of bearing:

- $K = 1$ for bearings, which may comply inclination of shaft,
- $K = 1,3$ for stiff bearings.

The range of $0.85 < n/n_{krit} < 1.25$ (Fig. 8.49b) must be categorically avoided, because undesirable high amplitudes of vibration arise.

Material Selection

It follows according to required strength, whereby for accessory applications usually S275JR, for normal load E295 to E360 and for higher stress tempered or case-hardened steel are used. For specific operating conditions non-metallic materials are also applied, e. g. ceramics for shafts in small canned motors, which serve for pump drives and which rotor turns in conveyed liquid (e. g. aquarium pumps) – see also advice on p. 503.

8.4 Bearings

The task to support parts moving towards each other with their force effects in drives and devices and to secure their spatial position is taken over by

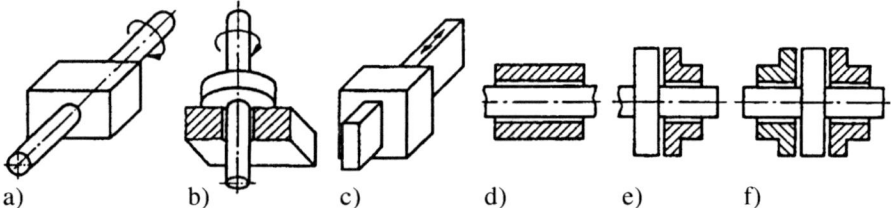

Fig. 8.50. Bearing and guidance. **a** Transverse or radial bearing. **b** Longitudinal or axial bearing. **c** Guidance. **d** Floating bearing. **e** Single-thrust bearing. **f** Fixed bearing

joints. Swivel joints for rotation are called bearings, prismatic pairs for translations are known as guidance (Fig. 8.50). Mostly the motion in guidances is linear; therefore they are called linear guidance (c) (see [8.1, 8.2]). At bearings transverse and longitudinal bearings differ depending on direction of load and support (radial and axial bearing, (a) and (b)). During construction of the bearing it must be considered that only intended motions are enabled and all the rest of them are blocked reliably. In floating bearings (Fig. 8.50d), besides rotation, linear motion may also appear. In single-thrust bearings (e) this displacement is blocked in one direction, in locating bearings (f) in both directions. Guiding and supporting of movable parts should happen without loss. This is antagonized by friction.

Depending on the kind of relative motion, dynamic friction from sliding of two bodies on each other, and rolling friction for a body rolling on another, sliding bearing and rolling bearing have respective merits.

8.4.1 Sliding Bearings

In sliding bearings, friction coefficient μ is mainly influenced by material pairing, surface character, and kind of lubrication as well as sliding speed.

For unlubricated bearings or bearings only lubricated once during assembling, μ may be assumed as constant as a first approximation with a value

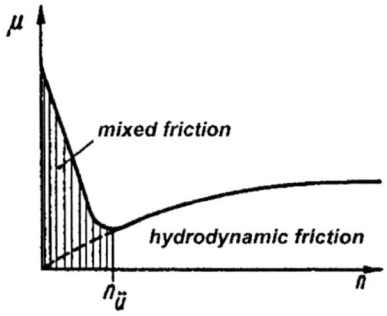

Fig. 8.51. Stribeck diagram

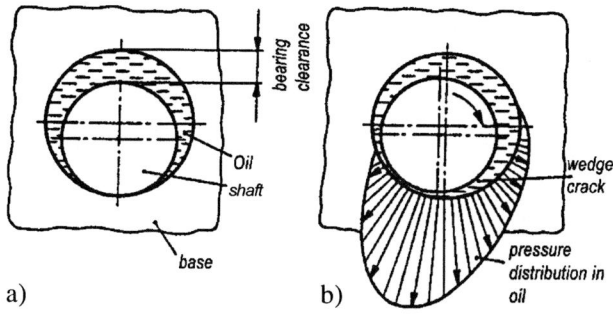

Fig. 8.52. Shaft in bore of bearing. **a** At $n = 0$. **b** At operational speed

of 0.1 up to 0.3 [8.1]. For bearings with permanent oil lubrication, Fig. 8.51 shows friction coefficient as a function of rotational speed n.

At $n = 0$ dry friction appears. With increasing rotational speed the share of fluid friction rises until a close and self-supporting lubrication film establishes at sufficiently high rotational speed. Rotational speed, at which complete fluid friction prevails, is called transient rotational speed $n_{\ddot{u}}$. Shaft then lifts off from bore of the bearing (Fig. 8.52). Abrasion in such bearings is avoided if operational speed is higher than transient speed $n_{\ddot{u}}$. It is then called abrasion-free or hydrodynamic bearing. Transient rotational speed may be calculated after *Vogelpohl* [8.1, 8.2]. Influencing variables are backlash and diameter of bearing as well as viscosity of lubrication oil.

Hydrodynamic lubrication may seldom be obtained and requires continuous operation in small drives and devices because of small bearing dimensions and relatively high backlash. At every start, stop or reversal of rotation zone of mixed friction is passed. Oscillating motion, switching motion, start-stop processes, tardy motions of adjustment etc. contradict conditions of hydrodynamic lubrication.

Thus sliding bearings in small drives are mainly wearing bearings. Their operational properties are basically defined by chosen material pairing and lubrication compatible design. In mechanical and electromechanical engineering hydrodynamic sliding bearings are used because of smooth running and high durability. In the following only wearing bearings are considered, which have to be calculated and designed for a maintenance-free operation.

Construction Forms

Pure axial bearings (Fig. 8.50b) are seldom required. Axial forces most often arise at lateral guidance of shaft and may be neglected compared to radially working bearing forces. Most often radial bearings are used. Bearing may be performed in two places (Fig. 8.53) or at sufficient width of connector in one place (Fig. 8.54). In most cases the shaft rotates and bearing connector stands still. Flat parts, e. g. toothed wheels, disks, sheet plate lever and others, may be supported by a fixed pivot (Fig. 8.55).

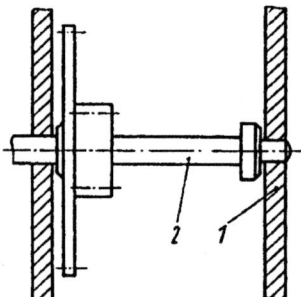

Fig. 8.53. Sliding bearing, two places – 1 blank; 2 axle

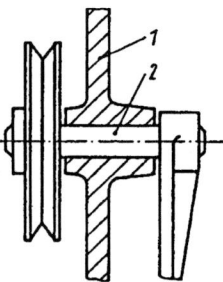

Fig. 8.54. Sliding bearing, single place – 1 blank; 2 axle

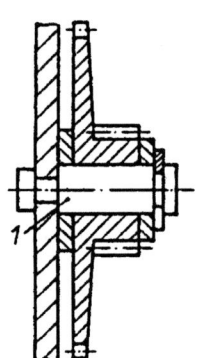

Fig. 8.55. Sliding bearing on riveted axle 1

Calculation

Friction moment is proportional to diameter of pivot. Low friction bearings will always have thin pivots. Danger of milling increases; thus pivot bending must be calculated (Fig. 8.56). From bending moment $T_b = Fa \leq W_b \sigma_{b\ zul}$ and bending section modulus $W_b = \pi d^3/32 \approx d^3/10$ the minimum diameter of pivot is

$$d_{\min} = \sqrt[3]{\frac{32Fa}{\pi \sigma_{b\ zul}}} \approx \sqrt[3]{\frac{10Fa}{\sigma_{b\ zul}}}\ . \tag{8.39}$$

Width must be chosen as $0.3 < b/d < 1.25$ for every bearing according to edge compression and in respect of acceptable surface pressure the appropriate material of bearing must be selected. Medium surface pressure for sliding

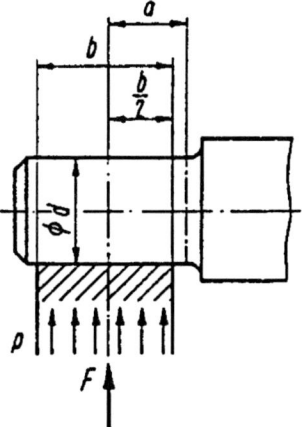

Fig. 8.56. Load of a pivot

bearings is

$$p_m = \frac{F}{bd} \leq p_{zul} \,. \tag{8.40}$$

Allowed values can be found in [8.1, 8.2].

Configuration

Pivots have a contact surface and, because of measures for limitation of axial backlash as well as compensation of axial forces, which have to be taken in radial bearings, a lay-on surface or track surface (Fig. 8.57). If low demands are made, the inclusion of pivot in simple hole bearing is sufficient (Fig. 8.58). For higher loads bushings are used.

While bearing width of hole bearing depends on thickness s of the part in which the bearing is mounted (plate, case, etc.), a higher bearing width b is

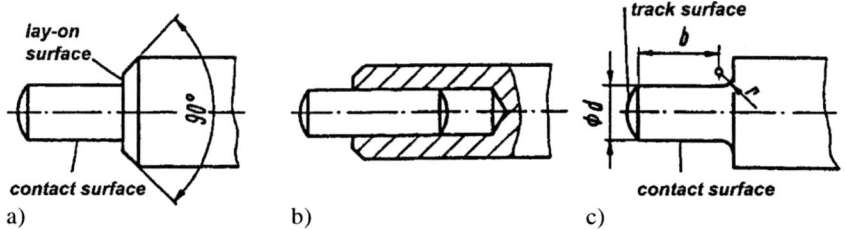

Fig. 8.57. Shapes of pivots. **a** Turned pivot with lay-on and contact surface. **b** Pressed in pivot. **c** Turned pivot with track and contact surface

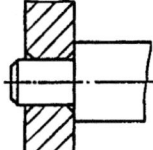

Fig. 8.58. Simple hole bearing

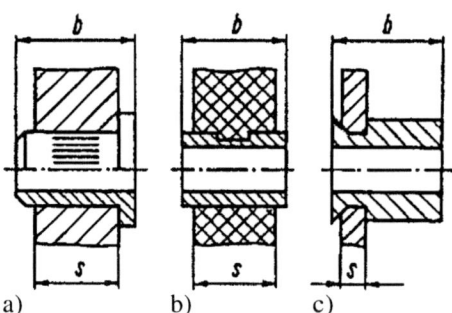

Fig. 8.59. Mounting of bushings

possible if bushings are used. It is mounted, e. g. by force fitting (Fig. 8.59a), surrounding (b) or riveting (c). For press-fitted plastic bushings the risk of slackening by stress relief or other effects of ageing exists. If bushings cannot be fixed additionally by form closure they should be stuck in.

Material Selection

Material of pivot should possess higher hardness and differ from material of bushing to reduce welding affinity. For pivots steel is almost exclusively used with a difference in Brinell hardness of $HB_{\text{pivot}} = (3\ldots 5)HB_{\text{bushing}}$. As bushing materials metals (brass, tin bronze and aluminium bronze, white alloy) and plastic (polyamide PA, polyoxymethylene POM, polytetrafluoroethylene PTFE) are used; examples are shown in Fig. 8.60.

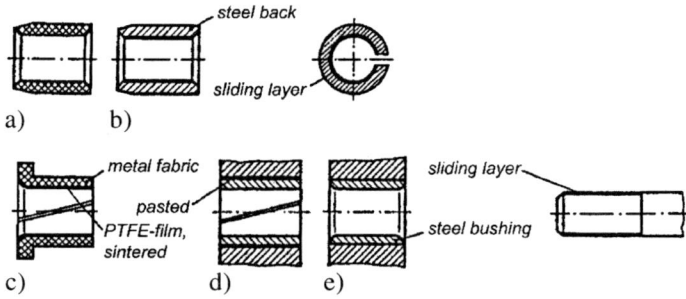

Fig. 8.60. Realizations of plastic sliding bearings. **a** Massive bearing. **b** Compound bearing. **c** Fabric or fiber bearing. **d** Film bearing. **e** Bearing with sliding layer

Lubrication

Function of lubrication process is to apply a sufficient volume of lubricant (oil, grease, sometimes also solid state lubricants like MoS_2). This requires access to the wear point (bore, flute, crack) and a repository for lubricant. Figure 8.61 shows some examples for wearing bearings.

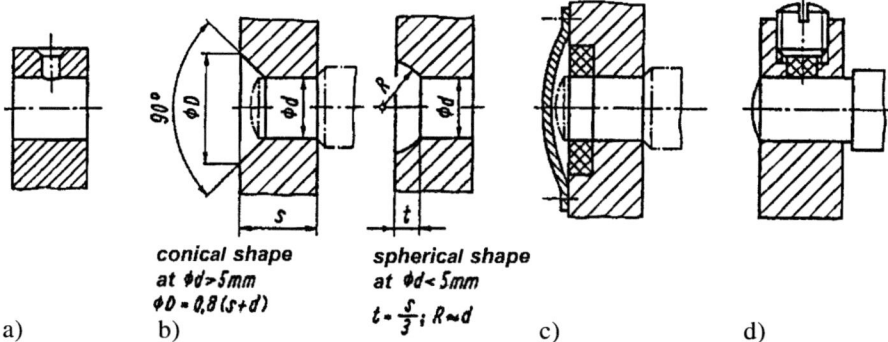

Fig. 8.61. Configuration of lubricating hole in wearing bearings

Already a simple oil hole (a) or oil cut (b) are effective, but oil should not have high leak affinity. For permanent lubrication felt ring lubrication (c) or felt pad lubrication (d) are more convenient.

Sintered Bearings

For maintenance-free operation sintered bearings were developed. Bearing materials are sintered brass and sintered iron with or without additional graphite. Sintered materials have a pore volume of 17–30% that is filled with low viscosity oil. Bushings of these materials are standardized in DIN 1850 and may be purchased pre-finished (Fig. 8.62).

To ensure extremely narrow backlash, as required for low noise bearing, mounting of cup and ball bearings is recommended to avoid alignment difficulties (Fig. 8.62c). Due to their spherical surface they may be brought into line exactly with alignment of shaft. A simple construction is also shown in Fig. 8.62c. One half of spherical calotte 2 is situated inside case 1, which is designed spherically according to calotte shape, while the other half is fixed by a snap ring 3 with spherical face.

If a felt ring is placed in bushing as additional oil repository (Fig. 8.63), certainty of function increases.

Stone bearings are mentioned, for which the material is gemstone (ruby or sapphire) and which are suitable for very small pivot diameters ($d \approx 0.1$–$2.5\,\mathrm{mm}$) as required in micro motors or measuring devices [8.1, 8.2].

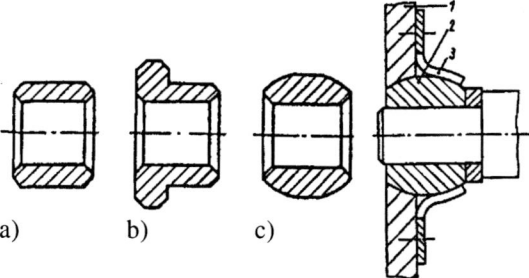

Fig. 8.62. Pre-finished bushings of sintered metal. a,b Bushing without and with collar. c Cup and ball bearing

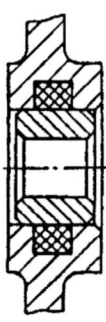

Fig. 8.63. Sintered bearing with additional lubrication

An important point for use of sliding bearings in minor motors may be costs. For comparable size they are only 20–30% in comparison with rolling bearings.

8.4.2 Rolling Bearings

Friction of rolling bearings is considerably lower than that of sliding bearings and practically independent of speed. Running surface and rolling body must be accomplished properly. This task is taken over by the producer of rolling bearings, who provides pre-finished bearings to customer. Challenge of constructing engineer is the choice of appropriate bearing type from offered product line as a result of calculation and to create opportune installation conditions.

Construction Forms

Rolling bearings consist of two rings (inner and outer ring), of which one is connected to rotating shaft and one to stationary case. Between the rings rolling bodies are arranged (spheres, cylinders, cone or barrel rollers or pins). For assured guidance and to avoid mutual contact rolling bodies are often fixed in so-called cages.

Ball and roller bearings differ and dependent on direction of peak load between radial and axial bearings. By far the major percentage of rolling bearings are radial bearings. Therefore the attachment "radial" is only used if special clarity is required. The attachment "axial" for axial bearings is always necessary.

In Table 8.10 rolling bearings are described, which are often used in small drives. Besides there are small rolling bearings with outer diameter $D < 2\,\text{mm}$, which are not characterized. They are produced as deep groove ball bearings with shaft diameters down to 1 mm. To simplify installation conditions, bearings are also produced with flanges (Fig. 8.64a). For smaller diameters (to 0.2 mm) the inner ring may be omitted, spheres run directly on hardened and polished shaft (b). For limitation of axial backlash of shaft, such bearings may be used with groove board (c).

Table 8.10. Configuration of rolling bearings (selection)

Configuration of rolling bearing	Properties and application
(Radial) deep groove ball bearing 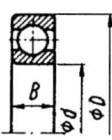	Deep groove ball bearings are the most often used rolling bearings for their universal applicability. Bearings have a high bearing strength in axial direction due to their deep grooves and good oscillation between balls and groove, even at high speed. Dimensions of smallest bearing are $d = 3$ mm, $D = 10$ mm, $B = 4$ mm. Deep groove ball bearings are self-supporting, that is they may not be resolved during mounting and demounting.
(Radial) separable ball bearing	In contrast separable ball bearings have a shoulder only on one side of outer ring. They function only one-sided, are not self-supporting and must be arranged in couples. Because of their easy breakdown, they are easy to mount. Radial bearing strength is lower and axial bearing strength is higher than in deep groove ball bearings. As a consequence of few osculation only slight friction arises.

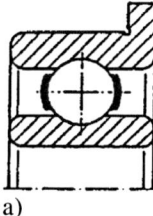
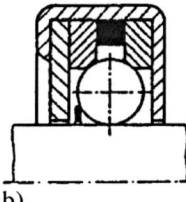
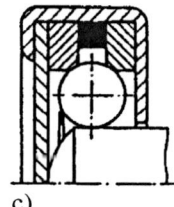

a)   b)   c)

Fig. 8.64. Small deep groove ball bearings. **a** With flange. **b** Without inner ring. **c** With groove board

Calculation

Required size of bearing is defined by existing load, operating conditions and requested durability. Every rolling bearing has a certain bearing strength. Dynamic bearing strength at rotating inner or outer ring and static bearing strength at rest or minor pivoting motion differ.

Dynamic bearing strength is connected with durability. Durability of a group of similar bearings is defined as number of rotation which is reached or exceeded by 90% of bearings before first occurrence of material fatigue. Therewith it is allowed that 10% of bearings fail before durability is reached. Nominal durability L in 10^6 revolutions is calculated with $p = 3$ for ball bearings and $p = 10/3$ for roller bearings according to Eq. (8.41) and may also be converted into hours of operation (L_h) for constant speed with Eq. (8.42):

$$L_{10} = \left(\frac{C}{P}\right)^p, \tag{8.41}$$

$$L_h = 10^6 \frac{L_{10}}{(60n)}; \tag{8.42}$$

C and P in N, L_{10} in 10^6 revolutions, L_h in h, n in rpm.

Dynamic load rating C of bearing is that load, at which nominal durability of L revolutions is reached. It may be found in roller bearing catalogue. The given definition of C assumes either a radial force F_r (radial bearing) or an axial force F_a (axial bearing) of constant value and direction. If both radial and axial load are existent, an equivalent load has to be calculated from both components. It corresponds to a pure radial load P_r for radial bearings or pure centric axial load P_a for axial bearings respectively, at which impact a rolling bearing would reach the same nominal durability as under real conditions.

Equivalent load is calculated $P = XF_r + YF_a$, whereby bearings factors X and Y may be found in rolling bearing catalogues [8.1, 8.2].

Mounting

If a shaft is supported by two or several rolling bearings, compensation of length variation (e. g. due to thermal expansion and mounting tolerance) must be ensured. In such cases (Fig. 8.65) one bearing is mounted fixed (fixed bearing) while the other bearing or bearings must be movable either on shaft or in casing (floating bearing).

Guidelines for fittings for mounting of rolling bearings (shaft-bearing, bearing-casing) are contained in DIN 5425 [8.1, 8.2].

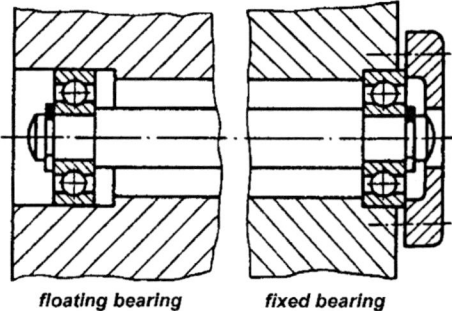

Fig. 8.65. General construction of rolling bearing with fixed and floating bearing

Lubrication and Sealing

Mostly grease is the lubricant which fills up all cavities of the bearing. Grease amount depends on rotational speed. Lowest friction, however, is reached by oil lubrication. Sealings ensure that lubricant does not leak out and pollutes environment and protects the bearing against dust and dirt. If bearings are not sealed another sealing has to be realized in construction. Contactless sealings are less effective than grinding sealings but they do not increase friction moment. For detailed presentation see [8.1, 8.2].

Advantages of rolling bearings compared with sliding bearings are low friction ($\mu \approx 10^{-3}$) also during start, simultaneous compensation of axial and radial forces, short axial overall length and standardized configurations. Disadvantages are larger diameter and costlier mounting, noisier run, impact sensitivity, danger of contamination and in most cases the higher price.

Please note: material indications (systems of indication, mostly used as abbreviation).

1. Abbreviations: with combinations of letters and numbers materials are indicated keyed.

Example: construction steel S185, minimal elastic limit 185 N/mm^2

Consider: at present DIN standards are converted to EN standards. This causes changes for abbreviations which are not completed yet (see also table below).

2. Material numbers: with up to seven-digit combinations of numbers, e. g. steels are characterized according to specific criteria (1st or 1st and 2nd digit: main material indicator; 3rd to 6th digit: sort number – composed of chemical conditions of production and application specifications; if required 7th and 8th digit for coding of special features, like, e. g. melting conditions and treatment status). For cast iron and non-iron metals special combinations are valid.

Examples: 1.1181 – heat-treated steel C35E, 1.0710 – machining steel 15S10.

Previous and present abbreviation and material numbers for selected steels:

Until now after DIN 17100	After DIN EN 10025	Material number
St33	S185	1.0035
St37-2	S235JR	1.0037
St44-2	S275JR	1.0044
St50-2	E295	1.0050
St60-2	E335	1.0060
St70-2	E360	1.0070

References

[8.1]   Krause, W.: Konstruktionselemente der Feinmechanik. 3. stark bearb. Aufl. München/Wien: Carl Hanser Verlag 2004
[8.2]   Krause, W.: Grundlagen der Konstruktion. Elektronik – Elektrotechnik – Feinwerktechnik. 8. Aufl. München/Wien: Carl Hanser Verlag 2002
[8.3]   Krause, W.: Plastzahnräder. Berlin: Verlag Technik 1985
[8.4]   Krause, W., Metzner, D.: Zahnriemengetriebe. Berlin: Verlag Technik 1988 und Heidelberg: Dr. Alfred Hüthig Verlag 1988

506   References

[8.5]   Krause, W.: Fertigung in der Feinwerk- und Mikrotechnik. Verfahren – Werkstoffe – Gestaltung. München/Wien: Carl Hanser Verlag 1996
[8.6]   Krause, W.: Gerätekonstruktion in Feinwerktechnik und Elektronik. 3. Aufl. München/Wien: Carl Hanser Verlag 2000
[8.7]   Luck, K.; Modler, K.-H.: Getriebetechnik. Analyse-Synthese-Optimierung. 2. Aufl. Berlin/Heidelberg: Springer-Verlag 1995
[8.8]   Krause, W.: Lärmminderung in der Feinwerktechnik. Düsseldorf: VDI-Verlag 1996
[8.9]   Kallenbach, E.; Bögelsack, G.: Gerätetechnische Antriebe. München/Wien: Carl Hanser Verlag 1991
[8.10] Krause, W.: Betriebsverhalten feinwerktechnischer Stirnradgetriebe. Teil I: Drehwinkeltreue; Teil II: Wirkungsgrad; Teil III: Lärmminderung. F&M 104 (1996) H. 10, S. 858; 105 (1997) H. 1–2, S. 50; 105 (1997) H. 4, S. 212
[8.11] Niemann, G.; Winter, H.: Maschinenelemente, Bde. I bis III. 4. Aufl., 2. Aufl., 2. Aufl. Berlin/Heidelberg: Springer-Verlag 2005, 2002, 2004
[8.12] Dresig, H.; Holzweißig, F.: Maschinendynamik. 6. Aufl. Berlin/Heidelberg: Springer-Verlag 2006

Standards and Directions

Section 8.1

[8.1.1] DIN 13,14 Metrisches ISO-Gewinde. 1986
[8.1.2] DIN 37 Darstellung und vereinfachte Darstellung für Zahnräder und Räderpaarungen. 1961
[8.1.3] DIN 103, 380 Metrisches Trapezgewinde. 1977
[8.1.4] DIN 111 Antriebselemente. Flachriemenscheiben. 1982
[8.1.5] DIN 654 Stahlbolzenketten; Maße, Befestigungsglieder. 1986
[8.1.6] DIN 695 Anschlagketten, Hakenketten, Ringketten. 1986
[8.1.7] DIN 780 Modulreihe für Zahnräder. 1977
[8.1.8] DIN 867 Bezugsprofil für Evolventenverzahnungen an Stirnrädern (Zylinderrädern) für den allgemeinen Maschinenbau. 1986
[8.1.9] DIN 868 Allgemeine Begriffe und Bestimmungsgrößen für Zahnräder, Zahnradpaare und Zahnradgetriebe. 1976
[8.1.10] DIN 2215 Endlose Keilriemen; Maße. 1975
[8.1.11] DIN 3960 Begriffe und Bestimmungsgrößen für Stirnräder (Zylinderräder) und Stirnradpaare (Zylinderradpaare) mit Evolventenverzahnung. 1987
[8.1.12] DIN 3961 bis 3967 Toleranzen für Stirnradverzahnungen. 1978
[8.1.13] DIN 3966 Angaben für Verzahnungen in Zeichnungen. 1978,1980
[8.1.14] DIN 3975 Begriffe und Bestimmungsgrößen für Zylinderschneckengetriebe mit Achswinkel 90°. 1976
[8.1.15] DIN 3990 Tragfähigkeitsberechnung von Gerad- und Schrägstirnrädern. 1987
[8.1.16] DIN 3992 Profilverschiebung bei Stirnrädern mit Außenverzahnung. 1964
[8.1.17] DIN 3998 Benennung an Zahnrädern und Zahnradpaaren. 1976
[8.1.18] DIN 7721 Synchronriemengetriebe (Zahnriemengetriebe), metrische Teilung, 1989

[8.1.19] DIN 58400 Bezugsprofil für evolventenverzahnte Stirnräder der Feinwerktechnik. 1984
[8.1.20] DIN 58405 Stirnradgetriebe der Feinwerktechnik. 1972
[8.1.21] DIN ISO 2203 Technische Zeichnungen; Darstellung von Zahnrädern. 1976
[8.1.22] ISO 5288, 5294 bis 5296 Synchronous belt drives. 2001, 1989, 1987, 1989 und ISO 13050 Curvilinear toothed synchronous belt drive systems. 1999
[8.1.23] VDI 2125, 2126 Ebene Gelenkgetriebe; Übertragungsgünstigste Umwandlung von Bewegungen. 1987, 1989
[8.1.24] VDI 2127 Getriebetechnische Grundlagen; Begriffsbestimmungen der Getriebe. 2001
[8.1.25] VDI 2130 Getriebe für Hub- und Schwingbewegungen; Konstruktion und Berechnung viergliedriger ebener Gelenkgetriebe für gegebene Totlagen. 2001
[8.1.26] VDI 2142 Auslegung ebener Kurvengetriebe; Bl. 1: Grundlagen, Profilberechnung und Konstruktion. 2002; Bl. 2: Rechnerunterstützte Profilberechnung. 1994
[8.1.27] VDI 2143 Bewegungsgesetze für Kurvengetriebe; Bl. 1: Theoretische Grundlagen. 2002; Bl. 2: Praktische Anwendungen. 2002
[8.1.28] VDI 2145 Ebene viergliedrige Getriebe mit Dreh- und Schubgelenken; Begriffserklärungen und Systematik. 1980
[8.1.29] VDI 2149 Bl. 1: Getriebedynamik - Starrkörper-Mechanismen. 1999
[8.1.30] VDI 2156 Einfache räumliche Kurbelgetriebe; Systematik und Begriffsbestimmungen. 2001
[8.1.31] VDI 2157 Planetengetriebe; Begriffe, Symbole, Berechnungsgrundlagen. 2002
[8.1.32] VDI 2158 Selbsthemmende und selbstbremsende Getriebe. 2002
[8.1.33] VDI 2159 Emissionskennwerte technischer Schallquellen; Getriebegeräusche. 1985
[8.1.34] VDI 2727 Konstruktionskataloge; Lösung von Bewegungsaufgaben mit Getrieben. 2000
[8.1.35] VDI 2741 Kurvengetriebe für Punkt- und Ebenenführung. 2002
[8.1.36] VDI 2758 Riemengetriebe. 2001

Section 8.2

[8.2.1] DIN 115 Antriebselemente; Schalenkupplungen. 1973
[8.2.2] DIN 116 Antriebselemente; Scheibenkupplungen. 1971
[8.2.3] DIN 740 Nachgiebige Wellenkupplungen. 1986
[8.2.4] VDI 2240 Wellenkupplungen; systematische Einteilung nach ihren Eigenschaften. 1971
[8.2.5] VDI 2241 Schaltbare fremdbetätigte Reibkupplungen und -bremsen; Bl. 1: Begriffe, Bauarten, Kennwerte, Berechnungen. 1982; Bl. 2: Eigenschaften, Auswahlkriterien. 1984
[8.2.6] VDI/VDE 2254 Bl. 1, 2: Feinwerkelemente; Drehkupplungen. 1976, 1978, 1990

Section 8.3

[8.3.1] DIN 471 Sicherungsringe (Halteringe) für Wellen. 1981
[8.3.2] DIN 748 Zylindrische Wellenenden für elektrische Maschinen. 1970
[8.3.3] DIN 3760 Radial-Wellendichtringe. 1972
[8.3.4] DIN 6799 Sicherungsscheiben (Haltescheiben) für Wellen. 1981
[8.3.5] DIN 24960 Gleitringdichtungen. 1992
[8.3.6] DIN 41591 Wellenenden für elektrisch-mechanische Bauelemente. 1976
[8.3.7] DIN 42020 Wellenenden mit Toleranzring bei elektrischen Kleinmotoren. 1982
[8.3.8] DIN 75532 Biegsame Wellen. 1976

Section 8.4

[8.4.1] DIN 615 Schulterkugellager. 1993
[8.4.2] DIN 620 Toleranzen der Wälzlager; Lagerluft. 1982
[8.4.3] DIN 625 Rillenkugellager. 1989
[8.4.4] DIN 1495 Gleitlager aus Sintermetall. 1983
[8.4.5] DIN 1850 Buchsen für Gleitlager. 1976
[8.4.6] DIN 5425 Wälzlager; Passungen, Toleranzfelder. 1984
[8.4.7] DIN 31698 Gleitlager; Passungen. 1979
[8.4.8] DIN ISO 281 Wälzlager; Dynamische Tragzahlen, nominelle Lebensdauer, Berechnungsverfahren. 1993
[8.4.9] VDI 2204 Bl. 1 bis 4: Auslegung von Gleitlagerungen. 1992
[8.4.10] VDI/VDE 2252 Bl. 1 bis 9: Feinwerkelemente; Führungen; Lager; Gelenke. 1970 bis 2003
[8.4.11] VDI 2541 Gleitlager aus thermoplastischen Kunststoffen. 1975

9
Project Design of Drive Systems

Christian Richter, Thomas Roschke

9.1 Requirements and Specifications

To find the solution of a technical problem it is at first necessary to define exactly the requirements. The more accurately the requirements are described, the better and more effectively the target can be met. First of all, the question is to answer which components of a drive system (cp. Fig. 1.1, Chap. 1) should be sized and parameterized. Accordingly, the interfaces must be specified. In any case the interface between drive and electric energy supply has to be described as well as the mechanical quantities of the interface to the driven load (active element).

The requirements of a drive system can be subdivided into six groups:

1. Control signals
   Among these are manual actuation by switches, direct operation by sensors, and control signals from the superior control hierarchy such as PLC (programmable logic control unit), computer or bus system; the level of signals, shape of signals etc.
2. Steady-state parameters
   This group includes the specification of torque, speed, direction of rotation, adjustment angles, forces, load cycle, supply voltage, line frequency and the number of line phases, permissible currents, service life, bearing loads, and stalling withstand.
3. Dynamic parameters
   Important parameters especially for servo drives are peak torque, acceleration, speed variations in synchronous operation, electrical and mechanical time constant, inertia, as well as current and voltage peaks.
4. Environmental requirements
   The drive should work faultless in its environment and likewise must not inadmissibly influence it. There exist safety-related regulations (e.g. European Union (EU) directives, standards) and approbations, which will be addressed in detail below. In this group, there are demands according

to noise and oscillations, stress by vibrations and shock, electromagnetic compatibility, system perturbations, degree of protection, product recycling, ambient temperature, air humidity, aggressive media and possible water exposure.

5. Design requirements
   Not only the visual appearance (including color and shape) can influence the impressions of the whole drive system but also general and mounting dimensions as well as mass could be given as an input.
6. Costs
   Finally, cost also determines the feasibility of a drive function. Here, not only the pure numerical value should be seen, but also the needed volume, the service, the technical reliability, the warranty conditions, and the reputation of the supplier.

Some important points as an example are listed above to give an idea of common requirements. Further specifications may be necessary depending on the type of drive. The more detailed the analysis of the drive function, the faster the optimal solution will be found.

For safety-related demands, the European Union released directives that define the same safety basis in all involved countries. These directives contain fundamental demands for technical products to make sure that with their operation no damage to humans or material assets will be caused. For small drive systems, the following directives are always valid. Sometimes more have to be respected depending on the type of product and its application.

- Low-voltage directive [9.1]
- Machinery directive [9.2]
- Electromagnetic compatibility directive [9.3].

Inside the EU, technical products must fulfill the demands of these directives (the responsibility of the manufacturer or vendor). It is not allowed to put products into operation that are not in accordance with the mentioned directives (the responsibility of the user). The compliance with appropriate directives is given when a CE-sign is attached to a final product or a declaration of manufacturer is supplied with a component or an incomplete machine. The attached documents contain applicable EC-directives and harmonized standards, which were used for the required tests. Some important standards for fractional-horsepower electric machines are listed in the references of Chaps. 9 and 10.

Approbations and marks of conformity are another way of presenting the compliance with safety-related demands. In this case, neutral and officially accredited test laboratories (the so-called notified bodies) check the safety-related features of a product. Examples of marks of conformity in Germany are the VDE mark (*V*erband *D*eutscher *E*lektrotechniker – German Association for Electrical, Electronic & Information Technologies) and the GS mark

(Geprüfte Sicherheit – evaluation according to the German Equipment and Product Safety Act GPSG). Only products with the UL mark (Underwriters Laboraties) can be delivered to the North American market. The UL mark is the most widely recognized and accepted evidence of a product's compliance with US and Canadian safety requirements, especially for fire and explosion protection as well as electromagnetic compatibility. From the start, a developer of equipment has to take care with the design and the choice of components such as motors regarding the requirements of his customer or target market respectively.

9.2 Approach to Drive Functions

To find solutions for drive functions, a systematic approach is advisable. This approach will be demonstrated with the treatment of the exercises in Sect. 9.7. In Fig. 9.1, the suggested approach is shown as a flowchart.

A thorough analysis of the requirements (cp. Sect. 9.1) is an essential prerequisite for the effective solution of drive functions. Thereafter, the active principle is chosen, the drive structure is defined, and the rough dimensioning is done. The check of the feasibility of the requirements and specifications is performed successively starting from step 5. Modeling and simulation methods [9.4, 9.5, 9.6, 9.7, 9.8, 9.9, 9.20, 9.21] can strongly support especially the analysis of the dynamic behavior (step 6) of sophisticated and extensive drive systems.

After each move from step 5 to step 10, the question is to answer if all given requirements can be fulfilled. If this is not the case, the following possibilities exist to still reach the target:

- the regarding requirement or parameter is rectifiable in certain limits,
- dedicated sections of the structure will be changed,
- a different active principle will be chosen.

Each question causes possible feedback, which is neglected in Fig. 9.1 to keep the chart concise.

Basically, the approach described above is only followed until step 6 during the treatment of the samples in Sect. 9.7, whereas the stepping drives are examined in detail. Comprehensive literature for the project design of drive systems with DC motors is already available [1.13, 9.10] to [9.12].

9.3 Classification of Common Drive Functions

9.3.1 Classification According to Motion

Regarding motion, four classes can be derived from the variety of drive tasks. Characteristic features of such a class are common parameters, which play a dominant role for this type of drive function.

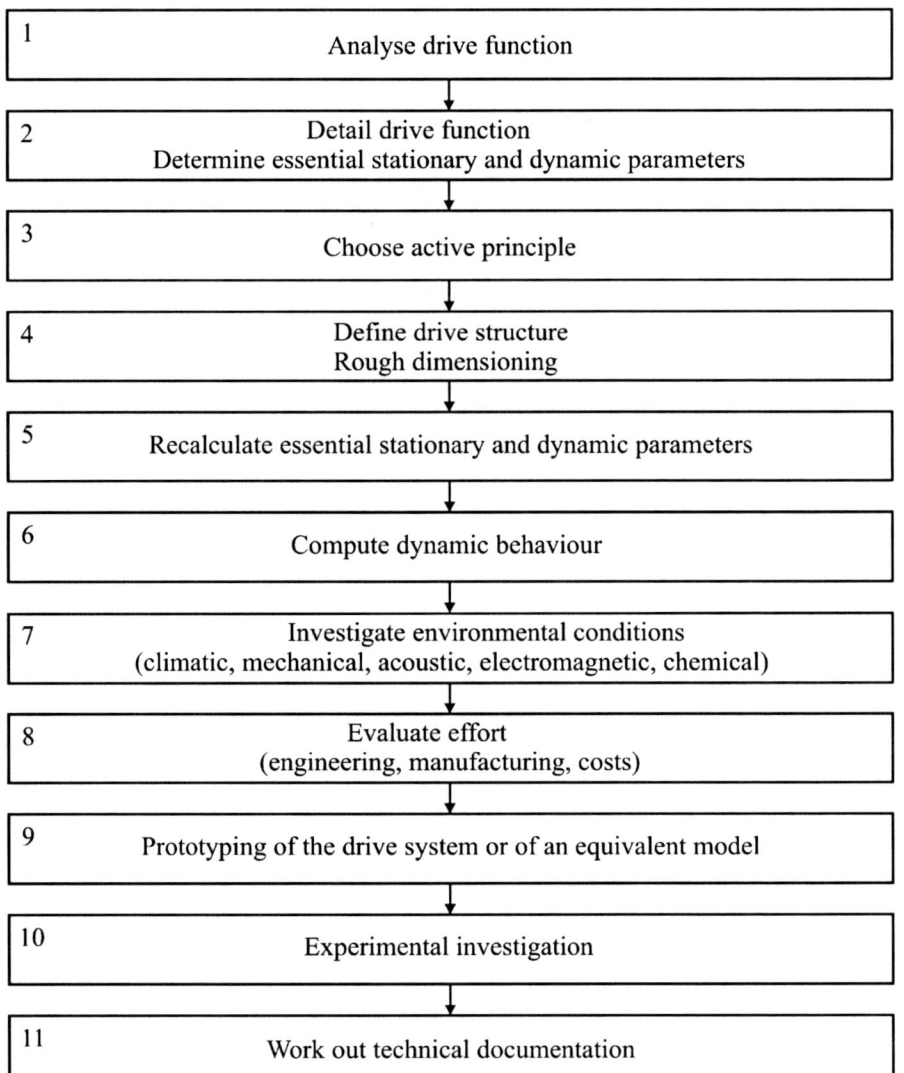

Fig. 9.1. Systematic approach to solve drive functions

In Table 9.1, the individual classes of drive functions are assigned with parameters independent from the type of motor used. Inside one class it does not matter if a rotary or a translational motion is realized. The examples below are mentioned to illustrate these statements.

The above-mentioned chapters described driving elements that can be divided with respect to their speed vs. torque characteristic into four groups (Fig. 9.2):

Table 9.1. Classification of drives according to the type of motion [1.13]

Class of drive function	Characteristic parameters	Operation mode	Examples
Constant speed	Slightly variable speed Operation at a defined speed ratio Crawl speed	Synchronous operation Asynchronous operation Closed-loop speed control Speed ratio control	Record or disk player drive Fan Pump Winding drive
Point-to-point control	Positioning accuracy Minimum of positioning time Maximum of speed	Positioning operation Group step operation of stepping drives	x–y-table Printer Dosing pump Movie cameras Bonding machines
Continuous-path control	Position accuracy during movement Maximum of acceleration	Time optimal positioning Servo control	x–y-recorder Robot drives Trip recorder Workpiece and tool positioning
Reversing operation	Reversing frequency Resonance frequency	Reversing operation	Sewing machine Photo plotter

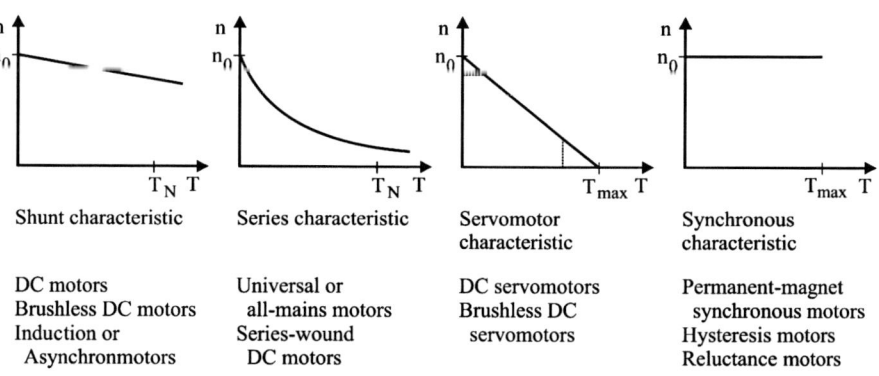

Fig. 9.2. Classification of speed vs. torque characteristic of rotary motors

- With increasing load, a shunt characteristic shows a low speed drop, e. g. from no-load to nominal-load operation less than 20%.
- Series characteristic represents a strong decline of the rotational speed with increasing load, e. g. from no-load to nominal-load operation more than 50%. Such motors are suitable for applications with high starting torque.
- Servomotor behavior exhibits a linear relationship to the speed and torque, whereas the characteristic can be used until standstill. Often, the max-

imum values of speed and torque are limited by special boundary conditions (dashed lines).
- Synchronous behavior is characterized by the independence of the speed from the load until to a maximum value of torque, e.g. the synchronous pull-out torque.

Equivalent relationships are valid for all linear or translational motors. In this case, the rotational speed has to be replaced by a translational velocity and the torque by the force.

Stepping motors are an exception, because their operational characteristics (cp. Fig. 3.27) have to be interpreted as limiting curves or margins. In the whole area limited by the torque curve each combination between stepping rate or speed, respectively, and torque is acceptable.

All these characteristics are valid for operation at constant mototr voltages. When the motors are operated by control electronics, quantities like voltage, current, and frequency can be altered. Each characteristic can be almost arbitrarily varied with such an influence on the input values of the electromechanical energy converter.

9.3.2 Classification According to Operating Mode

The thermal endurance as well as the wear, which is closely linked to the lifetime of the drive, are in the focus of such a consideration. Thereto regulations are made at least in three standards [10.21, 10.25] and [10.27], which are the most important for fractional-horsepower motors.

The rated or nominal duty defines the entirety of all electrical and mechanical quantities in duration and chronology, how they are defined from the manufacturer, and marked on the rating plate. Under this condition, the drive has to fulfill the agreed specifications. Starting from that, different duty types are defined. For fractional-horsepower motors, the most important duty types are continuous duty, short-time duty, and intermittent periodic duty.

Losses are generated with each transformation of energy. These losses are converted into heat, whereas the main heat sources are:

- Windings (Joule heat or ohmic loss due to electric current, $\sim I^2 R$),
- Magnetic circuit (hysteresis $\sim B^2 f$ and eddy-current losses $\sim B^2 f^2$ due to alternating magnetic flux and its harmonics),
- Bearings and rotor surface (bearing friction and windage loss),
- Commutator-brush system (electrical loss in brushes due to current transfer and arcing).

Components of an electric machine can be evaluated from a thermal point of view in the following way:

- Metal parts (copper or aluminium windings, iron circuit, housing) are thermally uncritical in the interesting temperature range of electric machines.

- Permanent magnets are limited in their temperature range. Above this limit, irreversible demagnetization will start. For instance, NeFeB-magnets show a maximum of allowed temperature between 100 and 180 °C.
- In bearings, the lubricating properties can be drastically altered at too high temperatures or the lubricant can be irreversibly changed.
- Sensitive against temperature stress are all plastic materials (e. g. winding insulation and bobbin), which show a drastic and irreversible change in their properties above the maximum of allowed operation temperature. This may lead to a complete breakdown of the electric machine.

Electric machines are subdivided into thermal classes of electrical insulation systems [10.29]. With a specified thermal class, the manufacturer guaranties that under rated conditions, the described heat sources will not impermissibly stress the motor, so that the specified product life can be reached. For instance, a "thermal class A" insulation will withstand a maximum winding temperature of 105 °C.

The power P_v dissipated by the energy conversion inside the electric machine leads *via* the heat storage inside the mass m with the thermal capacity c to a heating of the motor with a temperature rise of $\Delta \vartheta$. As soon as the machine is at a higher temperature compared to the environment it will dissipate heat through its surface O. The heat emission coefficient α characterizes this process. Therefore in Fig. 9.3, a simplified equivalent network is shown.

In a real machine, the dissipation is determined by the distribution of heat sources, the thermal conduction between the single parts of the machine, and the different levels of heat emission on separate surface sections. This mechanism is considered during the evaluation of the electric machine with the thermal class. Hence a simplification to a homogeneous body with a single heat source is permitted. The differential equation for the temperature rise in such a case is:

$$P_v \cdot dt = m \cdot c \cdot d\vartheta + O \cdot \alpha \cdot dt \ . \tag{9.1}$$

The time domain behavior of the temperature rise is exposed in Fig. 9.4a. The machine stores heat in its thermal capacity as long as the steady state with the overtemperature $\Delta \vartheta_{max}$ is reached. In this steady state, the generated power loss P_v and the heat emitted through the surface are equal.

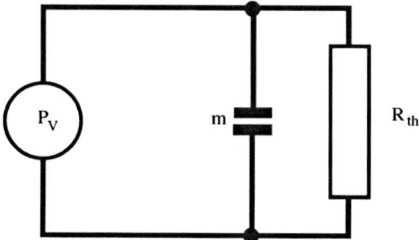

Fig. 9.3. Simplified equivalent network as a model of the thermal behavior of a fractional-horsepower motor

The continuous running duty (duty type S1) is defined with constant load and long duration, which lasts at least as long as the thermal steady state is reached (Fig. 9.4a).

During short-time duty (duty type S2), the run lasts too short to reach the thermal equilibrium. In the following break, the drive is switched off and the motor temperature decreases towards ambient temperature (Fig. 9.4b).

The intermittent periodic duty (duty type S3) is a sequence of identical load cycles where temperature increases during load and decreases during break time (Fig. 9.4c). Here it must be taken care of the relation between current and torque of the respective machine type. For example, with small synchronous motors, the current remains practically unchanged from no-load to full-load operation. In this case, the temperature rise is independent from the load stress.

From the comparison of the temperature characteristics in Fig. 9.4 it can be seen that for short-time and intermittent periodic duty a load period longer than the running time T_B will exceed the permissible maximum of temperature $\Delta\vartheta_{max}$. This is caused by higher losses compared to the continuous run of the same machine. Higher losses mean a larger loading and thereby a better utilization of the machine. It is important to ensure the duty factor or duty ratio ED (German: Einschaltdauer ED) and the whole duty cycle time T_S. The duty factor ED is defined as (T_B – operation-time):

$$ED = T_B/T_S . \tag{9.2}$$

Driving motors are classified from the manufacturer by operating mode. If no special information about the motor is given, a duty factor of $ED = 100\%$ must be assumed, which means continuous running.

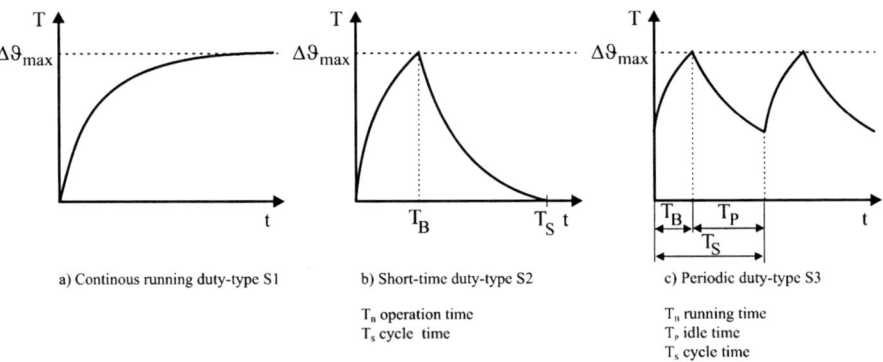

a) Continous running duty-type S1

b) Short-time duty-type S2

T_n operation time
T_s cycle time

c) Periodic duty-type S3

T_n running time
T_p idle time
T_s cycle time

Fig. 9.4. Temperature characteristics at different types of duty [10.21] (EN 60034-1)

9.4 Aspects of Motor Selection

To take better advantage of a motor, it is possible to increase its load during short-time or intermittent periodic duty compared to continuous running duty ($ED = 100\%$). For AC motors above approximately 10 W and all DC motors, the current increases with the torque. With a higher load, the demanded torque increases, too. The possible influence on the speed depends on the motor characteristic. The generated torque has nearly no influence on the value of the current for small AC motors below 10 W as well as stepper motors. Applying a higher voltage is here the only possible way to achieve a higher utilization of the motor. This is also possible for all larger motors. Figure 9.5 displays the qualitative interrelation of supply voltage and achievable torque vs. duty ratio. A precise quantitative statement can be only given for an individual motor.

A maximum of the ambient temperature of 40 °C can be assumed without any additional statement from the manufacturer. Higher temperatures shorten the loading capacity of the motor; lower temperatures allow a higher load as the rated one. The temperature rise of the motor is proportional to the power dissipation and inversely proportional to the heat dissipation capability. When the ambient temperature of a motor with unchanged cooling conditions is at 80 °C instead of 40 °C, and the motor is classified according to thermal class B (maximum winding temperature of 130 °C), the power dissipation has to be reduced with the coefficient $(130-80)/(130-40) = 0.55$. This means for instance that for a DC motor with permanent magnetic excitation the permissible torque is reduced to $0.75T_N$, because here the torque is directly proportional to the current, and the losses are proportional to the square of current.

Improved cooling allows a higher utilization of the motor. The heat can be better dissipated by forced cooling or by additional radiating surfaces or special heat sinks.

A simple check-up method for the thermal motor design is the measurement of the winding temperature and the comparison of the thermal class of

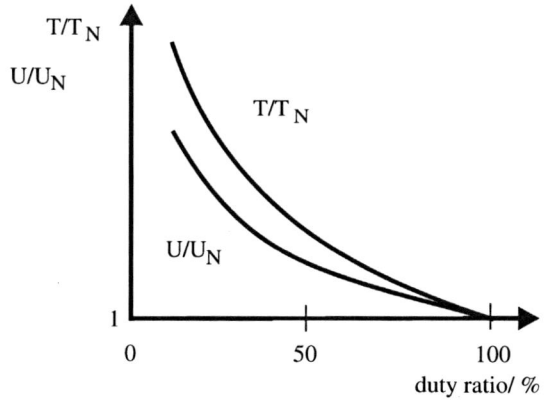

Fig. 9.5. Supply voltage and torque as a function of the duty ratio ED

insulation during practical tests under rated conditions (load, stress cycles, cooling, ambient temperature, etc.). Usually, the winding is the largest heat source of the motor.

The capacity value of a motor capacitor of single-phase asynchronous or synchronous motors does not need to be changed for operation with increased voltage. It is adapted to the winding with its inductivity and ohmic resistance. However, the dielectric strength or maximum permissible operating voltage of the capacitor might be increased, too.

The aspects of the selection of control electronics have already been discussed in Chap. 6. Here, the peak values of current and voltage are especially important, which can reach considerable magnitudes because of the inductivity of the motor windings. The chopper frequency should be higher than 13 kHz because then it is at the margin or outside of the human range of audibility.

The relationship between load and mechanical stress inside the components like gears, couplings, or clutches etc., and the influence on product lifetime was discussed in Chap. 8.

Noise is one of the most sensitive aspects of the selection of motors. The increasing sensitivity of people as well as the expansion of fractional-horsepower motors throughout further human spheres places permanently increasing demands for quieter drive systems. Therefore, an increasing number of authors deal with the causes of noise generation as well as noise prevention [8.8, 8.10, 9.14, 9.15, 9.16].

The following main causes of noise are:
- Laminated core hum caused by alternating magnetic fields with the double of the supply frequency, i.e. with 100 Hz at supply nets with 50 Hz.
- Rotating parts can develop noise at the bearings and with the fan. Usually, the noise is strongly increasing with the speed.
- The unbalance of rotating parts leads (especially at high speeds) to noticeable noise.
- The commutator-brush system of a DC motor causes besides friction very often an oscillation of the brushes, which emits noise.
- Stepping and synchronous motors miss the natural damping as a result of their stiff characteristics. Moreover, they very often tend to oscillate with their usually small inertia.
- Electronically commutated motors with up to three winding phases and brushed DC motors with a small number of slots [9.13] show strong torque pulsations, which can cause vibrations or noise, especially at low speed.
- The electronic control can for instance excite each motor to oscillations with unfavorable frequencies or pulse duty factors during pulse-width modulation or in chopping operation.
- Higher utilization of motors leads to increased noise.
- Gears (cp. Sect. 8.1) cause noise by colliding of the teeth. The type of material plays an important role, as well as the surface finish and the assembly tolerances. When gear wheels are off-center or not planar or

show pitch failures, the basic noise will be superposed by an additional alternating noise. Usually this is noticed as especially annoying.
- In general, fractional-horsepower drives are vibrating systems because they consist of rotating inertias, vibrating masses, and elasticities (electromagnetic field, springs). Therefore at least one natural frequency exists.

Some possibilities of noise reduction or prevention are:

- No operation near to a natural or resonant frequency,
- Fast passing of resonant ranges during speed or frequency changes,
- Use additional inertia to move natural frequencies towards smaller frequencies,
- Balancing of wheels of high-speed drives,
- Use of low-speed motors and direct drives (no noise caused by fan and gears),
- Integrate sliding or bush bearings instead of ball or roller bearings into motor design,
- Apply high control frequencies (>13 kHz) for pulse-width modulation or chopper operation,
- Avoid brushes using electronically commutated motors,
- Take motors with external rotor and consequently higher inertia,
- Reduce gear noise (cp. Chap. 8.1),
- Encapsulate with an housing or enclosure,
- Consider coupling conditions between the mechanical subunits of the drive and between drive system and device (usually the noise will be evaluated seriously different, if the drive system alone or the drive installed inside the device are considered).

Today, noise comes increasingly to the foreground and develops into one of the strongest sales arguments. However, there are no general statements for solving noise problems. The individual drive situation is to investigate and all components have to be considered as well. Favorable configurations of parts and subunits as well as adequate operational methods can avoid vibrations and noise, but also amplify remarkably. The effort for (and the costs of) such means of noise reduction should always be kept in mind.

9.5 Comparison of Position-Controlled DC Drives and Stepping Drives

Both active principles, the position-controlled DC motor and the stepping motor, can be used to realize positioning and servo drives. Therefrom the question arises, under which conditions which active principle is showing decisive advantages? Firstly, a comparison under the aspect of the technical and the commercial reachable parameters will be done. Secondly, determined position functions will be investigated [1.13, 9.17].

Table 9.2. Features and parameters of position-controlled DC drives and stepping drives

Feature/Parameter	Position-controlled DC drive	Stepping drive
Control	Closed-loop control	Open-loop control
Minimum of angular resolution	0.0025°	0.36° without additional means 0.0144° in micro stepping mode
Commercially reasonable angular resolution	$\geq 0.036°$	$\geq 0.36°$
Torque	For micro and small drives without limits	≤ 100 cNm (1000 cNm)
Speed/frequency	20 000 min^{-1} brushed motors 100 000 min^{-1} brushless motors	$\theta_s f_{A0\,max} = 5000°/s$ pull-in curve $\theta_s f_{B0\,max} = 45\,000°/s$ pull-out curve
Electromagnetic compatibility	Critical with brushed motors	
Operating life	5000 h for brushed motors 25 000 h for brushless motors	25 000 h

From the comparison of the parameters in Table 9.2, the following advantages of position-controlled DC motors can be found:

- up to ten times higher resolution,
- approximately five times better dynamic behavior,
- unlimited torque in the scope of micro and small drives.

On the other hand the following disadvantages exist:

- the closed-loop control needs an additional position sensor,
- the effort for electronics is higher (up to three control loops),
- with brushed motors the operating life is much lower,
- with brushed motors the electromagnetic compatibility (EMC) becomes critical; stepping motors and brushless motors are nearly similar due to EMC,
- the costs increase approximately linear with improved dynamic behavior.

The conclusion from the above is that stepping motors can be mainly used in a limited torque range with moderate dynamic demands and low to medium resolution. In this field, stepping drives lead to a lower effort for the whole drive system, which clearly shows cost advantages.

For a common positioning function such as point-to-point control, the following prerequisites are important in addition to the already above described points:

Prerequisites for a stepping drive:

- The stepping drive operates at a fixed stepping rate f_S = const. The consequence of this is a constant angular velocity $d\theta/dt = \theta_S f_S$.
- The starting frequency f_A is changed in dependency of the factor of inertia FI (Eq. 3.12).

Prerequisites for a position-controlled DC drive:

- The speed vs. torque and the current vs. torque characteristics show a linear relation.
- A steady-state approach is possible because the electrical time constant is very small compared to the mechanical time constant $\tau_{el} \ll \tau_m$. The transient electrical behavior is neglected.
- The highest available supply voltage and the permissible limits given by the motor manufacturer determine the maximum values of speed, current and torque.
- The positioning operation is subdivided into the three phases: acceleration, operation at maximum speed, and braking.

First of all, the case of a large angular movement is considered. Here the starting and braking phases do not influence the whole motion sequence considerably (Fig. 9.6).

The actuation angle of the stepping drive which is reached within the time T_{SM} is given by:

$$\theta_{0SM} = \theta_S \cdot f_R \cdot T_{SM} . \tag{9.3}$$

For the drive with the DC motor it results in:

$$\theta_{0DC} = 2 \cdot \pi \cdot n \cdot T_{DC} . \tag{9.4}$$

If the same positioning task is to solve, i.e. $\theta_{0SM} = \theta_{0DC}$ can be assumed, the ratio of the necessary actuating times is obtained by:

$$\frac{T_{DC}}{T_{SM}} = \frac{\theta_S \cdot f_B}{2 \cdot \pi \cdot n} \tag{9.5}$$

When an individual load case for the drives is considered, the operating frequency f_B related to the load torque T_L or the speed n, respectively, is to read from the motor characteristics and insert into Eq. (9.5).

To fulfill the inequality of the actuating times $T_{DC} < T_{SM}$, the required speed of the DC motor can be calculated from:

$$n > \frac{\theta_S \cdot f_B}{2 \cdot \pi} . \tag{9.6}$$

Since high-quality stepping drives achieve a product of $\theta_S f_{0B\,max} = 45\,000°/s$ in the no-load case, a DC motor has to offer a maximum speed above $n_{max} = 7500\,\text{min}^{-1}$ to realize a shorter actuating time.

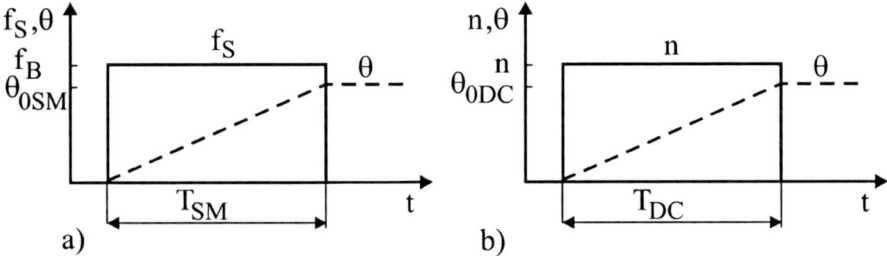

Fig. 9.6. Motion sequence with a large angular movement. **a** Stepping drive. **b** Position-controlled DC drive

For actuating tasks with small angular movement (Fig. 9.7), the stepping motor operates in the pull-in or start-stop range with the stepping rate f_A. For dynamic operation, the load inertia must be considered using Eq. (3.12). The actuation angle is given by:

$$\theta_{0SM} = \theta_S \cdot f_A \frac{1}{\sqrt{FI_{SM}}} T_{SM} \,. \tag{9.7}$$

The actuation angle is so small that the DC motor cannot reach its maximum speed. If in the considered short time an averaged constant torque T of the DC motor is assumed, it will constantly accelerate during the first phase T_a with an accelerating torque of:

$$T_{ba} = T - T_L \,. \tag{9.8}$$

Hence the speed increases linear, too. In the second phase, T_{Br}, the load torque, supports the braking. The braking torque can be calculated by:

$$T_{bBr} = T + T_L \,. \tag{9.9}$$

For the total motion sequence, the speed change must be $\Delta n = 0$. Therefore, the following equation is valid:

$$\int_{t_{DC}} T_b \cdot dt = 0 \,. \tag{9.10}$$

Consequentially, the following equilibrium can be obtained:

$$(T - T_L)T_a = (T + T_L)T_{Br} \,. \tag{9.11}$$

Since

$$T_{DC} = T_a + T_{Br} \tag{9.12}$$

is true, the equation

$$\frac{T_a}{T_{DC}} = \frac{1}{2}\left(1 + \frac{T_L}{T}\right) \tag{9.13}$$

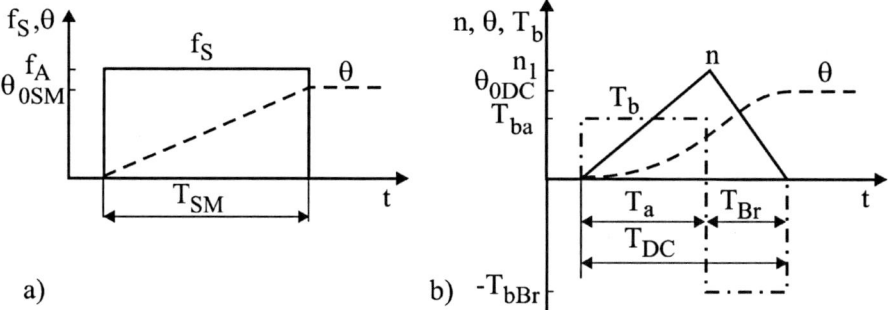

Fig. 9.7. Motion sequence with a small angular movement. **a** Stepping drive. **b** Position-controlled DC drive

results. The angles reached in each phase are:

$$\theta_a = \pi \cdot n_1 \cdot T_a \qquad \theta_{Br} = \pi \cdot n_1 \cdot T_{Br} \qquad (9.14)$$

or

$$\theta_a = \frac{T}{J_r \cdot FI_{DC}} \left(1 - \frac{T_L}{T}\right) \frac{T_a^2}{2} \qquad (9.15)$$

$$\theta_{0DC} = \frac{T}{4 \cdot J_r \cdot FI_{DC}} \left[1 - \left(\frac{T_L}{T}\right)^2\right] T_{DC}^2 . \qquad (9.16)$$

Using the mechanical time constant τ_m, where the factor of inertia FI is factored out, Eq. (9.16) can be formed:

$$\theta_{0DC} = \frac{\pi}{2} \cdot \frac{T}{T_a} \cdot \frac{n_0}{\tau_m} \cdot \frac{1}{FI_{DC}} \left[1 - \left(\frac{T_L}{T}\right)^2\right] T_{DC}^2 . \qquad (9.17)$$

From the comparison of Eqs. (9.7) and (9.17) the limit of the actuating time for equal actuation angles $\theta_{0SM} = \theta_{0DC}$ can be obtained:

$$T_{gr} = \frac{2}{\pi} \cdot \theta_S \cdot f_A \cdot \frac{T_a}{T} \cdot \frac{\tau_m}{n_0} \frac{FI_{DC}}{\sqrt{FI_{SM}}} \cdot \frac{1}{\left[1 - \left(\frac{T_L}{T}\right)^2\right]} . \qquad (9.18)$$

Respectively, the limit of the actuation angle for equal actuating times $T_{DC} = T_{SM}$ is given:

$$\theta_{gr} = \frac{1}{90°} (\theta_S \cdot f_A)^2 \frac{T_a}{T} \cdot \frac{\tau_m}{n_0} \cdot \frac{FI_{DC}}{FI_{SM}} \cdot \frac{1}{\left[1 - \left(\frac{T_L}{T}\right)^2\right]} . \qquad (9.19)$$

To explain the above statements further, a comparison is made between a high-quality stepping drive, which achieves in the pull-in range $\theta_S f_{A0\,max} = 5000°/s$, and a DC motor with a no-load speed of $n_0 = 6000\,\text{min}^{-1}$ and a mechanical time constant of $\tau_m = 10\,\text{ms}$. The operating point of the DC motor is given with $T/T_a = 3/4$. When equal factors of inertia $FI_{DC} = FI_{SM}$ and no-load operation ($T_L = 0$) are assumed, a limit of the actuation angle of $\theta_{gr} = 37°$ can be calculated. This means that for actuation angles smaller than $37°$ a stepping motor is superior to a DC motor under the described presumptions. For larger actuation angles, the opposite is true.

First of all, the required technical parameters must be compared with the achievable ones, when a decision for the best active principle has to be made for positioning tasks. As second criteria the typical limit values for the considered positioning task should be consulted. Thereby should be always kept in mind, that the better the realized dynamic behavior of the drive is, the higher the costs for the driving function will be.

9.6 Conversion of Mechanical Drive Parameters

It is not possible to design a tailored motor for each individual driving function. Often a gear for adaptation between driving motor and load is used. Moreover, very small angular velocities can be generated with a small high-speed driving motor and a reducing gear. This is most often helpful in costs and size compared to a direct drive. On the other hand, not only benefits are linked to the use of gears. Drawbacks are losses due to friction, backlash or inequality of the motion due to tolerances of the toothing [9.18].

A positioning drive often needs a translational motion. Usually such a motion is generated with a rotary motor and a mechanical motion converter between motor and load, such as a lead screw system (cp. Chap. 8).

Important conversions of the parameters from the mechanical load side towards the motor shaft are shown in Table 9.3. The appropriate structures of the drive systems are shown in Fig. 9.8.

The equations of Table 9.3 are written down for simple use as equation between related quantities, i. e. all quantities must be inserted necessarily with their units given in Table 9.4.

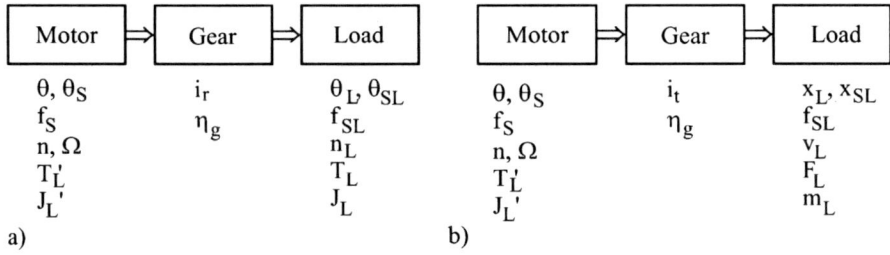

Fig. 9.8. Conversion of mechanical quantities. **a** Driving of a rotary moving load with a rotary motor. **b** Driving of a translational moving load with a rotary motor

9 Project Design of Drive Systems   525

Table 9.3. Conversion of mechanical drive parameters

Conversion rotation–rotation	Conversion translation–rotation
gear transmission ratio	gear transmission ratio
$i_r = \frac{n_L}{n}$	$i_t = \frac{x_{SL}}{\theta_S} = \frac{v_L}{6 \cdot n}$
angle	stroke
$\theta = \frac{\theta_L}{i_r}$	$\theta = \frac{x_L}{i_t}$
step angle	step width
$\theta_S = \frac{\theta_{SL}}{i_r}$	$\theta_S = \frac{x_{SL}}{i_t}$
stepping rate	stepping rate
$f_S = f_{SL} = \frac{6 \cdot n_L}{i_r \cdot \theta_S}$	$f_S = f_{SL} = \frac{v_L}{i_t \cdot \theta_S}$
angular speed	angular speed
$\Omega = \frac{0.1047 \cdot n_L}{i_r}$	$\Omega = \frac{1.745 \cdot 10^{-2} \cdot v_L}{i_t}$
revolutions per minute	revolutions per minute
$n = \frac{f_S \cdot \theta_S}{6} = \frac{f_{SL} \cdot \theta_{SL}}{6 \cdot i_r}$	$n = \frac{f_{SL} \cdot x_{SL}}{6 \cdot i_t}$
torque	force
$T'_L = i_r \cdot \frac{1}{\eta_g} \cdot T_L$	$T'_L = 5.73 \cdot i_t \cdot \frac{1}{\eta_g} \cdot F_L$
moment of inertia	mass inertia
$J'_L = i_r^2 \cdot \frac{1}{\eta_g} \cdot J_L$	$J'_L = 32.8 \cdot i_t^2 \cdot \frac{1}{\eta_g} \cdot m_L$

(A correct conversion of the inertia and the mass, respectively, on the basis of the kinetic energy does not consider the efficiency of the gear. However, in practice, the introduction of the gear efficiency into the conversion was useful for including unknown and undefined parameters of the energy conversion. When the gear efficiency is used, the calculation is always on the safe side.)

mechanical power at the load	mechanical power at the load
$P_L = 1.05 \cdot 10^{-3} \cdot T_L \cdot n_L$	$P_L = 10^{-3} \cdot F_L \cdot v_L$
mechanical power at the motor	mechanical power at the motor
$P = 1.75 \cdot 10^{-4} \cdot T'_L \cdot f_S \cdot \theta_S$	$P = 10^{-3} \cdot \frac{1}{\eta_g} \cdot F_L \cdot v_L$
$P = 1.75 \cdot 10^{-4} \cdot T_L \cdot \frac{1}{\eta_g} \cdot f_{SL} \cdot \theta_{SL}$	

Table 9.4. Units of the quantities used in the equations of Table 9.3

Quantity	Sign	Unit	Quantity	Sign	Unit
Angle	θ	°	Stroke	x	mm
Angular speed	Ω	s^{-1}	Stepping rate	f	Hz
Power	P	W	Translational transmission ratio	i_t	mm/°
Revolutions per minute	n	$\min^{-1}$	Speed	v	mm/s
Torque	T	cNm	Force	F	N
Inertia	J	gcm²	Mass	m	g

9.7 Samples of Drive Functions

In this chapter, individual drive functions are introduced to demonstrate the typical way of calculation as well as to explain essential criteria at representative examples.

9.7.1 Direct Drive of a Disk Storage Device

Hard disks belong to computer peripherals. The storage disks are directly fixed at the motor shaft. For a reliable writing and reading of the information at the disk a high running smoothness of the motor is required.

Requirements of the Drive

- Speed $n_\mathrm{L} = 2400\,\mathrm{min}^{-1}$
- Load torque $T_\mathrm{L} = 1\,\mathrm{cNm}$
- Supply voltage $U_\mathrm{N} = 12\,\mathrm{V} \pm 10\%$
- Load inertia $J_\mathrm{L} = 3500\,\mathrm{gcm}^2$
- Speed stability $\Delta n_\mathrm{L}/n_\mathrm{L} = 5\,\%\!o$ (long-term speed stability)
- Speed variation (deviation from constant velocity CV) $\mathrm{d}n_\mathrm{L}/n_\mathrm{L} = 0.1\,\%\!o$
  For the speed variation only one complete revolution is considered.
- Maximum starting time $T_\mathrm{a} = 15\,\mathrm{s}$ to nominal speed n_L
- No emission of electromagnetic fields, because the motor is placed in direct neighborhood to the stored data and the control electronics.
- Service life $L_\mathrm{B} = 25\,000\,\mathrm{h}$

Investigation to Find the Best Active Principle

Deviations from a constant velocity result from the changes of the load torque or the generated torque of the motor. Variations of the motor torque will result in acceleration or deceleration, therefore the permissible variation of the motor torque can be calculated from the required limit of the speed variation.

Each revolution will show a positive and a negative maximum of the speed variation, if a sinusoidal behavior is assumed. Therefore, the time of a quarter revolution should be inserted as the effective time for the speed variation. The following equation is valid:

This leads related to the rated load torque to:

$$T_\mathrm{b} = J_\mathrm{L} \cdot \frac{\mathrm{d}\omega}{\mathrm{d}t} = J_\mathrm{L} \cdot 2 \cdot \pi \cdot \frac{\mathrm{d}n_\mathrm{L}}{n_\mathrm{L}} \cdot n_\mathrm{L} \cdot \frac{4}{T} \quad (9.20)$$

$$= \frac{3500\,\mathrm{gcm}^2 \cdot 2\pi \cdot 0.1 \cdot 10^{-3} \cdot 2400\,\mathrm{min}^{-1} \cdot 4}{25\,\mathrm{ms}} = 0.14\,\mathrm{cNm}$$

$$\frac{T_\mathrm{b}}{T_\mathrm{L}} = \frac{0.14\,\mathrm{cNm}}{1.0\,\mathrm{cNm}} = 14\% \quad (9.21)$$

Decision

To ensure the required speed variation of less than 1 ‰, the generated torque of the motor should oscillate less than 14%, when a constant load torque is assumed. According to experience, this requirement cannot be fulfilled from a standard stepping motor. Therefore, the decision is made in favor of a DC motor.

Selection of the Structure of the Driving System

- Motor
  A brushed motor can never reach a service live of $L_B = 25\,000$ h and emits electromagnetic fields due to arcing of the brushes. Therefore, a brushless motor will be used. Figure 9.9 shows the operating characteristic of a suitable brushless DC (abbr. BLDC) motor for unregulated operation. The BLDC motor needs for one revolution a time of

$$T = \frac{1}{n_L} = \frac{1}{2400\,\text{min}^{-1}} = 25\,\text{ms}. \tag{9.22}$$

A complex closed-loop control would be necessary to realize the required low deviation from constant velocity with a motor with interior rotor, because the mechanical time constant of such motors (cp. Sect. 2.2) is of the same magnitude. In this case, it is better to use a motor with exterior rotor, which has a high inertia. The brushless motor A holds a momentum of inertia of $J_r = 2500\,\text{gcm}^2$. The permissible variation of the torque according to Eq. (9.21) is increased therewith to

$$T'_b = T_b \cdot \frac{J_r + J_L}{J_L} = 0.14\,\text{cNm} \cdot \frac{(2500 + 3500)\,\text{gcm}^2}{3500\,\text{gcm}^2} = 0.24\,\text{cNm}. \tag{9.23}$$

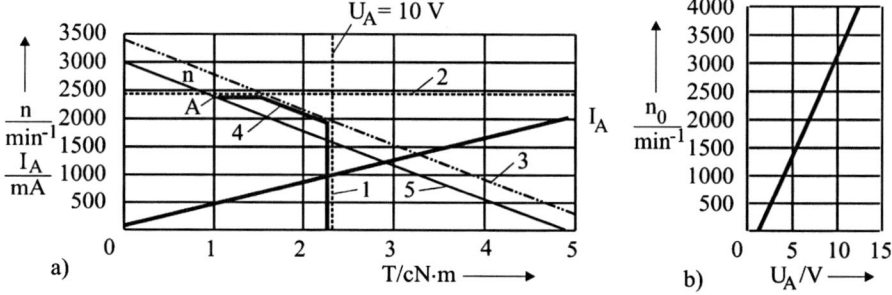

Fig. 9.9. Operating characteristic of brushless DC motor A. **a** Speed and current vs. torque characteristic. *1* characteristic for current limitation of $I_{max} = 1A$; *2* characteristic of speed limitation of $n_L = 2400\,\text{min}^{-1}$; *3* speed vs. torque characteristic for the minimum of supply voltage of $U_{min} = 10.8$ V; *4* acceleration curve; *5* speed vs. torque characteristic for $U = 10$ V; A operating point. **b** No-load speed vs. armature voltage

From this a ratio of 24% results related to the nominal torque. This can be assured from a high-quality brushless motor without any control.
- Structure
  The long-term speed stability can be only realized with a speed control loop. The structure of such a control circuit is shown in Fig. 9.10.
- Speed sensor
  It is sufficient to use the position sensor of the electronic commutation circuit and to generate an analog signal out of it, because only the long-term constancy of the speed is from interest [9.19].
- Speed controller
  The selection of the transfer function for the speed control is described in detail in [9.10] and [9.12]. Hence this way to achieve a determined behavior of the control circuit is not dwelled on in detail.
- Power amplifier with current control
  The choice of the transfer function can be made using [9.10] and [9.12]. Here it is important to consider the limitation of current. This is necessary to protect the power components of the amplifier against impermissible currents. Simultaneously, it helps to reduce costs. Without current limitation the starting current is $I_a = 2\,\text{A}$, whereas the rated current of the motor is only $I_N = 0.44\,\text{A}$. For instance, with the use of a current limitation of $I_{\max} = 1.0\,\text{A}$, the power components can be sized smaller and will be thereby cheaper. However, it must be checked if the required acceleration time T_a will be kept with this current limitation.

The supply voltage of the motor at the operating point has to be below the input terminal voltage, otherwise no control operation will be possible. The operating characteristics of the motor A are given for a voltage $U_A = 10\,\text{V}$.

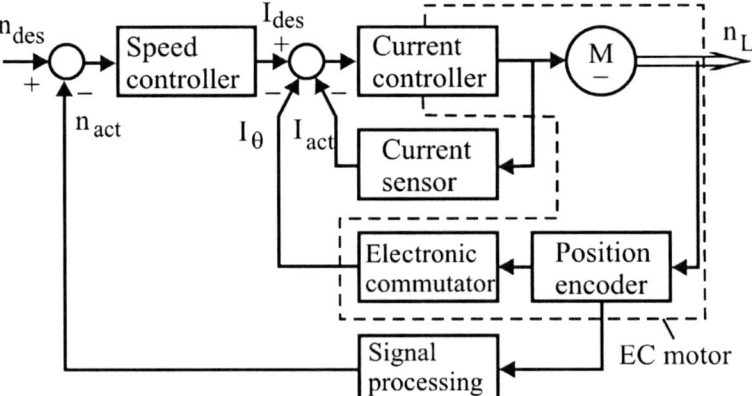

Fig. 9.10. Structure of the brushless motor A with speed-control loop

Verification of Important Drive Parameters

- Direct drive
  The storage disks are directly mounted on the motor shaft. Therefore, no conversion of parameters from load towards motor shaft is necessary.
- Mechanical power
  According to Tables 9.3 and 9.4, follows

$$P_L = 1.05 \cdot 10^{-3} \cdot T_L \cdot n_L = 1.05 \cdot 10^{-3} \cdot 1 \cdot 2400 = 2.52\,\text{W} \quad (9.24)$$

P_L	T_L	n_L
W	cNm	min^{-1}

- Electric power
  The electric power absorbed from the motor at the nominal working point amounts to

$$P_{el} = I_A \cdot U_A = 0.44\,\text{A} \cdot 10\,\text{V} = 4.4\,\text{W}\,. \quad (9.25)$$

- Efficiency
  The efficiency is given by

$$\eta = \frac{P_L}{P_{el}} = \frac{2.52\,\text{W}}{4.4\,\text{W}} = 57.2\%\,. \quad (9.26)$$

The electric power as well as the efficiency does not incorporate the power consumption of the control electronics. Both are related directly to the motor.

- Armature resistance
  The armature resistance is defining the armature current at standstill. When the starting current is taken from Fig. 9.9, the armature resistance can be derived:

$$R_A = \frac{U_A}{I_a} = \frac{10\,\text{V}}{2.0\,\text{A}} = 5.0\,\Omega\,. \quad (9.27)$$

- Mechanical time constants of motor and drive
  For the motor, the following equation can be applied

$$\tau_m = \frac{\Omega_0 \cdot J_r}{T_a} = \frac{2\cdot\pi\cdot 3030\,\text{min}^{-1}\cdot 2500\,\text{gcm}^2}{4.8\,\text{cNm}} = 1.65\,\text{s}\,, \quad (9.28)$$

and for the drive

$$\tau'_m = \tau_m \cdot FI = 1.65\,\text{s}\cdot\frac{(2500+3500)\,\text{gcm}^2}{2500\,\text{gcm}^2} = 3.97\,\text{s} \quad (9.29)$$

is valid.

- Tolerance of speed without closed-loop control
  The permissible voltage range is given with $U_N = 12\,\text{V} \pm 10\%$, which means that the minimum of the motor supply voltage is equal to $U_{min} = 10.8\,\text{V}$ and the maximum is $U_{max} = 13.2\,\text{V}$. From Chap. 2 it is known that a change of the armature voltage leads to a parallel shift of the speed vs. torque characteristic. From Fig. 9.9b, the no-load speeds related to the different voltages can be read:

$$U_A = 10\,\text{V} \qquad\qquad n_0 = 3030\,\text{min}^{-1}$$
$$U_{min} = 10.8\,\text{V} \qquad\qquad n_{0\,min} = 3285\,\text{min}^{-1}$$
$$U_N = 12\,\text{V} \qquad\qquad n_{0N} = 3670\,\text{min}^{-1}$$
$$U_{max} = 13.2\,\text{V} \qquad\qquad n_{0\,max} = 4050\,\text{min}^{-1}$$

The speed reduction from no-load to rated operational torque $T_L = 1.0\,\text{cNm}$ amounts to $\Delta n = (3030 - 2400)\,\text{min}^{-1} = 630\,\text{min}^{-1}$ according to Fig. 9.9a. The tolerance band of the speed for a load with the rated operational torque $T_L = 1.0\,\text{cNm}$ can be calculated for the permissible voltage tolerance of $U_N = 12\,\text{V} \pm 10\%$ under unregulated operation:

$$n_{L\,\text{unregulated}} = n_{0N} - \Delta n \pm \frac{n_{0\,max} - n_{0\,min}}{2}$$
$$= (3670 - 630)\,\text{min}^{-1} \pm \frac{(4050 - 3285)\,\text{min}^{-1}}{2} \qquad (9.30)$$
$$= (3040 \pm 382.5)\,\text{min}^{-1} = 3040\,\text{min}^{-1} \pm 12.6\% \,.$$

The result of $\pm 12.6\%$ is far from the required speed stability $\Delta n_L/n_L = 5\,\text{\textperthousand}$. This stability cannot be obtained without a closed-loop control. In general, a motor must be chosen that shows at the required load torque in unregulated operation a higher speed than the required one. Otherwise the controller would not be able to compensate the influence of disturbances.

- Maximum of acceleration time towards nominal speed
  As already discussed above under the point "power amplifier with current control", a current limitation of $I_{max} = 1.0\,\text{A}$ was introduced to use the advantage of smaller and cheaper power amplifier components. From this follows that the speed vs. torque characteristic will be limited too, which is shown in Fig. 9.9a with the dashed line 1. The running-up process towards nominal speed can be considered as a steady-state process, because the mechanical time constant $\tau_m = 3.97\,\text{s}$ is large compared to the electrical time constant of $\tau_{el} \approx 5\,\text{ms}$. In such a case, the transient behavior of the electric circuit can be neglected.
  The running-up process can be divided into two sections. The first part is the running up with a constant torque; the second part goes along the stationary characteristic. The conversion of the torque balance after time leads to

$$T = \int \frac{J_r + J_L}{T_{el} - T_L} d\Omega \,. \qquad (9.31)$$

The minimum of voltage $U_{\min} = 10.8\,\text{V}$ at the motor leads to a speed vs. torque characteristic, which is shown in Fig. 9.9a as a second characteristic line. In the case of a minimal voltage the motor accelerates with a constant torque towards a speed of $n_1 = 1770\,\text{min}^{-1}$ on this second characteristic line in the first phase of running up. For this reason, the acceleration time can be calculated as follows:

$$T_1 = (J_\text{r} + J_\text{L}) \cdot 2 \cdot \pi \int_0^{t_1} \frac{1}{T_\text{el} - T_\text{L}}\, \text{d}n$$

$$= (2500 + 3500)\,\text{gcm}^2 \cdot 2 \cdot \pi \int_0^{1770\,\text{min}^{-1}} \frac{1}{(2.4 - 1.0)\,\text{cNm}}\, \text{d}n = 7.95\,\text{s} \tag{9.32}$$

In the second phase ($n = 1770\ldots 2400\,\text{min}^{-1}$) is valid:

$$T_2 = (J_\text{r} + J_\text{L}) \cdot 2 \cdot \pi \cdot \frac{n_0}{T_\text{a}} \int_{n_1}^{n_\text{L}} \frac{1}{\frac{n_0}{T_\text{a}} \cdot (T_\text{a} - T_\text{L}) - n}\, \text{d}n$$

$$= (J_\text{r} + J_\text{L}) \cdot 2 \cdot \pi \cdot \frac{n_0}{T_\text{a}} \left[-\ln \left| \frac{n_0}{T_\text{a}} \cdot (T_\text{a} - T_\text{L}) - n \right| \right]_{n_1}^{n_\text{L}}$$

$$= (2500 + 3500)\,\text{gcm}^2 \cdot 2 \cdot \pi \cdot \frac{3285\,\text{min}^{-1}}{5.2\,\text{cNm}} \tag{9.33}$$

$$\cdot \ln \left| \frac{\frac{3285\,\text{min}^{-1}}{5.2\,\text{cNm}}(5.2 - 1)\,\text{cNm} - 1770\,\text{min}^{-1}}{\frac{3285\,\text{min}^{-1}}{5.2\,\text{cNm}}(5.2 - 1)\,\text{cNm} - 2400\,\text{min}^{-1}} \right| = 4.96\,\text{s}$$

Therefore, the total acceleration time to nominal speed sums up to:

$$T_\text{a} = T_1 + T_2 = 7.95\,\text{s} + 4.96\,\text{s} = 12.91\,\text{s} \tag{9.34}$$

and fulfils the requirement of $T_\text{a} \leq 15\,\text{s}$. Hence, it can be concluded that the current limitation is acceptable. If the supply voltage during running up is higher than the minimal voltage, the total acceleration time will be shorter according to the better speed vs. torque characteristic in the second phase of the running up.

Overall Evaluation

The verification of important parameters has shown that the brushless DC motor A can be used in this application. Moreover, it was demonstrated that a closed-loop speed control is needed. The disk drive is an example of a drive with high demands for speed stability of an accurately defined speed value. The requirement of a very small deviation from a constant velocity leads to a motor with a torque, which is to a large extent independent from the angular position of the rotor. A stepping motor does not easily fulfill this requirement.

9.7.2 Drive of a Dosing Piston Pump

A liquid is to dose in very small quantities with high accuracy by a piston pump. The filling and the flushing of the pump should be carried out in relatively short time. Furthermore, the drive system shall contain a motor that moves the piston with a screw thread (lead screw system).

Requirements of the Drive

- Filling volume of the pump $V_p = 60\,\text{cm}^3$
- Maximum of piston stroke $h_{k\,\text{max}} = 200\,\text{mm}$
- Piston diameter $d_k = 20\,\text{mm}$
- Minimum of dosing volume $V_{\text{min}} = 10\,\text{mm}^3$
- Dosing accuracy $|\Delta V_{\text{min}}/V_{\text{min}}| = 1.5\%$
- Maximum of dosing speed $v_{\text{dmax}} = 2000\,\text{mm}^3/\text{s}$
- Maximum of piston fill time $T_{\text{fmax}} = 4\,\text{s}$
- Maximum of force at the piston $F_{k\,\text{max}} = 40\,\text{N}$
- Piston operation with lead screw with a pitch of $h_s = 1.908\,\text{mm}$
- Mass of piston $m_k = 100\,\text{g}$
- Inertia of the lead screw $J_s = 65\,\text{gcm}^2$
- Density of the dosed liquid $\gamma_d = 1.5\,\text{g}/\text{cm}^3$
- Piston movement forward and backward
- Use in a chemical laboratory with aggressive gases

Investigation to Find the Best Active Principle

- Resolution
  From the minimum of dosing volume the step width h_{kS} of the piston can be calculated:

$$h_{kS} = \frac{V_{\text{min}} \cdot 4}{\pi \cdot d_k^2} = \frac{10\,\text{mm}^3 \cdot 4}{\pi \cdot 20^2\,\text{mm}^2} = 0.0318\,\text{mm} \, . \tag{9.35}$$

This step width is equal to an angular movement of the lead screw of

$$\theta_{sS} = \frac{h_{kS}}{h_s} = \frac{0.0318\,\text{mm} \cdot 360°}{1.908\,\text{mm}} = 6° \, . \tag{9.36}$$

- Maximum dosing frequency
  This results from the maximum of dosing speed together with the minimum of dosing volume:

$$f_{d\,\text{max}} = \frac{v_{d\,\text{max}}}{V_{\text{min}}} = \frac{2000\,\text{mm}^3/\text{s}}{10\,\text{mm}^3} = 200\,\text{Hz} \, . \tag{9.37}$$

- Maximum stepping rate during filling
  The piston must be filled completely within the maximum allowed fill time, i.e. the piston has to pass the full stroke. The required stepping rate is

defined by the piston stroke and the step width of the piston:

$$f_{f\,max} = \frac{h_{k\,max}}{T_{f\,max} \cdot h_{kS}} = \frac{200\,\text{mm}}{4\,\text{s} \cdot 0.0318\,\text{mm}} = 1572\,\text{Hz}\,. \qquad (9.38)$$

- Maximum lead screw speed
  The pitch of the screw and the step width of the piston give the number of steps for a complete revolution of the lead screw. The speed of the lead screw is defined by the maximum of stepping rate during filling divided by the number of steps for one revolution:

$$n_{s\,max} = \frac{f_{f\,max} \cdot h_{kS}}{h_s} = \frac{1572\,\text{Hz} \cdot 0.0318\,\text{mm}}{1.908\,\text{mm}} = 1572\,\text{min}^{-1}\,. \qquad (9.39)$$

- Load torque
  The load torque can be determined directly from the power (equation from Table 9.3 and 9.4) when the motor and the lead screw are directly coupled:

$$T_L = \frac{10^{-3} \cdot F_{k\,max} \cdot h_{k\,max}}{T_{f\,max} \cdot 1.05 \cdot 10^{-3} \cdot n_{s\,max}} = \frac{10^{-3} \cdot 40 \cdot 200}{4 \cdot 1.05 \cdot 10^{-3} \cdot 1572} = 1.21 \qquad (9.40)$$

T_L	$F_{k\,max}$	$h_{k\,max}$	$T_{f\,max}$	$n_{s\,max}$
$\text{cN} \cdot \text{m}$	N	mm	s	min^{-1}

- Dosing accuracy
  The required dosing accuracy is 1.5%, which cannot be directly fulfilled by a common stepping drive. Precision drives reach values of 2.2% without load. However, this requirement can be realized if two steps of a stepping motor are assigned to the smallest dosing volume:

$$\left|\frac{\Delta V_{min}}{V_{min}}\right| = \frac{1}{n} \cdot \left|\frac{\Delta \theta_S}{\theta_S}\right| = \frac{1}{2} \cdot 2.2\% = 1.1\%\,. \qquad (9.41)$$

With this assignment the maximum stepping rates are doubled, i.e. the stepping rate for dosing becomes $f_{d\,max} = 400\,\text{Hz}$ and the rate for filling $f_{f\,max} = 3144\,\text{Hz}$.

Decision

A stepping drive can realize the required parameters. Thereby the dosing will be done directly in pull-in or start-stop mode. It is necessary to use the pull-out range for the filling phase. The low dynamic requirements and the relatively small torque do not justify the use of an expensive position-controlled DC drive. A brushed motor cannot be used because of the application in a chemical laboratory with aggressive gases. Only a brushless motor would be useable, which avoids the arcing at the commutator completely.

Selection of the Structure of the Driving System

- Structure of the driving system
  The chosen structure of the driving system is shown in Fig. 9.11.
- Programmer
  The programmer selects the operating modes "Dosing" or "Filling", respectively, and controls the appropriate logic sequencer. It delivers the numerical dosing quantum to the counter and defines in the mode "Filling" the direction of movement. Both moving directions are necessary, when not only filling but also cleaning or rinsing of the pump is supported.
- Logic sequencer "Dosing"
  The logic sequencer "Dosing" generates a fixed frequency, which is equal to the maximum dosing rate. The destination signal is clearly defined, because dosing requires that the liquid is ejected. In Fig. 9.12a the frequency vs. time characteristic is shown. Different pulse width represents different dosing volumes.
- Logic sequencer "Filling"
  This logic sequencer delivers a variable frequency to implement a motion sequence for a "positioning operation within optimal time" (cp. Fig. 3.34). In Fig. 9.12b the motion is displayed with the current parameters. For the filling operation, the destination of movement is given, too. It is the opposite direction of the dosing operation. During rinsing of the pump the piston has to be repeatedly moved forward and backward.
- Stepping drive
  The operating characteristic of the chosen stepping drive B is shown in Fig. 9.13. Furthermore, the stepping drive offers the following important features:

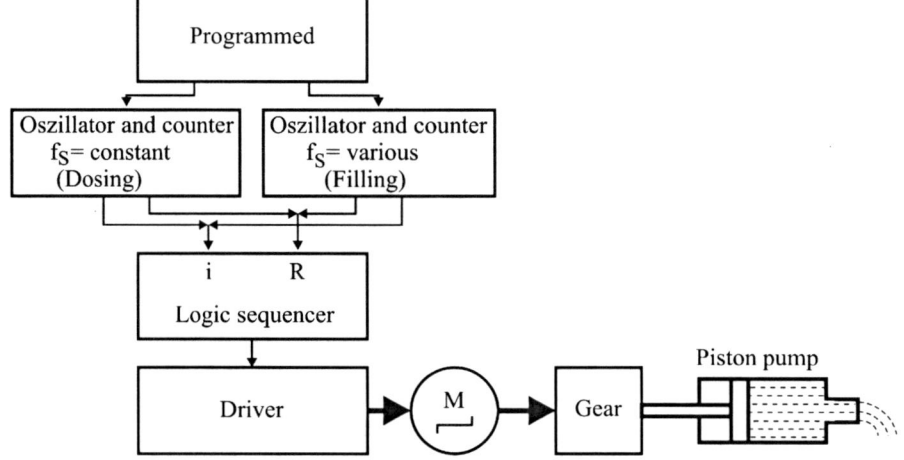

Fig. 9.11. Structure of a drive with a piston pump

step angle	$\theta_S = 7.5°$
number of steps per revolution	$z = 48$
inertia of the rotor in the motor	$J_r = 12.5\,\text{gcm}^2$
positioning accuracy	$\Delta\theta_S = 10'$

- Gear

  It is necessary to introduce a rotational gear, because of the given parameters of the lead screw pitch h_s, the required step width of the piston h_{kS}, and the step angle of the stepping motor θ_S. The rotational translation ratio of this gear between motor and lead screw is given as follows:

$$i_r = \frac{h_{kS} \cdot z}{h_s \cdot n} = \frac{0.0318\,\text{mm} \cdot 48}{1.908\,\text{mm} \cdot 2} = 0.4\,. \tag{9.42}$$

The total translation ratio between stepping motor and piston yields to:

$$i_t = \frac{h_{kS}}{n \cdot \theta_S} = \frac{0.0318\,\text{mm}}{2 \cdot 7.5°} = 2.12 \cdot 10^{-3}\,\text{mm}/°\,. \tag{9.43}$$

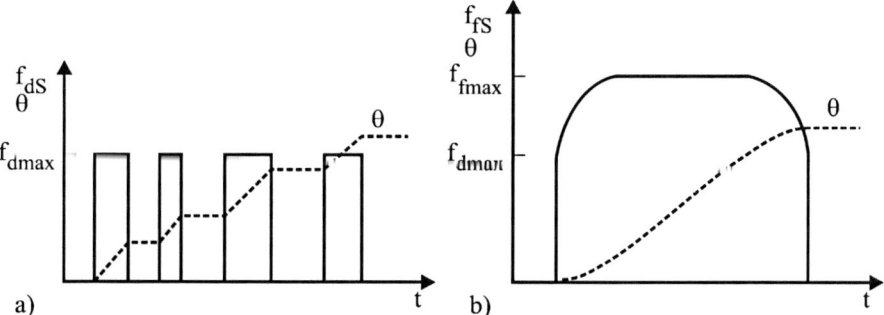

Fig. 9.12. Motion sequences of dosing piston pump. **a** Dosing. **b** Filling

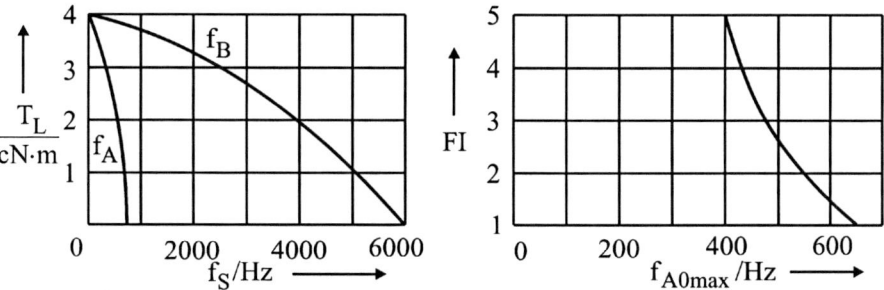

Fig. 9.13. Operating characteristics of stepping drive B. **a** Operating characteristic for factor of inertia $FI = 1$. **b** Correction curve of the pull-in range for factor of inertia $FI > 1$

Verification of Drive Parameters

- Dosing accuracy
  The positioning accuracy of the stepping motor is equal to:
  $$\left|\frac{\Delta\theta_S}{\theta_S}\right| = \frac{10'}{7.5°} = 2.2\% \ . \tag{9.44}$$
  With the assignment of two angular steps of a stepping motor to the minimum of the dosing volume, the required dosing accuracy of 1.5% can be met (cp. Eq. (9.41)). A prerequisite is an almost backlash-free coupling between the stepping motor, gear and piston.
- Maximum piston speed during dosing
  It is given by the maximum of the dosing speed and the dimensions of the piston pump:
  $$v_{kd} = \frac{v_{d\,max} \cdot 4}{\pi \cdot d_k^2} = \frac{2000\,\mathrm{mm}^3/\mathrm{s} \cdot 4}{\pi \cdot 20^2\,\mathrm{mm}^2} = 6.37\,\mathrm{mm/s} \ . \tag{9.45}$$
- Stepping rate during dosing
  During dosing the drive operates in the pull-in range. Its stepping rate results in:
  $$f_{d\,max} = \frac{v_{kd}}{i_t \cdot \theta_S} = \frac{6.37\,\mathrm{mm/s}}{2.12 \cdot 10^{-3}\,\mathrm{mm}/° \cdot 7.5°} = 400\,\mathrm{Hz} \ . \tag{9.46}$$
- Torque during dosing
  For the calculation of the torque according to Tables 9.3 and 9.4, a total efficiency of $\eta_g = 0.85$ is assumed, which leads to:
  $$T_L = 5.724 \cdot i_t \cdot \frac{1}{\eta_g} \cdot F_L = 5.724 \cdot 2.12 \cdot 10^{-3} \cdot \frac{1}{0.85} \cdot 40 = 0.57 \tag{9.47}$$

T_L	F_L
cNm	N

- Total moment of inertia of the load
  This consists of the individual inertia of the rotational gear, of the lead screw, of the piston and in the worst case of the liquid in a completely filled dosing cylinder:
  $$J_L = J_g + J_s' + J_k' + J_d' \ . \tag{9.48}$$
- Individual moments of inertia
  The individual moments of inertia have to be transformed to the motor shaft using the equation of Table 9.3 including the translation ratios according to Eq. (9.43).
  Rotational gear: assumptions $J_g = 0.2 J_r$; $\eta_g' = 0.95$
  $$J_g = 0.2 \cdot J_r = 0.2 \cdot 12.5\,\mathrm{gcm}^2 = 2.5\,\mathrm{gcm}^2 \tag{9.49}$$

lead screw

$$J'_s = i_r^2 \cdot \frac{1}{\eta'_g} \cdot J_s = 0.4^2 \cdot \frac{1}{0.95} \cdot 65\,\text{gcm}^2 = 10.95\,\text{gcm}^2 \qquad (9.50)$$

piston

$$J'_k = 32.79 \cdot i_t^2 \cdot \frac{1}{\eta_g} \cdot m_k = 32.79 \cdot (2.12 \cdot 10^{-3})^2 \cdot \frac{1}{0.85} \cdot 100\,\text{g} \cdot \text{cm}^2$$
$$= 0.02\,\text{gcm}^2 \qquad (9.51)$$

dosed liquid

$$J'_d = 32.79 \cdot i_t^2 \cdot \frac{1}{\eta_g} \cdot V_d \cdot \gamma_d = 32.79 \cdot (2.12 \cdot 10^{-3})^2 \cdot \frac{1}{0.85} \cdot 60 \cdot 1.5\,\text{g} \cdot \text{cm}^2$$
$$= 0.02\,\text{gcm}^2\,. \qquad (9.52)$$

Therewith the total moment of inertia of the load according to Eq. (9.48) results in $J_L = 13.5\,\text{gcm}^2$.
- Factor of inertia FI

$$FI = \frac{J_r + J_L}{J_r} = \frac{(12.5 + 13.5)\,\text{gcm}^2}{12.5\,\text{gcm}^2} = 2.1\,. \qquad (9.53)$$

- Maximum possible stepping rate for the pull-in region
According to Fig. 9.13, the pull-in stepping rate of stepping drive B amounts to approximately $f_A = 600\,\text{Hz}$ for a load torque of $T'_L = 0.57\,\text{cNm}$ and a factor of inertia of $FI = 1$. Using the correction curve of the pull-in range $FI = f(f_{A0max})$ for the current case of $FI = 2.1$ and Eq. (3.12), the maximum possible stepping rate for the drive system is given:

$$f_{A(2)} = f_A \cdot \frac{f_{A0\,\text{max}(2)}}{f_{A0\,\text{max}}} = 600\,\text{Hz} \cdot \frac{510\,\text{Hz}}{650\,\text{Hz}} = 470\,\text{Hz}\,. \qquad (9.54)$$

The maximum possible stepping rate for the start-stop region is higher than the required dosing rate of $f_{dmax} = 400\,\text{Hz}$. Thus the required dosing frequency is also feasible in the critical case.
- Total number of pulses for the maximum of piston stroke
It can be calculated from the complete filling volume of the pump, the minimum of dosing volume and the number of steps required for the minimum of dosing volume:

$$i_{max} = \frac{V_p \cdot n}{V_{min}} = \frac{60\,\text{cm}^3 \cdot 2}{10\,\text{mm}^3} = 12\,000\,. \qquad (9.55)$$

- Sequence of motions during filling

A motion sequence according to Fig. 9.12b is selected for the filling process. The acceleration and the braking period follow an exponential function. The function for the acceleration period is given as follows:

$$f_\mathrm{f}(t) = f_\mathrm{d\,max} + (f_\mathrm{f\,max} - f_\mathrm{d\,max}) \cdot (1 - e^{-t/\tau_\mathrm{a}}) \,. \tag{9.56}$$

The acceleration and deceleration period were not considered during the calculation of the already determined maximum stepping rate for filling according to Eq. (9.38). As a result of the required dosing accuracy, the smallest dosing volume was assigned with two steps. Taking into consideration these facts, a maximum stepping rate for filling of $f_\mathrm{fmax} = 5400\,\mathrm{Hz}$ is chosen and a time constant of $\tau_\mathrm{a} = 300\,\mathrm{ms}$ is assumed. The total number of pulses during the acceleration period can be determined to $i_\mathrm{a} = 1953$ by integration of Eq. (9.56).

The number of pulses i_B with the maximum stepping rate are given below, because acceleration and deceleration follow the same mathematical function:

$$i_\mathrm{B} = i_\mathrm{max} - 2 \cdot i_\mathrm{a} = 12000 - 2 \cdot 1953 = 8094 \,. \tag{9.57}$$

Therefore, the required time yields to:

$$T_\mathrm{B} = \frac{i_\mathrm{B}}{f_\mathrm{f\,max}} = \frac{8094}{5400\,\mathrm{Hz}} = 1.5\,\mathrm{s} \,. \tag{9.58}$$

Hence the total time for filling results in:

$$T = 2 \cdot T_\mathrm{a} + T_\mathrm{B} = 2 \cdot 1.2\,\mathrm{s} + 1.5\,\mathrm{s} = 3.9\,\mathrm{s} \,. \tag{9.59}$$

This time is smaller than the required time of $T_\mathrm{fmax} = 4\,\mathrm{s}$. The filling sequence has to be programmed as follows:
- start with a stepping rate of $f_\mathrm{dmax} = 400\,\mathrm{Hz}$
- enhance the stepping rate with a time constant of $\tau_\mathrm{a} = 300\,\mathrm{ms}$ during 1.2 s towards 5400 Hz
- initiate the braking period after

$$i = i_\mathrm{a} + i_\mathrm{B} = 1953 + 8094 = 10\,047\,\mathrm{pulses} \,. \tag{9.60}$$

- Advantageous for a safe stopping is an early start of the braking period, e.g. about 50 pulses earlier. Because the calculated time (9.59) is smaller than the required time, it is still ensured not to exploit completely the limit for the permissible fill time T_fmax. After arriving at the stop stepping rate, the drive moves slowly towards the end position.
- stop with the starting stepping rate $f_\mathrm{dmax} = 400\,\mathrm{Hz}$

Overall Evaluation

Stepping drive B fulfils the requirements. The dosing sequence is realized within the pull-in range of the motor. For filling and rinsing of the pump, a motion sequence with time optimal positioning is chosen to shorten the time spent. Therefore the drive is accelerated towards the pull-out range. The required dosing accuracy can be ensured by assignment of two angular steps of the motor to the smallest dosing volume.

9.7.3 Drive of a Drilling Machine for Printed Circuit Boards (PCBs)

Drill holes on a PCB are allocated to a fixed grid, whereas not all grid points have to be drilled. For machining, the PCB has to be moved underneath the drilling tool in x- and y-direction. In the example below, only one axis will be explained. The large number of holes in a PCB requires a very short cycle time. The cycle time includes transport time and processing time, where the workpiece rests. The driving motor moves the workpiece fixture with a lead screw system.

Requirements of the Drive

- Raster of drill grid $x_s = 2.5\,\text{mm}$
- Positioning accuracy $\Delta x_s = 0.05\,\text{mm}$
- Maximum of stroke $x_{\max} = 300\,\text{mm}$
- Lead screw pitch $h_s = 5\,\text{mm}$
- Cycle rate $c = 1000\,\text{min}^{-1}$
- Ratio of operation time (transport) to rest time (processing) $T_B{:}T_P = 2{:}1$
- Cycles per workpiece $c_w = 500$
- Total mass of work piece and its fixture $m_w = 2\,\text{kg}$
- Total inertia of lead screw including tacho-generator and incremental encoder $J_s = 50\,\text{gcm}^2$
- Maximum of force at the workpiece $F_w = 13\,\text{N}$

Investigation to Find the Best Active Principle

- Fragmentation of the drill grid
  The requirement of accuracy will be always fulfilled, if a single step of the drive is smaller than the required position accuracy. For the raster of the drill grid, i steps are necessary, which can be calculated as follows:

$$i = \frac{x_S}{\Delta x_S} = \frac{2.5\,\text{mm}}{0.05\,\text{mm}} = 50\,. \tag{9.61}$$

- Angular resolution of the lead screw
  The pitch of the lead screw and a single step together can be used to determine the angular movement of the lead screw per each single step.

If the single step is assumed with the required positioning accuracy, the angular movement per single step follows:

$$\theta_S = \frac{360° \cdot \Delta x_S}{h_s} = \frac{360° \cdot 0.05\,\text{mm}}{5\,\text{mm}} = 3.6° \ . \qquad (9.62)$$

- Cycle time
  The cycle time is given by the required cycle rate:

$$T = \frac{1}{c} = \frac{1}{1000\,\text{min}^{-1}} = 60\,\text{ms} \ . \qquad (9.63)$$

- Operation time and rest time
  Both can be determined from the cycle time and the given ratio of operation time to rest time:

$$T_B = \frac{2 \cdot T}{3} = \frac{2 \cdot 60\,\text{ms}}{3} = 40\,\text{ms} \qquad T_P = \frac{1 \cdot T}{3} = \frac{1 \cdot 60\,\text{ms}}{3} = 20\,\text{ms} \ . \qquad (9.64)$$

- Maximum of frequency
  The maximum of the stepping rate for the movement of the PCB within a raster of the grid can be determined from the number of steps per raster and the operation time:

$$f_{A\,\text{max}} = \frac{i}{T_B} = \frac{50}{40\,\text{ms}} = 1250\,\text{Hz} \ . \qquad (9.65)$$

- Angular movement of the lead screw per raster of the drill grid
  The angular resolution and the number of steps per raster of the drill grid give the angular movement:

$$\theta(x_S) = i \cdot \theta_S = 50 \cdot 3.6° = 180° \ . \qquad (9.66)$$

Decision

The possibility of using a stepping motor with a step angle of 3.6° or 1.8° directly coupled to the lead screw results from the determined angular resolution of the lead screw of $\theta_S = 3.6°$. The total stroke per raster of the drill grid will be executed with a relatively small angular movement of $\theta(x_S) = 180°$. This angle is clearly above the estimated limit for stepping motors, based on the comparison between stepping drive and position-controlled DC drive made in Sect. 9.5 that this angle is clearly above the estimated limit for stepping motors. Therefore, a position-controlled DC drive should be preferred in this application. Moreover, the product of $\theta_S f_{A\text{max}} = 4500°/\text{s}$ under load is very high. In Table 9.2, for high-quality stepping drives under no-load operation, a value of 5000°/s is given for the pull-in region. These points all together underline that a stepping drive is not best suited for the considered case. Hence a position-controlled DC drive is chosen.

Selection of the Structure of the Driving System

The chosen structure of the drive is shown in Fig. 9.14.

- Position controller
  The desired position x_{des} is fed as an input into the position controller. The desired value is the position of the next to process drill hole. It is compared with the actual position x_{act}. The output of the position controller is the analogue value of the desired speed n_{des}.
- Speed controller
  A motion sequence according to Fig. 9.15 without a period at constant speed is assumed, because the angular movement is small and the operation time is very short, too. The drive will accelerate until maximum speed n_{max} with a constant acceleration and afterwards brake again with a constant torque until standstill. The speed controller works with a large loop gain to reach short rise times. If not, the maximum of force F_w at the workpiece is needed, the maximum speed will be reached faster by the then larger acceleration. In this case, the motion sequence will switch over to a trapezoidal curve.
- Driver
  The driver takes over the supply of the servomotor with electrical power. Simultaneously it is a member of the current control circuit, which limits the current towards the maximum value I_{max}. This leads advantageously to smaller and cheaper electronic power components. The verification of the drive system will show which value of current limitation is permitted.

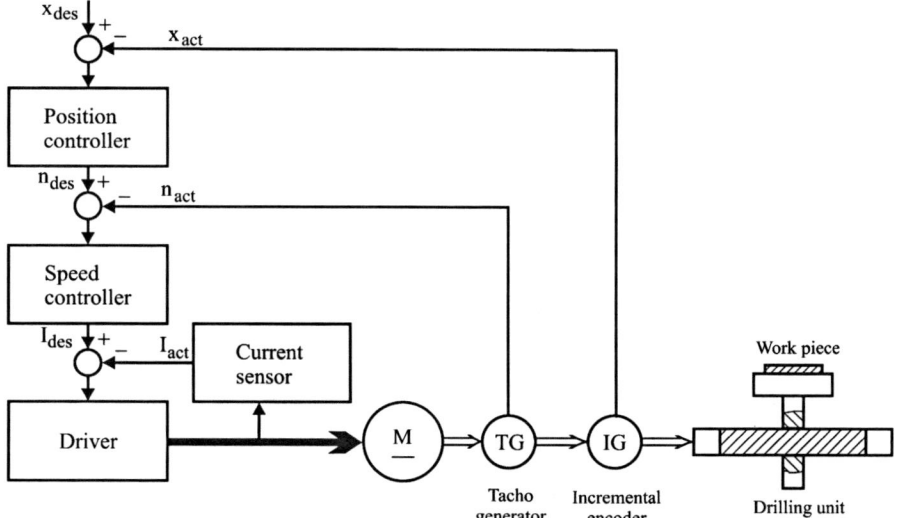

Fig. 9.14. Structure of the drive of a PCB drilling machine

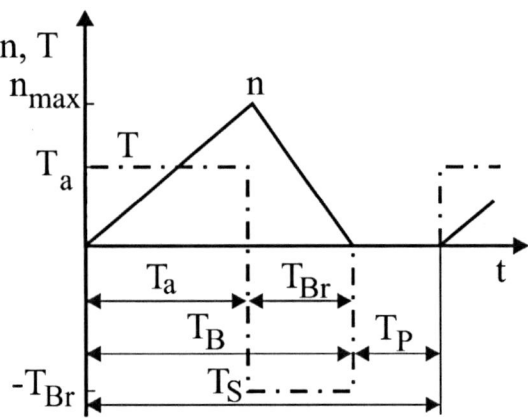

Fig. 9.15. Speed and torque vs. time characteristic during a single cycle time

- DC servo motor
  No special requirements with respect to lifetime or critical environmental parameters are demanded from the drive. Maintenance is always considered for devices or machines of this type. Therefore, a brushed DC motor can be used for this drilling unit. The required type of motion is a classical point-to-point control, i.e. the motion sequence between the points to approach do not need special requirements, e.g. a small tolerance regarding a constant velocity run. Consequently, the widespread used and economic type of a brushed DC motor with slotted armature can be chosen. The operating characteristic of the DC motor C is shown in Fig. 9.16. In the following section, other important parameters are given:
  - Rated voltage $U_N = 24$ V
  - Rated torque $T_N = 14.5$ cNm
  - Starting torque $T_a = 32.7$ cNm
  - No-load current $I_0 = 0.3$ A
  - Starting current $I_a = 6.0$ A
  - Inertia of the motor $J_r = 160$ gcm^2
  - Electrical time constant $\tau_{el} = 2$ ms

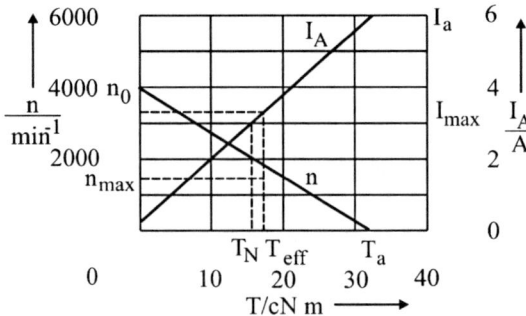

Fig. 9.16. Operating characteristic of the DC motor C

9 Project Design of Drive Systems

The tacho-generator as well as the incremental encoder could be integrated in the motor design.

- Tachometer generator
  The tacho-generator produces the actual speed value n_{act} and is carried out as a DC type.
- Incremental encoder
  The incremental encoder is used as a position transducer and delivers the actual position value x_{act}. The position x_{act} can be measured as an angular position of the rotor, if an almost backlash-free coupling between servo motor, incremental encoder and lead screw is assumed. Corresponding to the step angle of the lead screw of $\theta_S = 3.6°$, the transducer should purposefully deliver 100 pulses/2π or a multiple of it (increased positioning accuracy). If such a stiff and almost backlash-free coupling is not possible, then the linear movement has to be measured directly. Even in this case, the basic function of the position control circuit will remain unchanged.

Verification of Drive Parameters

- Translational transmission ratio
  In the case of a direct coupling between lead screw and motor the translational transmission ratio yields to:

$$i_t = \frac{\Delta x_S}{\theta_S} = \frac{0.05\,\text{mm}}{3.6°} = 0.01389\,\text{mm}/° \,. \tag{9.67}$$

- Load torque
  Assuming a gear efficiency of $\eta_g = 0.9$, the load torque converted to the motor shaft is given according to Tables 9.3 and 9.4:

$$T_L' = 5.724 \cdot i_t \cdot \frac{1}{\eta_g} \cdot F_w = 5.724 \cdot 0.01389 \cdot \frac{1}{0.9} \cdot 13 = 1.15\,\text{cNm} \tag{9.68}$$

T_L	i_t	F_w
cNm	mm/°	N

- Inertia of the work piece and its fixture
  It yields converted to the motor shaft to:

$$J_w' = 32.79 \cdot i_t^2 \cdot \frac{1}{\eta_g} \cdot m_w = 32.79 \cdot 0.01389^2 \cdot \frac{1}{0.9} \cdot 2000 = 14.1\,\text{gcm}^2 \,. \tag{9.69}$$

- Total inertia of the load
  The inertia of the lead screw is directly coupled to the motor shaft and must not be reduced by a conversion. The total inertia can be determined as:

$$J_L = J_s + J_w' = (50 + 14.1)\,\text{gcm}^2 = 64.1\,\text{gcm}^2 \,. \tag{9.70}$$

- Factor of inertia FI

$$FI = \frac{J_r + J_L}{J_r} = \frac{(160 + 64.1)\,\text{gcm}^2}{160\,\text{gcm}^2} = 1.4\ . \tag{9.71}$$

- Motor torque
  According to Fig. 9.15, a constant angular acceleration is assumed for the starting and the braking process. This presupposes a constant torque of the motor. Electrical transient phenomena will be neglected. The torque characteristic corresponds to Fig. 9.7. Therewith the required motor torque T can be calculated according to Eq. (9.16), if the following parameters are inserted:
  Total angle for a single motion sequence $\theta_{0\text{GM}} = i\theta_S$
  Total time for a single motion sequence $T_{\text{DC}} = T_B$
  Converting and dissolving of Eq. (9.16) leads to:

$$T^2 - \frac{\theta_S \cdot i \cdot J_r \cdot FI \cdot 4}{T_B^2} - T_L'^2 = 0 \tag{9.72}$$

$$T = 17.7\,\text{cNm}\ . \tag{9.73}$$

Thereto a current of $I_{\max} = 3.39 A$ belongs according to Fig. 9.16:

- Accelerating torque
  The torque corresponding to Fig. 9.15 can be calculated using Eqs. (9.8) and (9.9):

$$T_a = T - T_L' = (17.7 - 1.15)\,\text{cNm} = 16.55\,\text{cNm} \tag{9.74}$$

$$T_{Br} = T + T_L' = (17.7 + 1.15)\,\text{cNm} = 18.85\,\text{cNm}\ . \tag{9.75}$$

- Accelerating time
  According to Eq. (9.13) the following is effective:

$$T_a = 0.5 \cdot \left(1 + \frac{T_L'}{T}\right) \cdot T_B = 0.5 \cdot \left(1 + \frac{1.15\,\text{cNm}}{17.7\,\text{cNm}}\right) \cdot 40\,\text{ms} = 21.3\,\text{ms}\ . \tag{9.76}$$

- Braking time
  The braking time can be calculated as the difference between operation and acceleration time:

$$T_{Br} = T_B - T_a = (40 - 21.3)\,\text{ms} = 18.7\,\text{ms}\ . \tag{9.77}$$

- Maximum speed

$$\Delta n = n_{\max} = \frac{\Delta \Omega}{2 \cdot \pi} = \frac{T - T_L'}{J_r \cdot FI \cdot 2 \cdot \pi} \cdot T_a$$
$$= \frac{(17.7 - 1.15)\,\text{cNm}}{160\,\text{gcm}^2 \cdot 1.4 \cdot 2 \cdot \pi} \cdot 21.3\,\text{ms} = 1503\,\text{min}^{-1}\ . \tag{9.78}$$

- Operating mode
  According to [10.21], the motion sequence corresponding to Fig. 9.15 belongs to the duty type S5. With this duty type starting, running with constant torque (this part is zero in our special case), braking, and rest period influence the heating of the motor. When the required motor torque of $T = 17.7\,\mathrm{cNm}$ is compared with the thermal acceptable rated torque of $T_\mathrm{N} = 14.5\,\mathrm{cNm}$, it turns out that a continuous operation is not permitted (cp. Sect. 9.3.2). If no data about the different types of duty other than continuous running are given, the effective torque has to be identified, which determines the permitted temperature rise [9.10, 9.11, 9.12] and [10.21]. This is the case here. Current and torque are proportional with a DC motor. The losses and the temperature rise follow the square of the current. Therewith the subsequent equation results:

  $$T_\mathrm{N}^2 \cdot T = T_\mathrm{ef}^2 \cdot T_\mathrm{B} \tag{9.79}$$

  $$T_\mathrm{ef} = T_\mathrm{N} \cdot \sqrt{\frac{T}{T_\mathrm{B}}} = 14.5\,\mathrm{cNm} \cdot \sqrt{\frac{60\,\mathrm{ms}}{40\,\mathrm{ms}}} = 17.76\,\mathrm{cNm}\ . \tag{9.80}$$

  The effectively permitted torque T_ef is practically equal to the required torque $T = 17.7\,\mathrm{cNm}$. This approves the statement made above that the current should be advisably limited. The current limitation restricts at the same time the generated torque. The maximum speed of $n = 1835\,\mathrm{min}^{-1}$ related to this torque will be not reached during the positioning process (cp. Fig. 9.16).

- Mechanical time constant
  It can be determined from the operating characteristic. For the unloaded motor, the following is true:

  $$\tau_\mathrm{m} = \frac{\Omega_0 \cdot J_\mathrm{r}}{T_\mathrm{a}} = \frac{2 \cdot \pi \cdot 4000\,\mathrm{min}^{-1} \cdot 160\,\mathrm{gcm}^2}{32.7\,\mathrm{cNm}} = 20.5\,\mathrm{ms}\ , \tag{9.81}$$

  and for the complete drive

  $$\tau'_\mathrm{m} = \tau_\mathrm{m} \cdot FI = 20.5\,\mathrm{ms} \cdot 1.4 = 28.7\,\mathrm{ms} \tag{9.82}$$

  is valid.

- Current vs. time function
  For all the previous mentioned and the above-determined values it was assumed that the electrical transient behavior has no significant influence ($\tau_\mathrm{el} \ll \tau'_\mathrm{m}$). The ohmic resistance of the motor can be calculated from the starting current neglecting the voltage drop across the brushes:

  $$R = \frac{U_\mathrm{A}}{I_\mathrm{a}} = \frac{24\,\mathrm{V}}{6\,\mathrm{A}} = 4\,\Omega\ . \tag{9.83}$$

  Thereof the inductance is given:

  $$L = R \cdot \tau_\mathrm{el} = 4\,\Omega \cdot 2\,\mathrm{ms} = 8\,\mathrm{mH}\ . \tag{9.84}$$

At rated voltage, and neglecting the brushes voltage drop, i.e. $U'_A = U_N$, the motor current yields to:

$$i_A(t) = \frac{U_N \cdot \tau_{el} \cdot \tau_m}{L \cdot (\tau_m - \tau_{el})} \cdot (-e^{-t/\tau_{el}} + e^{-t/\tau_m}) \tag{9.85}$$

$$i_A(t) = \frac{24\,\text{V} \cdot 2\,\text{ms} \cdot 28.7\,\text{ms}}{8\,\text{mH} \cdot (28.7 - 2)\,\text{ms}} \cdot (-e^{-t/2\,\text{ms}} + e^{-t/28.7\,\text{ms}}) \tag{9.86}$$

$$= 6.45\,\text{A} \cdot (-e^{-t/2\,\text{ms}} + e^{-t/28.7\,\text{ms}}).$$

By dissolving Eq. (9.85) a time of 1.8 ms can be obtained, after which the current reaches its maximum value. Here the current limitation becomes effective and for this reason, the assumption is justified that the electrical transient phenomena could be neglected.

Overall Evaluation

The verification of the parameters has proven that the requirements could be fulfilled with the chosen DC motor. The high cycle rate of $1000\,\text{min}^{-1}$ can be realised using the duty type S5. A speed and a current control loop are subordinated to the position control circuit.

The selected drive is an example of a servo drive, which operates with a very short cycle time in a start-stop mode. The required high dynamics was decisive for the choice of a position-controlled DC motor.

9.7.4 Drive of a Light Pointer

When during production of printed circuit boards (PCBs) the components have to be placed manually, their correct position should be marked. A light is placed above the PCB emitting a focused beam. The light is moved with a lead screw system by a rotary motor. To mark the required PCB positions, a resolution of 2.5 mm in the x- and y-direction is needed. The location of the component should be shown with a flicker-free light band. Depending on the type of component, different light figures such as straight lines, triangles and rectangles are required. Two drives operate in the x- and y-direction in reversing duty, whereas the path length and speed vary. In the following section only one axis is examined.

Requirements of the Drive

- Raster of the PCB grid $x'_L = 2.5\,\text{mm}$
- Positioning accuracy $\Delta x'_L = 1,0\,\text{mm}$
- Reversing duty, for a distance covering 5 grid points ($4x'_L = 10\,\text{mm}$) a minimum frequency of $f_{min} \geq 10\,\text{Hz}$ is required
- Lead screw pitch $h_s = 5\,\text{mm}$
- Inertia of the screw spindle $J_s = 20\,\text{gcm}^2$
- Mass of the light fixture $m_L = 90\,\text{g}$
- Maximum of friction torque at the screw spindle $T_f = T_L = 0.5\,\text{cNm}$

Investigation to Find the Best Active Principle

- Reversing duty
  With these requirements it is not recommended to use an oscillating drive system, because of the strongly differing amplitudes and the operation at different positions. Oscillating drive systems are characterised by a single resonant operation frequency and preferably constant amplitudes.
  Nevertheless, the tendency of a positioning drive to oscillate should be used for an effective energy conversion as much as possible. The light band is composed from n path elements x'_L depending on the type of component. Therefore an optimal motion has to be found for the path elements x'_L as the fundamental movement. Depending on the required total path length, n, fundamental movements will be composed together.
- Time per raster step
  It can be calculated from the fundamental requirement for a flicker-free image:

$$T_L = \frac{1}{f_{SL}} = \frac{1}{4 \cdot f_{min}} = \frac{1}{4 \cdot 10\,\text{Hz}} = 25\,\text{ms} . \tag{9.87}$$

- Travel of the light fixture per raster step (travel of the nut)
  Figure 9.17 shows the structure of the drive system and its basic components.

$$x_L = x'_L \cdot \frac{200\,\text{mm}}{960\,\text{mm}} = 2.5\,\text{mm} \cdot \frac{200\,\text{mm}}{960\,\text{mm}} = 0.521\,\text{mm} . \tag{9.88}$$

- Angular movement of the screw spindle per raster step
  The travel of the light and the pitch of the screw determine the spindle movement. The angle per raster step is given by:

$$\theta_L = \frac{x_L}{h_s} = \frac{0.521\,\text{mm} \cdot 360°}{5\,\text{mm}} = 37.5° . \tag{9.89}$$

- Relative positioning accuracy
  It is not so critical with $1.0\,\text{mm}/2.5\,\text{mm} = 40\%$.

Decision

The use of a stepping drive is recommended according to the moderate requirements, e. g. the low positioning accuracy, the small torque and the slow movement (the product $\theta_L f_{SL} = 1500°/s$ is small). A strong advantage of the stepping drive is the open-loop operation directly controlled by the computer without any sensor during retracing of the grid points.

Selection of the Structure of the Driving System

- Microcontroller – Oscillator
  The stepping rate might be constant, so therefore only a simple oscillator is required or this function is carried out from the microcontroller itself. The

microcontroller determines in any case the position to approach and the length of the light beam with the number i of pulses as well as the reversing points with the direction signal R. It can take over all the functions inside the dashed frame of Fig. 9.17.

- Stepping drive
  It is not possible to design a stepping motor with a step angle of 37.5°. However, a stepping motor with a step angle of $\theta_S = 7.5°$ can be used. In this case, the fundamental movement x_L will comprise $n = 5$ steps. Stepping motor B is suitable for this task. Its data was already given in Sect. 9.7.2.

Verification of Drive Parameters

- Translational transmission ratio
  With a screw spindle movement of $\theta_L = 37.5°$ the nut in the light fixture will travel $x_L = 0.521$ mm. Out of it the translational transmission ratio can be determined:

$$i_t = \frac{x_L}{\theta_L} = \frac{0.521 \text{ mm}}{37.5°} = 0.01389 \text{ mm}/° \; . \tag{9.90}$$

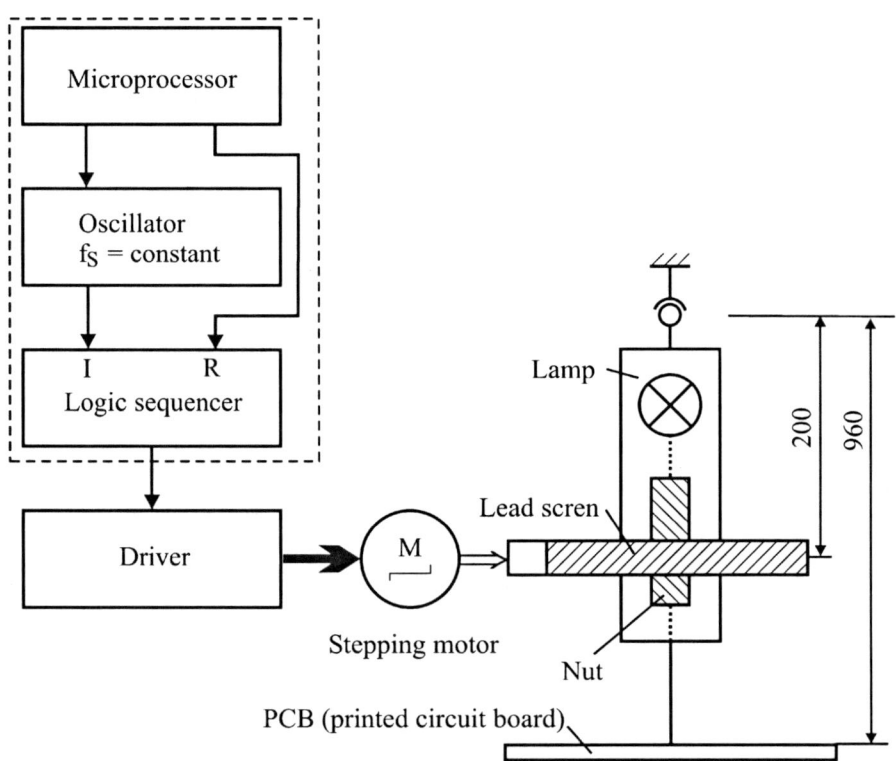

Fig. 9.17. Structure of the drive of the light pointer

- Inertia of the light fixture
  The light fixture rotates at the pivotal point. As a simplification is supposed that the light is moving linear with its total mass, because the angle of rotation per raster step is very small. Under the assumption of a gear efficiency $\eta_g = 0.85$ the inertia converted towards the motor shaft is given:

$$J'_L = 32.79 \cdot i_t^2 \cdot \frac{1}{\eta_g} \cdot m_L = 32.79 \cdot 0.01389^2 \cdot \frac{1}{0.85} \cdot 90 = 0.67 \, \text{gcm}^2 \,. \tag{9.91}$$

- Total inertia of the load
  The inertia of the screw spindle has not to be converted, because it is directly coupled to the motor shaft.

$$J_L = J_s + J'_L = (20 + 0.67) \, \text{gcm}^2 = 20.67 \, \text{gcm}^2 \,. \tag{9.92}$$

  Practically, the inertia of the light fixture converted to the motor shaft has no influence. Therefore the assumption made above about the linear movement of the light fixture proved to be allowed.
- Factor of inertia FI

$$FI = \frac{J_r + J_L}{J_r} = \frac{(12.5 + 20.67) \, \text{gcm}^2}{12.5 \, \text{gcm}^2} = 2.65 \,. \tag{9.93}$$

- Natural frequency
  According to Eq. (3.11), together with the torque constant of the motor in the point of origin of $k = 0.571 \, \text{cNm}/°$, the natural frequency is given:

$$\omega_0 = \sqrt{\frac{k}{J_r \cdot FI}} = \sqrt{\frac{0.571 \, \text{cNm}/°}{12.5 \, \text{gcm}^2 \cdot 2.65}} = 313.7 \, \text{s}^{-1} \,. \tag{9.94}$$

- Phase portrait
  Using digital simulation, the phase portrait of a free-oscillating stepping motor can be computed (cp. Sect. 3.5.2). The numerical solution will be not described in place. Free-oscillating or undamped stepping motor means, that the friction torque of the screw spindle of $T'_L = 0.5 \, \text{cNm}$ (12.5% related to $T_{\max}$) is not considered in the first instance. The result of this computation is shown in Fig. 9.18.
  Reference value for the axis of abscissa is the step angle $\theta_S = 7.5°$. For the axis of ordinates a value of $\theta_S \omega_0 = 2352.7°/\text{s}$ is given taking into account the factor of inertia $FI = 2.65$. For the purpose of clarity, the time as parameter was not marked out in the phase portrait.
- Sequence of motions
  First time period: At the time instant $t_0 = 0$ the stepping motor starts with the initial values $\theta/\theta_S(0) = -1$ and $d\theta/dt(0) = 0$. It follows the phase curve until time instant $t_1 = 3.8 \, \text{ms}$.

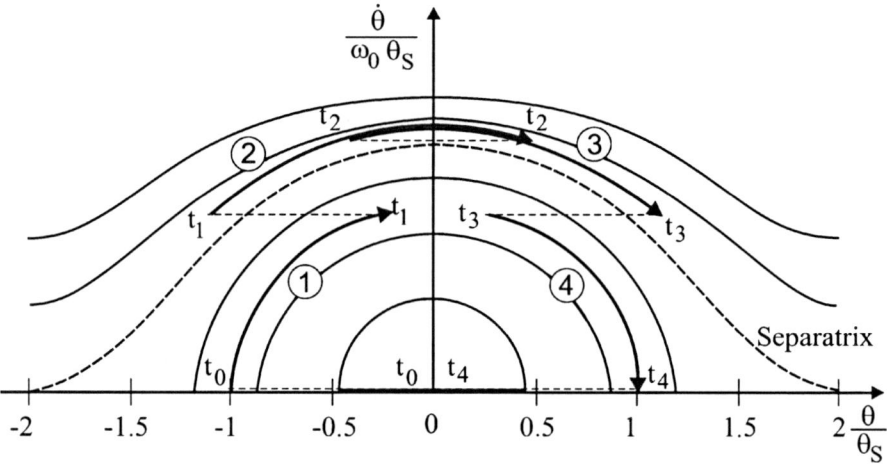

Fig. 9.18. Phase portrait of the undamped stepping motor B. The numbers 1...4 characterise the sequence of motions in the individual time periods during each raster step, t_n are the switching instants at arrival of a pulse i

Second time period: The instructions according to Eqs. (3.24) and (3.25) imply a shifting of the initial value of the deflection angle about the value of $\theta/\theta_S = -1$ to the left-hand side. Since the speed does not change, this results in a horizontal line. In this time period, the stepping drive runs beyond the separatrix until the instant $t_2 = 7.6$ ms.

Third and fourth time period: The stepping drive approaches again the abscissa axis, i.e. It will be decelerated. At the time instant $t_4 = 15.2$ ms, it reaches the angle $\theta/\theta_S = 1$. The speed will be zero.

Fifth time period: The stepping drive is already at zero speed. The switching towards the turbulence point means that it will stop without any oscillations.

In Fig. 9.19, all single sequences of motion are put together to the entire motion sequence. It can be realised, that the high velocity beyond the separatrix will cause that the travel of exactly five angle steps is already completed after four pulses. The fifth pulse secures that the stepping motor will stop without any oscillation. Without the fifth pulse it would execute undamped oscillations with the amplitude of $\theta/\theta_S = 1$ around the point of turbulence.

The backward motion in the phase portrait is a mirror-inverted characteristic below the abscissa axis.

According to Fig. 9.19 the achieved time per raster step yields to:

$$T_L = 5 \cdot T_i = 5 \cdot 3.8 \text{ ms} = 19 \text{ ms} , \tag{9.95}$$

so that it is smaller than the required. The precise verification considering damping and the current vs. time function during commutation of the windings resulted in a time of 21.5 ms per raster step.

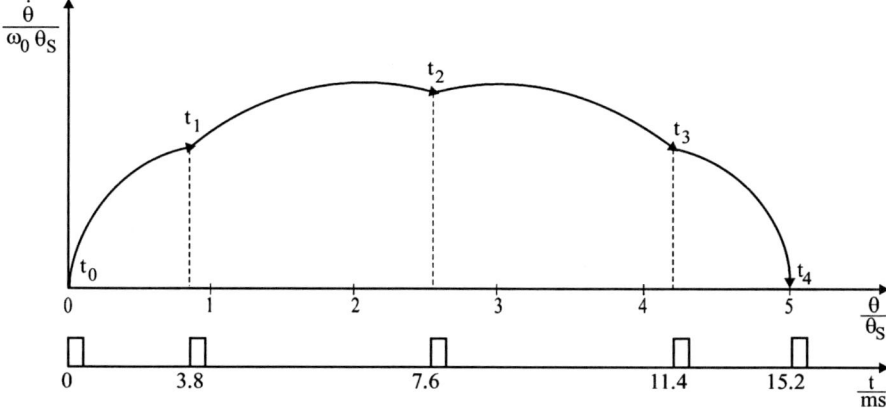

Fig. 9.19. Phase curves of a light pointer during a single raster step. The phase curve results from combining all segments of Fig. 9.18 with each other. The time scale is nonlinear and should help only for orientation

Overall Evaluation

The verification has shown that a reversing drive with strongly different amplitude can be advantageously realised with a stepping motor. The entire travel will be composed of absolute equal and self-contained motion sequences. Thereby any total amplitude or travel can be generated. The control is based on a fixed stepping rate. This example illustrates how a phase portrait can help to work out a descriptive solution.

9.7.5 Drive of a Flexible-Tube Pump or Peristaltic Pump

A liquid is applied on a strip material running with variable speed. The thickness of the coating is adjustable in grades. However, the thickness has to be independent of the changing speed of the tape, which is caused by the production technology. The strip material in some cases may be very long and therefore a continuous delivering pump is used for the coating liquid. The delivery volume or liquid quantity is precisely dosed with a flexible-tube pump, where revolving rollers deform a flexible hose in such a way that defined chambers transport a special amount of the liquid (see Fig. 9.21).

Requirements of the Drive

- Speed of the tape driving wheel $n_1 = 1000\ldots 3000\,\text{min}^{-1}$
- Speed of the tape or strip material, respectively, $v_\text{L} = 1\ldots 3\,\text{m/s}$
- Maximum of tape speed variation $\frac{dv_\text{L}}{dt}\big|_\text{max} = 0.9\,\frac{\text{m}}{\text{s}^2}$
- Tape width $b = 1\,\text{m}$
- Coating thickness $d = 0.05\ldots 0.25\,\text{mm}$
- Gradation of the coating thickness $\Delta d = 0.05\,\text{mm}$

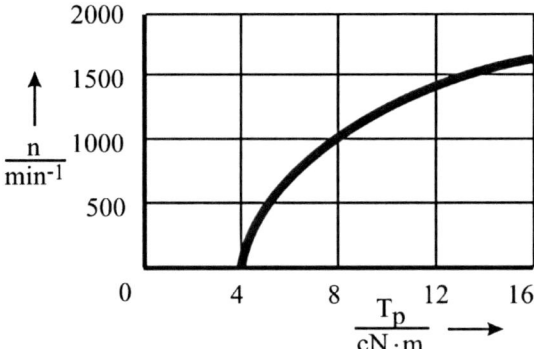

Fig. 9.20. Operating characteristic of the peristaltic pump

- Density of the coating material $\gamma = 1.2\,\text{g}/\text{cm}^3$
- Delivery volume of the peristaltic pump $V_p = 36\,\text{cm}^3/2\pi$
- Operating characteristic of the peristaltic pump (see Fig. 9.20)
- Inertia of the peristaltic pump with complete filling $J_p = 85\,\text{gcm}^2$
- One rotational direction of the peristaltic pump

Investigation to Find the Best Active Principle

- Speed adjusting range of the peristaltic pump
  It can be determined from the speed range of the tape and the span of the coating thickness:

$$\frac{n_{p\,\text{max}}}{n_{p\,\text{min}}} = \frac{v_{L\,\text{max}}}{v_{L\,\text{min}}} \cdot \frac{d_{\text{max}}}{d_{\text{min}}} = \frac{3\,\text{m/s}}{1\,\text{m/s}} \cdot \frac{0.25\,\text{mm}}{0.05\,\text{mm}} = 15\,. \qquad (9.96)$$

- Maximum delivery volume for the coating process
  It is resulting from the maximum speed of the tape and the maximum coating thickness to apply:

$$V_{\text{max}} = v_{L\,\text{max}} \cdot b \cdot d_{\text{max}} = 3\,\text{m/s} \cdot 1\,\text{m} \cdot 0.25\,\text{mm} = 750\,\text{cm}^3/\text{s}\,. \qquad (9.97)$$

- Maximum speed of pump
  It is given by the greatest amount of delivery volume for the coating process and the delivery volume of the pump per revolution:

$$n_{p\,\text{max}} = \frac{V_{\text{max}}}{V_p} = \frac{750\,\text{cm}^3/\text{s}}{36\,\text{cm}^3} = 1250\,\text{min}^{-1}\,. \qquad (9.98)$$

- Minimum speed of pump
  The smallest required speed of the pump can be calculated from the maximum speed and the speed adjusting range of the peristaltic pump:

$$n_{p\,\text{min}} = n_{p\,\text{max}} \cdot \frac{n_{p\,\text{min}}}{n_{p\,\text{max}}} = 1250\,\text{min}^{-1} \cdot \frac{1}{15} = 83.3\,\text{min}^{-1}\,. \qquad (9.99)$$

- Maximum power of the peristaltic pump
  For a speed of $1250\,\text{min}^{-1}$ a torque of $T_p = 10\,\text{cNm}$ can be readout from the diagram of Fig. 9.20. According to Table 9.3, and using the units of Table 9.4, the power in W can be calculated:

$$P_{p\,\text{max}} = 1.05 \cdot 10^{-3} \cdot T_p \cdot n_{p\,\text{max}} = 1.05 \cdot 10^{-3} \cdot 10 \cdot 1.25 = 13.1\,\text{W}\,. \tag{9.100}$$

Decision

Torque, speed itself, and speed adjustment range are relatively small in the given task. Therefore, the decision was made to use a stepping drive. A speed proportional signal for control of the stepping motor has to be derived from the driving wheel of the tape. Therewith the absolute synchronism is ensured between tape speed, speed of the peristaltic pump, delivery volume, and finally coating thickness.

Selection of the Structure of the Driving System

In Fig. 9.21, the chosen structure of the drive of the peristaltic pump is shown. The individual components are described in detail in the following.

- Incremental encoder
  The incremental encoder is rigidly coupled to the driving wheel and generates a signal, which is directly proportional to the tape speed. The selection of pulse numbers per turn is made in a way, that the signal can be used

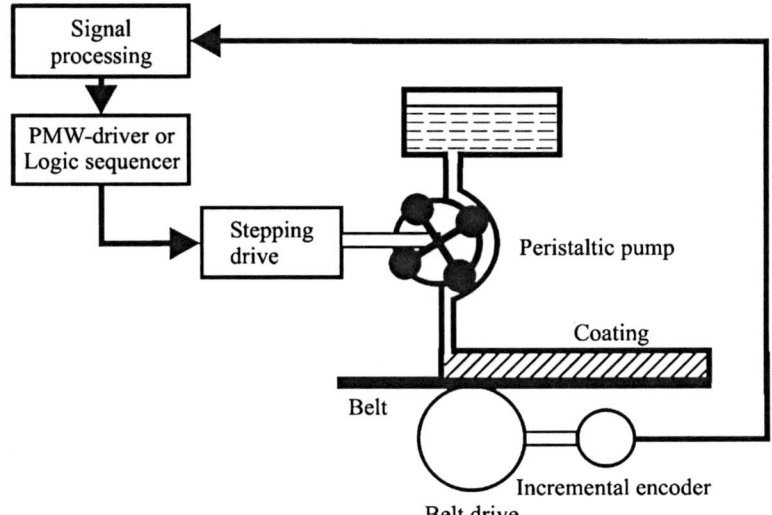

Fig. 9.21. Structure of the drive of the peristaltic pump

directly to control the stepping motor or at least only a simple pulse conditioning is needed, e.g. an additional frequency division. Only a single pulse sequence is necessary, because just one rotational direction of the tape and the pump is existent. A pulse number of $i = 400/2\pi$ was chosen. The number of pulses per turn and the maximum speed of the tape driving wheel give the maximum signal frequency:

$$f_{\text{meß max}} = i \cdot n_{1\,\text{max}} = 400 \cdot 3000\,\text{min}^{-1} = 20\,\text{kHz}\,. \tag{9.101}$$

- Stepping motor
  The number of steps of the stepping motor should have a fixed ratio to this maximum signal frequency. The natural starting point should be the maximum speed of the peristaltic pump. Firstly the frequency divider ratio i_{el} is arbitrary. Therefore, it is essential:

$$z = \frac{1}{i_{\text{el}}} \cdot \frac{f_{\text{meß max}}}{n_{\text{p max}}} = \frac{1}{i_{\text{el}}} \cdot \frac{20\,\text{kHz}}{1250\,\text{min}^{-1}} = \frac{960}{i_{\text{el}}}\,. \tag{9.102}$$

With a fixed ratio of the frequency divider of $i_{\text{el}} = 20$, a stepping motor can be used with $z = 48$ or $\theta_S = 7.5°$, respectively. The operating characteristic of the chosen stepping motor D is shown in Fig. 9.22. More important data of the stepping drive is given in the following:

Holding torque $\quad T_H = 20\,\text{cNm}$
Positioning accuracy $\quad \Delta\theta_S = 15'$
Inertia of rotor $\quad J_R = 70\,\text{gcm}^2$
Resonant frequencies $\quad f_{\text{res}} = 25/75\,\text{Hz}$

- Frequency divider
  The frequency divider has to support the fixed divider ratio of $i_{\text{el}} = 20$ as well as a frequency switching to achieve different values of the coating thickness. Therefore, the following divider ratios are necessary:

Coating thickness d (in mm)	0.05	0.10	0.15	0.20	0.25
Divider ratio f_{mess}/f_1	100	50	33.3	25	20

- Logic sequencer
  The direction signal R is not to switch, because only one direction of rotation is needed. Thus it is connected to a constant voltage potential.

Verification of Drive Parameters

- Maximum stepping rate
  The rotatory transmission ratio i_R is equal to 1, because of the direct coupling between stepping motor and peristaltic pump. According to the

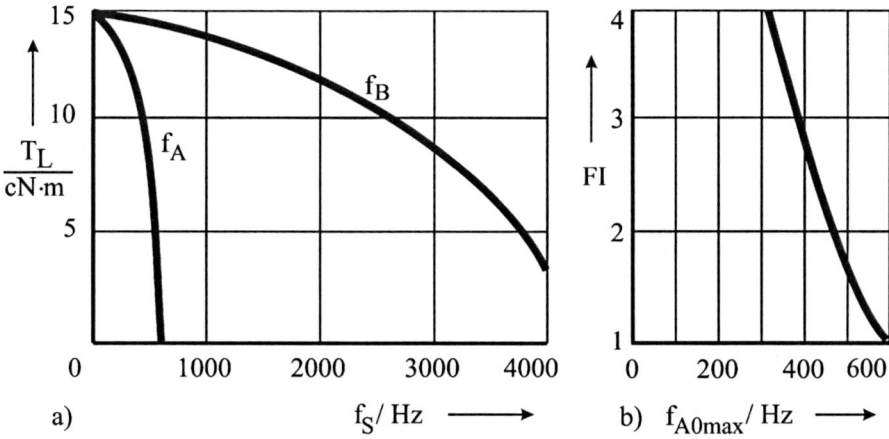

Fig. 9.22. Operating characteristic of the stepping motor D. **a** Operating characteristic for factor of inertia $FI = 1$. **b** Correction curve of the pull-in range for factor of inertia $FI > 1$.

equation of Table 9.3 and using the units of Table 9.4 the maximum of stepping rate in Hz is given:

$$f_{S\,\max} = \frac{6 \cdot n_{p\,\max}}{i_R \cdot \theta_S} = \frac{6 \cdot 1250}{1 \cdot 7.5} = 1000\ . \tag{9.103}$$

At a pump torque of $T_p = 10\,\text{cNm}$ this stepping rate is located in the pull-out range of the stepping drive D.

- Minimum stepping rate
  Likewise, the minimum of the stepping rate in Hz is calculated using the lowest speed of the pump:

$$f_{S\,\min} = \frac{6 \cdot n_{p\,\min}}{i_R \cdot \theta_S} = \frac{6 \cdot 83.3}{1 \cdot 7.5} = 66.7\ . \tag{9.104}$$

- Factor of inertia FI
  According to Eq. (3.10):

$$FI = \frac{J_R + J_P}{J_R} = \frac{(70 + 85)\,\text{gcm}^2}{70\,\text{gcm}^2} = 2.21\ . \tag{9.105}$$

- Natural frequencies
  The resonant frequencies of the stepping motor will be reduced by the external moments of inertia. The highest resonant frequency after this reduction can be calculated using Eq. (3.14):

$$f_{Res} = \frac{f_{Res0}}{\sqrt{FI}} = \frac{75\,\text{Hz}}{\sqrt{2.21}} = 50.5\,\text{Hz}\ . \tag{9.106}$$

This higher natural frequency shows a large enough distance to the smallest operating frequency f_{Smin}. Therefore, no instability is expected in the used frequency range.

- Maximum of frequency variation
  The maximum stepping rate f_{Smax} is located in the pull-out range. Since the frequency variation is strongly limited in the pull-out range, it must be examined closer. The most critical operating area is characterised by the maximum tape speed variation and maximum stepping rate. In this case the maximum of frequency variation of the stepping drive is given:

$$\left.\frac{\mathrm{d}f_S}{\mathrm{d}t}\right|_{\max} = \left.\frac{\mathrm{d}v_L}{\mathrm{d}t}\right|_{\max} \cdot \frac{f_{S\max}}{v_{L\max}} = 0.9\,\mathrm{m/s}^2 \cdot \frac{1000\,\mathrm{Hz}}{3\,\mathrm{m/s}} = 300\,\mathrm{Hz/s}\,. \quad (9.107)$$

A maximum pull-in rate $f_{\text{A0max}} = 430\,\text{Hz}$ can be derived from Fig. 9.22b for a factor of inertia $FI = 2.21$; for comparison the maximum pull-in rate f_{A0max} is $600\,\text{Hz}$ for $FI = 1$. According to Fig. 9.22a, the pull-in rate f_A is equal to $470\,\text{Hz}$ at a load torque of $T_\text{p} = 10\,\text{cNm}$ and $FI = 1$. Corresponding to Eq. (3.12), the maximum pull-in rate for the load torque $T_\text{p} = 10\,\text{cNm}$ and a factor of inertia $FI = 2.21$ can be determined to:

$$f_{A(2)} = f_A \cdot \frac{f_{A0\max(2)}}{f_{A0\max}} = 470\,\mathrm{Hz} \cdot \frac{430\,\mathrm{Hz}}{600\,\mathrm{Hz}} = 337\,\mathrm{Hz}\,. \quad (9.108)$$

The effectively permitted maximum pull-in rate $f_{A(2)}$ exceeds the maximum frequency variation according to Eq. (9.107). Hence the stepping drive will follow the change of the tape speed without faults, i.e. no step loss occurs.

Overall Evaluation

With the verification was shown that the stepping motor D could be used for this application. Furthermore, the presented drive is a sample of a path or trajectory control using a stepper motor. The stepping drive has to synchronously follow a given variable motion. At the same time this is an example of an electronic gear, which was already described in Sect. 3.5.1. Here an absolute synchronism prevails between two rotating shafts, although both are separated in space and can have any position to each other. Additionally, the transmission ratio can be changed electronically due to process requirements instead of using an expensive mechanical indexing gearbox.

9.7.6 Bowden Wire Drive of a Recording Device

Physical or computed quantities are often displayed in the form of a diagram to support an easy interpretation. In such a case, a plotter or x–y-recorder are used, which are generally equipped with voltage inputs. These inputs can be switched within several voltage ranges using internal amplifiers, to allow the adaptation towards the chosen diagram quantities. If the recording process is examined from a drive-dimensioning perspective, the input parameters of these drives will be within small limits.

In the following, only one axis of such a recording device is considered. The second axis can be handled alike. The given task is to develop a drive system that moves a recording carriage with pen using a Bowden wire. The input quantities of the drive are provided by a converter that conditions the input values of the recording device in an appropriate manner. The drive has to be controlled with a position control in the special shape of a trajectory control, because time-variant signals have to be registered. The variable input signal should be fixed each particular time as accurate as possible.

Requirements of the Drive

- Maximum recording length $x_L = 450$ mm
- Static accuracy $x_{SL} = 0.1$ mm
- Maximum recording speed $v_{Lmax} = 500$ mm/s
- Minimum recording speed $v_{Lmin} = 0.5$ mm/s
- Maximum dynamic deviation $\Delta x = 2$ mm at v_{Lmax}
- Mass of the recording carriage including the Bowden wire $m_w = 50$ g
- Maximum friction at the carriage $F_L = 1$ N
- Diameter of the driving and the return pulley $d_1 = d_2 = 6.37$ mm
- Width of the pulleys $b_1 = b_2 = 5$ mm
- Material of the pulleys: Aluminium

Investigation to Find the Best Active Principle

- Maximum of mechanical power
  When the motor has to overcome the highest friction force at maximum speed, the maximum of mechanical power according to the equation of Table 9.3 and 9.4 is reached. A gear will be not used to avoid problems with backlash of the gear, radial eccentricity of the gear wheels as well as noise. It is assumed that all existing friction forces are added up in the given value for the friction force at the carriage. With the gear efficiency $\eta_G = 1$ the maximum of power in Watts can be determined:

$$P = 10^{-3} \cdot \frac{1}{\eta_G} \cdot F_L \cdot v_{L\,max} = 10^{-3} \cdot \frac{1}{1} \cdot 1 \cdot 500 = 0.5 \text{ W} \; . \tag{9.109}$$

- Required resolution
  From the maximum recording length and the static accuracy, the required resolution can be calculated as:

$$z_g = \frac{x_L}{x_{SL}} = \frac{450 \text{ mm}}{0.1 \text{ mm}} = 4500 \; . \tag{9.110}$$

- Maximum number of motor revolutions
  The driving pulley for the Bowden wire is directly mounted at the motor shaft. Therefore the maximum number of motor revolutions U_{max} is

given with the maximum recording length and the perimeter of the driving pulley:

$$U_{\max} = \frac{x_L}{\pi \cdot d_1} = \frac{450\,\text{mm}}{\pi \cdot 6.37\,\text{mm}} = 22.5 \;. \tag{9.111}$$

- Number of steps of the motor
  The 22.5 revolutions of the motor can be subdivided into steps according to the required resolution. The motor has to perform z steps per revolution:

$$z = \frac{z_g}{U_{\max}} = \frac{4500}{22.5} = 200 \;. \tag{9.112}$$

These number of steps leads to a step angle of:

$$\theta_S = \frac{360°}{z} = \frac{360°}{200} = 1.8° \;. \tag{9.113}$$

- Maximum stepping rate
  It can be calculated from the maximum recording speed and the required resolution:

$$f_{S\,\max} = \frac{v_{L\,\max} \cdot z_g}{x_L} = \frac{500\,\text{mm/s} \cdot 4500}{450\,\text{mm}} = 5\,\text{kHz} \;. \tag{9.114}$$

- Minimum stepping rate
  In analogy to the equation above, the minimum recording speed is used for the minimum stepping rate:

$$f_{S\,\min} = \frac{v_{L\,\min} \cdot z_g}{x_L} = \frac{0.5\,\text{mm/s} \cdot 4500}{450\,\text{mm}} = 5\,\text{Hz} \;. \tag{9.115}$$

Decision

Based on the required resolution, a stepping motor with a step angle of $\theta_S = 1.8°$ could be used. However, the dynamic requirements are in contrast to that. The maximum stepping rate amounts to 5 kHz and the maximum dynamic deviation at this speed should be less than 2 mm. This means that the maximum stepping rate must be obtained after approximately 4 ms, which is equivalent to an operation at this frequency within the pull-in range. The product $\theta_S f_{A0\max} = 9000°/\text{s}$ is much higher than common values of stepping motors (cp. Chap. 9.5). Therefore, the decision is made to use a position-controlled drive with a DC servo motor.

Selection of the Structure of the Driving System

Figure 9.23 presents the chosen structure of the drive for a single axis of a recording device. The individual subassemblies and components are explained in the following in detail.

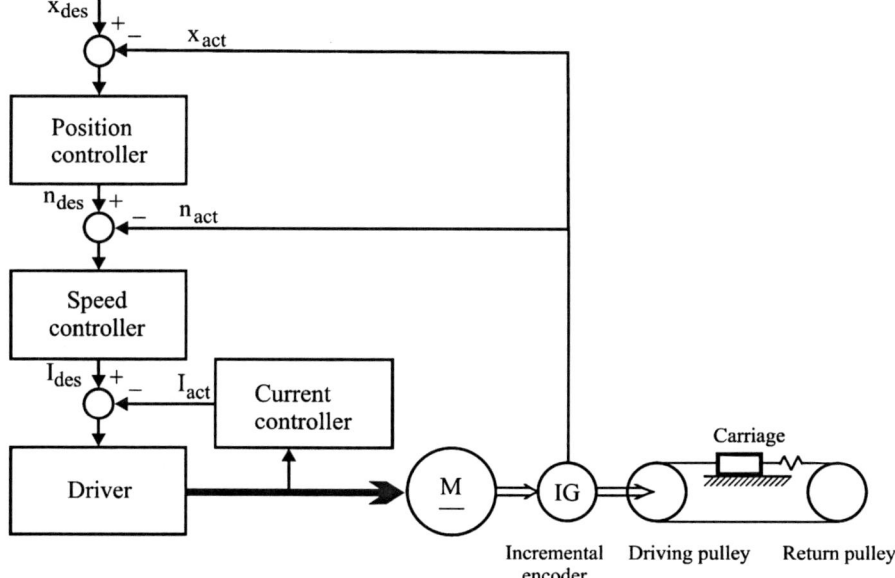

Fig. 9.23. Structure of the drive of a recording device

- Driving motor
  A DC motor with a bell-shaped rotor is chosen (cp. Sect. 2.1) based on the requirements for a good dynamic behavior, on a speed variation range of 1:1000, and on a relatively small mechanical power of 0.5 W. The main parameters of this DC motor E are as follows:

Nominal voltage	$U_N = 24\,\text{V}$
Nominal speed	$n_n = 3200\,\text{min}^{-1}$
Nominal torque	$T_N = 1\,\text{cNm}$
Nominal current	$I_N = 280\,\text{mA}$
Efficiency	$\eta_N = 48\%$
Armature resistance	$R_A = 34\,\Omega$
Mechanical time constant	$\tau_m = 20\,\text{ms}$
Electromagnetic time constant	$\tau_e = 0.1\,\text{ms}$
Thermal time constant	$\tau_{th} = 0.2\,\text{min}$
Starting voltage	$U_a = 0.5\,\text{V}$
Maximum permissible current	$I_{max} = 1.4\,A$
Maximum permissible speed	$n_{zul} = 10\,000\,\text{min}^{-1}$
Total inertia including the pulse encoder	$J_R = 10.7\,\text{gcm}^2$

The operating characteristics of the DC motor E are shown in Fig. 9.24.
- Incremental encoder
  The encoder is integrated together with the motor in one physical unit. It delivers 100 basic pulses and has two scanning signals, which show an

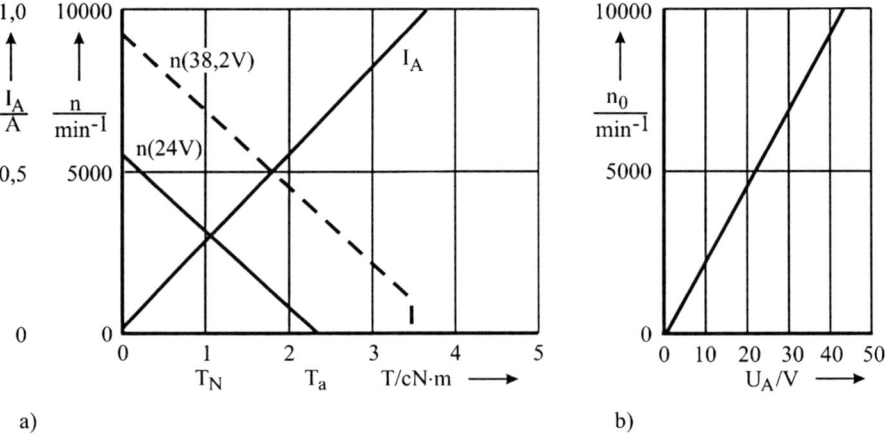

Fig. 9.24. Operating characteristics of the DC motor E. **a** Speed and current vs. torque for different voltages. **b** No-load speed vs. armature voltage

electric phase shift of $\pi/2$. These two scanning signals are required for the position control to sense different directions of rotation. The 100 basic pulses per scanning track can be quadruplicated with common procedures. Finally, it results in a resolution of $z = 400$ per revolution.

- Control circuit

  A position control circuit with inner speed and current control loops will be used (cp. Chap. 6) to reach the required static accuracy as well as a highly dynamic behavior. The incremental encoder combines an extraordinary static position resolution with a very small load for the motor (no additional friction torque, very small inertia, no additional power demand). It is practically applied as a sensor for both, the position and the speed.

Verification of Drive Parameters

In the following section, an overview of the development and parameterization of the control circuit is given. Specific requirements of the driving motor are investigated and explained based on references for the dimensioning of position control circuits [6.14, 6.15], which treat this subject much more detailed and give a comprehensive overview about all criteria. Multi-loop feedback control circuits are sized starting from inside and going outwards.

- Driver

  The servo motor has to move in both directions of rotation. Thereby a four-quadrant driver circuit has to be applied (see Sect. 6.2.2.2). The current control is designed for current limitation. A constant current stands for a constant torque according to Eq. (2.1). This means in our case a constant acceleration torque during running up. The parametrization of the current limit is made according to the maximum permissible current of the motor

$I_{gr} = I_{max} = 1.4$ A. The verification of the drive system will show which value of current limitation is permitted and necessary.

To reach the maximum current during standstill, according to Eq. (2.9), the following voltage has to be supplied to the motor:

$$U_H = I_{max} \cdot R_A = 1.4\,\text{A} \cdot 34\,\Omega = 47.6\,\text{V} \,. \tag{9.116}$$

Corresponding to Eq. (3.2), the inductance will define the current slope of a winding in the point of origin. Using Eq. $\tau_e = L_A/R_A$ the current slope is given:

$$\frac{di_A}{dt} = \frac{U_H}{L_A} = \frac{U_H}{\tau_e \cdot R_A} = \frac{47.6\,\text{V}}{0.1\,\text{ms} \cdot 34\,\Omega} = 14\,\text{A/ms} \,. \tag{9.117}$$

In our case, the current will be at its permissible maximum value of 1.4 A already after $3\tau_e = 0.3$ ms.

- Speed control circuit
  The design of the current controller as current limiter leads towards the complete neglection of the already small electromagnetic time constant of $\tau_e = 0.1$ ms. In this case, the transfer function of the DC motor Eq. (6.16), see Sect. 2.1.2.6 as well) is reduced to the following form:

$$\frac{\omega}{u'_A} = \frac{K_S}{1 + p\tau_m} \,. \tag{9.118}$$

- Translational transmission ratio
  According to the equation in Tables 9.3 and 9.4, the translational transmission ratio can be determined using the step angle $\theta_S = 0.9°$ and the number of pulses $z = 400$ of the encoder. This leads towards the double of the required static accuracy and the following translational transmission ratio:

$$i_t = \frac{0.5 \cdot x_{SL}}{\theta_S} = \frac{0.5 \cdot 0.1}{0.9} = 0.0556\,\text{mm}/° \,. \tag{9.119}$$

- Maximum speed
  The maximum recording speed defines the maximum speed of the motor. According to the equation in Tables 9.3 and 9.4, the maximum speed in min^{-1} is given by:

$$n_{max} = \frac{v_{L\,max}}{6 \cdot i_t} = \frac{500}{6 \cdot 0.0556} = 1500 \,. \tag{9.120}$$

- Inertia of the load
  It sums up from the individual inertias of both pulleys and the mass of the carriage:

$$J_L = J_1 + J_2 + J'_w \,. \tag{9.121}$$

- Individual moments of inertia
  Driving pulley
  Assuming a massive disc the inertia of the driving pulley is given:

$$J_1 = \frac{m_1 \cdot d_1^2}{8} = \frac{d_1^4 \cdot \pi \cdot b_1 \cdot \gamma}{32} = \frac{6.37^4 \,\mathrm{mm}^4 \cdot \pi \cdot 5\,\mathrm{mm} \cdot 2.7\,\mathrm{g/cm}^3}{32} \qquad (9.122)$$
$$= 0.02\,\mathrm{gcm}^2 .$$

  Return pulley
  Since the return pulley exhibits the same diameter and runs at the same speed as the driving pulley, a conversion towards the motor shaft is not necessary. Because of identical dimensions and materials, the following is true:

$$J_2 = J_1 = 0.02\,\mathrm{gcm}^2 . \qquad (9.123)$$

  Carriage
  Based on the equation in Tables 9.3 and 9.4, the conversion of the mass of the carriage towards the shaft leads to the following inertia in gcm^2:

$$J'_\mathrm{w} = 32.8 \cdot i_t^2 \cdot \frac{1}{\eta_\mathrm{G}} \cdot m_\mathrm{w} = 32.8 \cdot 0.0556^2 \cdot \frac{1}{1} \cdot 50 = 5.06 . \qquad (9.124)$$

- Total moment of inertia
  Using Eq. (9.121) the total inertia is obtained from the individual inertias:

$$J_\mathrm{L} = J_1 + J_2 + J'_\mathrm{w} = (0.02 + 0.02 + 5.06)\,\mathrm{gcm}^2 = 5.1\,\mathrm{gcm}^2 . \qquad (9.125)$$

- Factor of inertia
  The factor of inertia is given according to Eq. (3.10):

$$FI = \frac{J_\mathrm{R} + J_\mathrm{L}}{J_\mathrm{R}} = \frac{(10.7 + 5.1)\,\mathrm{gcm}^2}{10.7\,\mathrm{gcm}^2} = 1.48 . \qquad (9.126)$$

- Maximum torque of the motor
  The motor generates its maximum of torque during operation with the maximum of current. Initially the output constant of the electrical machine k_M is determined according to Eqs. (2.1) and (2.2) using the diagram in Fig. 9.24:

$$k_\mathrm{M} = \frac{T_\mathrm{N} + T_\mathrm{f}}{I_\mathrm{N}} = \frac{T_\mathrm{N}}{I_\mathrm{N}} \cdot \frac{1}{1 - \frac{I_0}{I_\mathrm{N}}} = \frac{1\,\mathrm{cNm}}{280\,\mathrm{mA}} \cdot \frac{1}{1 - \frac{15\,\mathrm{mA}}{280\,\mathrm{mA}}} = 3.77 \cdot 10^{-3} \frac{\mathrm{cNm}}{\mathrm{mA}} . \qquad (9.127)$$

  Therewith, the friction torque is calculated:

$$T_\mathrm{f} = k_\mathrm{M} \cdot I_0 = 3.77 \cdot 10^{-3} \frac{\mathrm{cNm}}{\mathrm{mA}} \cdot 15\,\mathrm{mA} = 0.057\,\mathrm{cNm} . \qquad (9.128)$$

The difference of the maximum of the electromagnetically generated torque and the friction torque leads to the maximum available torque at the motor shaft:

$$T_{\max} = k_M \cdot I_{\max} - T_f = 3.77 \cdot 10^{-3} \frac{\text{cNm}}{\text{mA}} \cdot 1.4\,\text{A} - 0.057\,\text{cNm} = 5.22\,\text{cNm}\,. \tag{9.129}$$

- Load torque
  The friction force at the carriage has to be converted to a torque at the shaft of the motor. According to the equation in Tables 9.3 and 9.4, the load torque in cNm is given:

$$T'_L = 5.73 \cdot i_t \cdot \frac{1}{\eta_G} \cdot F_L = 5.73 \cdot 0.0556 \cdot \frac{1}{1} \cdot 1 = 0.318\,. \tag{9.130}$$

- Acceleration time
  For constant current, the equation of motion (Eq. 2.19) can be transformed into:

$$k_M \cdot I_{\max} = T'_L + T_f + FI \cdot J_R \cdot \frac{\Delta \omega}{\Delta t}\,. \tag{9.131}$$

Hence the acceleration time ΔT_b from zero to maximum speed can be calculated:

$$\Delta T_b = \frac{2\pi \cdot n_{\max} \cdot FI \cdot J_R}{k_M \cdot I_{\max} - T'_L - T_f} \tag{9.132}$$

$$= \frac{2\pi \cdot 1500\,\text{min}^{-1} \cdot 1.48 \cdot 10.7\,\text{gcm}^2}{3.77 \cdot 10^{-3} \cdot \frac{\text{cNm}}{\text{mA}} \cdot 1.4\,\text{A} - 0.318\,\text{cNm} - 0.057\,\text{cNm}} = 5.1\,\text{ms}\,. \tag{9.133}$$

The maximum dynamic deviation at maximum recording speed $v_{L\max}$ is required with equal or less than $\Delta x = 2\,\text{mm}$. Therewith, the permissible acceleration time until the maximum speed with a constant acceleration can be derived:

$$T_{\text{bzul}} = \frac{2 \cdot \Delta x}{v_{L\,\max}} = \frac{2 \cdot 2\,\text{mm}}{500\,\text{mm/s}} = 8\,\text{ms}\,. \tag{9.134}$$

The loading of the DC motor can be reduced below the maximum permissible current of 1.4 A, because the permissible acceleration time is higher than the achievable acceleration time during operation with the maximum permissible current. A new current limit of $I_{\max} = 950\,\text{mA}$ is chosen. Therewith the motor can develop a maximum torque according to Eq. (9.129):

$$T_{\max} = k_M \cdot I_{\max} - T_f = 3.77 \cdot 10^{-3} \frac{\text{cNm}}{\text{mA}} \cdot 950\,\text{mA} - 0.057\,\text{cNm}$$
$$= 3.52\,\text{cNm}\,. \tag{9.135}$$

The acceleration time is increased towards the value (Eq. 9.132):

$$\Delta T_\mathrm{b} = \frac{2\pi \cdot 1500\,\mathrm{min}^{-1} \cdot 1.48\,\mathrm{A} \cdot 10.7\,\mathrm{gcm}^2}{3.77 \cdot 10^{-3}\frac{\mathrm{cNm}}{\mathrm{mA}} \cdot 950\,\mathrm{mA} - 0.318\,\mathrm{cNm} - 0.057\,\mathrm{cNm}} = 7.76\,\mathrm{ms}\,. \tag{9.136}$$

This is still below the required one of 8 ms. Thus the requirements can be fulfilled as well with the newly made assumptions based on the shown estimations.

- Armature voltage for maximum acceleration

  The motor has to be supplied with the following armature voltage to work at the operating point of $T_\mathrm{max} = 3.52\,\mathrm{cNm}$ at a speed $n_\mathrm{max} = 1500\,\mathrm{min}^{-1}$. Then the maximum acceleration can be realized and with the torque M_max the maximum speed is reached as fast as required. By transformation of Eq. (2.5) and under neglection of the total brush voltage drop, the needed armature voltage is given:

  $$U_\mathrm{A} \approx 2\pi \cdot n_\mathrm{max} \cdot k_\mathrm{M} + \frac{R_\mathrm{A} \cdot (T_\mathrm{max} + T_\mathrm{f})}{k_\mathrm{M}} \tag{9.137}$$

  $$= 2\pi \cdot 1500\,\mathrm{min}^{-1} \cdot 3.77 \cdot 10^{-3}\frac{\mathrm{cNm}}{\mathrm{mA}} + \frac{34\,\Omega \cdot (3.52 + 0.057)\,\mathrm{cNm}}{3.77 \cdot 10^{-3}\,\mathrm{cNm/mA}}$$

  $$= 38.2\,\mathrm{V}\,. \tag{9.138}$$

  According to Fig. 9.24b, the DC motor will arrive with this voltage at a no-load speed of $n_0 = 8832\,\mathrm{min}^{-1}$. The steady-state speed vs. torque characteristic is shown with a dashed line in Fig. 9.24a. This no-load speed is below the maximum permissible speed of this motor of $n_\mathrm{zul} = 10\,000\,\mathrm{min}^{-1}$. The knee of the curve on the bottom is caused by the current limitation.

- Current slope

  The current slope in the point of origin Eq. (9.117) is changed by the smaller voltage of 38.2 V, compared to the voltage of 47.6 V, which is needed to reach the maximum permissible current as follows:

  $$\frac{\mathrm{d}i_\mathrm{A}}{\mathrm{d}t} = \frac{U_\mathrm{A}}{\tau_\mathrm{e} \cdot R_\mathrm{A}} = \frac{38.2\,\mathrm{V}}{0.1\,\mathrm{ms} \cdot 34\,\Omega} = 11.23\,\mathrm{A/ms}\,. \tag{9.139}$$

  The final value of the current $I_\mathrm{max} = 950\,\mathrm{mA}$ will be already achieved after 0.19 ms. The above statements about the neglection of the electrical transient phenomena are still valid.

- Temperature rise of the motor

  The steady-state load torque of $T_\mathrm{L} = 0.318\,\mathrm{cNm}$ is below the nominal value of the motor torque. This means that the caused temperature rise of the motor will be below the nominal value, too.

  The maximum current $I_\mathrm{max} = 950\,\mathrm{mA}$ exceeds 3.4 times the nominal current. Out of it follows 11.5 times higher losses compared to the nominal ones without consideration of the changed resistance by the temperature rise. Simultaneously, the limit of temperature rise will be 11.5 times higher compared to the nominal situation. In the worst case these losses have to be expected during a time of $T_\mathrm{b} \approx 8\,\mathrm{ms}$, when the recording carriage is accelerated towards the maximum speed. The temperature rise follows an exponential function with the time constant τ_th (cp. Sect. 9.3.2). Therefore, the ratio between the temperatures at maximum and nominal torque is known:

$$\frac{\Theta_\mathrm{max}}{\Theta_\mathrm{N}} = \frac{\Theta_\mathrm{max\,e}}{\Theta_\mathrm{Ne}} \cdot (1 - \mathrm{e}^{-T_\mathrm{b}/\tau_\mathrm{th}}) = 11.5 \cdot (1 - \mathrm{e}^{-8\,\mathrm{ms}/0.2\,\mathrm{min}}) = 0.0077\;. \tag{9.140}$$

This value is uncritical. If the device is operated with the maximum recording speed of $v_\mathrm{Lmax} = 500\,\mathrm{mm/s}$, it will need only a time of $0.9\,\mathrm{s}$ to travel the maximum recording length of $x_\mathrm{L} = 450\,\mathrm{mm}$. It is expected that enough breaks will occur and the averaged loading will be that low that the motor is not overloaded thermally.

- Minimum speed
  The minimum speed is $1.5\,\mathrm{min}^{-1}$ according to the variation range of 1:1000. Motors with bell-shaped rotors allow (even at such low speeds) a smooth running without cogging. The required armature voltage for this speed and the given load torque amounts in analogy to Eq. (9.137) to:

$$\begin{aligned} U_\mathrm{A\,min} &\approx 2\pi \cdot n_\mathrm{min} \cdot k_\mathrm{M} + \frac{R_\mathrm{A} \cdot (T'_\mathrm{L} + T_\mathrm{f})}{k_\mathrm{M}} \\ &= 2\pi \cdot 1.5\,\mathrm{min}^{-1} \cdot 3.77 \cdot 10^{-3}\frac{\mathrm{cNm}}{\mathrm{mA}} + \frac{34\,\Omega \cdot (0.318 + 0.057)\,\mathrm{cNm}}{3.77 \cdot 10^{-3}\,\mathrm{cNm/mA}} \\ &= 3.4\,\mathrm{V}\;. \end{aligned} \tag{9.141}$$

- Position control loop
  The position control circuit should be adjusted according to the optimum of absolute value for undelayed input signals, which are present for the examined path or trajectory control [9.4, 9.10]. The position controller has to provide the reference variable for the speed control loop and should include simultaneously a speed limitation.

Overall Evaluation

The required good dynamic behavior and the smooth running at low speed leads to the use of a position-controlled DC drive. It is possible to ensure the required dynamic behavior and to reduce substantially the effect of the time constants of the motor with an underlayed speed control loop with high amplification and an inner current control loop with current limitation. The precise sizing of the controller has to consider the dead band in connection with the no-load angular velocity and the armature voltage (cp.

Sect. 2.1.2). Motors with bell-shaped rotors are preferably equipped with metal brushes and their low voltage drop in the range of $U_{\text{Bü}} = 0.1\ldots0.5\,\text{V}$ to keep this dead band marginal. The heavy thermal stress at maximum acceleration is temporally that short that it has almost no negative influence on the lifetime. The presented drive is an example for the path control of a DC motor.

9.7.7 Drum Drive

(in cooperation with Steffen Schnitter)

A drum has to be driven in a device operated with extra-low voltage. The life expectancy of the device is very high and the constancy of the speed should be long-term stable. In this drive design project, for special aspects computer simulation will be used to find a solution.

Requirements of the Drive

- Voltage $U = 24\,\text{V} \pm 10\%$
- Frequency $f = 50\,\text{Hz}$
- Speed of the drum $n_T = 50\,\text{min}^{-1}$
- Torque at the drum $T_T = 3.3\,\text{cNm}$
- Inertia of the drum $J_T = 150\,\text{gcm}^2$
- Service life $L_B = 25\,000\,\text{h}$

Investigation to Find the Best Active Principle

- Motor
  The high life time requests lead to a motor almost without wearing parts. AC motors are best suitable for such demands. A synchronous motor can fulfil the requirement for a long-term stable constancy of speed, because its speed is directly proportional to the frequency of the supply net and independent from voltage variations inside the accepted tolerance (cp. Sect. 2.4.2).
  The mechanical power to drive the drum is given according to the equations from Tables 9.3 and 9.4:

$$P_T = 1.05 \cdot 10^{-3} \cdot T_T \cdot n_T \qquad (9.142)$$
$$= 1.05 \cdot 10^{-3} \cdot 3.3 \cdot 50 = 0.173\,\text{W}\,. \qquad (9.143)$$

In this power range, synchronous motors based on the claw-pole principle are normally used. Their advantage is a very good ratio between power and costs. However, claw-pole synchronous motors are not available for such a small speed as it is required from this drum drive. Common speeds are 250 rpm and more with a supply frequency of 50 Hz. Therefore, a synchronous motor with a speed of 500 rpm and an additional downstream gear were chosen.

- Gear

  The required transmission ratio of the gear according to Table 9.3 is:

  $$i_\mathrm{r} = \frac{n_\mathrm{T}}{n} = \frac{50}{500} = 0.1 \ . \tag{9.144}$$

  There are two main advantages when the drum is driven with the inserted gear. First of all, the inertia at the motor shaft will be reduced remarkably, and secondly, the gear backlash relieves the starting of the motor. Claw-poled synchronous motors in this power range are used without special starting aids. This means that such motors must have synchronised with the supply net within a single period of the alternating voltage. The factor of inertia FI for a trouble-free starting of the motor should not exceed 2.5...3.

Decision

A claw-poled synchronous motor is chosen for driving of the drum. In series connected to the motor is a gear with the transmission ratio of $i_\mathrm{r} = 0.1$. The drum is directly mounted on the output shaft of the gear. With such a drive system, the required long service life and the long-term stable constancy of the speed can be fulfilled.

Selection of the Structure of the Driving System

The structure of the drive is shown in Fig. 9.25.

- Motor

  The chosen claw-poled synchronous motor F is characterised by the following parameters:

Rated voltage	$U_\mathrm{N} = 24\,\mathrm{V} \pm 10\%$
Rated frequency	$f_\mathrm{N} = 50\,\mathrm{Hz}$
Rated torque	$T_\mathrm{N} = 0.5\,\mathrm{cNm}$
Rated speed	$n_\mathrm{N} = 500\,\mathrm{min}^{-1}$
Motor inertia	$J_\mathrm{r} = 2.1\,\mathrm{gcm}^2$
Running capacitor	$CB = 5.6\,\mu\mathrm{F}$

- Gear

  The chosen gear is featured with a transmission ratio $i_\mathrm{r} = 0.1$ and an efficiency of $\eta_\mathrm{g} = 0.95$.

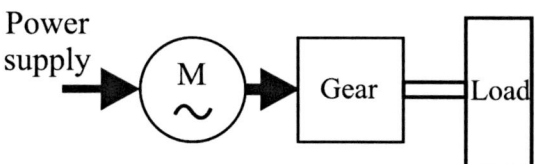

Fig. 9.25. Structure of the drive with synchronous motor, gear and load

Verification of Drive Parameters

- Reduced inertia of the drum
  The inertia of the drum reduced to the motor shaft can be calculated with the equation from Table 9.3 using the units given in Table 9.4:
  $$J'_L = i_r^2 \cdot \frac{1}{\eta_g} \cdot J_T = 0.1^2 \cdot \frac{1}{0.95} \cdot 150 = 1.58 \, \text{gcm}^2 \,. \tag{9.145}$$
  Therewith the factor of inertia FI is given according to Eq. (3.10):
  $$FI = \frac{J'_L + J_r}{J_r} = \frac{(1.58 + 2.1) \, \text{gcm}^2}{2.1 \, \text{gcm}^2} = 1.75 \,. \tag{9.146}$$
  The factor of inertia FI is situated inside of the claimed range. Hence the motor should start without difficulties.

- Torque
  The required torque for the drum has to be converted to the motor shaft. Using the equation from Table 9.3, the torque is given:
  $$T'_L = i_r \cdot \frac{1}{\eta_g} \cdot T_T = 0.1 \cdot \frac{1}{0.95} \cdot 3.3 = 0.347 \, \text{cNm} \,. \tag{9.147}$$
  The motor with its rated torque of $T_N = 0.5 \, \text{cNm}$ can drive the drum together with the gear with a safety factor of 1.44.

- Noise
  The advantage of using a gear in the drive system is the reduction of torque and inertia of the motor. Therefore, also a very small motor could be used. At the same time the backlash of the gear relieves the starting of the motor. On the other hand, the backlash of the gear is disadvantageous and can cause noise, especially when the driving motor tends to oscillate.

Claw-poled synchronous motors are usually built with two stators (cp. Fig. 2.133). One stator is directly supplied from the mains, the second one from a capacitor. Both stators are not magnetically coupled and behave like single-phase AC machines. The alternating field generated of each stator can be divided in its effect into two inversely rotating fields. The supply of the second stator from the capacitor aims at eliminating both of the inversely rotating fields in its superposed effect by a phase shift of $\pi/2$ of the currents in both stators. An entire compensation of the inversely rotating fields will be not possible in reality due to asymmetries in the stators, due to capacitor values which are only stepwise available, and due to load-dependent operating points. This means that always a certain addiction to oscillations has to be expected. In spite of this, such motors are produced in a large volume worldwide due to their simple structure, their robustness and their competitive price. The accurate investigation of the oscillating characteristic of such drives is not possible with approximate calculations. Difficulties are linked for example to the backlash of the gear and the stiffness of the components of the drive system. Therefore, computer simulation will be applied for this investigation.

Computer Simulation

The software package SIMPLORER [9.24] was used to obtain the simulation results shown here. However, the basic operation methods of such dynamic simulation systems are comparable for different software packages. Here the solution will be revealed using a realistic example.

Each simulation package provides a model library, where often-required components are available and usually graphically represented. Each model offers at least one input and one output. The parameters of these models are usually assigned in dedicated parameter windows. Such models of the electrical domain are for example resistors, inductors and voltage sources.

Moreover, new models can be created and the relation between in- and output can be defined with mathematical equations or characteristic curves. Inside of the simulation software, such models and their equations are joined with each other by connecting the different inputs and outputs of their graphical representation.

For the project design of drive systems, in many cases it is recommended to cooperate directly with the manufacturer of the drives. Two facts support this statement: firstly, the software for computer simulation is expensive and requires experienced users; secondly, specific parameters are often required, which cannot be found in regular datasheets and product documentation. This will be indicated in the following example at the appropriate points. Furthermore, many manufacturers own verified and proven models for different drive structures. In such a case, often only minor modifications are necessary to find reliable solutions in a fast way.

The software package SIMPLORER is based on physical quantities. The values of these quantities have to be entered in SI-units, e.g. the current in A, the voltage in V, the time in s. Appropriate appendices can be used for smaller and larger values, e. g. a current $I = 10\,\text{mA}$ can be entered as "0.01" or "10 m".

- Electrical domain

    Figure 9.26 shows the electrical circuit of the motor. The components resistor R, inductor L, capacitor CB, and voltage source U are standard models out of the library. The input values for each component are summarized in Table 9.5.

Table 9.5. Input parameters of the electric circuit

Component	Parameter	SI-Unit	Value
Voltage source	Voltage (U)	V	24
	Frequency (f)	Hz	50
Resistor	Resistance (R)	Ω	230
Inductor	Inductance (L)	H	391 m
Running capacitor	Capacitance (CB)	F	5.6 u (u is used for μ)

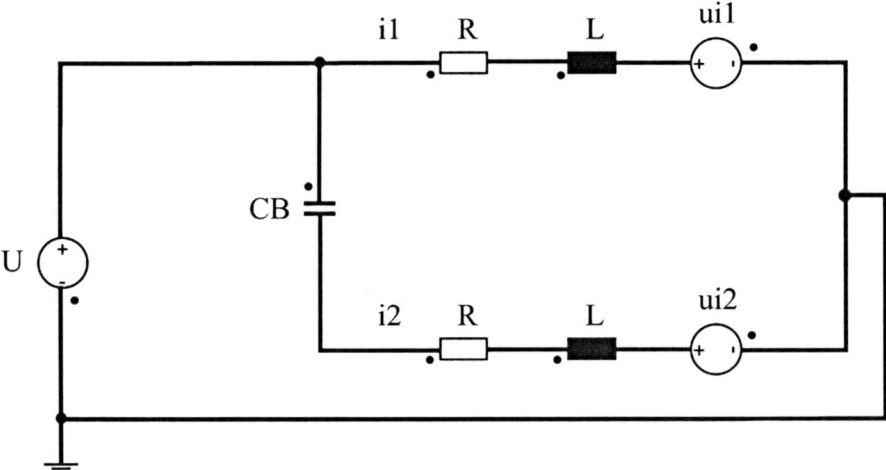

Fig. 9.26. Schematics of the electrical domain of the computer model

The inductance is a typical parameter, which can only seldom be found in product datasheets. More often the electrical time constant τ_{el} is given. The synchronous motor F has a time constant of $\tau_{el} = 1.7\,\text{ms}$. The advantage of such a declaration is the independence of the individual winding data of the motor. Using Eq. $\tau_e = L_A/R_A$ the inductance of the motor can be calculated:

$$L = \tau_{el} \cdot R = 1.7\,\text{ms} \cdot 230\,\Omega = 391\,\text{mH} \, . \tag{9.148}$$

New components will be defined for the induced voltages u_i of the motor. Therefore, the motor constant k and the number of pole pairs p are used. Either the motor constant is determined using Eqs. (2.1) and (2.3) respectively, or the manufacturer is asked for it:

$$k = \frac{T}{I} = \frac{u_i}{\Omega} = 0.16\,\text{Nm/A} \, . \tag{9.149}$$

The number of pole pairs of a synchronous motor can be calculated using the speed:

$$p = \frac{f}{n} = \frac{50\,\text{Hz}}{500\,\text{min}^{-1}} = 6 \, . \tag{9.150}$$

Both induced voltages in Fig. 9.26 are described with the following equations:

$$ui1 = -k \cdot \text{RotorOMEGA} \cdot \sin(p \cdot \text{RotorPHI}) \tag{9.151}$$
$$ui2 = -k \cdot \text{RotorOMEGA} \cdot \sin(p \cdot \text{RotorPHI} - pi/2) \, . \tag{9.152}$$

$pi/2 = \pi/2$ considers the spatial rotation of both stators to each other by a half pole pitch. RotorOMEGA is the mechanical angular velocity of the rotor in s^{-1}, RotorPHI the angular position in radians.

Therewith the electrical circuit of the motor is described and the numerical simulation delivers both currents $i1$ and $i2$, which are required for the following physical domains of the model.

- Mechanical domain

  Figure 9.27 presents the mechanical domain of the drive system. In each stator of the motor a torque will be generated. The rotor is featured with an inertia of $J_r = 2.1\,\text{gcm}^2$ and a friction torque of $T_f = 0.3\,\text{cNm}$.

  Both generated torque components are described with the following equations:

  $$T1 = k \cdot i1 - \sin(p \cdot \text{RotorPHI}) \tag{9.153}$$
  $$T2 = k \cdot i2 - \sin(p \cdot \text{RotorPHI} - pi/2) \,.$$

The gear is a provided standard model and needs the parameters shown in Table 9.7.

The total stiffness results from the elasticity of the teeth of the gear wheels. For the input of the transmission its definition has to be taken into account. Here in the computer model it is inversely described to Table 9.3. The stiffness inside the backlash model stands for the contact behavior, e. g. characterised by the grease at the tooth flank. These parameters of the gear are again very special values, which can be only provided by the manufacturer or must be extensively investigated. The input parameters of the load, i.e. the drum are shown in Table 9.8. Again, a model of the standard library could be used.

For the investigation, the speed of the rotor and the load are from special interest. It is given using the angular velocity:

$$n = \frac{\omega}{2\,\pi} \,. \tag{9.154}$$

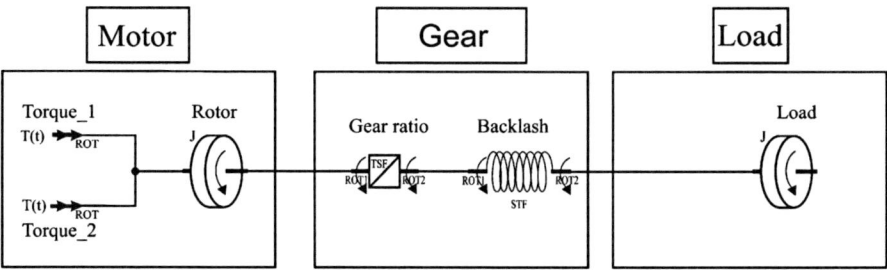

Fig. 9.27. Computer model of the mechanical domain

Table 9.6. Input parameters of the rotor

Component	Parameter	SI-unit	Value
Rotor	Inertia J_r	kgm^2	2.1e-7
	Friction torque T_f	Nm	3 m

Table 9.7. Input parameters of the gear

Component	Parameter	SI-unit	value
Gear	Total stiffness	Nm/rad	10 000
	Transmission ($1/i_r$)		10
	Efficiency η_g		0.95
	Backlash	°	±1
	Contact stiffness of backlash	Nm/rad	2.5

Table 9.8. Input parameters of the load

Component	Parameter	SI-unit	Value
Load	Load inertia J_t	kgm²	150 e-7
	Load torque T_T	Nm	33 m

This is written in the modelling language as follows:

$$\text{nRotor} = \text{RotorOMEGA}/(2 \cdot pi) \cdot 60 \quad (9.155)$$
$$\text{nLast} = \text{LastOMEGA}/(2 \cdot pi) \cdot 60 \,.$$

The coefficient 60 is used for the conversion into $\min^{-1}$, the common unit for the revolutions per time.

- Simulation results

In Fig. 9.28, the speed of the motor vs. time is shown, and in Fig. 9.29 the speed of the drum vs. time with a running capacitor $CB = 5.6\,\mu\text{F}$ is shown. After switching on the supply voltage, the motor remains in its rest position for approximately 4 ms. Initially, an electric current has to be established in order that a torque can be generated. Thereafter a strong

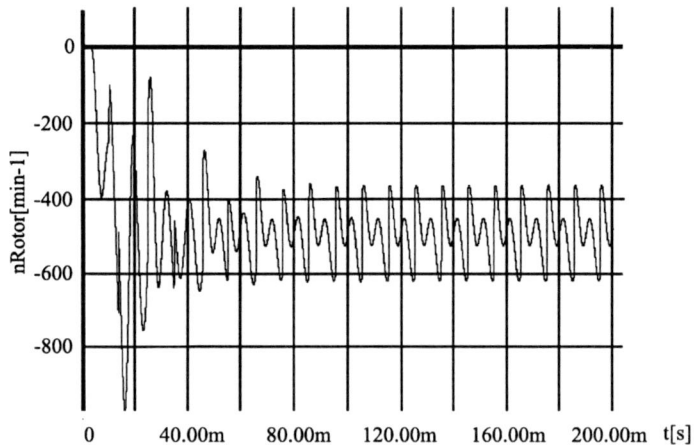

Fig. 9.28. Speed of the motor with a running capacitor $CB = 5.6\,\mu\text{F}$

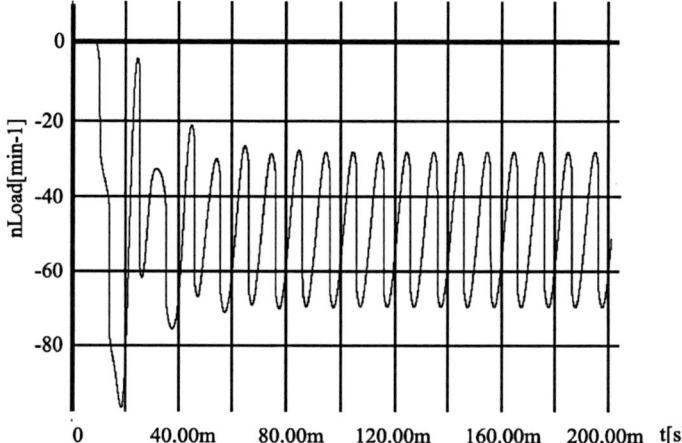

Fig. 9.29. Speed of the drum with a running capacitor $CB = 5.6\,\mu\text{F}$

acceleration period starts. However, the motor is not able to reach inside of the first half period of the supply frequency the synchronous speed of $n_N = 500\,\text{min}^{-1}$. This occurs only in the second half of the first period. The bouncing and the very high acceleration afterwards are caused by the backlash inside the gear. After a transient period, the oscillations of the motor speed swing around an averaged speed of $n_N = 500\,\text{min}^{-1}$.

When the steady state after $t > 60\,\text{ms}$ is examined, the strong content of higher harmonic components of the speed oscillations attracts attention. That means that the pinion of the motor strikes the first gear wheel with different high frequencies and generates noise. This leads to strong oscillations of the drum speed, too. In Fig. 9.29 it is obvious that the acceleration flank of the drum speed has a clearly stronger slope compared to the "braking flank". Always when the motor pinion hits the gear wheel a strong impulse is generated. Afterwards the drum runs without contact to the driving pinion and slows down. The speed variations of the drum range around a value of $n_T = (50 \pm 20)\,\text{min}^{-1}$.

The strong higher harmonic components of the motor speed as well as the speed oscillations of the drum cause unacceptable noise. Therefore, a second simulation run is done, which only differs in the value of the running capacitor of $CB = 6.8\,\mu\text{F}$ from the first one.

Figures 9.30 and 9.31 show that the waveform of the motor speed is almost harmonically and the drum speed variates only in the range of $n_T = (50 \pm 9)\,\text{min}^{-1}$. Therefore, the time domain behavior of the drive could be substantially improved.

A FFT (Fast Fourier Transformation) analysis can be easily added, when the simulation results are available in the computer. The results of this analysis for the motor speed are shown in Fig. 9.32 using the time interval $t = (100\ldots 200)\,\text{ms}$ in the steady state. Especially the motor speed is

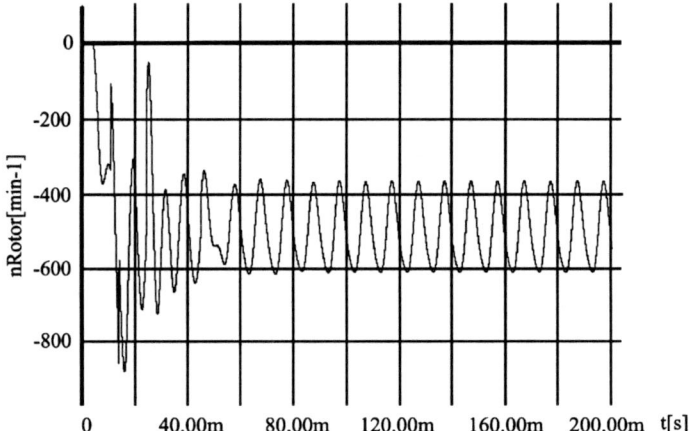

Fig. 9.30. Speed of the motor with a running capacitor $CB = 6.8\,\mu\text{F}$

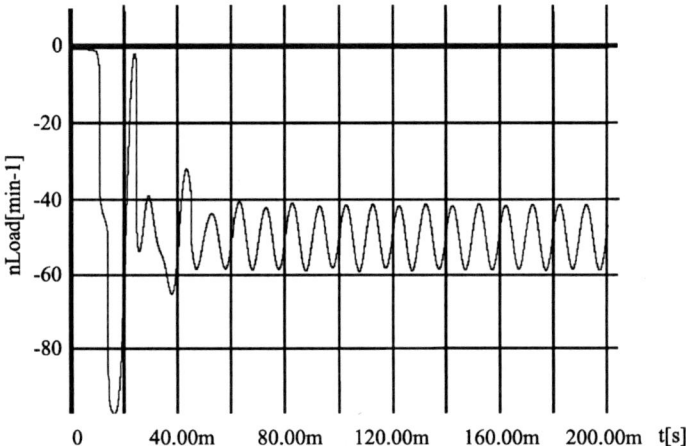

Fig. 9.31. Speed of the drum with a running capacitor $CB = 6.8\,\mu\text{F}$

interesting, because this is the source of potential noise. With the specified running capacitor of $CB = 5.6\,\mu\text{F}$, a large spectrum of upper harmonics in the motor speed is obtained. Even the highest amplitude is not obtained with the fundamental wave, but with the first harmonics at $f = 200\,\text{Hz}$. Besides the fundamental wave only a considerably alleviated first harmonic at $f = 200\,\text{Hz}$ is existent in the second simulation run with the improved running capacitor of $CB = 6.8\,\mu\text{F}$. Already the inspection in the time domain (cp. Fig. 9.30) turned out an almost harmonic oscillation.

The fundamental wave with $f = 100\,\text{Hz}$ can normally not completely be avoided. It results from the inversely rotating fields of the magnetically uncoupled stators of the claw-poled synchronous motor, as already described above.

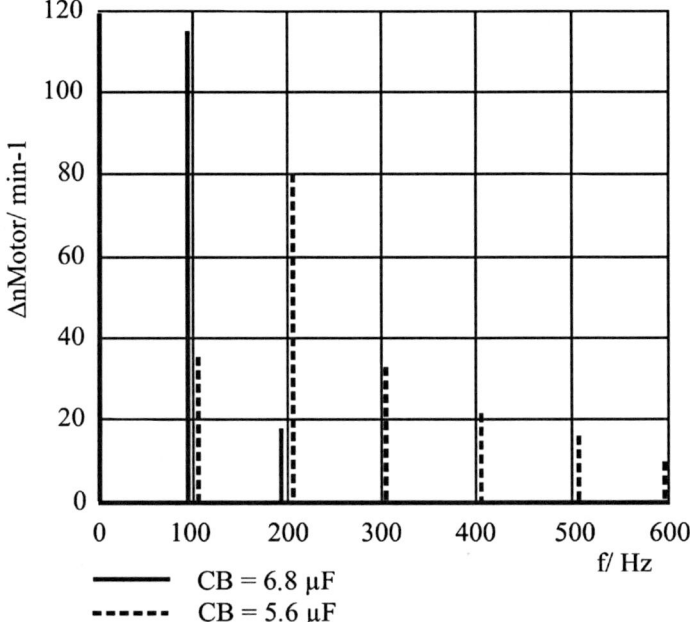

Fig. 9.32. FFT analysis of the motor speed in the time interval (100...200) ms

Overall Evaluation

The claw-poled synchronous motor with series-connected downstream gear can drive the drum according to the given requirements. Experience shows that this is an economic solution as well.

Concerning noise, the interaction between the synchronous motor and gear can be critical. A considerable improvement of the vibration behavior of the drive could be reached with the change of the running capacitor using computer simulation. The dynamic behavior of such a drive train with a vibrating motor as a source of mechanical energy, the gear with backlash, and the stiffness of the load cannot be described with differential equations in a closed form. The example showed which special conclusions can be gained using computer simulation compared to simple conventional calculations. From a parameter variation could be derived, which parameters show a serious influence on the system behavior and against which parameters the system is insensitive. Such a method can remarkably reduce the experimental effort in the testing phase of a drive design as well as can give direction for targeted investigations. However, the ultimate proof of the solution for a driving project design can only be made with real experiments.

The shown outline of the solution has turned out, how important a close cooperation with the drive manufacturer is. One point is the computer modelling of the drive system. Not all required models are available in libraries of simulation packages. Some have to be created by oneself. Drive manufacturers often have complete models of their products available. Furthermore, the specific parameters required for the simulation are often not contained in datasheets and product catalogues. Cooperation usually leads to the fastest, most effective and unerring solution.

Standards and Directions

[9.1]   European Commission (EC)-directive on Low Voltage 2006/95/EC, repeals former 73/23/EEC, more details see http://ec.europa.eu/enterprise/newapproach/standardization/harmstds/reflist.html
[9.2]   EC-directive 2006/42/EC on Machinery, repeals former 98/37/EC
[9.3]   EC-directive 2004/108/EC on Electromagnetic Compatibility, repeals former 89/336/EEC
[9.4]   Schönfeld, u.a. Bewegungssteuerungen, Springerverlag, Abschnitt 10
[9.5]   Holl, E. Mathematische Modellierung von Kraftfahrzeug-Servoantrieben zum Zwecke der Entwurfsoptimierung, Fortschrittsbericht VDI, Reihe 21, Nr. 246, 1998
[9.6]   Eidam, J. Beurteilung und Simulation des Betriebsverhaltens von lagegeregelten Direktantrieben als „Elektronische Kurvenscheibe", Fortschrittsberichte VDI, Reihe 1, Nr. 279, 1997
[9.7]   Kalender, T. Neue Wege bei der Steifigkeitsmodellierung und Diagnose von spielfreien Präzisionsgetrieben, Tagungsband SPS/IPC/DRIVES 93, VDE Verlag 1993, S. 451–460
[9.8]   The-Quan Pham Modellierung, Simulation und Optimierung toleranzbehafteter Mechanismen der Feinwerktechnik, Dissertation TU Dresden, Institut für Feinwerktechnuik, 1998
[9.9]   Maas, S. u.a. Auslegung eines Schrittmotorantriebs mit einem Modell hoher Ordnung, antriebstechnik 1996, Nr. 7, S. 52–54 und Nr. 8, S. 57–60
[9.10] Schönfeld, R., Habiger, E. Automatisierte Elektroantriebe, Berlin, Verlag Technik, 1981
[9.11] VEM-Handbuch: Die Technik der elektrischen Antriebe- Grundlagen, 8. Auflage Berlin: Verlag Technik 1986
[9.12] Schönfeld, R. Grundlagen der automatischen Steuerung, 2. Auflage Berlin: Verlag Technik; Dr. Alfred Hüthig Verlag 1984
[9.13] Braun, H. u.a. Polfühligkeit pemanentmagnetisch erregter Gleichstrommotoren, F + M Carl Hanser Verlag, München 1996, Heft 7–8, S. 562–566
[9.14] Enzmann, B. Sonderverzahnungen aus Kunststoff für geräuscharme Getriebe, antriebstechnik 1990, Heft 5, S. 42–44
[9.15] Hirn, H., u.a. Leistungsstark und leise, Maschinenmarkt 1989, Heft 19, S. 46–50
[9.16] DIN-Fachbericht 72: Erfassung und Dokumentation der Geräuschqualität von Elektromotoren für Kfz-Zusatzantriebe, Beuth Verlag GmbH Berlin Wien Zürich 1998
[9.17] Richter, C. Gleichstrom- oder Schrittantrieb – wer hat wo seine Stärken? Tagungsband SPS/IPC/Drives 93, VDE Verlag 1993, S. 407–416

[9.18] Richter, H. Versuchsstand zur Analyse von Kleingetriebemotoren, Diplomarbeit TU Dresden, Institut für Feinwerktechnik 1996
[9.19] Hanitsch, R, u.a. Bürstenloser Gleichstrommotor digital geregelt, Elektronik 1980, Heft 20, S. 67–71
[9.20] Gospodaric, D. u.a. Parameterbestimmung von Schrittmotoren und SR-Motoren unter Zuhilfenahme der Komponenten- und Systemsimulation, VDI-Berichte 1269 „Innovative Kleinantriebe", 1996, S. 133–150
[9.21] Joneit, D. Der Drehschubmotor – rechnergestütze Simulation des kompletten Antriebs für die Projektierung, VDI-Berichte 1269 „Innovative Kleinantriebe", 1996, S. 205–210
[9.22] Europäische Zeichen für elektrotechnische Erzeugnisse, elektronik industrie, Nr. 10, 1998, S. 126
[9.23] Buschmann, H., Jung, R.: Funktionsorientierte Systematik elektrischer Kleinmotoren. F & M 91 (1983) H. 7, S. 345–350
[9.24] Ansoft Corporation, 225 West Stadion Square Drive, Pittsburgh, PA 15219-1119

Standards

[10.1] DIN 1495-1: 1983 Sintered metal plain bearings which meet specific requirements for fractional and subfractional horsepower electric motors - Part 1: Spherical bearings: Dimensions (German)
[10.2] DIN 1495-2: 1983 Sintered metal plain bearings which meet specific requirements for fractional and subfractional horsepower electric motors - Part 2: Cylindrical bearings: Dimensions (German)
[10.3] DIN 1495-3: 1996 Sintered metal plain bearings which meet specific requirements for fractional and subfractional horsepower electric motors - Part 3: Requirement and Testing (German)
[10.4] DIN 42014: 1987 Earthing conductor terminals for small motors, requirements, tests, design examples (German)
[10.5] DIN 42016: 1974 Built-in motors for apparatus, connecting dimensions (German)
[10.6] DIN 42017: 1984 Marking and coordination of the connecting leads for small motors (German)
[10.7] DIN 42026-1: 1977 Permanent magnet segments: directives for selection of dimensions (German)
[10.8] DIN 42027: 1984 Servo motors: classification and survey (German)
[10.9] DIN 42028-1: 1980 Connectors with receptacles and tabs for small motors: forms and dimensions (German)
[10.10] DIN 42961: 1980 Rating plates for rotating electrical machinery: design (German)
[10.11] DIN VDE 0580: 2000 Electromagnetic devices and components - General specifications (German)
[10.12] EN 10106: 2007 Cold rolled non-oriented electrical steel sheet and strip delivered in the fully processed state
[10.13] EN 10341: 2006 Cold rolled electrical non-alloy and alloy steel sheet and strip delivered in the semi-processed state
[10.14] EN 60034-1: 2005 Rotating electrical machines - Part 1: Rating and performance (IEC 60034-1: 2005)

[10.15] EN 60034-5: 2007 Rotating electrical machines - Part 5: Degrees of protection provided by integral design of rotating electrical machines (IP code) - Classification (IEC 60034-5: 2006)
[10.16] EN 60034-6: 1993 Rotating electrical machines - Part 6: Methods of cooling (IC code) (IEC 60034-6: 1991)
[10.17] IEC TS 60034-20-1: 2002 Technical Specification: Rotating electrical machines - Part 20-1: Control motors - Stepping motors
[10.18] EN 60068-1: 1994 Environmental testing - Part 1: General and guidance (IEC 60068-1: 1992)
[10.19] EN 60068-4: 1996 Environmental testing - Part 4: Information for specification writers - Test summaries (IEC 60068-4: 1994)
[10.20] EN 60085: 2004 Electrical insulation - Thermal classification (IEC 60085: 2004)
[10.21] EN 60335-1: 2007 Household and similar electrical appliances - Safety - Part 1: General requirements (IEC 60335-1: 2006)
[10.22] EN 60529: 2000 Degrees of protection provided by enclosures (IP code)
[10.23] EN 60664-1: 2003 Isolation coordination for equipment within low-voltage systems - Part 1: Principles, requirements and tests (IEC 60664-1: 2002)
[10.24] EN 60730-1: 2007 Automatic electrical controls for household and similar use: Part 1: General requirements (IEC 60730-1: 2003)
[10.25] EN 60950-1: 2006 Information technology equipment - Safety - Part 1: General requirements (IEC 60950-1: 2005)
[10.26] EN 61000-4-1: 2007 Electromagnetic compatibility (EMC) - Part 4-1: Testing and measurement techniques - Overview of IEC 61000-4 series (IEC 61000-4-1: 2006)
[10.27] EN 61000-6-1: 2007 Electromagnetic compatibility (EMC) - Part 6-1: Generic standards - Immunity for residential, commercial and light-industrial environments (IEC 61000-6-1: 2005)
[10.28] EN 61000-6-2: 2005 Electromagnetic compatibility (EMC) - Part 6-2: Generic standards - Immunity for industrial environments (IEC 61000-6-2: 2005)
[10.29] EN 61000-6-3: 2007 Electromagnetic compatibility (EMC) - Part 6-3: Generic standards - Emission standard for residential, commercial and light-industrial environments (IEC 61000-6-3: 2006)
[10.30] EN 61000-6-4: 2007 Electromagnetic compatibility (EMC) - Part 6-4: Generic standards - Emission standard for industrial environments (IEC 61000-6-4: 2006)
[10.31] EN 60404-1: 2000 Magnetic materials - Part 1: Classification
[10.32] EN 60404-8-1: 2004 Magnetic materials - Part 8-1: Specification for individual materials - Magnetically hard materials
[10.33] EN 60404-8-5: 1989 Magnetic materials - Part 8-5: Specification for individual materials - Specification for steel sheet and strip with specified mechanical properties and magnetic permeability
[10.34] EN 60404-8-6: 2007 Magnetic materials - Part 8-6: Specification for individual materials - Soft magnetic metallic materials

Important Symbols

A	cross section
B	magnetic flux density
D_i	bore diameter
E_A	back electromotive force (back-e.m.f.)
F	force
H	field intensity
I	current
I_0	no-load current
I_a	armature current
I_a	starting current
J	(moment of) inertia (torque)
J	current density
L	inductance
l	stack length
L_p	principal inductance
$L_{\sigma 1}$	stator leakage inductance
$L_{\sigma 1}$	rotor leakage inductance
m	number of phases
n	speed (revolutions per minute, rpm)
n_0	no-load speed
n_s	synchronous speed
n_N	nominal speed
P	power
p	number of pole pairs
P_{Cu}	copper losses
P_{Fe}	iron losses
P_N	nominal power
q	number of slots per pole and phase
R_1	stator resistance
R_2	rotor resistance
R_a	armature resistance

R_{Sh}	shunt resistor (resistance)
R_S	series resistor (resistance)
T	torque
T_0	no-load torque
T_s, T_H	starting torque, holding torque
T_N	nominal torque
U	voltage
U_i	induced voltage
v	velocity
W	energy
w_1	number of turns, stator
w_2	number of turns, rotor
δ	width of air gap
η	efficiency
μ	(effective) permeability
μ_0	free-space permeability
μ_r	relative permeability
Φ	magnetic flux
Ψ	flux linkage
ω	angular velocity

Biographies

Professor Dipl.-Ing. Dr. Wolfgang Amrhein received his Dipl.-Ing. degree from the Technical University of Darmstadt, Germany, and his doctor's degree from the Swiss Federal Institute of Technology Zurich, all in electrical engineering. In 1990, he joined Papst-Motoren, St. Georgen, Germany, where he led the department for drive development. From 2000 until 2007 he served as Scientific Head of the Linz Center of Competence in Mechatronics (LCM), Austria. Currently, he is Director of the Institute of Electric Drives and Power Electronics, Johannes Kepler University, Linz, and Area Coordinator of the research area Mechatronic Design of Machines and Components at the Austrian Center of Competence in Mechatronics (ACCM). Further, he leads the committee Electrical Devices and Servo Drives of the VDE/VDI (GMM), Germany.

Prof. Dr.-Ing. habil. Hans-Jürgen Furchert lectured at the University of Applied Sciences in Gießen-Friedberg (Germany) in subjects electrical engineering, electrical fractional-horsepower motors, printed circuit technique, opto-electronic systems as well as applied precision engineering and guided there the laboratory for actuaters. He studied at the further College for electrical engineering, faculty precision mechanics and optics, took one's doctor's degree and university lecturing qualification at the Technical University Ilmenau (Germany). He worked at the Carl Zeiss Jena Company as a development engineer and consultant.

Prof. Dr.-Ing. Friedrich Wilhelm Garbrecht worked mostly in the area of control engineering at the Fachhochschule (University of Applied Sciences) Gießen-Friedberg. He studied mechanical engineering at the Technical University of Braunschweig and received his Dr.-Ing. at the University of Erlangen-Nuremberg. After studying, he worked first in the test department of the aerospace industry in Northern Germany. After this he was in the department for measuring and control engineering of a company in the chemical industry in the Rhine region.

Dr.-Ing. Elmar Hoppach received his PhD degree from Darmstadt University. After being with automotive electronics suppliers as head of R&D and CEO for almost a decade, he is now working as a consultant on fractional-horsepower drives.

Prof. Dr.-Ing. habil. Hartmut Janocha studied electrical and mechanical engineering at the Technical University (today: Leibniz University) of Hannover. Since 1989 he has been full professor at Saarland University, where he holds the chair at the Laboratory of Process Automation (LPA). His main fields of work include unconventional actuators, 'intelligent' structures, new signal-processing concepts, and machine or robot vision.

Prof. Dr.-Ing. habil. Eberhard Kallenbach was longstanding (from 1990–2002) director of the Institute for Microsystems Technology, Mechatronics and Mechanics of the Technical University Ilmenau. Since 1992 he has been the leader of the Steinbeis-Transferzentrum Mechatronik Ilmenau, since 1996 member of Saxony Academy of Science, Leipzig, and since 2002 member of acatech – the Council for Technical Science of the Union of German Academies of Science and Humanities. In addition he is a member of the American Society for Precision Engineering (ASPE).

Prof. Dr.-Ing. habil. Dr. h.c. Werner Krause is emeritus professor for construction of precision engineering at the Technical University Dresden. Until 2002, he was director of the Institute of Precision Engineering. At the same time he led the study course Precision and Micro Technology at the Faculty of Electrical Engineering and Information Technology. He is member of the Saxony Academy of Science, Leipzig, and member of acatech – the Council for Technical Science of the Union of German Academies of Science and Humanities.

Dr.-Ing. Andreas Möckel is a lecturer at the Institute of Electrical Energy Conversion and Automation of the Technical University of Ilmenau, where he also received his academic education and doctorate.

Prof. Dr.-Ing. habil. Dieter Oesingmann was the chairman of the Department of Electrical Machines of Small Power within the Faculty of Electrical Engineering and Information Technology and the director of the Institute of Electrical Energy Conversion and Automation of the Technical University of Ilmenau. He studied at the Faculty of Electrical Power Engineering of the Technical University of Dresden. Prof. Oesingmann made his thesis and habilitation at the Technical University of Ilmenau and worked several years as faculty member and lecturer there. He was a team manager at the State Combine for Electrical Appliances in Suhl for 3 years. He is chairman of the Association for Promotion of Education and Research in the Field of Electrical Machines at the Faculty of Electrical Engineering and Information Technology.

Dr.-Ing. habil. Christian Richter is a specialist of small stepper and synchronous motors and is recently retired from the management of development of small electric machines of Saia-Burgess Dresden GmbH. He studied electric machines at Dresden University of Technology, where he gained his PhD and his postdoctoral qualification (habilitation), too. Thereafter he worked on the development of miniature and fractional-horsepower motors in the company VEB Kombinat Elektromaschinenbau and later as the head of department in VEM Antriebstechnik AG, Dresden.

Dr.-Ing. Thomas Roschke is Manager of Research & Development of the Saia-Burgess Dresden GmbH, a member of Johnson Electric, which develops small stepper motors including electronics as well as solenoids. He studied electrical engineering and precision mechanics at Dresden University of Technology, where he did his PhD at the Institute of Solid State Electronics. Later he led the research team "Electromagnetic Actuators" at the Institute of Precision Mechanics of Dresden University, which worked mainly for switchgear and aerospace industry.

Prof. Dr.-Ing. Wolfgang Schinköthe, born in 1955, studied electrical engineering with a main focus on precision engineering at the Technical University of Dresden. From 1981 to 1989 he worked there as a temporary, later as a permanent scientific assistant at the Institute of Electronics Technology and Precision Engineering, where he attained a doctorate in 1985. Starting from 1989 he was first project manager at Robotron-Elektronik Dresden and afterwards chief designer at Feinmeß Dresden. In 1993 he got a reputation to the University of Stuttgart, where he is since then chair holder and director of the Institute of Design and Production in Precision Engineering. The main points of his research are, among others, linear direct drives as well as piezoelectric travelling wave motors.

Dipl.-Ing. Dr. Siegfried Silber is Associate Director of the Institute of Electric Drives and Power Electronics at the Johannes Kepler University of Linz and manager of the Business Unit Electrical Drives of LCM (Linz Center of Competence in Mechatronics), Austria. He received his Dipl.-Ing. degree in electrical engineering from the University of Technology Graz and the PhD degree from the Johannes Kepler University Linz.

Prof. Dr.-Ing. Hans-Dieter Stölting studied electrical engineering at the Universities of Aachen and Stuttgart. At the Institute of Electrical Machines of the latter university he became a scientific assistant and received his doctorate. Afterwards he joined Siemens, Würzburg, where he worked on the field of fractional-horsepower motors. In 1980, he got a reputation to the Institute of Drive Systems and Power Electronics of the Leibniz University of Hannover where his points of research are small and micro electrical drives. From 1993 to 2002 he led the committee Electrical Devices and Servo Drives of the VDE/VDI (GMM).

Prof. Dr.-Ing. Heinz Weißmantel is working at the Institute of Elektromechanical Constructions (EMK) at the Technical University Darmstadt (TUD). He was working some years in Fa. HELLA, Lippstadt, and director of the Faulhaber Group, Schönaich. For the last 5 years he worked on the project "Design for Elderly".

Index

absolute angle encoders 106
absolute incremental encoders 289
AC controller 48
AC magnet 227
AC synchronous linear motors 262
AC tachometer generator 51
AC voltage divider 291
acceleration time 544, 563
acceleration torque 522, 544
active magnetic bearing 426
active power 131
active system damping 426
actuating time 523
actuation angle 521, 523
adjustment of controller parameters 376
aerostatic guide 269, 276, 290, 298
air-gap power 131
air-gap reactances 41
aligned position 117
AlNiCo magnets 15
α, β-reference frame 361, 381
alternating field 131
aluminium-nickel-cobalt magnets 15
amplification factor 379, 380
analogous power amplifier 291
analogous power control elements 290
analogue amplifier 339
analogue CBSM 365
analogue position encoders 290
analogue techniques 365
analogue voltage signals 364
analytic computation methods 242

angular frequency of the slip 411
angular speed 411, 452
anisotropic magneto-resistive effect 109
anisotropic rotors 153
anode potential 355
anti-twist mechanism 282
armature branch 57
armature coil 57
armature resistance 529
armature winding 35, 74
armed brush 52
asymmetrical armature field 81
asymmetrical winding 81, 130
automotive application 357
auxiliary winding 129
axial bearing 420, 430, 431
axial stiffness 424
axis transformation 381
axles 492

Bach's relation 464
backup bearings 430
balance operation 129
balanced motor 145
balanced rotor 21
ball monorail guidance 269
band gearings 474
barium-ferrite 147
Barkhausen circuit 46
bearing strength 464, 467, 469, 472
bearingless motors 434, 437
bearingless pumps 418, 440, 441

bearingless segment motors   442
bearings   23, 423, 424, 495
bell rotors   23
belt drive   472
bending moment   464, 492
bevel gears   470
bias current   427
bias flux   427, 431
bimodal resonator   333
bipolar control   163, 171, 179
bipolar operation of stepping motors   358
block commutation   121, 359, 390
block diagram   292
   of closed-loop controlled EC-motors   394
   of controlled DC-motors   379
   of drive electronics for three-phase AC-motors   399
   of the field-oriented controlled asynchronous motors   389
block voltage commutation   121
bootstrap circuit   348, 349
boundary surface   256
boundary surface forces   267
box coils   257, 268
braking time   544
breakdown torque   149, 152, 371
brush bridge   41
brush contact resistance   39
brush holder   51
brush springs   51
brushed DC motors   518
brushless AC motor   68, 81, 94, 106
brushless DC motor   68, 81, 106, 116
brushless permanent magnet motors   67, 68
butterfly curve   320

cam mechanisms   480
cam step mechanism   482
canned motor   146
capacitor motor   132
capacitor voltage   134
capacitor-start motor   129
cardan joint coupling   487
carrier frequency   291
carrier-based sinusoidal modulation (CBSM)   360, 365

cascade control   282, 376, 379
cascaded closed loops   288
cathode potential   355
CCD line scanning   295
CCD lines   294
CE-sign   510
centrifugal force   490
centrifugal governors   46
characteristic frequencies   188
characteristic motor   287
charge control   338
charged capacitor   349
chopper amplifier   292
chopper circuits   8, 177
chopper frequencies   177
chopping operation   370
chorded windings   69, 76, 77
circular arc toothing   460
circular current   76
circular field   130
clamping mechanism   344
Clarke's transformation   96
classification of electromagnets   226
claw-poled PM motors   163
claw-poled principle   146
claw-poled stepping motor   179
clockwork gearing   460
closed loops for electrodynamic linear motors   288
closed-loop control   11, 340, 347
   of DC motors   377, 380
   of universal motors   376
closed-loop position control   10, 11
closed-loop rotational speed control
   of electronically commutated motors   391
closed-loop speed control   10, 11
closed-loop torque control   11
clutches   487
cogging forces   263
cogging torque   76, 79, 84, 88, 94, 100, 104
colossal magneto-resistive effect   109
commutating winding   35
commutation   36, 262, 356
commutation capacitor   355
commutation circuit   59
commutation signals   284
commutation zone   36

Index 587

commutator motors   68, 125, 131
commutator series motors   33
commutator-brush system   518
comparator   364
compensated half step mode   182
compensating winding   35
$I \times R$ compensation   372
complex power   131
compound motor   13
compressors   418
continuous running duty (duty type S1)   516
control algorithms   398
control circuit   560
control electronics   334
control of bipolar windings   177
control signals   509
control system   282, 284
controller parameters   395
controller structure   390
core losses   131
cost-minimization   49
coupling drives   477
couplings   482
cross-pole connection   144
crossed helical gears   466
crown wheel gears   470
Curie temperature   320, 322
current control   181, 189, 289
cycle time   193, 359, 363, 540
cyclo-gear   460
cycloidal toothing   460
cylindrical coil system   267
cylindrical coils   257
cylindrical commutator   21
cylindrical gears   456, 459
cylindrical magnets   278
cylindrical stand gear   461
cylindrical worm   468, 469

d, q-reference frame   381
damping   197
data acquisition   293
DC controller   49
DC electromagnet   208
DC intermediate circuit   348
DC motor   14, 352, 517, 542
DC permanent magnet motors   13, 517
DC-servo mode   262

DC/DC-converters   351
dead band generator   401
dead time generation   347
declaration of manufacturer   510
delta connection   75, 128
demagnetising curve   15
depolarization   320, 326
detent torque   182
diagrams   196
diametric pitch (modulus)   455, 456
differential equation for the temperature rise   515
differential equation of elastic line   494
differential equation of motion   429
differential operator   387
differential throttle measurement systems   291
differential throttle principle   290
differential transformer   290
digital circuits   365
direct current magnet   207
direct drive   113, 529
direct piezoelectric effect   317
discrete CBSM   366
disk motors   126
disk rotor   52, 86
disk rotor design   89, 90
dosing accuracy   533, 536
driver   541, 560
duty cycle time   516
duty factor   516
duty ratio   359, 516
dynamic behaviour   214, 233, 395
dynamic equations   287
dynamic load rating   504
dynamic mathematical model of the motor   401
dynamic operation   31
dynamic parameters   509

eddy current damping   425
eddy current losses   36, 256
eddy current sensors   278
efficiency   27, 44, 318, 339, 459, 467, 469, 470, 474, 478, 529
eigenfrequency   322, 323, 330, 332, 334
eigenmode   335
elastic compliance coefficient   317
elastic structures   430

electric arcs 61
electric magnet 117
electric polarization 317
electric power tools 33
electrical damping 425
electrical degrees 128
electrical equivalent circuit 93
electrical magneto-mechanical structures 270
electrical partial system 253, 256
electrical time constant 288, 289
electrodynamic drives 251
electrodynamic MCM 271
electromagnet 205
electromagnetic bearings 427
electromagnetic compatibility 511
electromagnetic compatibility directive 510
electronic commutation 261
electronic control 518
electronic frequency change 193
electronic gears 192
electronically commutated motors 68, 518, 519
electrostrictive effect 318
elliptical rotating field 146
elliptical rotating field 131
embedded magnets 83
EMC problems 289
enable signal 354
epicyclic gear 452, 459, 461
equation of motion 195
equivalent circuit
  of asynchronous motors in the α, β-reference frame 388
  of controlled DC-motors 380
  of EC-motors in the d, q-reference frame 394
  of the asynchronous motor in the α, β-reference frame 384
  of the asynchronous motor in the d, q-reference frame 410
  of the field-oriented asynchronous motors 388
equivalent control circuit 379
equivalent network 41
exciter winding 34
expanders 419
external inertia 31

external rotor 81, 84, 86, 126
external rotor design 88
external rotor motor 69

factor of inertia 184, 187, 537, 544, 549, 555, 562
feed-forward control 340
feedback system 347
ferrite magnets 17
ferroelectric materials 318, 319
ferromagnetic cylinder 267
ferromagnetism 421
field computation methods 242
field control 67
field orientation 386
field weakening 99, 139
field winding 13
field-effect transistor (FET) 349
field-oriented control 380
final control element 289, 291
finer tooth pitch 169
firing point 48
firm couplings 484
flange coupling 484
flank clearance 462
flat and box coil arrangements 268
flat belt gearings 474
flat coils 256, 257, 289
flat commutator 21
flat windings 268
flexible drives 472
flexural mode 333
floating bearing 504
floating emitter potential 351
flux expanding arrangements 257
flux observer 411
flux oriented d, q-reference frame 409
flux-oriented coordinate system 97
fold bellow coupling 487
force constant 252
force locus 436, 437, 443
forces on boundary surfaces 253, 264, 267, 286
four-coordinate motor 271
four-quadrant operation 348, 353
fractional pitch windings 76
free-oscillating stepping motor 195
freewheeling diode 177, 353, 358
frequency control 67

Index 589

frequency converter 138, 148, 357, 395
frequency divider 554
frequency ramp 372
frequency-proportional drives 145
friction coefficient 497
friction forces 266
friction motors 333
friction torque 31, 43, 183
full bridge or H-bridge 171, 353
full bridge with thyristors 355
full step mode 170, 181
full-bridge mode 9
full-pitch windings 69
full-wave control 10
full-wave phase-angle control 368

galvanic separation 398
gate input power 352
gear 518, 519, 524, 535, 567
gear trains 452, 454
gearing steps 453
gearings 450
Geneva mechanism 482
giant magnetic inductance effect 109
giant magneto-resistive effect 109
glass cylindor 284
globoid wheel 468
globoid worm 468
Grashof 477
grinding spindles 418
groove ball bearings 503
group step operation 193
GS mark 511
guide gearings 450
gyroscopic effects 430

H-bridge mode 9
half step mode 171, 180
half-bridge 349
half-wave phase-angle control 367
Hall sensor 106, 109, 262, 289
hard disk drives 258
harmonic drive 462
harmonic fields 132
heteropolar bearing 427, 431, 433
heteropolar configurations 287
high control frequencies 519
high grade motors 19
high ripple content 359

high-energy magnet materials 265, 268, 277
high-performance profiles 476
high-resistance cage rotor 127
high-speed drives 419
highly dynamic drives 194
holding torque 170, 181, 185
homopolar bearing 427, 431, 432
homopolar configurations 287
household appliances 33, 367
hybrid motor 113, 163
hybrid stepping motor 179
hydrodynamic lubrication 23, 497
hysteresis 36, 131, 256, 318, 326, 340–342
hysteresis loop 49, 150
hysteresis loss 150, 267
hysteresis model 341
hysteretic behaviour 319

impressed current 288
inchworm motor 330, 342–344
incremental angle encoders 106
incremental encoder 543, 553, 559
incremental optical encoders 108
indirect coupling 348, 351, 355, 358
indirect-coupled signal 355
individual moments of inertia 536, 562
inductive pulse transformation 348
inductive transmitter 355
inertia of the light fixture 549
inertia of the load 561
inertia of the work piece and its fixture 543
inertia J 31
input power 37
input signals 170
insulated gate bipolar transistor (IGBT) 349
integrated power modules 357
intelligent power modul (IPM) 358, 398
interfering forces 271, 282, 296
interferometric position encoders 290
intermitted operation 254
internal rotor 81, 84, 86, 126
internal torque 387, 409
interpoles 61
interrupt service routine 399

590   Index

inverse control   338, 340–342
inverse piezo effect   329
inverter   119
inverter states   362
involute gearing   459
involute profile   455
irreversible partial degaussing   285

kinetic energy   197

lap winding   57
Laplace transformations   287
large-scale integration   398
large-signal operation   340
laser-interferometric   293
lead-zirconate-titanate   319
leaf springs   56
leakage fluxes   58, 257, 259
leakage inductances   383
life time problems   260
lifetime   14, 35
linear code   295
linear commutation   60
linear direct drives   250
linear displacement   165
linear hybrid stepping motor   299
linear measuring systems for movement tasks   289
linear stepping motor   164, 303, 306, 308
linear voltage regulator   347
load angle   145, 149, 152
load capacity   467, 469, 472
load torque   187, 533, 543, 563
locked-rotor torque   36
logic circuit   364
logic sequencer   162, 554
logic sequencer "Dosing"   534
logic sequencer "Filling"   534
long-term constant-speed   192
long-term stability   319
longitudinal effect   318, 321
longitudinal mode   333
Lorentz forces   435
low-voltage directive   510
low-voltage motor   353
lubrication   463, 467, 500, 504

magnet torque   154

magnet-rotor motor   151
magnetic bearing technology   419
magnetic coupling   489
magnetic drive design   241
magnetic effectiveness   210
magnetic integrated circuit sensors   109
magnetic partial system   253, 256
magnetic spin moments   421
magnetization characteristic   40
magnetization losses   38
magneto-resistive sensors   106, 109
manufacturing tolerances   181
mass performance relation   251
material selection   463, 467, 469, 472, 495, 500
mathematical model
  of the asynchronous machine   387
mathematical motor model   93
mathematical pendulum   184
matrix equation   386
  of asynchronous motors   386
maximum dosing frequency   532
maximum lead screw speed   533
maximum piston speed during dosing   536
maximum possible stepping rate for the pull-in region   537
maximum pull-out rate   185
Maxwell force equation   253
Maxwell forces   427, 435
mechanical commutation   260
mechanical damping   425
mechanical partial system   254, 255
mechanical power   529
mechanical time constant   31, 288, 529, 545
mechanical torsion block   292
mechanical torsion blocking   281
mechanical transfer units   449
mechanical transformers   449
Merrill motors   153
micro computer controller   290
micro stepping mode   161
microcontroller   119, 364, 547
micromotor   149
microprocessor   162
microscope focusing   282
microscope image clarity analyser   282

microstepping 309
mid-point tap 358
miniaturized motors 265
minimum stepping rate 555, 558
modular design 269
modulation methods 359
momentary transmission 454
monomodal resonator 333, 334
moonie transducer 328
motion converters 298
motor constant 25, 252
motor field sensor 374
motor selection 517
motor with spring counter-force 288
motor-voltage changing 137
movement on two axes 273
movements along a path 297
moving coil system 262
moving iron armature 256
moving permanent magnets 267
multi-axis path generators 297
multi-coordinate drive (MCD) 271, 303
multi-coordinate measurement system 271, 292
multi-coordinate regulation 273
multi-variable control 272
multilayer ceramics 319, 321, 322
multilayer stack 331, 335
multiple ring arrangements 423
multivariable control 292
mutual inductance 41

natural frequency 494, 519, 549
negative stiffness 424
neodymium iron boron magnets 17
no-load speed 36, 132
nodal equation 362
noise performance 465
noise reduction 519
non-commutated homopolar configurations 285
non-ferrous cores 263
non-ferrous winding 261
non-involute gear 460
non-magnetic spacers 152
non-saturated part 39
normal forces 263, 266
normal pitch 455

numerical simulations 195

ohmic losses 131
ohmic voltage drop 41
one-phase equivalent stator circuit of EC-motors 392
one-phase motors 146, 147
one-quadrant operation 352
open-loop control 11, 347
open-loop operation 188
open-loop speed control 10
operating air gap flux density 286
operation at variable speed 192
operation voltage monitoring 289, 290
operation with one transistor 8
optimal current waveforms 100
optimization problem 262
opto-coupler 351, 355
oscillating drives 193
oscillating system 188
oscillation equation 184
oscillator 547
output stages for the current 295
over-commutation 62
overall transmission 459
overlaid commutation 262

P-controllers 289
pantographs or cross-push guides 273
passive magnetic bearings 419
passive stabilization 437
path measurement 290
path measurement signal 262
path proportional signal 290
path-transfer-functions 287
perforated air supply elements 277
periodic duty (duty type S3) 516
peripheral speed 459
permanent magnet bearings 420, 421
permanent magnet DC motors 14
permanent magnet design shapes 84
permanent magnet designs 81
permanent magnetic excitation 252
permanent magnetic excitation field 285
permanent magnets 15
permanent-split capacitor motor 129
permanently excited stepping motors 182

Index

perpetual forces 36
phase angle controller 368
5-phase hybrid stepping motor 170
phase portrait 183, 195, 202, 549
five-phase stepping motor 182, 188
phase-angle control 10
phase-shifter capacitor 367
phasor diagram 42
photodiode array 293
physical measurement system 271
PI or PID controllers 289
PI-controller 395
PID-controller 296
piezo actuators 319
piezo constant 323
piezoelectric charge constant 317
piezoelectric effect 317
piezoelectric materials 318
piezoelectric polymers 321
pivot drives 258, 290
planar bevel gear 471
planar spur gearing 471
planar stepping motor 305
plugging construction 461
point-to-point control 520
polarized electromagnet application 239
polarized electromagnet shape 236
pole- or ring-core impedances 66
polychloroprene 475
polycrystalline ceramics 318
polyurethane 475
polyvinylidene fluoride 318, 321
position control loop 289, 565
position controller 289, 541
position encoders 123, 290, 359
positioning accuracy 189, 250, 278, 290
positioning and servo drives 519
positioning dynamic 250
positioning operations with time optimum 194
positioning processes 288
positioning system 341–343
potential energy 197
power amplifier with current control 528
power amplifiers 338, 339
power electronics 119
power factor 43, 131

power modules 347
power rate 31
power switches 352
Prandtl-Ishlinskii hysteresis operator 341
pre-selected rotational speed 358
pre-selected rotor position 358
precious metal brushes 24
Preisach hysteresis operator 341
prismatic systems 257
problem-specific drives 298
profile overlap 454
proportional controller (P-controller) 378
proportional-integral controllers (PI controllers) 378
protective function 358, 379
protective functions 398
PTFE steel 276
pull out of synchronism 145
pull-in range 187
pull-out range 187
pulsating torque 131, 145
pulse duty factor 370
pulse-width modulated amplifiers 289
pulse-width modulation (PWM) 347, 364
punching-bending-fitting process 146
pyroelectricity 320
PZT 319
PZT ceramics 319

quasi-static mode 322
quasi-symmetrical winding 129
quasistatic operation 285

rack-and-pinion gearings 460
radial bearing 420, 422, 430
radial bearing forces 438
radial stiffness 422–424
radio shielding 35
rail seek and tracking mode 290
ramp generator 368
rare-earth magnets 17
rated or nominal duty 514
rating torque 36
reactive power 131, 155
reciprocal or inverse piezoelectric effect 317

Index 593

rectilinear translation   452, 479
reductor principle   167
reference detectors   294
reference marks   284, 294
reference potential   358
reference profile   457
reference voltage   364
reference voltage space vector   363
relative physical measures   292
relative positioning accuracy   547
reluctance forces   253, 256, 263, 264,
    279, 286, 437, 442
reluctance rotor   153
reluctance stepper   116
reluctance stepping motor   179
reluctance torque   100, 151, 436
remanent flux   39
required resolution   557
requirements of a drive system   509
reset time   380
resistance-start motor   129
resolver   106
resonant frequency   494
resonant ranges   519
resonator   333–337
resonators   333
rest positions   196
reverse recovery time   353
reversing duty   136, 547
rhombic-wound winding   90
ring coils   164
ring magnet   164
ripple torque   21
roller screw drive   478
roller-guidance systems   276
rolling bearings   502
rotary stepping motor   303
rotating field   144
rotating field angle   381
rotating transformation   382
rotation diagram   53
rotation translation movement
    converters   250
rotational speed   356, 367, 368, 449,
    452, 459, 488, 495
rotational speed control
  of asynchronous motors   371
  of DC-motors   370
  of EC-motors   392

  of stepping motors   376
  of synchronous motors   374
  with frequency converters   371
rotor imbalances   427
rotor position sensor   374
rotor resonances   427
rotor-current frequency   130
rotor-dynamic aspects   430
running capacitor   134

saddle point   202
safety-related demands   510
salient-pole motor   151
samarium cobalt magnets (SmCo)   17
sampling controller   380
scanned optoelectronically   272
screw mechanism   451, 477
sealing   504
selection of control electronics   518
self-aligning couplings   485
self-locking feature   165
self-supporting windings   91
self-switching clutches   489
self-switching coal brushes   56
self-ventilation   36
sensor signal   273
sensorless closed-loop control   401
  of DC-motors   403
sensorless control   124
sensorless control in the d, q-reference
    frame
  of asynchronous motors   411
  of synchronous motor   408
sensorless loadangle control
  of synchronous motor   408
sensorless positioning control
  of synchronous motor   408
separable ball bearings   503
separatrix   197, 199, 202
sequence of motions during filling   537
series characteristic   513
series motor   40
series wound motor   14
servo technique   374
servomotor behavior   513
sets of twin permanent magnets   271
shaded-pole motors   126, 148
shading coils   140
shear effect   321

short-circuit monitoring   290
short-circuit winding   373
short-coil systems   259
short-time duty (duty type S2)   516
shunt characteristic   513
shunt motor   40
shunt wound motors   13
sickle-shaped coils   167
signal delay   379
signal flow diagram   287, 288
signal generator   162
signal processing errors   294
signal processor   119
signal processors or transputers   292
signal transmission by opto-couplers   351
simulation of a rotational speed-control loop   397
simulation of closed-loop rotational speed control   395
simulation of closed-loop rotational speed-control   397
single ring arrangements   423
single-phase equivalent circuit   386
   of asynchronous motors   386
single-phase motor   132
single-stack   167
sintered bearings   501, 502
sintered metal   501
sinus evaluation   361
sinusoidal commutation   71
sinusoidal evaluated PWM commutation   390
sinusoidal voltages   359
skew-wound winding   90
skewing   78, 79
slide screw drive   478
slider   256, 260, 262
slider-crank chains   479
sliding bearing   496, 498
sliding or bush bearings   519
slip changing   137
slip frequency   372
slip rings   50
slip-frequency compensation   373
slot skewing   128
slot to pole number ratio   80
slot-commutator   53
slotless armature design   89

slotless cylindrical winding design   90
slotless discoidal winding design   89
slotless motors   94
slotless skew winding   148
small-signal operation   324
smoothing capacitor   355
soft magnetic   431
space vector   95
space vector modulation   361, 400
space-vector diagram
   for synchronous motors   405
   for synchronous motors in the α, β-reference frame   406
spark suppression capacitors   66
sparks   61
speed adjusting range of the peristaltic pump   552
speed control   30, 45, 136
speed control circuit   561
speed controller   528, 541
speed dependent clutch   490
speed proportional damping   183
speed reversal   50
speed sensor   528
speed torque characteristic   30, 36, 512
spherical joint coupling   487
split-cage motor   126, 148
split-phase motor   129
spring constant   494
spring ring coupling   487
spring-mass system   494
springs   56
spur gears   455, 459
square-wave commutation   74, 106
squirrel-cage rotor   153
squirrel-cage winding   127, 151
stand gears   452, 460
standard motor   125
standards for fractional-horsepower electric machines   510
standing wave   335, 336
star connection   75, 128, 357
star wheel gearing   482
start-stop region   187
starting cage   151
starting capacitor   134
state controller   296
stator coordinate system   97
stator current vector   405

Index   595

stator inductance   403
stator voltage vector   405
steady-state approach   184
steady-state parameters   509
steel   463, 505
Steinmetz-connections   128
step angle   161, 169, 180
step errors   181
step width   532
stepper motor   358, 517
stepping drive   534, 548
stepping gears   481
stepping motor   554
stepping rate   175, 187
stepping rate during dosing   536
Stribeck diagram   496
strontium-ferrite   147
structure of the driving system   534
subcontrollers   376
subordinated current control   289
surface-mounted magnets   83
surface-mounted permanent magnets   94
switch mode power supply (SMPS)   398
switch-on time   366, 391
switched reluctance synchronous motor   116
switching amplifier   339, 340
switching cycle   400
switching Hall sensors   106
switching noises   75
symmetrical winding   129
synchronous behavior   514
synchronous belt gearings   475
synchronous speed   130, 134
synchronous torque   154
system design   241
system of control   295

tachometer generator   543
tang-commutator   53
tangential force   464
tapped winding   137
tapping of the exciter winding   47
temperature coefficient T_k 17
temperature rise of the motor   564
temperature-dependent resistance   412
thermal behaviour of drives   515

thermal partial system   254, 256
three-coordinate drives   293
three-level controller   340
three-phase a,b,c-system   362
three-phase bridge   356, 365, 398
three-phase induction motors   130
three-phase motor   134, 362
three-phase windings   128
thrust demands   260
thyristor converter   356
time per raster step   547
time-controlled switching   362
timer
   clock frequency   400
   with symmetric output signals   400
tin-can motor   164
tolerances of toothing   462
tooth belts   475
tooth coil windings   436
toric magnets   257
torque dependent clutch   490
torque during dosing   536
torque fluctuations   77
torque pulsations   128
torque ripple   77, 88, 99, 100, 104
torque saddles   134
torsionally flexible couplings   486
total acceleration time   531
total dynamic behaviour   225
total inertia of the drive system   187
total inertia of the load   543, 549
total moment of inertia   562
total translation ratio   535
trailing   55
trailing cable   256
trailing hoses or cables   271
transfer behaviour   288
transfer function   287, 376, 450
   of asynchronous motors   387
   of DC-motors   378
   of EC-motors   393
   of PI-controllers   395
z-transfer function   396
   of PI-controller   396
transfer gearings   450
transformation ratio   193
transistor for motor control   49
transition resistance   352
translational motion   524

596   Index

translational transmission ratio   543, 548, 561
transmission (transmission ratio)   452, 459, 464, 466, 468, 473
transmitting and reflecting cross grid   284
transputer system   282
transversal effect   318, 321, 326, 328
transverse flux motor   113
traveling wave   336–338
traveling wave motors   338
triangle wave generator   364
trimorph   327
tube cores   66
tubular piezo transducers   327
turbo-compressors   419
turbo-molecular pumps   418
turn-off signal   364
turn-on delay   364
turn-on signal   364
turn-on time   350, 353
two-axis reference frame   362
two-coordinate measuring   293
two-coordinate motors   282
two-phase equivalent motor model   95
two-phase stepper motor   358
two-quadrant operation   352
two-value capacitor motor   129
types of stepping   162
typical motion sequences   191

ultra-centrifuges   418
ultrasonic motors   330, 332–334, 336, 337
UL mark   511
unaligned position   117

under-commutation   62
unipolar control   163, 171, 175, 179
unipolar operation   358
unipolar stepper motor   358
universal motor   33, 367
upper phase plane   202

V-belts   474
variable-reluctance motor   163
VDE mark   510
vector control   297, 387
velocity control   289
ventricular assist device   441
vibration-damping elements   425
voice coil motor driver   290
voice coil motors   251, 252, 265
voltage control   46, 338–340, 342
voltage differential equation   429
voltage lifting   372
voltage limitation   363
voltage space vector   362

wave mode   170
wave winding   57
wear limit   56
Weiss' domains   319
wheel chain   453
wire-coupled disturbance   351
wolfrom gear   460
worm gears   468

x–y–z–Φ-module   306
x–y-hybrid stepping motor   304
z–Φ-hybrid stepping motor   304
zero crossings   182, 368

Printed in the United States
116960LV00001B/40-60/P